PSAM 12
Probabilistic Safety Assessment and Management
22-27 June 2014 • Sheraton Waikiki, Honolulu, Hawaii, USA

Proceedings of the Probabilistic Safety Assessment and Management (PSAM) 12 Conference

Volume 8 - Wednesday PM I

PSAM 12

Probabilistic Safety Assessment and Management

22 - 27 June, 2014

Sheraton Waikiki, Honolulu, Hawaii USA

CONFERENCE PROCEEDINGS

Volume 8

Wednesday PM I

Foreword

It is was our honor to welcome you to Honolulu, Hawaii, for the twelfth rendition of the Probabilistic Safety Assessment and Management (PSAM) Conference. The planning for PSAM Honolulu began back in 2007 (before PSAM 9 in Hong Kong), when we looked at several locations around the United States, included Arizona, California, Boston, and even considered locations in Oceania. Based upon the feedback both during and after the conference, PSAM 12 proved to be a great success.

We would like to thank all of the volunteers, those that served before, during, and after the Conference. Members of the Technical Program Committee, the Organizing Committee, the session chairs, and the presenters have our gratitude for making PSAM 12 the most memorable PSAM yet.

This publication represents the technical proceedings for the Conference. Due to the large number of published papers (a total of 391), we have subdivided the technical content (papers) into five volumes, one for each day of the conference.

On behalf of the International Association for Probabilistic Safety Assessment and Management Board of Directors, we hope that this publication will provide a valuable technical resource in addition to a reminder of the memorable stay in the Hawaiian Islands.

Dr. Curtis Smith
Technical Program Chairs

Dr. Todd Paulos
General Chair

Sponsors

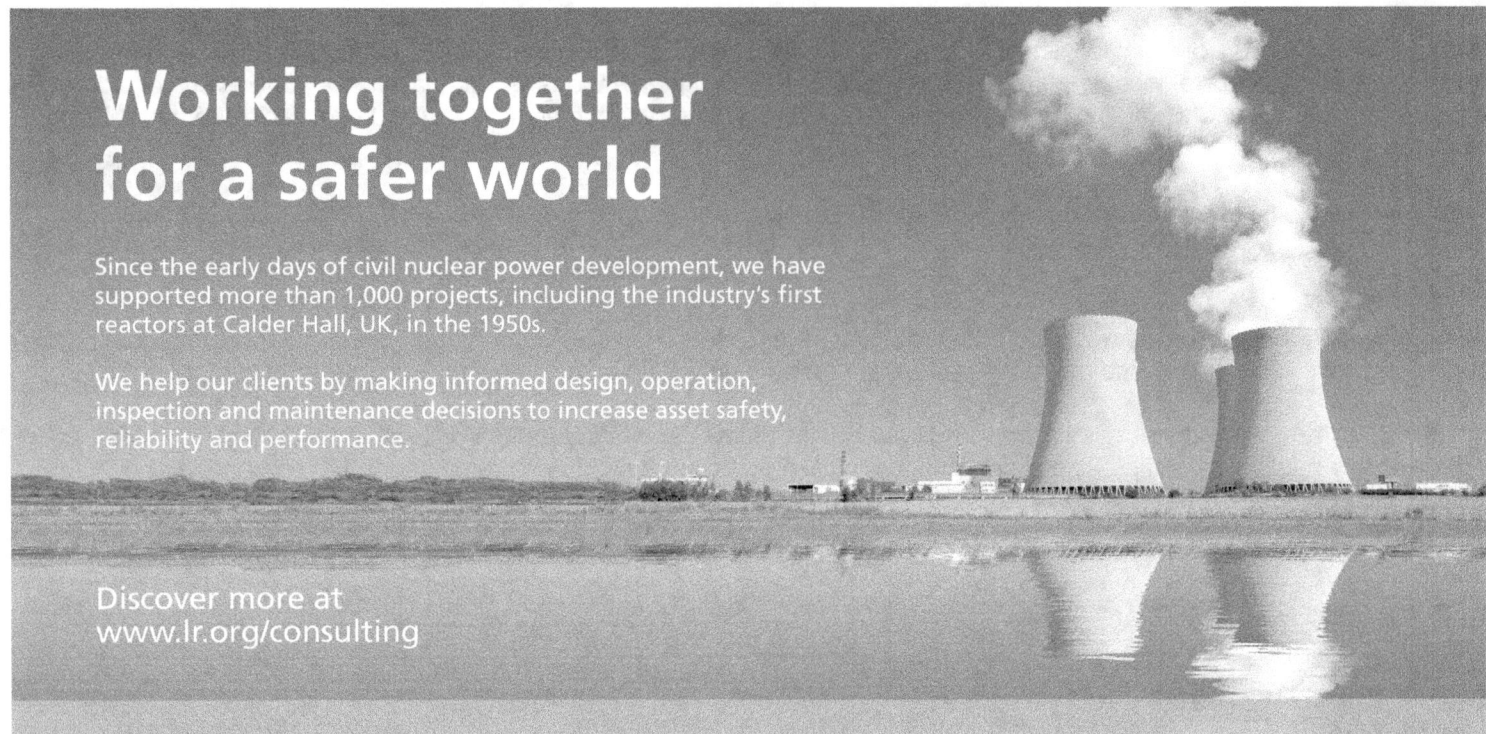

Sponsors

Technical Program Committee

Technical Program Chair: Curtis Smith, INL USA
Assistant Technical Program Chairs: Steve Epstein, Lloyd's Register Japan
Vinh Dang, PSI Switzerland
Ted Steinberg, QUT Australia

We would like to thank the members of the PSAM 12 Technical Program Committee. These individuals helped to make PSAM 12 a success by reviewing abstracts, technical papers, organizing sessions, and providing technical leadership for the conference.

Technical Committee Members:

- Roland Akselsson
- S. Massoud (Mike) Azizi
- Tito Bonano
- Ronald Boring
- Roger Boyer
- Mario Brito
- Kaushik Chatterjee
- Vinh Dang
- Claver Diallo
- Nsimah Ekanem
- Steve Epstein
- Fernando Ferrante
- Federico Gabriele
- Ray Gallucci
- S. Tina Ghosh
- David Grabaskas
- Katrina Groth
- Seth Guikema
- Steve Hess
- Christopher J. Jablonowski
- Moosung Jae
- Jeffrey Joe
- Vyacheslav S. Kharchenko
- James Knudsen
- Zoltan Kovacs
- Ping Li
- Harry Liao
- Francois van Loggerenberg
- Jerome Lonchampt
- Soliman A. Mahmoud
- Diego Mandelli
- Donoval Mathias
- Zahra Mohaghegh
- Thor Myklebust
- Cen Nan
- Mohammad Pourgolmohammad
- Marina Roewekamp
- Clayton Smith
- Shawn St. Germain
- Ted Steinberg
- Kurt Vedros
- Smain Yalaoui
- Robert Youngblood
- Enrico Zio

Organizing Committee

General Chair: Dr. Todd Paulos
General Vice Chair: Prof. Stephen Hora, USC
Technical Program Chair: Curtis Smith, INL USA
Webmaster, Registration, Support for Papers/Abstracts Submission and Review: Hanna Shapira, TICS

Table of Content

Page   Paper

10   *148*   **Experimental Approach to Evaluate the Reliability of Digital I&C Systems in Nuclear Power Plants**
Seung Jun Lee (a), Man Cheol Kim (b), and Wondea Jung (a)
a) Korea Atomic Energy Research Institute, Daejeon, Korea, b) Chung-ang University, Seoul, Korea

18   *170*   **The Contribution to Safety of a Diverse Backup System for Digital Safety I&C Systems in Nuclear Power Plants, a Probabilistic Approach**
W. Postma, J.L. Brinkman
NRG, Arnhem, the Netherlands

30   *196*   **Modeling of Digital I&C and Software Common Cause Failures: Lessons Learned from PSAs of TELEPERM® XS-Based Protection System Applications**
Robert S Enzinna (a), Mariana Jockenhoevel-Barttfeld, Yousef Abusharkh (b), and Herve Bruneliere (c)
a) AREVA Inc. Lynchburg, VA, USA, b) AREVA GmbH, Erlangen, Germany, c) AREVA SAS, Paris, France

41   *283*   **Methodology for Safety Assessment of the Defense-in Depth and Diversity Concept of the Digital I&C by Modernization of an NPP in Finland**
Ewgenij Piljugin (a), Jarmo Korhonen (b)
a) Gesellschaft fuer Anlagen und Reaktorsicherheit (GRS) mbH, Garching, Germany, b) Fortum, Power and Heat, Helsinki, Finland

51   *31*   **Scrum, documentation and the IEC 61508-3:2010 software standard**
Thor Myklebust(a), Tor Stålhane (b), Geir Kjetil Hanssen (a), Tormod Wien (c) and Børge Haugset (a)
a) SINTEF ICT, b) IDI NTNU, c) ABB

63   *378*   **A Software Package for the Assessment of Proliferation Resistance of Nuclear Energy Systems**
Zachary Jankovsky, Tunc Aldemir, Richard Denning (a), Lap-Yan Cheng and Meng Yue (b)
a) The Ohio State University, Columbus, Ohio, USA, b) Brookhaven National Laboratory, Uptown, New York, USA

74   *241*   **Risk Estimation Methodology for Launch Accidents**
Daniel J. Clayton, Ronald J. Lipinski (a), and Ryan D. Bechtel (b)
a) Sandia National Laboratories, Albuquerque, NM, USA, b) Office of Space and Defense Power Systems, U.S. Department of Energy, Germantown, MD, USA

82   *296*   **Development of Online Reliability Monitors Software for Component Cooling Water System in Nuclear Power Plant**
Yunli Deng, He Wang, Biao Guo
Fundamental Science on Nuclear Safety and Simulation Technology Laboratory, College of Nuclear Science and Technology, Harbin Engineering University, Harbin, P.R. China

93   *340*   **A Parallel Manipulation Method for Zero-suppressed Binary Decision Diagram**
Jin Wang, Shanqi Chen, Liqin Hu, Rongxiang Hu, Fang Wang, FDS Team
Institute of Nuclear Energy Safety Technology, Chinese Academy of Sciences, Hefei Anhui, China

99   *89*   **Realistic Modelling of External Flooding Scenarios - A Multi-Disciplinary Approach**
J. L. Brinkman
NRG, Arnhem, The Netherlands

109   *149*   **Insights from the Analyses of Other External Hazards for Nuclear Power Plants**
James C. Lin
ABSG Consulting Inc., Irvine, California, United States

121   *245*   **The Next-Generation Risk Assessment Method About the Effect of a Slope And Foundation Ground on a Facility in a Nuclear Power Plant**
Susumu Nakamura (a), Ikumasa Yoshida (b), Masahiro Shinoda (c), Tadasi Kawai (d), Hidetaka Nakamura (e), and Masaaki Murata (f)
a) Dept. of Civil & Environmental Eng., College of Engineering, Nihon University, Koriyama, Japan, b) Tokyo City University, Tokyo, Japan, c) Railway technical research institute, Kunitachi, Japan, d) Tohoku University, Sendai, Japan, e) Japan nuclear regulation authority, Tokyo, Japan, f) Mitsubishi heavy industry, Takasago, Japan

130   *289*   **Probabilistic Tsunami Hazard Analysis for Nuclear Power Plants on the East Coast of Korean Peninsula**
In-Kil Choia, Min Kyu Kim, Hyun-Me Rhee
Korea Atomic Energy Research Institute, Daejeon, Korea

138   *305*   **External Events PSA for the Spent Fuel Pool of the Paks NPP**
Attila Bareith (a), Jozsef Elter (b), Zoltan Karsa, Tamas Siklossy (a)
a) NUBIKI Nuclear Safety Research Institute, Budapest, Hungary, b) Paks Nuclear Power Plant Ltd., Paks, Hungary

Table of Content

Page	Paper	
148	132	**Ramifications of Modeling Impact On Regulatory Decision-making - A Practical Example** Ching Guey *Tennessee Valley Authority, Chattanooga, TN, U.S.A.*
152	138	**A Fresh Look at Barriers from Alternative Perspectives on Risk** Xue Yang, Stein Haugen *Norwegian University of Science and Technology, Trondheim, Norway*
164	140	**Monitoring Major Accident Risk in Offshore Oil and Gas Activities by Leading Indicators** Helene Kjær Thorsen (a) and Ove Njå (b) *a) Safetec Nordic AS, Oslo, Norway, b) University of Stavanger, Stavanger, Norway*
178	185	**The Role of NASA Safety Thresholds and Goals in Achieving Adequate Safety** Homayoon Dezfuli (a), Chris Everett (b), Allan Benjamin (c), Bob Youngblood (d), and Martin Feather (e) *a) NASA, Washington, DC, USA, b) ISL, Rockville, MD, USA, c) Independent Consultant, Albuquerque, NM, USA, d) Idaho National Laboratory, Idaho Falls, ID, USA, e) Jet Propulsion Laboratory, Pasadena, CA, USA*
190	195	**Improving Consistency Checks between Safety Concepts and View Based Architecture Design** Pablo Oliveira Antonino, Mario Trapp *Fraunhofer IESE, Kaiserslautern, Germany*
202	379	**Uncertainty Evaluation in Multi-State Physics Based Aging Assessment of Passive Components** Askin Guler, Tunc Aldemir, and Richard Denning *Nuclear Engineering Program The Ohio State University, Columbus, OH, USA*
212	537	**Passive System Evaluation by Using Integral Thermal-Hydraulic Test Facility in Passive NPP(nuclear power plant) PSA(probabilistic safety assessment) Process** Ruichang Zhao, Huajian Chang, Yang Xiang *State Nuclear Power Technology Research & Development Center, Beijing, China*
220	576	**Probabilistic Assessment of Composite Plate Failure Behavior under Specific Mechanical Stresses** Somayeh Oftadeh, Mohammad Pourgol-Mohammad, and Mojtaba Yazdani *Sahand University of Technology, Tabriz, Iran*
229	216	**Development of Feedwater Line & Main Steam Line Break Initiating Event Frequencies for Ringhals Pressurized Water Reactors** Anders Olsson, Erik Persson Sunde (a), and Cilla Andersson (b) *a) Lloyd's Register Consulting, Stockholm, Sweden, b) Ringhals NPP, Väröbacka, Sweden*
241	158	**Improvement of the Reliability and Robustness of Variance-Based Sensitivity Analysis of Final Repository Models by Application of Output Transformation** Dirk-Alexander Becker *Gesellschaft fuer Anlagen- und Reaktorsicherheit (GRS) mbH, Braunschweig, Germany*
251	177	**Bayesian Approach Implementation on Quick Access Recorder Data for Estimating Parameters and Model Validation** Javensius Sembiring, Lukas Höhndorf, and Florian Holzapfel *Institute of Flight System Dynamics TUM, München, Germany*
259	194	**Comparative Assessment of Severe Accidents Risk in the Energy Sector: Uncertainty Estimation Using a Combination of Weighting Tree and Bayesian Hierarchical Models** M. Spada, P. Burgherr and S. Hirschberg *Laboratory for Energy Systems Analysis, Paul Scherrer Institute (PSI), Villigen PSI, Switzerland*
271	234	**Investigation of Different Sampling and Sensitivity Analysis Methods Applied to a Complex Model for a Final Repository for Radioactive Waste** Sabine M. Spiessl, and Dirk-A. Becker *Gesellschaft fuer Anlagen- und Reaktorsicherheit (GRS) mbH, Braunschweig, Germany*
283	253	**Importance Analysis for Uncertain Thermal-Hydraulics Transient Computations** Mohammad Pourgol-Mohammad (a), Seyed Mohsen Hoseyni (b) *a) Department of Mechanical Engineering, Sahand University of Technology, Tabriz, Iran, b) Department of Basic Sciences, East Tehran Branch, Islamic Azad University, Tehran, Iran*

Table of Content

Page Paper

294 304 **Insights from an Integrated Deterministic Probabilistic Safety Analysis (IDPSA) of a Fire Scenario**
M. Kloos, J. Peschke (a), B. Forell (b)
a) GRS mbH, Garching, Germany, b) GRS mbH, Cologne, Germany

306 430 **Uncertainty Propagation in Dynamic Event Trees - Initial Results for a Modified Tank Problem**
Durga R. Karanki, Vinh N. Dang, and Michael T. MacMillan
Paul Scherrer Institute, Villigen PSI, Switzerland

320 460 **An Approach to Physics Based Surrogate Model Development for Application with IDPSA**
Ignas Mickus, Kaspar Kööp, Marti Jeltsov (a), Yuri Vorobyev (b), Walter Villanueva, and Pavel Kudinov (a)
a) Royal Institute of Technology (KTH), Stockholm, Sweden, b) Moscow Power Engineering Institute, Moscow, Russia

331 315 **A Toolkit for Integrated Deterministic and Probabilistic Risk Assessment for Hydrogen Infrastructure**
Katrina M. Groth (a), Andrei V. Tchouvelev (b,c)
a) Sandia National Laboratories, Albuquerque, NM, USA, b) AVT Research, Inc., Canada, c) International Association for Hydrogen Safety, HySafe

Experimental Approach to Evaluate the Reliability of Digital I&C Systems in Nuclear Power Plants

Seung Jun Lee[a], Man Cheol Kim[b], and Wondea Jung[a]
[a] Korea Atomic Energy Research Institute, Daejeon, Korea
[b] Chung-ang University, Seoul, Korea

Abstract: Owing to the unique characteristics of digital instrumentation and control (I&C) systems, the reliability analysis of digital systems has become an important element of probabilistic safety assessments. In this work, an experimental approach to estimate the reliability of digital I&C systems is considered. A digitalized reactor protection system was analyzed in detail, and the system behavior was observed when a fault was injected into the system using a software-implemented fault injection technique. Based on the analysis of the experimental results, it is possible to not only evaluate the system reliability but also identify weak points of fault-tolerant techniques by identifying undetected faults. The results can be reflected in designs to improve the capability of fault-tolerant techniques.

Keywords: Digital I&C system, PSA, fault injection, fault-tolerant technique, failure detection coverage

1. INTRODUCTION

The probabilistic risk/safety assessment (PRA/PSA) has been widely used in the nuclear industry for licensing and identifying vulnerabilities to plant safety since 1975. PSA techniques are used to assess the relative effects of contributing events on system-level safety or reliability. They provide a unifying means of assessing physical faults, recovery processes, contributing effects, human actions, and other events that have a high degree of uncertainty [1,2].

Recently, instrumentation and control systems (I&C) in nuclear power plants (NPPs) have been changed into digitalized systems. Deterioration and an inadequate supply of components of analog I&C systems have caused inefficiency and high maintenance costs. Moreover, since the fast evolution of digital technology has made it possible to design more reliable functions for NPP safety, the transition from analog to digital has been accelerated. Owing to the unique characteristics of digital I&C systems, a reliability analysis of the digital systems has been introduced as one of the important issues in the PSA field [3,4].

The report published in 1997 by US National Research Council states that appropriate methods for assessing safety and reliability are key to establishing the acceptability of digital I&C systems in safety-critical plants such as NPPs [3]. The HSE's guide also pointed out the importance of the PSA for software-based digital applications as a demonstration of safety [5]. However, there is no widely accepted method for digital I&C PSAs. Conventional PSA techniques cannot adequately evaluate all features of digital systems. Failure coverage, common cause failures, and software reliability are the three most critical factors in the safety assessment of digital systems [6].

This work suggests an experimental approach to evaluate the reliability of digital I&C systems.

2. CHARACTERISTICS OF DIGITAL I&C SYSTEMS

Digital I&C systems are designed based on software and have unique characteristics utilizing software. The following should be considered in digital I&C system reliability evaluations:

- Failure coverage: Digital I&C systems have various fault-tolerant techniques for enhancing the system reliability. In the reliability evaluation, the fault-tolerant techniques and their failure coverage must be considered. A fault is a source that has the potential of generating failures. Fault-tolerance is the system's capability to help the system perform correctly the specific required functions in spite of the presence of faults. In a fault tolerance evaluation, failure detection coverage is a crucial factor [7]. Failure detection coverage is a measure of the system's ability to perform failure detection, failure isolation, and failure recovery. For evaluating the failure detection coverage, it is important to exclude the duplicated effect of fault-tolerant techniques since various fault-tolerant techniques are implemented simultaneously at each level of the system hierarchy, such as component-level fault detection algorithms (e.g., memory checksum, watchdog timer for detecting microprocessor halt), board-level self-diagnostics (e.g., loop back check for input and output module), and system-level error detection mechanisms (e.g., automatic periodic test, state comparison algorithm of redundant modules). In addition, a different inspection period and range of each fault-tolerant technique should be considered [8].

- Common cause failure (CCF): The issues related to a system are the risk concentration and diversity (including CCF), the failure coverage of a self/peer monitoring, the effectiveness of an automated periodic system testing, and the network communication failures. The use of a single microprocessor module for multiple safety-critical functions will cause a severe concentration of risk in a single microprocessor. Safety-critical applications have adopted a conservative design strategy, based on functional redundancies. However, the software programs of these functions are executed by one microprocessor sequentially. Therefore, the level of redundant design of digital systems is usually higher than those of conventional mechanical systems. This higher redundancy will clearly reduce the risk from a single component failure, but raise the severity of CCF consequence. This higher level of redundancy exponentially increases the number of CCF events modeled in a fault tree, if conventional CCF modeling methods are applied. In some nuclear power plants, there are four signal processing channels for the safety parameters, and each channel consists of two or four microprocessor modules for the same function. For example, in the RPS of the OPR-1000 plant, there are 16 processors that do the identical function of local coincidence logic. In this case, the system model will have 65519 events for representing the CCFs of the local coincidence logic processors [3].

- Software reliability: The prediction of software reliability using a conventional model is generally much harder than for hardware reliability. It is notable that there has been a lot of discussion among software engineering researchers about whether a software failure can be treated in a probabilistic manner. Software faults are design faults by definition. That is, software is deterministic and its failure cannot be represented by a 'failure probability'. However, software can be treated based on a probabilistic method because of the randomness of the input sequences [3].

In this work, a digitalized reactor protection system (RPS) was tested for evaluating its reliability using one of the fault injection techniques. A fault injection is a technique for validating the reliability by observing the system behavior when a fault is injected. It consists of the accomplishment of controlled experiments where the observation of the system's behavior in the presence of faults is induced explicitly. The target system is tested without decomposition, thus problems in the system such as CCF or software flaws are reflected in the test results. That means the fault injection method threatens not the components of a system, but the whole system as it is, and the experiment results include all possible effects of problems existing inside the system. A limited software-implemented fault injection technique in which faults can be injected into memory and register was used based on an assumption of that all faults in a system are reflected on the faults in the memory and register. To reduce the necessary fault injection experiments and obtain reliable results, the memory map of the target software was analyzed. An unnecessary fault injection can be eliminated and the importance of specific memory area can be identified based on the analyzed memory map.

3. FAULT INJECTION EXPERIMENTS

3.1. Target Digital Reactor Protection System

We propose an experimental approach to evaluate the reliability of digital I&C systems. For a more realistic evaluation, the prototypes of digital I&C systems that have been adopted in a real digitalized NPP were used for the experiment. The target digital I&C system is the Integrated Digital Protection System (IDiPS) Reactor Protection System (RPS), which was developed in Korea [9,10] during the Korea Nuclear Instrumentation and Control System (KNICS) research and development project. The IDiPS RPS has four independent channels, where each channel consists of bistable processors (BPs), coincidence processors (CPs), an automatic test and interface processor (ATIP), a cabinet operator module (COM), and other hardware components [11].

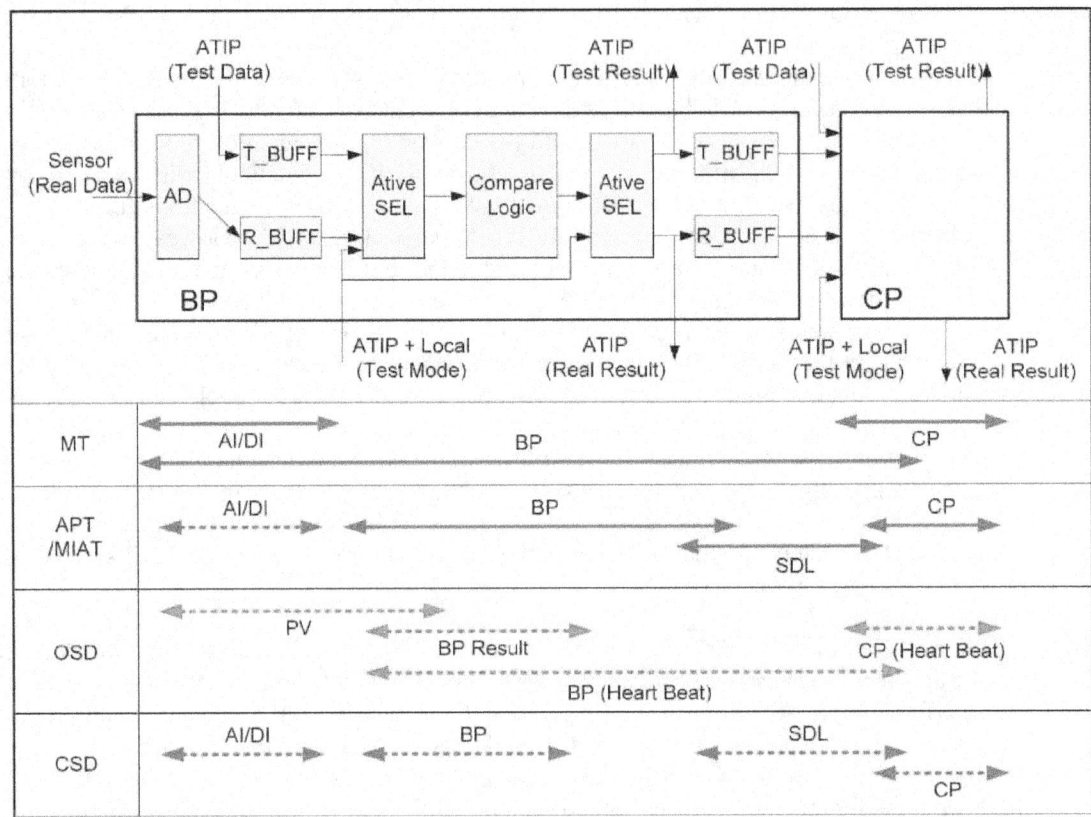

Figure 1: Fault-tolerant techniques in the IDiPS RPS [11]

IDiPS RPS tests can be classified into two categories: active tests and passive tests. Figure 2 shows four types of tests that have different types of coverage and periods [11].

- Active tests consist of automatic periodic tests (APTs), manual initiated automatic tests (MIATs), and manual tests (MTs) [9]. An APT is periodically initiated by the ATIP without any human intervention. An MIAT is almost the same as an APT except for the operator initiation and tested trip parameter selection. An MT is generally performed once per month.
- A passive test partially checks the system's integrity. This test consists of component self-diagnostics (CSD) and online status diagnostics (OSD).

In our work, the BP of the IDiPS-RPS was selected as a target system. Among the failure detection functions of the target system, three were considered: OSD, CSD, and APT.

3.2. Fault Injection Techniques

We used a software-implemented fault injection technique in which faults can be injected only into the memory [12,13]. Our fault injection experiment was conducted based on the assumption that all faults in a system are reflected in the faults in the memory because a fault should affect the memory related to the calculation process or variables and cause a wrong output. A fault of any component in a system may have an effect on the calculation process, reading input variables, generating output variables, and so on. A wrong calculation, program halt, variable changes, or wrong execution path may be caused by the fault. Conversely, the fault may have no effect on the output. If a fault does not have any effect on the output, then it is impossible to detect the fault because there are no observable consequences from the fault. If a variable related to the system output is changed by an inappropriate value for the current situation, then the fault may be detectable [14].

The fault injection experiment was performed for a system, not for a single component of the system. The different inspection period and range problem is not available because the behavior of the system against the injected fault was observed.

A fault injection experiment was performed based on the following three steps. First, fault types were identified according to the effects of injected faults. Based on the fault types, the failure detection coverage was defined. Second, a memory map of the target system was analyzed to perform efficient experiments. Unnecessary experiments were eliminated to reduce the number of experiments required. Finally, fault injection experiments were performed, and the results were analyzed.

3.3. Definition of Failure Detection Coverage

Faults in digital I&C systems are categorized into seven types according to their consequence and detection potential, as shown in Table 1.

Table 1: Categorization of faults into seven types

	Changed and used			Unused or unchanged
	Correct output	Wrong output	No output	
Detected	A	C	E	G
Undetected	B	D	F	

- Correct output (Fault types A and B):
    - Even when a bit is changed by a fault, and the changed bit is used to generate a system output, there may not be any effect on the output because the changed bit is not directly related to the output generation. For example, a stuck-at-1 fault changes "variable A" from 16 (binary: 10000) to 24 (binary: 11000). In this case, if the set point for "variable A" is 10, then the output is not changed, because both 16 and 24 are greater than the set point. This type fault is categorized as a safe fault.

- Wrong output (fault types C and D):
    - The bit changed by a fault may cause a wrong system output. For example, "variable A" has a value of 16 (binary: 10000), and the set point is 10. If the highest bit of "variable A" is changed by a stuck-at-0 fault, then "variable A" becomes 0 (binary: 00000) and a wrong output is generated.

- No output (fault types E and F):
    - The bit changed by a fault may cause a program halt or infinite loop, and thus the program does not generate an output. In this case, nothing is written on bits for the output, and the previous output is not updated.

- Unused or unchanged (fault type G):
    - A memory area is not assigned to any program code or variables. Even though some memory area is assigned and used, there will be no effect on the output unless a fault

changes a memory bit. For instance, if a stuck-at-0 fault is injected on a bit that was already 0, then nothing is changed. These unused or unchanged bits do not have any effect on the output generation, and it is impossible to detect such faults.

If a system works correctly despite the presence of a fault, the fault is called a "safe fault." A "correct output" (fault types A and B) and "unused or unchanged" (fault type G) fault types are classified as a "safe fault." Even if a malicious fault causes a "wrong output" or "no output" (fault types C, D, E, and F), if it is detected, the system will remain in a safe state. Such detectable malicious faults (fault types C and E) are also classified as a "safe fault" in terms of safety. If a malicious fault is not detected, the fault is classified as an "unsafe fault." The fault types are categorized as shown below:

- No-effect faults: A, B, G
- Malicious faults: C, D, E, F
- Safe faults: A, B, C, E, G
- Unsafe faults: D, F

We define the failure detection coverage as the probability of detecting malicious faults. The equation of failure detection coverage is defined as

$$(C + E) / (C + D + E + F) \qquad (1)$$

3.3. Fault Injection Experiments

We performed fault injection experiments on the memory area of the BP application. Faults were injected into the memory of the BP application using the Code Composer tool [15], and an automatic fault injection program was developed for the experiment. Figure 2 shows the environment of the fault injection experiment. Two types of memory faults, stuck-at-0 and stuck-at-1, were considered because a memory bit has a binary value.

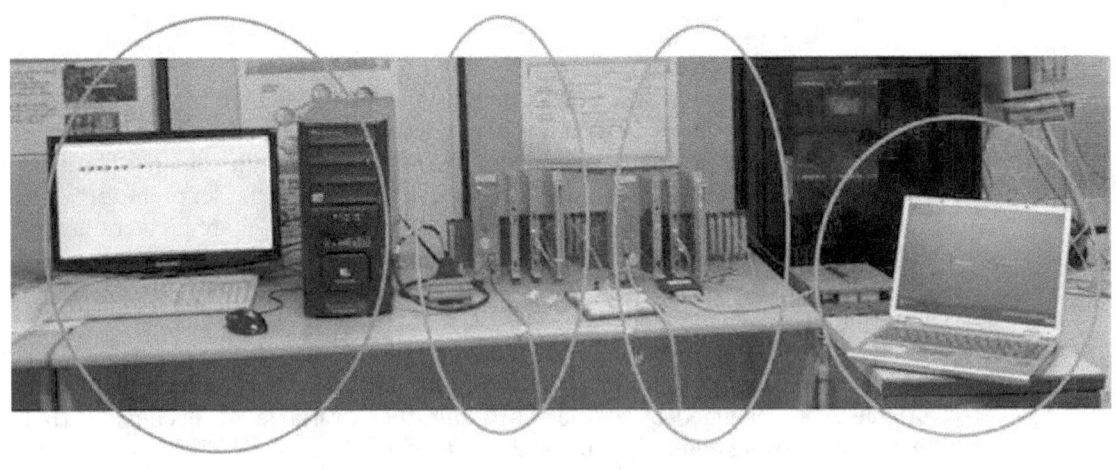

Data acquisition and storage   Automatic Test Processor   Bitable Processor   Fault Injector

Figure 2: The environment for the fault injection experiment [11]

Because this experiment was for a feasibility study, the fault injection experiments were performed under limited conditions as follows in order to reduce the experimental time.

- A limited memory area was examined. A fault injection experiment for every single bit requires a large amount of time because of the large memory size, and each fault injection experiment takes approximately 1 min. For example, a total of approximately 8 million experiments are necessary just for the memory of the BP OS code. Moreover, the memory size of the BP application is much greater than that of the BP OS. Therefore, fault injections were performed on 3% of used memory area, and only two bits of each assembler line (the first and last bits) were examined. Usually, the first and last bits have a more significant effect than the other bits, and thus this limited condition may result in a more conservative output.

- The environment for the fault injection experiments is not exactly the same as the actual operating environment. The fault injection conditions differ from plant operating conditions even though actual digital I&C systems are examined, because the fault injection environment is implemented using only BP and ATIP. If other components are connected, then different behaviours can be observed. However, in terms of failure detection, it is expected that the results will differ little from those of the actual operating environment.

3.4. Experiment Results

A total of 55,752 fault injection experiments were then performed excluding the unused memory area, and the following observations were made.

- Faults resulting in no effect (fault types A, B, and G): (90.77% of injected faults)
- Faults resulting in no trip (fault types C, D, E, and F): 5,144 (9.23% of injected faults)
- Detected faults (fault types C and E): 5,028 (9.02% of injected faults)
- Undetected faults (fault types D and F): 116 (0.21% of injected faults)

Among the faults that caused a trip signal generation failure (C + E + D + F), the undetected faults (D + F) occupied 2.26%. Therefore, the failure detection coverage of the target system was 97.74%, based on Equation. 1.

3.5. System Reliability

The failure probability of an analog I&C system is calculated with the failure probabilities of the components. If an analog system consists of four relays and a failure of any relay causes a system malfunction, then the system failure probability is *p(relay failure probability) x 4*. However, since digital I&C systems consist of hardware and software, software failure probability should be considered in addition to the hardware failure probability. Moreover, in spite of a system failure, a system malfunction is prevented if the failure is detected. The equation for the failure probability of a digital I&C system is as follows:

*p(digital I&C system failure) = (p(HW failure) + p(SW failure)) * (1-p(failure detection))*   (2)

Systems applied in NPPs are highly reliable and examined through strict validation/verification processes. In fact, no failure was observed when a fault was not injected in the experiments. Since only systems that do not have any flaw can be adopted in NPPs, it is not possible to estimate the system failure probability through this experiment. Usually, the failure probability of a component in analog I&C systems is about 1E-5 – 1E-6. If it is assumed that the failure probability of a digital RPS including hardware and software failure probability is 1E-5 and the failure detection coverage is 90%, then the system failure probability is 1E-6. If the failure detection coverage is 99%, then the system failure probability reduces to 1E-7. The failure detection coverage of fault-tolerant techniques is a very important factor to enhance the reliability.

4. CONCLUSION

The unique characteristics of digital I&C systems should be considered to estimate the reliability of digital I&C systems. In the present work, the reliability of digital I&C systems was estimated through fault injection experiments. A software-implemented fault injection technique in which faults are injected into the memory was used based on the assumption that all faults in a system are reflected in the faults in the memory. The fault injection experiment was performed based on the following three steps. First, fault types were identified according to the effects of the injected faults. Based on the fault types, the failure detection coverage was defined. Second, the memory map of the target system was analyzed to perform efficient experiments. Unnecessary experiments were eliminated to reduce the required number of experiments. Finally, fault injection experiments were performed, and the results were analyzed. For a feasibility study, a limited number of fault injections were performed using two digital I&C components. Based on the experimental results and analyzed memory map, the number of faults for each fault type was estimated.

Based on the experiment result analysis, it is possible not only to evaluate the reliability of digital I&C systems but also to point out the weakness of fault-tolerant techniques by identifying the undetected faults. The result can be reflected to the design to improve the capability of fault-tolerant techniques.

Acknowledgements

This work was supported by Nuclear Research & Development Program of the National Research Foundation of Korea grant, funded by the Korean government, Ministry of Science, ICT & Future Planning (Grant Code: 2012M2A8A4025991).

References

[1] S. Authen and J. Holmberg. "*Reliability Analysis of Digital Systems in a Probabilistic Risk Analysis for Nuclear Power Plants*". Nuclear Engineering and Technology. 44, pp. 471-482, (2012).
[2] T. Aldemir, et al. "*Dynamic Reliability Modeling of Digital Instrumentation and Control Systems for Nuclear Reactor Probabilistic Risk Assessments*". NUREG/CR-6942, (2007).
[3] H. G. Kang, et al. "*An overview of risk quantification issues of digitalized nuclear power plants using static fault tree*". Nuclear Engineering Technology. 41, pp.849-858. (2009).
[4] M. Douglas, et al. "*Digital Instrumentation and Control Systems in Nuclear Power Plants*", National Academy Press, Washington, D.C, (1997).
[5] HSE, "*The use of computers in safety-critical applications*", London, HSE Books, 1998.
[6] H. G. Kang and T. Sung. "*An analysis of safety-critical digital systems for risk-informed design*". Reliability Engineering and System Safety. 78, pp. 307-14, (2002).
[7] J. B. Dugan and K. S. Trivedi. "*Coverage Modeling for Dependability Analysis of Fault-Tolerant Systems*. IEEE Transactions on Computer. 38(6), pp.775-787, (1989).
[8] S. J. Lee, et al. "*Reliability assessment method for NPP digital I&C systems considering the effect of automatic periodic tests*", Annals of Nuclear Energy. 37, pp.1527-1533, (2010).
[9] K. C. Kwon and M. S. Lee. "*Technical Review on the Localized Digital Instrumentation and Control Systems*". Nuclear Engineering and Technology. 41, pp. 447-454, (2009).
[10] S. Hur, D. H. Kim, I. K. Hwang. "*A New Automatic Periodic Test Method for the Digital Reactor Protection System*". NPIC&HMIT, Knoxville, Tennessee, USA, (2009).
[11] J. G. Choi, et al. "*Fault Detection Coverage Quantification of Automatic Test Functions of Digital I&C System in NPPs*", . Nuclear Engineering and Technology. 44, pp. 421-428, (2012).
[12] S. J. Kim, et al. "*A method for evaluating fault coverage using simulated fault injection for digitalized systems in nuclear power plants*". Reliability Engineering and System Safety. 91, pp. 614-623, (2006).
[13] M. Hsueh, T. K. Tsai, R. K. Iyer. "*Fault injection techniques and tools*". IEEE Transaction on Computer. 30, pp. 75-82, (1997).

[14] J. S. Lee, et al. *"Evaluation of error detection coverage and fault-tolerance of digital plant protection system in nuclear power plants"*. Annals of Nuclear Energy. 33, pp. 544-554, (2006).
[15] Texas Instruments, 1994. Code Composer, User's Guide.

The contribution to safety of a diverse backup system for digital safety I&C systems in Nuclear Power Plants, a probabilistic approach

W. Postma[a*], J.L. Brinkman[a]

[a] NRG, Arnhem, the Netherlands

Abstract: NRG performed a research project on the influence on safety of diverse backup systems next to the existing digital I&C safety systems in Nuclear Power Plants (NPPs). As part of this research project a probabilistic approach has been used to evaluate the basic options to connect a diverse backup system logically to the existing digital I&C systems.

One can distinguish four different basic design options: (1) no backup system is used; (2) the backup system is used only if the digital system has failed, the switch-over to the backup system is automatic; (3) the backup system is used only if the digital system has failed, the switch-over to the backup system is manual; (4) the backup system is in continuous operation, with an equal vote as the digital system.

Design (2) and (4) have been modeled and compared with the situation without backup system (design 1). This paper will discuss the model and the results, including the sensitivity analyses, in order to reflect on the probabilistic impact of a diverse backup system.

Keywords: Digital I&C, Backup system, PRA, CCF

1. INTRODUCTION

NRG performed a research project on the influence on safety of diverse backup systems next to the existing digital I&C safety systems in Nuclear Power Plants (NPPs). As part of this research project a probabilistic approach has been used to evaluate the basic options to connect a diverse backup system logically to the existing digital I&C systems. The probabilistic evaluation and its results will be discussed in this paper.

Digital I&C systems, compared with analogue I&C systems, provide many important technical benefits. They are basically drift-free and system performance has been improved in terms of accuracy and computational capabilities. Since digital I&C systems have outstanding capabilities of data handling and storage, the plant operating conditions can be effectively measured and displayed.

However, digital systems are vulnerable to common cause failure (CCF) that may cause redundant safety systems to fail in their functions. Avoiding CCF is of vital importance for a safe and reliable operation of digital safety systems.

With this background most regulators require a diverse backup system to complement the digital safety systems in NPPs. The main goal of the diverse backup system is to enhance the safety further. The system has to bring the plant in a safe state if all digital systems have failed due to a CCF.

One can distinguish four different basic design options: (1) no backup system is used; (2) the backup system is used only if the digital system has failed, the switch-over to the backup system is automatic; (3) the backup system is used only if the digital system has failed, the switch-over to the backup system is manual; (4) the backup system is in continuous operation, with an equal vote as the digital system.

Design (2) and (4) have been modeled and compared with the situation without backup system (design 1). Design 3 is very similar to design 1 and mainly limits the number of functions suitable for backup, because of the time frame that is needed to carry out the manual action to switch to the backup system.

*w.postma@nrg.eu

2. MAIN DESIGN OPTIONS IN BACKUP SYSTEM LOGIC
An important aspect in assessing the impact of the diversity strategy on safety is how the backup/diverse system "knows" when to operate. In this section the three main design options to incorporate the backup system in the architecture are discussed:

- **Automatic switchover (design 2)**: The system is actuated when failures in the other systems are detected by the monitoring system;
- **Manual switchover (design 3)**: The system is actuated manually if the operators notice a loss of the computerized systems;
- **Continuous operation with an equal vote to the digital system (design 4)**: The system functions completely in parallel to the other systems, but incorporates a limited number of functions.

The manual switchover and the automatic switchover are very similar. In both cases the backup system will only be used if failures in the normal systems are <u>detected</u>. As it is of course impossible to react on <u>undetected</u> failures of the normal systems, the fraction of undetected failures will have a large impact on the effectiveness of the backup system. To make a best estimate of the reliability of a system it is therefore important to know the fraction of undetected failures as compared to the detected failures.

Every option has its advantages:
- The automatic switchover and the manual switchover will not likely contribute to the unavailability and the back-up-system is only used when really necessary, consequently the reliability requirements can be less stringent than for the digital system. Also, the automatic switchover is fast and reliable.
- The manual switchover does not need additional computerized elements to carry out the switchover, so it will be more robust against a complete loss of the computerized I&C.
- Continuous operation with an equal vote to the digital system always contributes to the safety.

The disadvantage of the back-up in continuous operation is that the backup system should be classified in the same safety category as the standard systems, as it carries out the same, although limited number of safety functions. Another disadvantage is that the reliability requirements for spurious actuations should also be equal to those of the standard digital systems in order to prevent an increase in spurious trips and actuations.

In digital I&C systems there are numerous possibilities to mitigate the effects of failures. One of the mitigative measures is to switch the outputs of the failed module to a predefined status or notifying the operators, if possible. If it is possible to define a failsafe status of a module the additional effect on safety of a backup system might be small, as will be shown in this paper.

In all cases it is necessary to ensure the independence of the computer based systems and the backup system in order to prevent that failures in the computer based systems can affect the operation of the backup and vice versa. Additionally the diversity between the digital I&C system and the backup system is needed to eliminate potential CCF.

3. SYSTEM ARCHITECTURE
The architecture that is used for the present study is an example system that is based on information in publically available literature on digital I&C for Nuclear Power Plants. The schematic of the architecture is shown in Figure 1.

The input of the system is carried out 4-fold redundant, the output is 2-fold redundant. Every redundant channel has two diversities, named diversity A and B. There is no communication between the two diversities.

The data acquisition is done by four sensors. The signal of a sensor is – per sensor - divided over diversity A and B. A/D converters convert the signal into a digital signal.

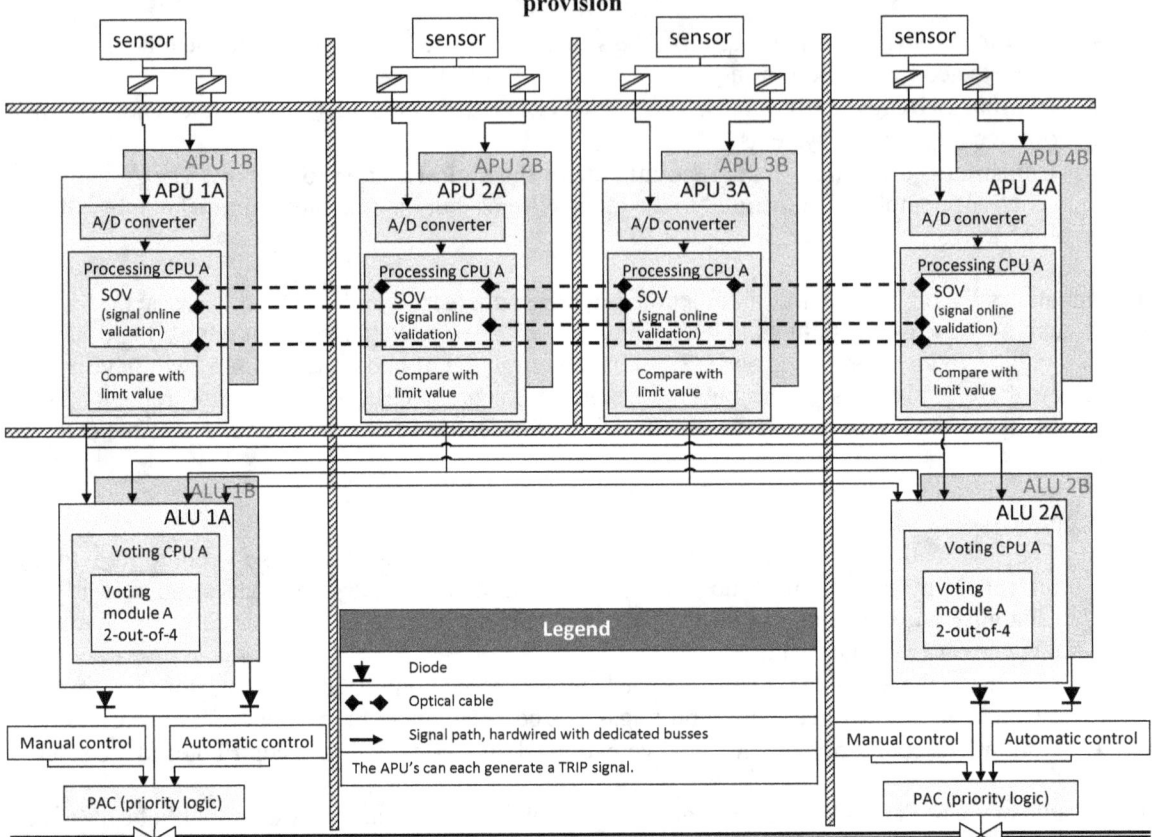

Figure 1: I&C structure that is the basis for the model. The architecture is shown without backup provision

In the acquisition and processing units (APUs) the sensor values are processed and validated. The digitized input is computed to get the corresponding physical value of the measurement that is processed.

The digitized values from all four channels are distributed to all the redundant channels for validation purposes. Each input is compared to the measuring range limits. The digitized inputs of the other channels are used for comparison.

The Actuation Logic Units (ALUs) receive the result of the validation process of every APU of the corresponding diversity. In the ALU a 2 out of 4 voting process takes place to decide whether or not to send an actuation signal to the Priority Actuation and Control (PAC).

The PAC sends an actuation command to the actuator (for example a valve). Also, the PAC ensures that safety commands will always get priority over non-safety commands. The PACs in both trains are assumed to be identical.

For all communication optical cables are used to ensure independency, using dedicated buses.

Three designs are modeled:
- Design 1: No backup provision
  - The architecture of design one corresponds to the architecture presented in Figure 1.

- Design 2: Switchover
  - In this case the backup system is used when the normal system has failed detectably. An automatic switchover is modeled (design 2). The architecture of design two is shown in Figure 2. The backup system is coupled to the standard system by using an inverse AND-gate (hardwired). If the backup system sends activation signal AND the monitoring of ALU A AND B gives an error message (sending 0, fail safe), the inverse AND-gate gives an activation signal. In normal operation the monitoring of ALU A and B gives an active high signal and the output of the backup systems is blocked.
- Design 4: Continuous operation with an equal vote to the digital system (design 4)
  - The architecture of design 4 is shown in Figure 3. The backup system is connected logically to the standard system via an OR-gate. This means that either subsystem A, subsystem B or the backup can initiate an actuation signal.

Figure 2: Backup system logic configuration in case of an automatic switchover to the backup system, when the normal system has failed detectably (design 2)

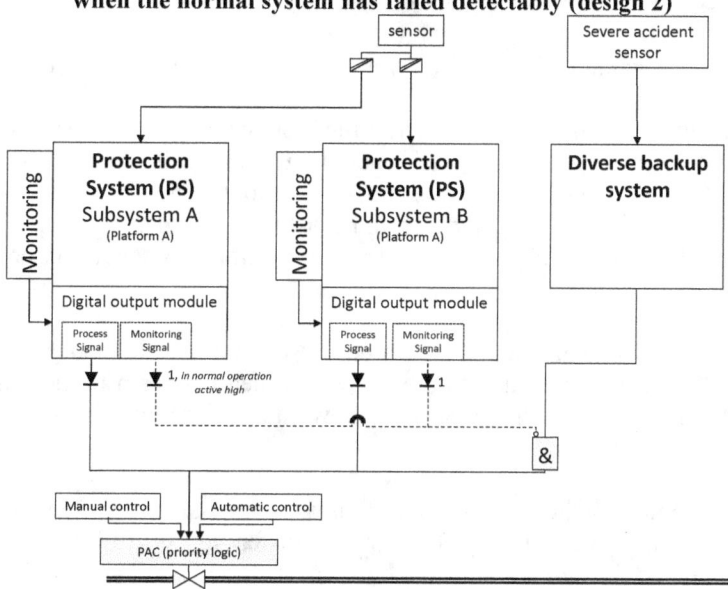

Figure 3: Backup system logic configuration in case of equal and continuous operation of the backup (design 4)

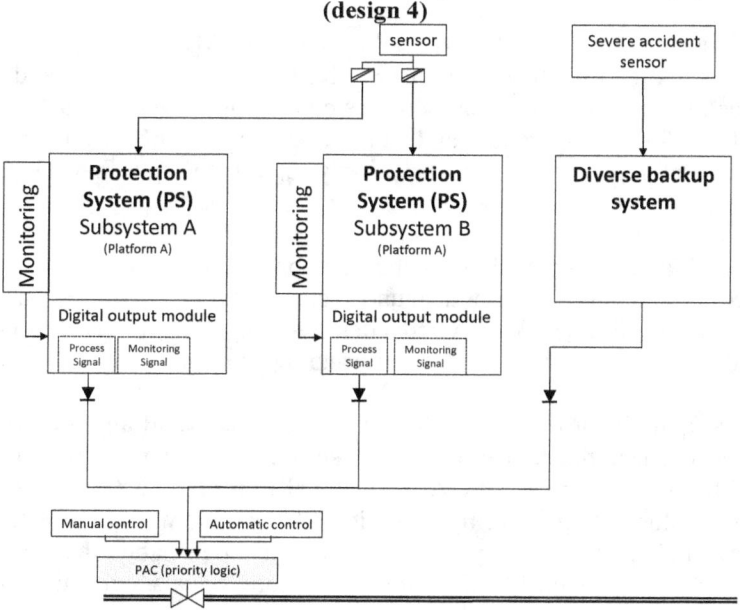

4. FAULT TREE MODEL
4.1. General description of the model and the approach
Given the available information in publically available literature the most appropriate level of abstraction is the I&C unit level. An I&C unit has a specific task in the process of executing the function. An example of an I&C unit is the ALU, which has the specific task of sending an actuation signal after carrying out a voting process. If more details are available, the I&C units can be further decomposed in CPU, I/O-modules, racks etc. (module level). In order to make the I&C unit level an accurate description of the module level, appropriate assumptions have to be made on for example diagnostic coverage, common cause failures and predefined outputs.

Three top-events are considered:
1. Failure to actuate without backup system which corresponds to design 1;
2. Failure to actuate with a backup system that is used only when the normal system has failed detectably which corresponds to design 2;
3. Failure to actuate with a backup system in continuous operation with an equal vote to the digital system which corresponds to design 4.

Design 3 is not modeled separately, because it is very similar to design 2 (see section 1).

The model considers only one automatic protection function, carried out by the protection system of a plant. The model reflects the system up to and including the actuation signal and does not include failure of the actuator itself. Additionally, the model does not include manual actions and support systems. The power supply is not modeled assuming a high redundancy of power supply systems. Based on experience, the impact of loss of power on the failure to actuate is negligible for highly redundant power supply systems.

The standard top-down ("upstream") approach has been used to make the model. Starting with the top-event, failure to send an actuation signal, to the valve and then upstream to the sensors. To take into account all relevant failures for all components, all flow paths are followed consistently, starting with the priority logic and ending at the sensors.

In case of design 2, automatic switchover to the backup after the standard system has failed detectably, the backup is built into the system in way that it reacts differently to different types of faults. If the standard system fails detectably the backup will be used, if the standard system fails undetectably the backup cannot be used. This defeats the "good practices" way of making fault trees, because it forces one to think in combinations of failures, which is very error prone.

In this model it is assumed that the backup is used only if both ALUs have failed detectably. If there are detectable failures in an ALU, the output of the ALU is marked as invalid and the output goes to the predefined value (=0, no actuation signal) and the control signal goes from 1 (active high) to 0. If there is a combination of a detected failure and an undetected failure of the ALUs, the backup system is not coupled in. Therefore, the event that the priority logic receives a faulty input signal from the ALUs is split into four options:

1. A detected failure in both ALUs → the backup is used;
2. A detected failure in ALU A and an undetected failure in ALU B → backup is not used;
3. An undetected failure in ALU A and a detected failure in ALU B → backup is not used;
4. An undetected failure in both ALUs → backup is not used.

The ALUs can get a faulty input from the APUs. A detected failure of an APU will lead to a "high" signal (= partial trigger) due to the defined fail safe position. The voting logic will now see at least one high signal and will need only one additional high signal to send an actuation signal to the priority logic. The result is that the voting logic, by definition, changes from 2 out of 4 (2oo4) to 1 out of 3 (1oo3) in the corresponding diversity. If yet another APU fails detectably, the output is again set to 1 (=partial trigger). The voting module (ALU) will now see 2oo4 signals high and will send an actuation signal to the PAC. So, detected failures of the APUs will not contribute to a failure to actuate. Only the undetected failures of the APUs will contribute to a failure to actuate. If three or more APUs in a diversity fail undetected, the ALUs in that diversity will not be able to send an actuation signal.

It is assumed that the backup system is completely independent and diverse from the digital I&C systems. For the backup system different sensors are used than for the digital I&C system. The backup system cannot be set to a predefined value. No CCF has been assumed between the backup system and the standard systems in order to determine the maximum impact of the backup system.

3.2. Failure modes
Five failure modes are included in the model:
1. **Detected failures (DF0) programmable logic (APU/ALU/PAC)**: In case of detected failures, the I&C units will go to the predefined outputs, which are as follows:
   - APU = 1 (partial trigger)
   - ALU = 0 (no actuation)
   - PAC = 0 (no actuation)
   - The failure of a module can be detected by internal means and by external means. For example, if an APU fails, this can be detected by the APU itself (internal means). Another possibility is the detection of a failure at the next I&C unit in the processing line, as it is the case for the sensors. If there is a failure of a sensor, the sensor itself will not detect that, but the APU is capable of validating the signal. If the signal is lost or corrupted and the APU can detect that (external means), the output of the APU is set to 1.
   - Examples: loss of power supply, loss of input, crash of a microprocessor, etc.
2. **Undetected failures (UFB) programmable logic (APU/ALU/PAC)**: under the undetected failures, the failure mode "undetected-blocking" is considered, because this is a fail to danger. In case of undetected blocking the output of an I&C unit does not generate a (partial) trigger, although a partial trigger is should be generated. Because the failure mode is not detectable, the output cannot be set to the predefined output value. Examples: saturation of a system network, progressive saturation of a memory block, etc.
3. **Detected (DF) and undetected (UF) failures backup system**: In both cases, the backup system fails to send an actuation signal. It is assumed that it is not possible to set the output of the backup system to a predefined state. In case of a detected failure, repair can be started. In case of an undetected failure, the failure will be discovered after testing.
4. **Out of range (OOR) sensor**: A sensor can indicate a value that is outside a 5% bound of the value read by the other sensors. In this case an error message is sent to the control room. However, because this type of validation is dependent on communication with other sensors, it is conservatively assumed that this type of error does not set the APU to the predefined output value.
5. **In range but low (IRL) sensor**: This failure mode is a calibration error. All sensors lie within a 5% bound of each other, but all sensors indicate a too low value. This failure mode is undetected.

3.2. Data
Failure rates
The data that is used, is obtained from publically available sources and is shown in Table 1. Table 1 shows the total failure rate. Based on the diagnostic coverage the failure rate is split into a detected and undetected fraction. The failure rates for the hardware modules APU, ALU and the backup system are the sum of the theoretical failure rates of the power supply, backplane, CPU, memory, carrier board, digital input piggyback, digital output piggyback and serial I/O piggyback (MIL-HDBK-217F [1]). For the backup system similar reliability data is assumed as for the digital I&C system. Also for the sensors are similar data used as for the sensors of the digital I&C system.

Table 1: Data used for the fault tree model

Component	Failure rate [h^{-1}]	Remarks	Data source
Sensor out of range	$1.0 \cdot 10^{-6}$		[2]
Sensor in range but low	$1.0 \cdot 10^{-6}$	CCF all sensors: 5.0E-9 [h^{-1}]	-
APU/ALU/ Backup system	$1.6 \cdot 10^{-5}$	It is assumed that all these modules consist of the same hardware, but have different software modules with the same reliability.	[3]
PAC	$4.4 \cdot 10^{-5}$	This value corresponds to the failure rate of the AV42, which is a priority and actuation control module.	[4]

Diagnostic coverage

The ratio between detected and undetected failures is largely determined by the coverage of the monitoring. In other words, what percentage of failures will be recognized by monitoring? The diagnostic coverage used in the model is 99%. This is the highest diagnostic coverage that is claimed in the IEC 61508 standard for requirements for electrical/electronic/programmable electronic safety-related systems [5].

Test intervals

Test intervals are an important measure to limit the impact of undetected failures on the system failure probability. At this point it is not known what the test intervals are. Therefore the test intervals are assumed to be the same for all modules and are set to 1 year (8760h), which corresponds to the yearly outage of most nuclear power plants.

Repair times

Repair times depend on the kind of failure, accessibility and availability of spares. For example, an I/O module can be replaced with power on the backplane bus, but if processing units have to be replaced, the power of the backplane bus has to be shut down.

For the repair times a realistic assumption has been made, namely 8h. Assuming that repair can be carried out in one shift. The replacement time reflects the repair time from detection until the notification that the system is repaired.

Common cause failure (CCF)

CCF is the failure of two or more identical or similar components as the result of a common cause which has not been explicitly modeled, for example design errors or errors during maintenance. In redundant systems, common cause failure (CCF) can lead to loss of more than one channel or train. Modeling CCF is important, because CCF can be a dominant contribution to system unavailability.

The following groups have been considered for CCF analysis:

1. Sensors;
2. APUs (Acquisition and Processing Units);
3. ALUs (Actuation Logic Units);
4. PACs (Priority and Actuation and Control).

The Binomial Failure Rate model (BFR model) has been used for the quantification of the CCF. In the BFR model both dependent and independent failures are considered. For the modeling of CCF one is interested in the dependent failures. For this study the generic data from the report "European Utilities requirements for LWR Nuclear Power Plants" [6] have been used. Which leads to a percentage of common cause failure of 14% (=common cause failure probability/total failure probability). This does not mean that 14% of the failure on demand is caused by CCF. For example 3 out of 4 APUs have to fail to contribute to the failure on demand. The undetected common cause failure of 3 out of 4 APUs is in the order of 10^{-8}, while a combination of three APUs failing independently is in the order of 10^{-21}, which is much less than the failure rate for common cause failure. Therefore common cause failure is a dominant contribution to the total failure on demand.

For the CCF of sensors other CCF values have been chosen, based on the values known from analog systems. The parameters of the BFR model are chosen such that the probability of a CCF of the sensors (for example due to calibration errors) is 0.6% of the total failure rate of one sensor.

4. RESULTS
4.1. Dominant cutsets

The dominant cutsets provide valuable information on the model. In Table 2 the dominant cutsets are shown. The first 10 cutsets represent in each case more than 95% of the total failure rate. The first order cutsets, which are all common cause failures, provide the largest contribution to the PFD. In all three models, the dominating cutset is an undetected common cause failure of the priority modules (PAC), representing more than 60% of the total probability of failure on demand (PFD).

Compared to design 1 (without backup system) the following is observed for design 2 and 4:
- Design 2, automatic switchover, addresses only one failure mode: a detected CCF of 4oo4 ALUs (0.8% of PFD), because in case of a detected failure of all ALUs the system will switch over to the backup system. The other cutsets of design 2 are the same as the cutsets of design 1.
- Design 4, continuous backup, also addresses calibration errors of the sensors (14% of PFD), because two independent sensors are available for the backup system, and undetected CCF of the ALUs (4.6% of PFD). The new dominant cutsets compared to design 1 represent combinations of CCF in the backup-system together with CCF in the digital systems.

Table 2: Dominant cutsets. Cutsets that show up in every model are marked grey.

	no backup (design 1)	backup that is coupled in, in case of failure of the normal system (design 2)	backup in equal and continuous operation (design 4)
1	Undetected CCF of 2oo2 PACs	Undetected CCF of 2oo2 PACs	Undetected CCF of 2oo2 PACs
2	Calibration error of 4 out of 4 sensors	Calibration error of 4 out of 4 sensors	Detected CCF of 2oo2 PACs
3	Detected CCF of 2oo2 PACs	Detected CCF of 2oo2 PACs	Combination of independent undetected failures of 2oo2 PACs
4	Undetected CCF of 4oo4 ALUs	Undetected CCF of 4oo4 ALUs	Combination of undetected failure of PAC1 and detected failure of PAC2
5	Combination of independent undetected failure of 2oo2 PACs	Combination of independent undetected failures of 2oo2 PACs	Combination of undetected failure of PAC2 and detected failure of PAC1
6	Undetected CCF of 6oo8 APUs	Undetected CCF of 6oo8 APUs	Combination of detected independent failure of 2oo2 PACs
7	Detected CCF of 4oo4 ALUs	Undetected failure of 7oo8 APUs	Calibration error of the sensors of the backup system in combination with calibration error of the sensors of the standard system
8	Undetected failure of 7oo8 APUs	Combination of undetected failure of PAC1 and detected failure of PAC2	Calibration error of the sensors of the backup system in combination with undetected CCF of 4oo4 ALUs
9	Combination of undetected failure of PAC1 and detected failure of PAC2	Combination of undetected failure of PAC2 and detected failure of PAC1	Undetected CCF of 2oo2 backup systems and a calibration error of the sensors of the standard systems
10	Combination of undetected failure of PAC2 and detected failure of PAC1	Combination of detected independent failure of 2oo2 PACs	Undetected CCF of 2 oo2 backup systems and an undetected CCF of 4oo4 ALUs

4.1. Failure to actuate on demand

The quantification results (PFD) of the three models are summarized in Table 3. The results show that the impact of a switchover backup system (design 2) on safety is very small from a probabilistic point of view (1% lower PFD). This can be reasoned from the cutsets, since the dominant cutset represent failures that are not addressed by the backup system, in particular, undetected failures and failures of the priority logic.

The backup that is in continuous operation with an equal vote as the digital system does have a significant effect on the PFD, because also the undetected failures of the ALUs and APUs, and the calibration errors are addressed by this configuration.

Table 3: Summary of the quantification results of the three models

	no backup (design 1)	backup that is coupled in, in case of failure of the normal system (design 2)	backup in equal and continuous operation (design 4)
1st order	$1.5 \cdot 10^{-4}$	$1.5 \cdot 10^{-4}$	$1.1 \cdot 10^{-4}$
2nd order	$6.9 \cdot 10^{-6}$	$6.8 \cdot 10^{-6}$	$6.6 \cdot 10^{-6}$
3rd order	$7.9 \cdot 10^{-9}$	$7.9 \cdot 10^{-9}$	$6.5 \cdot 10^{-10}$
Higher order	-	-	-
Total	$1.6 \cdot 10^{-4}$	$1.6 \cdot 10^{-4}$	$1.2 \cdot 10^{-4}$

4.2. Sensitivity analysis common cause failure

The impact of the percentage of CCF on the PFD has been analyzed by adjusting the ratio between the common cause failure rate and the total failure rate of a component. The results are shown in Figure 4. The response to the percentage of CCF is the same for all three architectures. The higher the percentage of CCF the larger the difference is between the PFD of the model with and without backup.

The difference between design 2 (switch-over backup) and design 1 (no backup) remains very small: 4% lower PFD for a CCF of 20%. As the percentage of CCF increases the contribution of CCF to the PFD also increases. Consequently the impact of detected CCF of the ALUs increases, which is addressed by the backup system. However this effect is masked by the increased contribution of cutsets that are not addressed by the backup system, for example CCF of the priority logic.

The system with a backup in continuous operation (design 4) performs better for all values. The difference increases as the percentage of CCF increases.

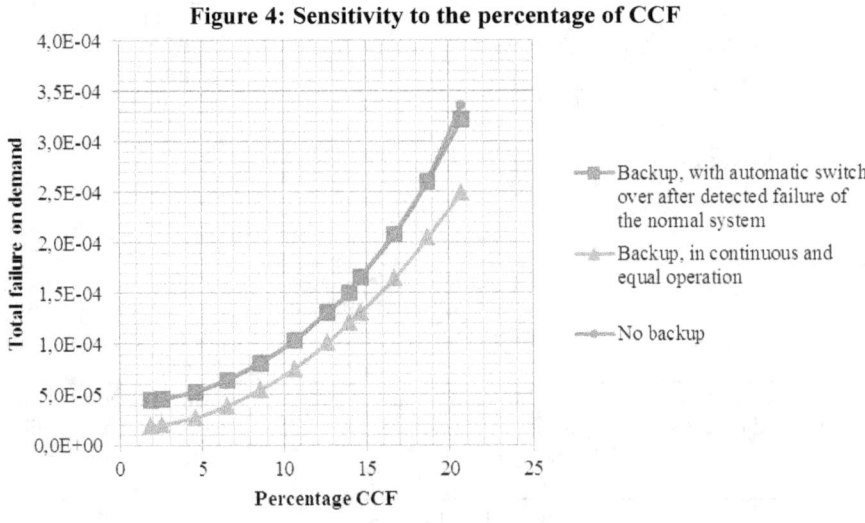

Figure 4: Sensitivity to the percentage of CCF

4.4. Sensitivity to the diagnostic coverage

The value of the diagnostic coverage is uncertain. Additionally it is known that more detailed models (lower level of abstraction) show a higher importance of detected failures, i.e. the diagnostic coverage is higher than could be expected from the model with a high level of abstraction. Therefore it is important to know the sensitivity of the results to the value of the diagnostic coverage. The diagnostic coverage has been varied from 80% to 100%; the effect on the PFD of the three models is shown in Figure 5.

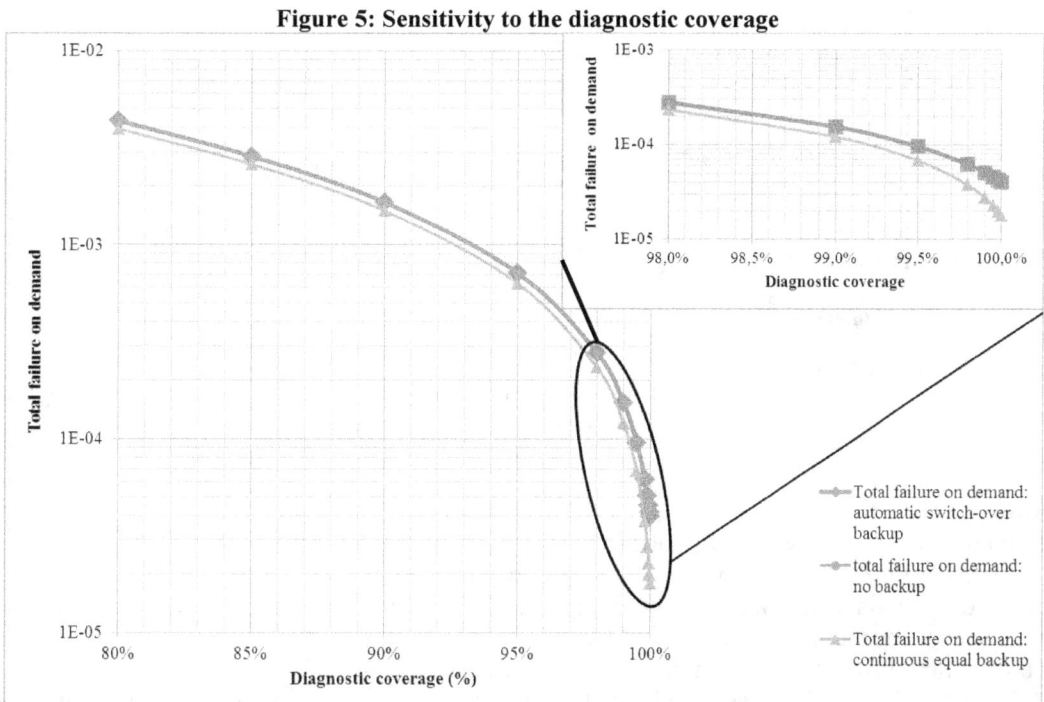

Figure 5: Sensitivity to the diagnostic coverage

The results show that the PFD is very sensitive to the diagnostic coverage. The PFD changes almost two orders of magnitude between 90% detected failures and 100% detected failures. Also, the higher the diagnostic coverage, the more sensitive to the diagnostic coverage the model will be.

However the differences between the models are very small as the diagnostic coverage is changed. In the limit of 100% diagnostic coverage the dominating cutsets of design 1 (no backup) are from CCF of the sensors and CCF of the PAC, neither are addressed by a switchover backup. The failure that is addressed by the switchover backup, detected CCF of the ALUs, presents only 3% of the total PFD. Therefore the system with an automatic switchover backup system will only perform 3% better than the system without backup. In the limit of 100% diagnostic coverage a continuous backup will perform 57% better than the system without backup, because also the CCF of the sensors is addressed.

Note that if all detected failures that are addressed by the backup system would lead to a predefined failsafe value, the models would give the same result. However in the present study it is assumed that the detected failures of the ALUs and the sensors will not lead to a failsafe output.

4.5. Impact of the test interval

The test interval has a high impact on the PFD. As the test interval is decreased, the PFD also decreases. This can be explained by the fact that the undetected failures have a less pronounced impact on the PFD in this case. The results are shown in Figure 6. The dominating cutsets of design 1 (no backup) for a chosen limit value for the test interval of 1h are: 1) detected CCF of the PAC (91% of the PFD), 2) detected CCF of the ALU (7% of the PFD). From these failures only the detected CCF of the ALUs is addressed by a backup system (both design 2 and 4) because a backup system cannot address CCF of the priority logic (PAC). Therefore, for both designs, switchover backup and continuous backup, a decrease in the total PFD of ~7% results when a test interval of 1h is chosen (PFD without backup ≈ $1.8 \cdot 10^{-5}$).

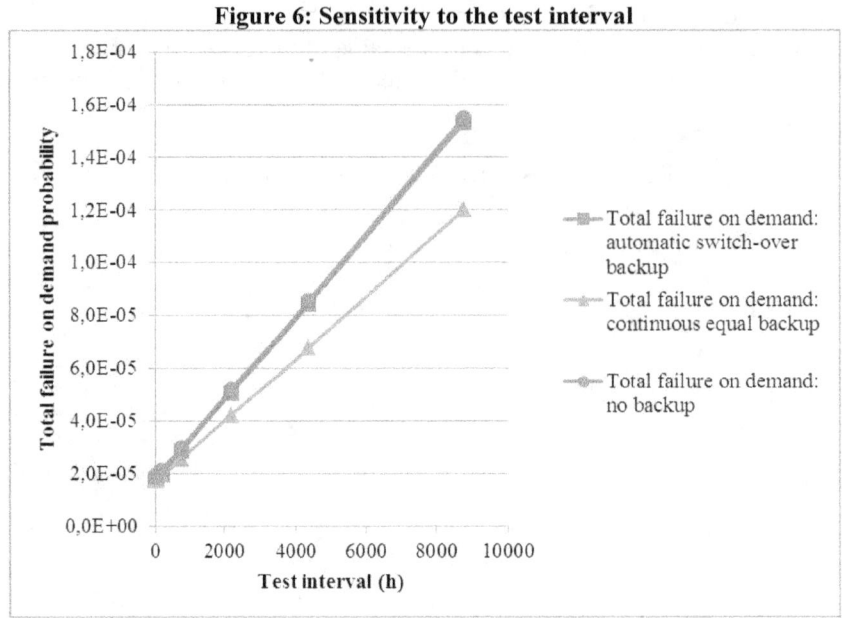

Figure 6: Sensitivity to the test interval

4.6. Impact of the failure rate of the priority logic

Since common cause failure of the priority logic has the highest contribution to the PFD, the model is very sensitive to changes of common cause failure probability of the priority logic. For all three models, the common cause failure probability has been changed in order to observe whether the backup systems have a more pronounced effect on the PFD in case of a lower common cause failure probability of the PAC. As is shown in Table 4 the effect of the backup systems is more pronounced if no CCF between the PACs is assumed (for example, because the PACs are diverse). Especially the backup in continuous operation shows significantly better results than the system without backup. The switch-over backup shows slightly better results (~3%) than the system without backup.

Table 4: Impact of the CCF probability of the PACs on the PFD

CCF of 2oo2 PACs	no backup	switch over backup	continuous backup
• Detected: $4 \cdot 10^{-4}$ • Undetected: $4 \cdot 10^{-6}$	$5.6 \cdot 10^{-04}$	$5.6 \cdot 10^{-04}$	$5.3 \cdot 10^{-04}$
• Detected CCF: $4 \cdot 10^{-5}$ • Undetected CCF: $4 \cdot 10^{-7}$	$1.6 \cdot 10^{-04}$	$1.6 \cdot 10^{-04}$	$1.2 \cdot 10^{-04}$
no CCF	$4.1 \cdot 10^{-05}$	$4.0 \cdot 10^{-05}$	$6.6 \cdot 10^{-05}$

5. CONCLUSION

How the backup is implemented does have an influence on the impact of the backup system on the probability of failure on demand (PFD). In general a backup system in continuous operation and with an equal vote as the digital system has a more pronounced effect than a backup system that is used when the standard systems have failed detectably.

Several parameters have been changed to determine the influence on the model: the percentage of common cause failure, the diagnostic coverage, the test interval and the common cause failure rate of the PAC. The results are summarized in Table 5.

Table 5: Summary of the results in terms of % lower PFD with backup compared to the system without backup

	No backup PFD	Switch-over backup (design 2) (% lower PFD)	Continuous backup (design 4) (% lower PFD)
Values as described in section 3.2	$1.6 \cdot 10^{-4}$	1%	23%
20% of CCF	$3.4 \cdot 10^{-4}$	4%	26%
2% of CCF	$4.5 \cdot 10^{-5}$	2%	58%
Diagnostic coverage of 100%	$4.1 \cdot 10^{-5}$	3%	57%
Diagnostic coverage of 80%	$4.3 \cdot 10^{-3}$	0%	9%
Test interval of 1h	$1.9 \cdot 10^{-5}$	7%	8%
No CCF between PACs	$4.1 \cdot 10^{-5}$	3%	84%

If the backup system is only used when the other systems have failed detected, the impact on the PFD is very small. The dominating cutsets are: 1) failure of 2 out of 2 priority logic modules (PAC), either CCF or independent, either detected or undetected, 2) CCF of 4 out of 4 sensors, 3) undetected CCF of the ALUs, 4) detected failure of the ALUs (0.8% of the total PFD). A switchover backup can only address the detected failure of the ALUs, therefore the maximum decrease of the PFD with a switchover backup is 0.8%. The continuous backup can also address CCF of the sensors (if other, independent sensors are used) and undetected failure of the ALUs, which leads to a decrease of the PDF of 23%.

Furthermore it is shown that if the percentage of CCF, diagnostic coverage, the test interval or the failure rate of the priority logic is changed, the decrease of the PFD of a switch-over backup system is relatively small (maximum 7%). The relatively small gain needs to be balanced to the increased complexity and possibilities for spurious failures. Additionally, another possibility to mitigate detected failures is to set the failed module to its predefined output value, and if possible to a failsafe value. In this study a failsafe value is only assumed for the APUs and not for the ALUs and PACs. In the limit that all failures are detected and lead to a failsafe status, the backup system will not have any impact.

If the backup system is used in continuous operations with an equal vote to the digital system the performance is significantly better than for the system without backup. The calculated impact is for one function, not for the complete digital I&C system. The impact on the complete system will be less, since the standard I&C systems are already diverse and implement diverse functions.

Other possibilities to improve the safety of a digital protection function are: 1) increasing the diagnostic coverage; 2) investing in timely detection for example by periodic testing; and 3) diversifying the priority logic.

Acknowledgements
This research has been carried out under contract with the Dutch Ministry of Economic Affairs and the Kernfysische Dienst (KFD).

References
[1] Department of Defense, *"Military Handbook Reliability Prediction of Electronic Equipment (MIL-HDBK-217F)"*, 2 December 1991, Washington DC.
[2] J. Schüller, *"Methods for determining and processing probabilities"*, NRG, 1997, Den Haag.
[3] NRG, *"Faalkansanalyse NSTA, NSTA - Vernieuwen NSS en Startautomaat"*, CMG Public Sector B.V., 2002.
[4] Stefan Authen, Kim Björkman, Jan-Erik Holmberg, Josefin Larsson, *"Guidelines for reliability analysis of digital systems in PSA context - Phase 1 Status Report"*, NKS, 2010.
[5] "IEC 61508-2: requirements of electrical/electronic/programmable electronic safety-related systems", CEI/IEC 61508-2:2000, first edition, 2000, Geneva.
[6] "European Utilities Requirements for LWR Nuclear Power Plants", Volume 2: Generic Requirements, Chapter 17, PSA methodology, appendix C, Method and Data for Treatment of Common Cause Failures, Revision B, November 1995.

Modeling of Digital I&C and Software Common Cause Failures: Lessons Learned from PSAs of TELEPERM® XS-Based Protection System Applications

Robert S Enzinna[a], Mariana Jockenhoevel-Barttfeld[b], Yousef Abusharkh[b], and Herve Bruneliere[c]

[a]AREVA Inc. Lynchburg, VA, USA
[b]AREVA GmbH, Erlangen, Germany
[c]AREVA SAS, Paris, France

Abstract: The authors have created probabilistic safety assessment (PSA) models of TELEPERM® XS (TXS)-based digital protection systems for a variety of nuclear power plant applications in the USA and around the world. This includes PSA models for digital protection system upgrades, and protection systems for new reactor builds. The PSA models have involved detailed digital instrumentation and control (I&C) fault tree models that have been fully integrated with the full plant PSA model. This paper discusses lessons learned, insights, and modeling recommendations gleaned from this experience.

The paper discusses recommended level of modeling detail, development of failure rate and fault coverage data, treatment of fault tolerant design features and common cause failure (CCF) defenses, fault tree modularization/simplification, and other topics of interest. Practical suggestions for PSA modeling are made based on experience gained from actual digital I&C PSA models built for several internal and external customers.

Modeling of CCF for the TXS hardware modules and for the software is highlighted, especially focusing on the quantification of software common cause failures (SWCCF). The authors describe the methodology used for quantification of SWCCF in the PSA studies, the definition of realistic software CCF modes, and estimation of failure probabilities.

Keywords: PSA, PRA, Digital I&C, Software Common Cause Failure.

1. INTRODUCTION

The guidance provided herein draws upon lessons learned from previous probabilistic safety analysis (PSA) work for digital I&C designs, including the EPR™ advanced nuclear power plant projects in Europe and the U.S., and digital reactor protection system (RPS) and engineered safety features actuation system (ESFAS) upgrades such as for the Oconee nuclear plant. The Oconee RPS/ESFAS upgrade was the first full-scale digital RPS/ESFAS replacement in the US nuclear industry, and the reliability analysis performed for that project was a first-of-a-kind (FOAK) for a U.S. PSA. The purpose of that analysis was to satisfy customer and regulatory requirements for reliability analysis of the safety related I&C system.

The reliability model was created as a stand-alone fault tree model but with the intent that it could be integrated into the customer's plant PSA at a later time. The model was subsequently integrated into the Oconee PSA by the customer after the system was installed.

The lessons learned from the Oconee digital I&C model were then used and developed further in the U.S. EPR™ PSA. For the U.S. as well as for the European EPR™ projects, the digital I&C fault tree models were fully integrated with the overall PSA development from the start. Since previous PSA's for the existing nuclear plant fleet usually contain neither fully digital control rooms, nor fully digital safety-related I&C systems, the EPR™ projects were also unique with respect to the level of regulatory scrutiny and review that the PSA models have received.

Parallel efforts by AREVA teams in the U.S, France and Germany have built digital I&C models for TELEPERM® XS (TXS) customers around the world. These PSA models have reflected the varying needs and dictates of customers and regulatory authorities in several different countries. The lessons learned that are summarized in this paper encompass that broad and diverse experience.

Much of the digital I&C experience at AREVA revolves around the TELEPERM® XS digital I&C, which is the system platform developed at AREVA GmbH to implement safety I&C systems with highest safety responsibility. The first TXS systems were put into operation more than 20 years ago and have an excellent reliably record. TXS I&C systems have been installed in over 60 units at 28 plant sites located in 11 countries and utilizing 10 different reactor designs. The TXS components have clocked billions of hours of operating experience without a common cause failure (CCF) of either software or hardware. Indeed even random failures are extremely rare. This track record is largely due to the efforts of our colleagues at AREVA GmbH who designed the TXS platform from the start to be highly reliable and immune from software common cause failure. The designers of the system studied the failure modes and failure causes of conventional computer-based systems, and then built the TXS platform to exclude those failure modes. Features were built into the platform to reduce latent software defects, eliminate failure triggers, eliminate failure propagation, and minimize failure consequence. One of the challenges of the PSA modeling effort has been to reflect those CCF defenses in a fair and realistic way, while creating a methodology that is sufficiently robust yet practical in its application.

The following sections discuss the nuances and unique issues associated with PSA modeling of modern digital I&C systems, which have surfaced during the collective experience of the authors' work.

2. LESSONS LEARNED FROM TELEPERM® XS PSA STUDIES

2.1. Lesson 1: Operating Experience shows that SWCCF is Rare in a Well-Designed System

Over the last 20 years, TELEPERM® XS systems have been installed in over 60 nuclear units. These systems contain two versions of the main TXS processing module. These TXS processing modules have clocked 250 million hours of operation without experiencing a SWCCF. Indeed, even processor module hardware failures during this accumulated field experience are very rare (fewer than 20). Some of the reasons for this outstanding reliability will be explained in the paper.

There is a tendency, especially by regulators, to be conservative or "bounding" and ascribe SWCCF probabilities to hypothetical failure modes such as "failure of all computerized I&C." However, taking diversity requirements into account that are usually included between the different reactor protection subsystems, no relevant/realistic CCF mode of the software can be identified for TXS that would cause the complete failure of the system (e.g., both subsystems).

It is important that the failure modes and effects to be included in the PSA model are credible and are assigned realistic probabilities.

It is cautioned that if the assumed CCF is overly conservative, that it may disguise more meaningful insights. We are not suggesting that SWCCF be omitted from the PSA model, or be overly optimistic. However, if the assumed SWCCF dependency is unrealistic, then its inclusion may mask the importance of other failures modes and the value of corresponding design countermeasures. Therefore, the primary emphasis should be to also include software (SW) failure modes and effects in the PSA model that are realistic relative to the design features of the system.

2.2. Lesson 2: Understand why the CCF Defenses in the Platform Design are Important

As a vendor of safety-related computer systems for nuclear power plants (NPPs), AREVA has studied the failure modes of SW in standard computer systems. This research has resulted in features and defenses in the TXS design to rule out many common SW failure modes and reduce the frequency and consequence of others [1].

An example is the so-called "data-storm" event. This common failure mode, which afflicts standard computer systems, occurs when "special loading" taxes the operating system (OS) capacity. This is a failure mechanism where the communication bus is bogged down by excessive network traffic. It is therefore important for the reliability analyst to understand whether the system uses networks with variable loading (and examine associated loading analysis), or whether the system is the type that uses cyclic processing and invariable bus loading, like TXS. Strictly cyclic operation and constant loading of communication and processing buses involved in TXS prevent this failure mode and ensure that an actual system demand puts no more stress on the OS than any other cycle.

Operating experience in standard digital systems also indicates that interference between application program data (e.g., due to dynamic memory allocation) and faults in releasing system resources (e.g., time dependencies due to internal system clock) are leading causes of failure. In TXS-based systems these failure modes are eliminated by static memory allocation and asynchronous operation. As a general rule, interference from the application SW on the OS and hardware resources is forbidden, and consequences such as process-driven interrupts are not allowed. These features alleviate leading failure causes (such as OS lockup due to memory conflict) that plague standard computer systems. Therefore it is important for the reliability analyst's understanding to know whether the system in question uses dynamic or static memory allocation.

OS features such as invariant cyclic processing, and invariance of process and communication bus load, are designed to reduce failures due to external influences and ensure that the stress during a demand is the same as during a normal non-demand cycle. These features remove dynamic interaction failure mechanisms from the design. A primary reason for the use of deterministic program execution and cyclic operation in the OS platform is to disconnect the OS from the signal trajectories and establish a pattern of predictable system behavior. Deterministic program execution limits the opportunity for failure due to untested software paths and data sets because there is only one path through the software instructions, which does not change in response to input state changes or plant initiating events.

The platform and OS design have an important role in limiting SW failure triggers and failure consequences. Additional detail on failure modes and defenses is available in the referenced document [1], as well as in industry consensus documents such as IEC-62340 [2].

2.3. Lesson 3: SWCCF Recommendations for Application SW

From a PSA perspective, the authors favor a SWCCF quantification methodology that is realistic, and practical to apply. From the design perspective, the methodology must recognize the value of the defensive features that are built into the platform design and into the system architecture.

In a TXS-based system, defense against SWCCF involves four constituent parts:
- A software lifecycle (SLC) process that reduces latent errors,
- A platform design that reduces failure triggers,
- Platform features that eliminate failure propagation (minimize failure consequence), and
- Functional diversity.

The SWCCF methods that are described below achieve the goal of realistically reflecting the design features that influence SWCCF defense, without requiring excessive PSA resources. Hence the PSA analyst's attention is focused on the design features that most influence safety, and this helps to inform his/her interactions with the design team.

A probability estimate can be obtained via expert elicitation, given that the experts used have a good understanding of both the features of the SLC development process, and of the digital I&C platform design and its OS defensive measures. The analyst should understand the degree of customization that is allowed in the application software. Features such as the exclusive use of function block libraries (reusable software functional blocks that are simple, fully tested, verified, and rigorously controlled), automated development tools, and automatic code generation help reduce SW errors. It is also important for the PSA analyst get an appreciation for the functional specification process. (Is it a formal process? Is it "user friendly" for both the process and I&C engineers? How is it checked, verified, tested?) The PSA analyst should be familiar with the V&V methodology, the associated tools (e.g., simulation, inverse checking), and how the process conforms to applicable standards of good practice.

For example, TELEPERM® XS uses a functional specification process based upon functional diagrams. These function diagrams are difficult to distinguish from those used for traditional analog designs, and are deliberately designed to be familiar to both the process design engineers and the I&C engineers. The "components" on the functional diagram (bistable, summer, etc.) are software function blocks that mimic their analog counterparts. The executable code is generated automatically from the diagrams, and is checked via simulation, and other tools. TXS has no custom SW development in the traditional sense.

The platform and OS design also have an important role in limiting triggers of application software failure. With deterministic program execution there is just one path through the program instructions, and all of the application code is executed on each cycle. The objective of this type of design is to limit the opportunity for failure due to untested software paths and data sets.

We are wary of SWCCF estimation methods that ignore the platform design. Much of the published research tends to over-emphasize the software development aspect and ignore the profound effect that the platform design characteristics have on reducing SW failure and CCF. Research that is biased towards customized made-from-scratch software development is less valuable for platforms like TXS which restrict use of customized SW and employ a simple predictable operating system. We are suspicious of methodologies that attribute all of the failure probability to the likelihood of a SW defect because they ignore the second aspect of SW failure probability, which is the likelihood of a failure trigger. The defenses built into the I&C platform to reduce triggers have a marked effect on SW failure probability. CCF also requires propagation to redundant trains or diverse functions. And so the defenses built into the platform to reduce failure propagation and consequence are very important as well.

Another importance aspect of the platform design that is often overlooked in the research is the configuration control. When evaluating the software quality and V&V process, the entire SW lifecycle is important, not just the initial development. A good defense involves the whole life-cycle, because approximately 36% of the failures in generic digital I&C operating experience are introduced during maintenance and update activities occurring after product installation [3].

An expert elicitation process can compare the features of the system and process in question with the features typically associated with other high-reliability applications. IEC 61508 [4] and IEC-62340 [2] are suggested as a guideline for this engineering judgment. These are an industry good practices documents and IEC 61508 provides consensus estimates of reliability targets that can be achieved for differing safety integrity levels (SIL). Rigorous guidelines for compliance with each SIL are provided for both hardware and systemic (SW development) aspects of the design process.

Safety Integrity Level Targets From IEC-61508

SIL	Low demand mode (Probability of failure on demand)
4	$\geq 10^{-5}$ to $< 10^{-4}$
3	$\geq 10^{-4}$ to $< 10^{-3}$
2	$\geq 10^{-3}$ to $< 10^{-2}$
1	$\geq 10^{-2}$ to $< 10^{-1}$

In PSA studies, we have modified the failure probability within the target range, much like a performance shaping factor (PSF) would be used in human reliability analysis (HRA). The value of the PSF used is based on functional complexity.

In early studies we used a simple complexity adjustment that was based on application SW function. A simple one-parameter trip signal would be assigned a failure probability at the low end of the range, for example 1E-5/demand for a simple trip on high pressure in a SIL-4 system. A more complex trip that used two parameters would be assigned the base value for each parameter in the function, for example 2E-5/demand for a trip that combined high pressure and temperature. Redundant channels with the same software are conservatively treated with complete dependence.

Since the safety system designs being analyzed have extensive features to protect against propagation of failure between diverse functions, it was reasonable to assign application SWCCF probabilities to individual software functions, or groups of software functions (characterized by having the same functional requirement specifications, such as plant parameter inputs, algorithms and/or data trajectories), that are common to multiple processors. In the example mentioned above, if the pressure sensor input for the two functions was the same, then this introduced a SWCCF dependency as well (CCF trigger: same signal). Hence functions that were truly diverse (no common input parameters) got more credit in the PSA than functions that shared input.

In more recent work we are exploring the use of a more sophisticated complexity metric to shape the base SWCCF probability. This metric is under development, and analyzes the actual source code that is created by the TXS automatic code generator.

In our example PSA application, we also assigned a beta factor between any diverse SW functions that may appear in the same minimum cut set of the PSA. We used the beta factors when it was necessary to judge the coupling between similar, but not identical, SW functions.

Design standards such as IEC 62340 [2] provide strong endorsement of functional diversity as a defense against SWCCF. When coupled with the other defenses (reducing defects, reducing triggers, preventing propagation), function diversity provides an effective defense against specification errors, and reduces the probability of a common failure trigger by employing different signal trajectories. The functional diversity may be achieved within the digital system by using different input parameters, algorithms, and data trajectories, as well as by using diversity inherent in the plant process systems. The OS defensive features, discussed above, provide assurance that software failures do not propagate to diverse functions. The functional diversity is even more effective if it is implemented on independent computers.

Since the functional diversity may involve using some common SW elements, the PSA analyst may desire to model a dependency between the two digital functions using the familiar beta-factor approach. Quantification of these beta factors using hard data or analytical methods is difficult. Therefore, assignment of beta factor values will require the use of expert judgment, based on a qualitative assessment of the similarities between the functions. The recommended beta factor is between 0.1 and 0.001 depending on the similarity of the software functions (input parameter, algorithm, and data trajectory). Some suggestions are available in the referenced papers [5, 6].

2.4. Lesson 4: SWCCF Recommendations for Operating System SW - It is Helpful if the Platform has a Proven Track Record

If the OS used in the NPP I&C platform is supported by a mature operating history, then this may allow statistical inference methods to be used to assess this part of the software failure probability. For example, the AREVA TXS platform has performed for more than twenty years in dozens of plants worldwide. The computer processor modules have 250 million cumulative operating hours of service, without an OS failure. The authors have used this experience to generate upper bound failure probabilities using 95%-chi-squared statistics. This treatment is possible because the OS has features to ensure its independence from the application SW and from the plant process, and therefore the OS failure rate is not influenced by variations in the application SW or by interference from transient loading.

The fundamental contributor to OS reliability is the features that limit the propagation and the severity of application SW failures. Specific features of the OS platform such as strictly cyclic operation, constant bus loading, static memory allocation, and prohibition of process-driven interrupts are used to ensure predictable OS performance and behavior that is free of interference from the application program. These features are designed to ensure that application SW failures caused by special loading, unanticipated input signal trajectories, or other application program design errors will not affect the OS, and hence propagate a failure to other functions.

In safety-related NPP applications, there are also requirements for independence between redundant channels. Simultaneous OS failure in independent safety-related computers is rare, and not observed in the field data. Therefore, bounding statistical treatment (given sufficient failure-free operating experience) and/or expert judgment will be necessary to quantify the probability of CCF of the OS in redundant channels. However, even a very small probability assumed for system-wide CCF failure of the OS may dominate the PSA results. Therefore, it is cautioned not to be overly conservative with these estimates, as that may mask the sensitivity of the PSA to more realistic failure modes, and the design features (hardware and software) that influence them.

2.5. Lesson 5: There is no Substitute for Vendor Failure Rate Data for I&C Modules

AREVA has accumulated an extensive failure rate data base for the various modules that are used in the TXS platform. When modules are new, theoretical estimates are generated using the part-stress analysis methodology of the Siemens SN29500 standard. Once modules are in service, field data is collected and updated on a quarterly basis. Since TELEPERM® XS installations have been operating for 20 years, the operating experience is extensive. The most common TXS modules, such as processing modules and input/output (I/O) modules, have accumulated hundreds of millions of operating hours. The field data for these modules is processed to produce best estimate as well as 95%-chi-squared upper bounds. As the field data accumulates, the theoretical estimates from the part stress analysis invariably prove to be conservative, even compared to the upper bound field data. Since the design and reliability of digital I&C modules is very much vendor specific, the authors cannot recommend a generic data base that is a good substitute.

2.6. Lesson 6: Adjust Level of Modeling Detail to Availability of Data and Supporting Analysis

For digital I&C, the failure data (field data or theoretical estimates) is usually generated at the module or board level of detail. This provides a convenient level of detail for the PSA model.

However, the I&C design team will usually produce other useful analysis that is at a much finer level of detail. This may include failure modes and effects analysis (FMEA) at the piece part level, fault coverage analysis aligned to the failure modes of specific circuits, and failure mode taxonomy for specific types of triggering events.

This is contrasted with the PSA model which is typically aligned with functional failure of the associated process system. The I&C failure modes are reflected at the functional level of the actuated component or system, regardless of how the digital system itself may fail. However, understanding of the failure mode taxonomy is important for other reasons, namely allocating parameters for failure likelihood and detectability (fault tolerance), identifying common dependencies between functions, and for guiding the PSA analyst in identification of which components (hardware or software) can contribute to loss of safety function.

The PSA model should also be a tool that drives the design to improve. IEC standard 62340 [2], provides useful insights on the leading causes of latent defects (e.g., specification errors), and failure triggers (e.g., environmental stress, input signal trajectory). The standard provides recommendations for reducing latent defects, reducing failure triggers, and for reducing consequences to other channels and functions. Paramount in these recommendations is the use of functional diversity in the design. This includes functional diversity within the digital system design, as well as potential diversity that exists in the plant process systems. Functional diversity provides double protection because it safeguards against functional specification errors as well as triggers in the data trajectory. The focus of the standard suggests that an effective level of detail for the SW failure contribution in the PSA model is one that recognizes and encourages functional diversity in the design.

Diverse functions are helpful to some degree whether they reside on the same processors, on separate processors, or on an entirely different system. In any case, the PSA should make an assessment of the effectiveness of the OS design features that are supposed to prevent software failures from propagating to other functions and assign a reasonable contribution for OS CCF (appropriate to the degree of separation) to capture the probability that this objective is not achieved. If the assigned CCF probability is too conservative, then it may have the undesired effect of discouraging functional diversity.

2.7. Lesson 7: Fault Coverage Data is as Important as the Failure Rate Data

Each component or module has a parameter called "fault coverage." Fault coverage is an estimate of the percentage of the failure rate for each module that represents self-monitored (SM) versus non-self-monitored (NSM) failure modes. Failure modes that are self-monitored, or "covered," are those faults that can be detected and compensated for by the components downstream. To the PSA analyst, the coverage represents an estimate of the effectiveness of the fault-tolerant features and fault-propagation barriers in the integrated hardware/software design.

Fault coverage has an important role in the PSA model because it drives which mathematical unavailability model (repair-time model, test-interval model, or both) is used for each component. It determines if the reliability is modeled with a short or long mean-time-to-repair (MTTR). In a digital system, known failures can typically be repaired quickly via replacement of a rack-mounted module.

Undetected failures on the other hand may stay in the system for a relatively long time, for example until a scheduled surveillance test (periodicity is generally from few months to 2 years).

Because of the fault tolerant design, the system may compensate instantly for a "covered" failure. For example, in a protection system application of TXS, a certain module may perform a 2-out-of-4 coincidence logic. If the module senses that an input is faulted, then it is programmed to change the coincidence. As bad inputs are recognized, the coincidence can be programmed to transition from 2-out-of-4 to 2-out-of-3, then to 1-out-of-2 or even 1-out-of-1 if necessary (degradation). It can also be programmed to go to a pre-defined safe state, if desired.

Consequently, the postulated failures involving NSM failure modes will almost always dominate over the SM failure modes. This is true even if the NSM percentage of the failure rate is very small relative to the SM percentage. Therefore, the results are sensitive to the fault coverage parameter. Because of the importance of fault coverage, detailed FMEA of the TXS modules is performed to determine fault coverage.

2.8. Lesson 8: Failure Mode Taxonomy

The definition of realistic failure modes for SWCCF is an important input to the PSA study. In highly-redundant NPP safety systems, the specific values assigned to the SW reliability are less important to the overall system PSA model than the choice of which SW failure modes and effects (e.g., fails a single function, or fails multiple functions) to include in the model. In a multi-channel safety system, the PSA results are easily dominated by any SWCCF that is assumed to affect the function of redundant trains or diverse functions. It is therefore critically important that the SWCCFs that are included in the model are realistic, credible and representative.

2.9. Lesson 9: Use Fault Tree Modularization to Simplify the Analysis

Fault tree modularization is an effective means for simplifying the analysis. This may not be easy to accomplish given the integrated nature of digital I&C design applications. Modularization tends to increase model conservatism, but is a trade off with simplification of fault tree displays and presentation of minimum cut set results.

The referenced paper [7] has described the methodology developed by AREVA for the modeling of probabilities of failure per demand of I&C functions in the PSAs of Nuclear Power Plants for which the I&C detailed design (allocation of functions in the units) is clearly defined. The principle of the method (except the need for modeling software failures) remains applicable for analog platforms.

Sensors required for the elaboration of each I&C signal are individually modeled as well as their related conditioning components. Common cause failures are applied for sensors as well as for conditioning modules on a case by case basis. The elementary components used for the modeling of I&C processing parts are the single processing units. A processing unit typically consists of a sub rack with one or more processing modules, I/O modules and communication modules. A detailed model of the unit is developed separately and inserted into the fault tree as a modular basic event. CCF for the processing parts (or automation parts) are modeled at the functional level as well as the platform level.

This methodology has been successfully implemented in PSAs for new builds and existing plants.

The following advantages have been identified:
- The links between I&C and support systems are easy to implement in the model,
- The hazards analyses integrates I&C,
- A detailed modeling of units allows the detection of asymmetries or imbalances in the I&C design (inadequate allocation of signals in the processing units),
- This modeling is easily understandable with respect to the PSA cut sets analysis,
- The I&C architecture is accurately represented in the PSA.

3. CONCLUSION

Final Lesson: Always Remember that the Objective is to Improve the Design
The most important feature of a PSA methodology for digital I&C is that it represents credible failure modes and realistic (but not necessarily precise) likelihood estimates. SWCCF probabilities that are too conservative, or which represent hypothetical failure modes that are not credible, will drive the design function in directions that may not be productive. SWCCF estimates that are subjective, but well founded will better serve the design.

Since SW failures require both a latent defect and a trigger, the desirable PSA methodology must account not only for the characteristics of the software life cycle development process, but also for the characteristics of the platform and OS design that work to reduce the failure triggers. Also, since the NPP PSA study is primarily sensitive to CCF as opposed to failure of individual functions or channels, the desirable SW reliability method also addresses the likelihood that the SW failure propagates to redundant channels and/or diverse functions. Therefore, it is desirable that the methodology also consider the platform and OS design features that are intended to limit failure consequence. The best quantitative methodology is one that accounts for the features of both the design and of the SLC development process. This is the most important characteristic of a useful methodology, even if it results in a methodology that involves a high degree of engineering judgment and qualitative insight.

To represent the design fairly, it is important for the PSA analyst to understand the behavior of the system being modeled. This is accomplished through a close working relationship with the design activity, including the hardware and SW design as well as the SLC process. A FMEA from the design activity is especially helpful. Other useful material from design engineering may include functional diagrams, architecture diagrams, function block library definitions, fault coverage analysis, operating history, description of platform CCF defenses, and other information. It is also important for the PSA analyst to investigate the quality of the SLC process, to get an appreciation of where errors may be introduced (e.g., functional specification, SW maintenance and update), how they are avoided (e.g., formal specification methodology, reusable SW/function blocks, automatic code generation), where errors may be caught (e.g., V&V, testing, simulation), and how the process conforms to applicable standards of good practice. It is through these activities that the PSA engineer gets an appreciation for the effectiveness of the design and process defenses against defects, failure triggers, and failure propagation.

Acknowledgements
The authors acknowledge Dr. Arnold Graf, Dr. Christian Hessler, and others at AREVA GmbH who have devoted their careers to the study of digital I&C failure modes, and whose research has resulted in the

defenses of the TELEPERM® XS design that eliminate many software failure modes, and reduce the likelihood and consequence of postulated common cause failures.

References

[1] EMF-2110(NP)(A) Revision 1, *"TELEPERM XS: A Digital Reactor Protection System,"* May 2000, AREVA NP/Siemens Power Corporation. (ADAMS Accession Number ML003732662)

[2] IEC-62340, *"Nuclear Power Plants – Instrumentation and Control Systems Important to Safety – Requirements to Cope with Common Cause Failure,"* International Electrotechnical Commission.

[3] *"IEC 61508 Overview Report,"* version 2.0, exida, January 2006.

[4] IEC 61508, *"Functional Safety of Electrical/Electronic/Programmable Electronic Safety-Related Devices,"* International Electrotechnical Commission.

[5] Bob Enzinna, Li Shi, & Steve Yang (AREVA NP Inc.), *"Software Common-Cause Failure Probability Assessment,"* 6th ANS International Topical Meeting on Nuclear Plant Instrumentation, Control, and Human-Machine Interface Technology, Knoxville Tennessee, April 2009.

[6] *"Modeling of Digital I&C in Nuclear Power Plant Probabilistic Risk Assessments,"* white paper by Nuclear Energy Institute Digital I&C Working Group, July 2007.

[7] Hervé Brunelière, Caroline Leroy, Laurent Michaud (AREVA NP SAS), Nissia Sabri (AREVA NP Inc.), Peter Otto (AREVA NP GmbH), *"Finding the best approach for I&C modeling in the PSA in the different design phases"* 11th International Probabilistic Safety Assessment and Management Conference (PSAM11) & The Annual European Safety and Reliability Conference (ESREL12), Helsinki Finland, June 2012.

Methodology for Safety Assessment of Defense-in-Depth and Diversity Concept for Modernization of the Digital I&C of a NPP in Finland

Ewgenij Piljugin[a*], Jarmo Korhonen[b],

[a]Gesellschaft fuer Anlagen und Reaktorsicherheit (GRS) mbH, Garching, Germany
[b]Fortum, Power and Heat, Helsinki, Finland

Abstract: A new automation concept based on digital instrumentation and control (I&C) systems will be implemented in the Loviisa NPP plant in Finland within a modernization project. The new automation concept was developed by Fortum under consideration of the Defense-in-Depth and Diversity (3D) strategy. Different methodologies are used in several tasks of the design verification such as safety evaluation of the I&C functions, failure mode and effect analysis (FMEA) for identifying the relevant failure modes of the I&C hardware. The results of the analysis present generic and design specific issues. The generic issues primarily concern methodological aspects and design specific issues concern identifying failure modes of the I&C equipment, evaluation of the failures effects and the propagation paths, identification of candidates for common cause failure analysis (CCF), and identification of appropriate countermeasures to prevent or mitigate hazardous failure effects.

This paper presents some selected insights from the evaluation of the safety significant aspects of the reliability of the new digital I&C systems and discusses the results of V&V tasks from a methodological point of view. The identified issues should also support consideration of safety relevant aspects of digital safety important I&C systems in the probabilistic reliability analysis (PRA) of modernized nuclear power plants.

Keywords: Defense-in-Depth, diversity, CCF, FMEA, I&C, NPP, hardware, software.

1. INTRODUCTION

The Loviisa nuclear power plant (NPP) in Finland consists of two units with VVER-440 type pressurized water reactors. The first unit started operating in 1977, and the second in 1980. A new automation concept based on digital instrumentation and control (I&C) systems will be implemented in the Loviisa NPP within a modernization project. The Loviisa NPP was built to meet the most developed Western safety standards. Safety and operability is continuously improved by major modification projects. Most of the present I&C systems and as well as human-machine interfaces (HMI) of the main control room will be replaced. The new automation concept was developed by Fortum under consideration of the Defense-in-Depth and Diversity (3D) strategy control of transients and accidents. A first rough estimation of the reliability of the new automation concept was made at the early stage of the unfinished I&C systems design. Therefore some relevant aspects of the new digital I&C systems were not analyzed in depth, e.g., dependencies between digital I&C systems of different safety categories, potential common cause failures (CCF) of hardware and software. In the next stage of the design development a large amount of work was performed to analyze and systematically identify potential shortcomings in the new architecture of I&C systems and in the specification of I&C equipment. GRS experts have provided technical support to Fortum within the framework of verification of the design of selected I&C systems.

Different approaches and methodologies have been used in several tasks of the verification process taking into account the specific objectives of each task of the design verification such as top-down evaluation of the I&C functions, bottom-up approaches for identifying the relevant failure modes of the I&C hardware. The results of the analysis present different types of issues: generic and design specific ones. The generic type of issues primarily concern methodological aspects, e.g., appropriate grade of decomposition (functional breakdown) of the I&C systems in to functional units (e.g.

modules or basic components of the hardware, modules or elementary functions of the software) for performing an FMEA of an adequate level of abstraction [1, 2], consideration of self-monitored and self-revealed failures (failure modes), comprehensibility of the analysis, plausibility of assumptions. The design specific type of issues concern identifying probable failure modes of the I&C equipment, evaluation of the failures effects, identification of candidates for CCF analysis, recognition of appropriate countermeasures to prevent or mitigate failure effects.

2. DEFENSE-IN-DEPTH CONCEPT OF THE NEW LOVIISA I&C SYSTEMS

In the Loviisa Automation Renewal project (LARA) I&C systems and control room human-machine interfaces (HMI) of the plant units will be replaced gradually. All I&C systems including safety related protection systems will be implemented with digital platforms. This requires improvements of accident management principles and modifications of the plant´s defense-in-depth concept. The defense-in-depth principles of the LARA concept will be implemented by defining task categories and equipment belonging to those categories, which shall perform required safety functions. In this way the acceptance criteria given in Finnish safety requirements (YVL Guides) shall be met.

The conceptual design plan of the LARA project prescribes the design basis for different systems and procedures used in specific task categories (see Figure 1). Task categories are defined for the accident management (e.g. measures for control, prevention, mitigation) in different events of the plant. Each task category used for accident management consists of a functional entity formed by automation and process systems and the control room operations. Control room operations refer to emergency and abnormal operating procedures, control room ergonomics and shift operations (such as the accident management organization) utilized in accident situations. Automation systems include different levels, such as measurements, platforms and individual actuator controls. Process systems can be divided into main and auxiliary systems.

Figure 1: General description of a task category

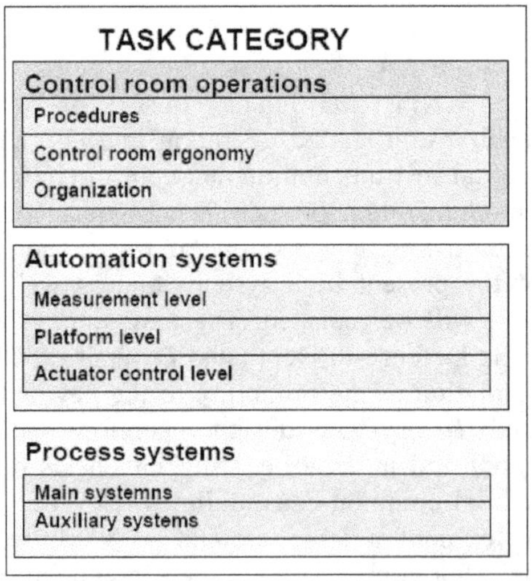

TASK CATEGORY
Control room operations
Procedures
Control room ergonomy
Organization
Automation systems
Measurement level
Platform level
Actuator control level
Process systems
Main systemns
Auxiliary systems

The event categories defined for the plant include the following:

Design basis categories (DBC)
- Normal operation (DBC 1)
- Anticipated operational occurrences (DBC 2)
- Postulated (design basis) accidents Class 1 (DBC 3) and Class 2 (DBC 4)

Design extension conditions (DEC)
- Class A (DEC A)

- Class B (DEC B)

In control of transients and accidents the defense-in depth principle is applied both in short and long term accident management (see Figures 2 and 3). 'Controlled state' refers to a state in which the plant is brought and safety functions performed by automatic actuations or by short-term manual operator actions. In the controlled state the reactor is shut-down and kept subcritical and its residual heat is removed or its heat is removed in the beginning of the safety demand caused by the event. 'Safe state' refers to a state in which the plant can remain for the duration of the required corrective measures, wherein the additional risks caused by the state are minimized and from which it is possible to return to normal operation – or from which the plant is brought to the final safe state. In the safe state the reactor is kept subcritical and its residual heat is removed with a good margin as long as the safety demand caused by the event exists. 'Final safe state' refers to a state in which the plant can be kept in the long term. In the final safe state the reactor is kept subcritical and its residual heat is removed with a good margin after the safety demand caused by the event no longer exists. In the course of the event while reaching the final safe state the safe state is kept and possible radioactive releases are controlled and the safety of the public is ensured. The reactor can be depressurized and the reactor core can be removed.

Figure 2: Defense-in-depth principle in short-term accident management

	Preventive protections		Reactor protections		
	Normal process control	Preventive protection	Reactor protection	Manual backup of reactor protection	Automatic backup of reactor protection
Small disturbance, normal operation (Production goes on)	Objective: Controlled state	Objective: Controlled state	Objective: Controlled state	Objective: Controlled state	Objective: Controlled state
Large, relatively frequent disturbance, transients (No safety systems actuated)					
Large, infrequent disturbance, Design basis accidents (Shutdown, safety systems actuated)					

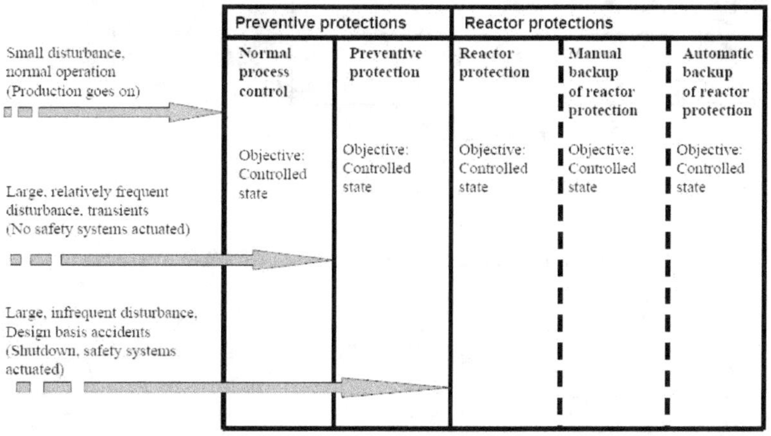

Figure 3: Defense-in-depth principle in long-term accident management

	Ordinary operations	Accident management			
	Ordinary shutdown	Safe Shutdown	Backup of manual accident management	Manual accident management	Diverse backup of manual accident management
Small disturbance	Objective: Safe state/ final safe state	Objective: Safe state	Objective: Safe state	Objective: Safe state	Objective: Safe state
Large, relatively frequent disturbance, transients (no safety systems actuated)					
Large, infrequent disturbance, design basis accidents (shutdown, safety systems actuated)					

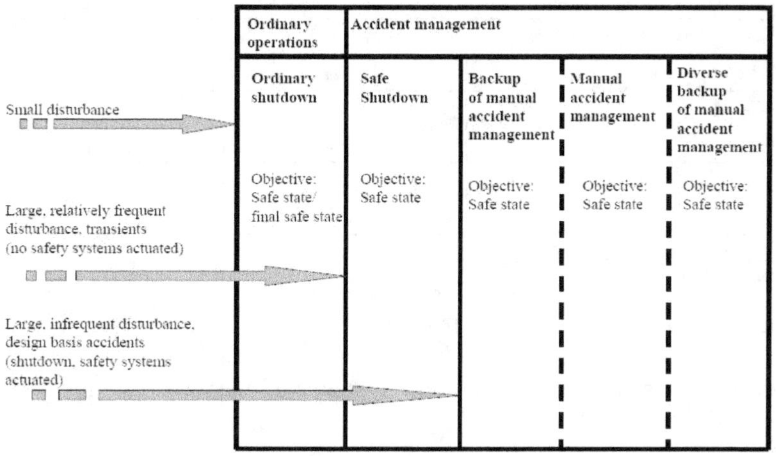

Safety classification of the equipment of the I&C systems is defined through the task categories (see Figures 4 and 5). Redundancy principle is generally applied to all safety classified task categories designed for accident management in different event categories to meet the reliability targets given for the safety classified systems. Diversity principle is applied to task categories performing actual safety functions in short- and long-term accident management to meet the risk targets given for the plant. Separation principle is applied to task categories performing preventive or actual safety functions in short- and long-term accident management categories to meet the reliability targets given for the safety classified systems.

Figure 4: Short term task categories of I&C systems

Event category		Task category				
		Normal Process Control NPC	Preventive protection PREV	Reactor protection RPS	Manual backup of RPS RPSMBU	Automatic backup of RPS RPSABU
DBC 2	Anticipated operational occurences	(SP)	RP (SP)	RP SP DP		
DBC 3	Postulated (design basis) accidents, class 1			RP SP DP		
DBC 4	Postulated (design basis) accidents, class 2			RP SP DP		
DEC A	Design extension conditions, class A		RP (SP) (DP)	RP SP DP	RP (SP) (DP)	
DEC B	Design extension conditions, class B		RP (SP)	RP SP DP	RP (SP) (DP)	(SP) (DP)

Quality principle:
- Safety class 2
- Safety class 3
- Safety class NS

RP Redundancy principle
SP Separation principle
DP Diversity principle

Figure 5: Long-term task categories of I&C systems

Event category		Task category				
		Ordinary shutdown OSD	Safe shutdown SSD	Backup of MAM MAMBU	Manual accident management MAM	Diverse backup of MAM MAMDBU
DBC 2	Anticipated operational occurences					
DBC 3	Postulated (design basis) accidents, class 1	(SP)	RP SP DP	RP SP DP		
DBC 4	Postulated (design basis) accidents, class 2	(SP)	RP SP DP	RP SP DP		
DEC A	Design extension conditions, class A	(SP)	RP (SP) (DP)	RP (SP) (DP)	(SP) (DP)	(SP) (DP)
DEC B	Design extension conditions, class B	(SP)	(SP) (DP)	(SP) (DP)		

Quality principle:
- Safety class 2
- Safety class 3
- Safety class NS

RP Redundancy principle
SP Separation principle
DP Diversity principle

For system design of the safety important I&C systems of the task category RPS the following principles will be considered:

- Single failure tolerance,
- Robustness against common cause failures,
- Fail-safe behavior for activation of the reactor trip.

The further I&C systems of other task categories will be implemented by using graded requirements of the defense-in-depth concept regarding application of redundancy, separation and diversity principles (see Figures 4 and 5). The figure 6 presents an overview of design basis of the new I&C system in which the different design principles and I&C equipment (e.g. I&C platforms) is assigned to the different task categories in short term and long term.

Figure 6: Design basis for assignment of the I&C equipment to the task categories

I&C Design	Short-term AM task categories					Long-term AM task categories				
	NPC	PREV	RPS	RPSMBU	RPSABU	OSD	SSD	MAMBU	MAM	MAMDBU
Measurements (1)	Transm NS	Transm 3	Transm 2	Transm 3	Transm NS	Transm NS	Transm 3	Transm 3	Transm NS	Transm 3
Measurements (2)			Transm 2D	Transm 3D						Transm 3D
Platform	SPPA-T2000				SPPA-T2000	SPPA-T2000			SPPA-T2000	
Platform	Teleperm XS	Teleperm XS	Teleperm XS	diverse I&C modules		Teleperm XS	Teleperm XS	Teleperm XS	Teleperm XS	diverse I&C modules
Actuator control (1)	Type NS				Type NS	Type NS			Type NS	bypass
Actuator control (2)	Type 1	Type 1	Type 1	Type 1	Type 1	Type 1	Type 1	Type 1	Type 1	Type 1
Design Basis										
Safety Class	NS	SC3	SC2	SC3	NS	NS	SC3	SC3	NS	SC3
Redundancy principle	not required	required	required	required	not required	not required	required	required	not required	not required
Separation principle	required	required	required	required	required	required	required	required	required	required
Diversity principle	not required	required	required	required	required	not required	required	required	required	required

The design of I&C systems of the LARA project provides usage of different types of measurement devices and of actuator control modules for different task categories:

- safety classified equipment Transm 2, 2D, 3, 3D and Type 1
- non-classified equipment Transm NS, Type NS.

The architecture of I&C systems of the LARA project comprises also different I&C platforms:

- Teleperm XS is a I&C system platform (AREVA) for safety I&C in the nuclear power plant. It comprises all the necessary hardware and software components, including the software tools required for engineering, testing and commissioning, operation and troubleshooting [3].
- SPPA-T2000 (Siemens) is an universal process control system for power plants and consists of following sub-systems (platforms): automation system, process control and management system, engineering system, diagnostics system, communication and bus system [4].

3. VERIFICATION AND VALIDATION OF THE DEFENSE-IN-DEPTH CONCEPT

3.1. Scope and objectives of the verification and validation

The verification and validation of the concept for the new I&C system has been performed in several steps in which the experts of the GRS have carried out the following tasks:

- Analysis of the overall I&C architecture concerning potential impact of common cause failures,
- A review of the FMEA of selected modules of the equipment of the safety-important I&C systems,
- An analysis of the separation of trains and redundancies at which a single failure requirement shall be applied,
- An analysis of the separation of different safety classes at which a single failure requirement shall be applied,
- An analysis of the separation of different lines of defense which addresses to the prevention of the propagation of the impacts of potential common cause failures (CCF) from one line of defense to another.

Important objectives of the work comprised identification of potential <u>active failures</u> of the hardware and software and identification of potential impacts and propagation paths:

- from one train/redundancy to the other trains/redundancies of the I&C systems of the safety classification SC2 (reactor protection system),
- from systems of the safety class 3 to the trains/redundancies of I&C systems of the higher safety class SC2.

Further the V&V analysis establishes also a relationship with other methods of reliability analysis (e.g. Fault Tree Analysis of the PRA) in which the FMEA will subsequently be integrated.

For the scope of these V&V tasks the <u>active failure</u> has been defined as follows:

- Active failure means a defective control that leads to a function that is unnecessary considering the demand mode or that differs from its specification, but which is not a lack of functionality. For instance, incorrect function sequences and starting or termination functions at the wrong time are active failure modes, as is producing incorrect data that lead to a wrong action by the operator.

3.2. Methodology

The main approach of the V&V assessment of the LARA I&C system has been developed on the basis of the Failure Mode and Effect Analysis (FMEA) methodology [5, 6]. The FMEA methodology can be applied principally to the assessment of digital I&C systems [7] and can help to identify functional dependencies between hardware and software of the overall I&C architecture of the plant. The challenge of the application of the FMEA methodology was to select a proper strategy for the functional decomposition of the I&C systems and to consider software faults with potential functional impact. The FMEA of a complex system or of a complex hardware module is usually an iterative analysis process. The traditional FMEA applies mainly the bottom-up approach, but in some cases of the V&V work it was necessary to apply a top-down approach for identification of functional impacts (e.g. identification of all dependent and independent failure causes of a specific failure effect). The key difference between both methods is that a bottom-up approach identifies failure modes of single components (e.g. hardware modules) and deduces the corresponding effect on the performance for the appropriate system level while a top-down approach identifies failure effects on system level and analyses at lower levels probable failure modes of the hardware and of the software that could result in the identified failure effect. The evaluation of the failure modes of the hardware and of the application software in the framework of the V&V project was made generally in the following manner, see Figure 7.

Figure 7: Main steps of the evaluation of the FMEA and CCF

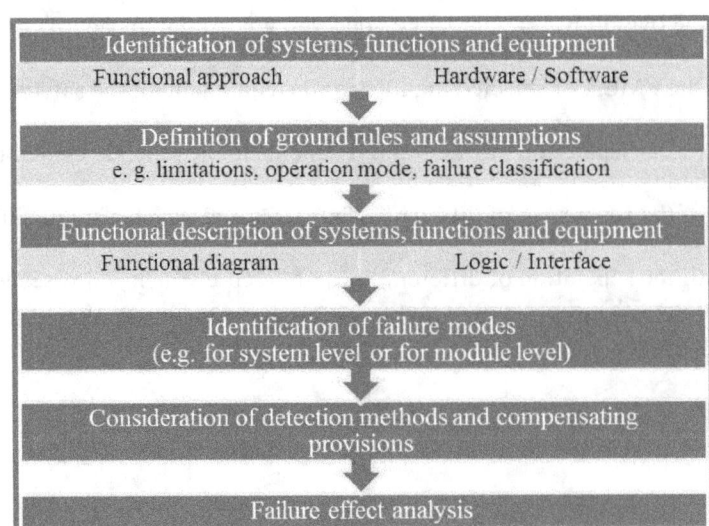

One step of the V&V process was the analysis of the propagation of a potential active single failure of the hardware and software from systems of safety class SC3 to systems of safety class SC2. Therefore the following assumptions were made for the definition of failure modes for an active failure:

- The influenced module or item of one I&C system generates an erroneous, but valid signal (e.g. a logical "1" signal) to the impacted module or item of another I&C system.
- Signals without a valid logical state are not considered.
- Distinction of permanent and non-permanent erroneous signals:
  Permanent erroneous signals.
  In this case it is assumed that the impacted module or item of the I&C system (SC2) receives an unintended, permanent and valid "1" input signal which does not change during the postulated event sequence.
  Non-permanent erroneous signal.
  In this case it is postulated, that a single, unintended, but valid "1" signal originating from the influenced module or item of one I&C system is received for a limited time by the impacted module or item of the I&C system (SC2). It is furthermore assumed, that this signal turns to "0" immediately afterwards, but that it is present long enough to trigger logical functions or other inadvertent actions.

Potential consequences of failure modes of both types of erroneous signals were en analysed for different plant states. For the purpose of this analysis two plant states were defined:

- Normal Operation: The plant is in normal power operation without initiation criteria for any safety important functions.
- Abnormal operation or accident conditions: initial state of the plant is normal operation when some kind of not further defined transient occurs, that leads to a significant change in the parameters which are relevant as input signals for the analysed safety important function of the I&C system SC2 (e.g. primary circuit pressure, steam generator level, containment pressure etc.).

Based upon the aforementioned definitions and assumptions it was analysed if and how signals generated by active failures of a module or item of an I&C system may affect the output signals of the module or item of the safety important I&C system (SC2).

Within the framework of the V&V project also the actuator control modules were assessed regarding fulfilling the requirements of the priority concept of actuation of safety important function and components.

The potential impact of a postulated CCF of a selected I&C function in all divisions of an I&C system was considered on module level for instrumentation (transmitter), on system level (e.g. for I&C platforms: Teleperm XS, SPPA-2000) and on module level for actuators (e.g. priority modules Type 1 and 2). On system level, the potential impacts of the postulated CCFs were evaluated separately for safety logic, voting logic and signal distribution. The evaluation for the instrumentation, safety logic and actuation logic was performed on the basis of simplified functional diagrams of the logic structures of the I&C functions.

In a further step of V&V process the CCFs of identical functional elements in all I&C functions of the safety important I&C system have been analyzed and presented in MS Excel worksheets (see Figure 7). The analysis was made based on the following assumptions:

- Instrumentation: consideration of CCF of identical transmitters only,
- Application software: consideration of CCF of identical functional blocks with changeable parameters (e.g. functional blocks of the software as set-point blocks, timing blocks),
- Application software: consideration of CCF of "Reset-Set" flip-flop blocks of the software,

- Voting logic: consideration of a CCF of the functional blocks "1oo2" or "2oo4" in all I&C functions at the same time as a "worst-case" path for further Defense-in-Depth analysis of whole safety concept,
- Analysis does not consider a potential CCF of simple elementary functions of the application software (e.g. "AND", "(N)AND", 2. MIN and 2. MAX) explicitly.

3.3. Results

The results of the FMEA and of the CCF analysis are documented in a large number of worksheets (MS Excel) containing detailed information regarding all relevant aspects of the analysis. For the analysis of the impact of a potential CCF on the safety important I&C functions detailed information about their logical design and signal processing was also evaluated.

The evaluation of potential CCF failures was limited to the identification of the potential failure modes and effects on the actuation signals of the safety important I&C functions based on system level FMEA methodology. The CCF failure modes have been defined for the functional elements of instrumentation, safety logic and actuation logic. The results of the CCF analysis are documented in worksheets (see also Figure 8) including the following information:

- CCF analysis level: Reference to measurement level, safety logic on platform level, voting logic on platform level, signal distribution between platform level and actuation level,
- Functional element: Items of hardware and of software affected by a postulated CCF,
- Postulated failure mode (permanent): Failure mode of the postulated CCF; all failures are considered as permanent failures from the time of failure manifestation onwards,
- Preconditions / plant state,
- Time scale: Information on the point in time / time span at which the failure effect is analysed,
- Description of failure mode propagation: Probable effects of the assumed CCF on relevant functional elements in the signal path,
- Affected components: single component or group of components which might be affected by the assumed CCF,
- Manual action concerning Set-Reset flip-flops (elementary functions of the application software): Distinction between operator actions relevant for the signal path like resetting of the flip-flops,
- Undisturbed input signal: Input signal which the affected components will get without occurrence of any failure,
- Input signal for components under consideration of CCF: Input signal which the affected components will get with occurrence of the assumed CCF,
- Impact of CCF: Distinction whether a CCF has an impact (indicated by red color in the table) or not depending on preconditions and current plant state,
- Failure effect of CCF: Type of failure effect of a CCF.

Figure 8: Example (screen shot of a worksheet) for CCF evaluation

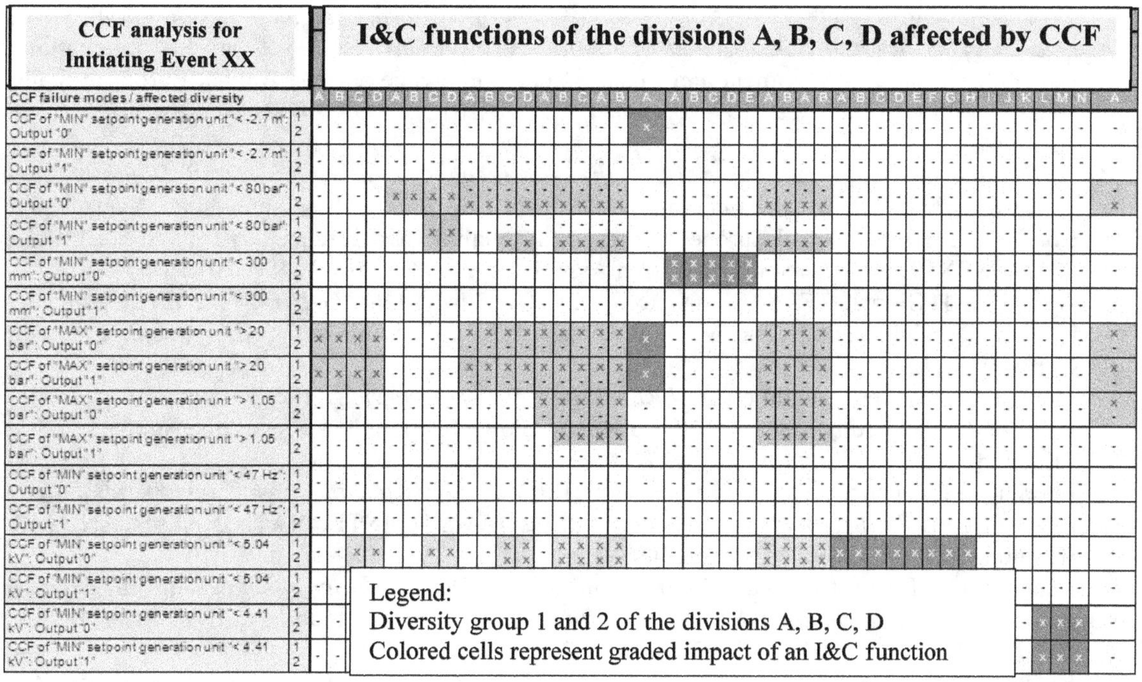

[Worksheet table: "CCF analysis for Initiating Event XX" with columns showing "I&C functions of the divisions A, B, C, D affected by CCF". Rows list CCF failure modes including various setpoint generation unit failures (MIN/MAX) with different thresholds (-2.7 m³, < 80 bar, < 300 mm, > 20 bar, > 1.05 bar, < 47 Hz, < 5.04 kV, < 4.41 kV) and Output "0" or "1", with diversity groups 1 and 2 indicated. Cells marked with "x" or shaded to represent graded impact.]

Legend:
Diversity group 1 and 2 of the divisions A, B, C, D
Colored cells represent graded impact of an I&C function

4. CONCLUSION

This paper presents some preliminary results regarding methodological aspects of the evaluation of the defense-in-depth concept of new digital I&C systems in the NPP Loviisa. The main approach of the design verification process has applied primarily failure mode and effect analysis (FMEA) for identifying relevant failure modes of I&C systems, functions and equipment and also for evaluation of probable failure effects. Some generic and design specific issues are identified and will be considered in the next phase of the V&V process.

The generic issues mainly concern methodological aspects such as choosing of an adequate level of abstraction (e.g. system level vs. module level) for the FMEA purpose and consideration of potential software faults in the CCF model. Design specific issues concern comprehensibility and completeness of failure mode identification for I&C equipment, evaluation of failure effects, identification of candidates for common cause failure analysis (CCF), and identification of appropriate countermeasures to prevent or mitigate hazardous failure effects.

The FMEA methodology provides a good basis to perform a comprehensible assessment of a defense-in-depth concept (architecture) of digital I&C systems. The CCF analysis requires in addition a coherent logical model of all relevant functional interrelations (e.g. signal processing/linking in the hardware and in the application software, and also in the priority and actuation logic) of the I&C system or/and of whole I&C architecture.

The identified issues are also intended to support the consideration of safety relevant aspects of digital safety important I&C systems in the probabilistic reliability analysis (PRA) of nuclear power plants, e.g. in a probabilistic model of digital I&C systems.

References

[1] J.-E. Holmberg, S. Authén, A. Amri, *Development of best practice guidelines on failure modes taxonomy for reliability assessment of digital I&C systems for PSA,* in: 11th International Probabilistic Safety Assessment and Management Conference and the Annual European Safety and Reliability Conference 2012 (PSAM11 ESREL 2012), ISBN: 978-1-62276-436-5, Curran Associates, Inc., Red Hook, NY, 2012,.

[2] E. Piljugin, S. Authén,, *Proposal for the Taxonomy of Failure Modes of Digital System Hardware for PSA,* in: 11th International Probabilistic Safety Assessment and Management Conference and the Annual European Safety and Reliability Conference 2012 (PSAM11 ESREL 2012), ISBN: 978-1-62276-436-5, Curran Associates, Inc., Red Hook, NY, 2012

[3] AREVA NP GmbH, *TELEPERM® XS System Overview, ANP: G-101-V1-12-ENG,* published by and copyright (2012) AREVA NP GmbH – Paul-Gossen-Strasse 100, 91052 Erlangen, Germany, 2012.

[4] Siemens AG Power Generation, *TELEPERM XP System Overview. A96001-S90-A772-V1-7600.* Published by and copyright (2002) Siemens AG Power Generation, Freyeslebenstraße 1, 91058 Erlangen, Germany, 2012.

[5] International Electrotechnical Commission (IEC), *Analysis techniques for system reliability – Procedure for failure mode and effects analysis (FMEA),* INTERNATIONAL STANDARD IEC 60812:2006(E), Second edition 2006-01.

[6] International Electrotechnical Commission (IEC), *Dependability management, Application guide – Analysis techniques for dependability –* Guide on methodology, INTERNATIONAL STANDARD IEC 60300-3-1:2003, Part 3-1, 2003.

[7] P. Haapaneni, A. Helminene *Failure Mode and Effects Analysis Of Software-Based Automation Systems,* STUK-YTO-TR 190, Finland, August 2002.

Scrum, documentation and the IEC 61508-3:2010 software standard

Authors: Thor Myklebust[1,a], Tor Stålhane[b], Geir Kjetil Hanssen[a] Tormod Wien[c] and Børge Haugset[a]
[a] SINTEF ICT
[b] IDI NTNU
[c] ABB

Abstract

Agile development, and especially Scrum, has gained increasing popularity.
IEC 61508 and several related standards for development of safety critical software has a strong focus on documentation, including planning, which shall show that all required activities have been performed. Agile development on the other hand, has as one of its explicit goals to reduce the amount of documentation and to mainly produce and maintain working software.
The problem created by the need to develop a large amount of documents when developing safety critical systems is, however, not a problem just for agile development – it has been identified as a problem for all development of safety critical software. In some cases up to 50% of all project resources has been spent on activities related to the development, maintenance and administration of documents. Thus, a way to reduce the amount of documentation will benefit all developers of safety critical systems.
By going systematically through all the documentation requirements in IEC 61508-1 (general documentation requirements) and IEC 61508-3 (software requirements) and by using the combined expertise of the five authors, we have been able to identify documents that are or can be generated by tools used in the requirement and development process, e.g. logs from requirement and testing tools and documents that can be made as part of the planning and discussions, e.g. snap shots of whiteboards. We have also identified documents that normally can be reused when issuing a new version of the software and identified documents that can be combined into one document.

Keywords: Scrum, safety-critical software, documentation, IEC 61508, Certification

1. Introduction

Agile development, and especially Scrum [1], has gained increasing popularity and has also been applied in the development of safety critical software, for instance in aviation and automotive [2, 3]. IEC 61508 [4] and several related standards for development of safety critical software has a strong focus on documentation, including planning, which shall show that all required activities have been performed. Agile development on the other hand, has as one of its explicit goals to reduce the amount of documentation and to mainly produce and maintain working software. Assessment of compliance with standards like IEC 61508 is outside the scope of agile methods.
The problem created by the need to develop a large amount of documents when developing safety critical systems is, however, not a problem just for agile development – it has been identified as a problem for all development of safety critical software. In some cases up to 50% of all project resources has been spent on activities related to the development, maintenance and administration of documents [5]. Thus, a way to reduce the amount of documentation will benefit companies that develop safety critical systems. We are, however, motivated by the focus on simplicity and pragmatism in agile methods and believe that adapting principles from agile software development to the development of safety critical systems will help to simplify the work with the documentation and thus to reduce costs.
Our work in this paper has been guided by the following research question: How can information from an agile software development process be used to reduce the documentation costs imposed by IEC61508?

---

[1] thor.myklebust@sintef.no

The authors have already published papers on how to adapt the agile development process to conform to the standards ISO 9001 (quality systems) [6], IEC 61508 (functional safety systems) [7] and IEC 60880 (nuclear systems) [8]. Some companies have been reluctant to adapt an agile approach due to the perceived risk of having to redo a large amount of documentation for each of the frequent and short iterations in the development cycle. How we have solved this problem is described in chapter 3 and 4 below.

This work has been performed as part of the SUSS[2] project, financed by The Norwegian Research Council.

2. Background

As Scrum and other agile processes are introduced also into the part of the software industry that develops safety critical systems, the industry is caught between the relevant standards that are pre-agile and mostly document driven and the agile concept which tries to avoid producing documents that does not directly relate or contribute to the development of working software. This is based on the agile manifest (http://agilemanifesto.org/) that states "Working software over comprehensive documentation ".

In our opinion the relevant standards overdo their focus on documents, mostly because they overdo their focus on process documentation. It is our experience that a large part of this documentation will only be used for proof of conformance (PoC) which is needed in two cases – for certification and in case the product will be drawn into a court case.

Using an agile approach will reduce the amount of in-process document needed. Another factor that will reduce lead time and cost is to tap the large potential for reuse of whole or parts of important documents. This can, however, only be achieved if they are written with reuse in mind.

SafeScrum

We have earlier attacked similar problems related to standards that were, often implicitly, intended for a document driven, waterfall process such as ISO 9001 [6] and IEC 61508 [7]. Our conclusion is the same in both cases: most of the standards' requirements are met without much ado, some requirements can be solved with a little flexibility from the developers and assessors while there are a few stumbling blocks that need new thinking. Of the 50 top-level requirements in ISO 9001, only four fall into this category. For the IEC 61508, we had to develop a new Scrum process – Safe Scrum – in order to cater to the identified problem areas.

SafeScrum [ibid.] is motivated by the need to make IEC 61508 more flexible with respect to planning, documentation and specification, as well as making Scrum a practically useful approach for developing safety critical systems.

Our model has three main parts. The first part consists of the IEC 61508 steps of developing first the environment description and then the SSRS (Software Safety Requirement Specification) phases 1-4 (concept, overall scope definitions, hazard and risk analysis and overall safety requirements). These initial steps result in the initial requirements of the system that is to be developed and is the key input to the second part of the model, which is the Scrum process. The requirements are documented in a *product backlog.* A product backlog is a list of required features and functions of the system prioritized by the customer.

Due to the focus on safety requirements, we propose to use two related product backlogs, one *functional product backlog*, which is typical for Scrum projects, and one *safety product backlog*, to handle safety requirements. We will keep track of how each item in the functional product backlog relates to the items in the safety product backlog, i.e. which safety requirements that are affected by which functional requirements.

---

[2] Norwegian: Smidig utvikling av Sikkerhetskritisk Software. English: Agile Development of safety Critical Software

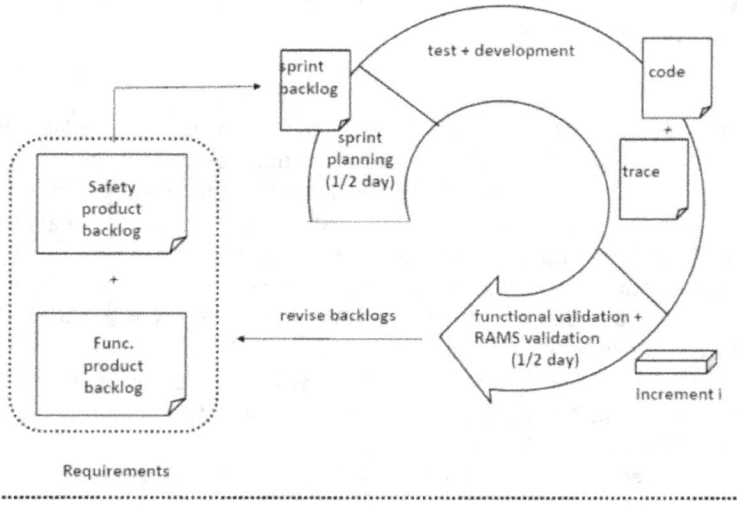

Figure 1: The SafeScrum model

Each Scrum iteration can be considered as a mini waterfall project or a mini V-model, and consists of planning, development, testing, verification and also validation. For the development of safety critical systems, traceability between system/code and backlog items, both functional requirements and safety requirements, is needed. The documentation and maintenance of trace information is introduced as a separate activity in each sprint – see Figure 1. In order to be performed in an efficient manner, traceability requires the use of a supporting tool. There exist several process-support tools that can manage traceability in addition to many other process support functions. Two out of many examples are Jira (www.atlassian.com/software/jira) and Rally software (www.rallydev.com).

An important practice in many Scrum projects is *test-driven development*, where the test of the code is written *before* the code is developed. Initial, this test is simple, but as the code grows, the test is extended to continuously cover the new code. The benefits of test-driven development are that the developer needs to consider the behaviour of the code, based on the requirements, before implementation, it enables regression testing, and it provides documentation of the code.

A sprint should always produce an *increment,* which is a piece of the final system, for example design, test rig or executable code. The sprint ends by demonstrating and validating the developed code to assess whether it meets the requirements in the sprint backlog. Some items may be found to be completed and can be checked out while others may need further refinement in a later sprint and goes back into the backlog. To make Scrum fit with IEC 61508, we propose that the final validation in each iteration is done both as a validation of the functional requirements and as a RAMS validation, to address specific safety issues. If appropriate, an assessor may take part in this validation for each sprint. The assessor could also take part in the retrospective after each sprint to help the team to keep safety consideration in focus. Running such an iterative and incremental approach means that the development project can be continuously *re-planned* based on the most recent experience with the growing product. Between the iterations, it is the duty of the customer or product owner to use the most recent experience to re-prioritize the product backlogs.

As the final step, when all the sprints are completed, a final RAMS validation will be done. Given that most of the developed system has been incrementally validated during the sprints, the final RAMS validation will be less extensive than when using other development paradigms. This will also help us to reduce the time and cost needed for certification.

Test Driven Development
TDD is a popular practice in agile development and is often used to supplement Scrum. We see TDD as a natural part of SafeScrum as well and believe that this may produce documentation being useful as PoC. TDD is a practice where all new code at the procedure or method level first needs to be described as a suite of mock objects and assertions (expected results from given inputs). The total

collection of unit tests grows as the code grows and is automatically executed frequently to test if the code works as defined after each change.

Trust
We have checked requirements related to "Trust" in several IEC and ISO standards [9-17]. During assessment work, we have observed that the level of trust that the assessor have in the manufacturer may affect the level of documentation needed for the approval of the product. In the standards evaluated, only ISO/IEC 17021 [13] mentioned the level of trust the assessor have in the manufacturer. Quote"*Familiarity (or trust) threats: threats that arise from a person or body being too familiar with or trusting of another person instead of seeking audit evidence*". This standard is also the only standard that mentions the requirements for trust related to the assessor (third party).The level of trust the assessor have in the manufacturer is a subjective issue so it is important to discuss the level of details, possible excessive bureaucracy and pragmatism with the assessor at the beginning of the certification process. The important issue is that the manufacturer has the information they need to do their job and the assessor to do his job.
Trust as a topic in this respect is closely linked to the level of competence and experience of the personnel.
In practice trust is mainly related to people, not organizations. This has been experienced by one of the SUSS participating companies when the certification body changed several of their assessors and as a result, trust was decreased.

Industrial challenges
The development of safety-critical systems is guided by document-driven and process-heavy standards. The safety-standard, IEC61508, assumes extensive documentation and strictly defined processes for the product safety certification including risk analysis, change control and traceability. Therefore the speed of change is lower in such projects, making them less flexible with respect to changing requirements from customers and markets.
The safety process being mostly a document driven process, where each step from planning and specification to design, coding, testing and validation and verification need to be documented as a Proof of Concept, put a lot of emphasis on the project organization and the competence and experience of the people involved. Furthermore, the requirement that there shall be unique traceability all the way from requirements to design, implementation, testing and validation and verification, complicates the picture and assumes the use of labor extensive procedures to be able to cope with often large amount of data. Data, which over the lifetime of a project that can span several years, is not necessarily static. Compared to a "normal" software development project, testing is without doubt the task requiring most additional effort. This is due to the rigid requirements on the documentation process to verify that the required functionality is implemented as specified. Tests must be implemented at all levels (unit, functional, system) with unique traceability, covering normal, exceptional and erroneous operation.

3. Requirements related to documentation
In search of a potential reduction of the necessary documentation we believe that proper adoption of agile software development principles from the Scrum methodology may reduce the costs of documentation. We expect to see two cost saving effects: 1) it will reduce lead time and increase the development process flexibility, thus reducing development costs and, 2) it will reduce the number of new documents. When doing modification of an already certified product, only a few documents are new e.g. test reports. Furthermore these documents can be based on templates or reuse (see IEEE std 1517:2010 [18] for more information related to reuse) or be automatically generated to further reduce documentation costs. See table I.

The challenge with this solution is to keep the process and available documentation in line with the IEC 61508 requirements while at the same time gaining the benefits from an agile development process. As described below, we can achieve this through a systematic walkthrough of the IEC 61508 requirements and only keep the minimum of documents or information that are needed to meet the standard's requirements.

Method when evaluating IEC 61508-1 documentation requirements
We have used the same method for the work reported here as we have used earlier – see [6, 7]. The process consists of the following two steps:
1. Check each relevant part of the standard (Part 1 ch. 5) and for each requirement ask "If we use Scrum, will we still fulfil this requirement?" this check is used to move the requirements into one out of three parts of an issues list – "OK", no further action requirement, "?", needs to be discussed further and "Not OK", will require changes to Scrum and, in a long term perspective, to IEC 61508. In addition to the issues list we will also get a lot of input to how to modify the Scrum process in order to reduce the amount of conflicts.
2. Check all requirements that are in the categories "?" and "Not OK" against a modified Scrum process model – in our case Safe Scrum. This will reduce the number of problematic requirements further. In addition, the accompanying discussions will enable us to identify new ways of tackling some of the problems discovered.

This process is used on the case at hand in the section "IEC 61508 walkthrough". Most of the categorizations done on the standard's requirements are to a certain degree subjective. For this reason we have included all relevant roles in the assessment: one assessor, two Scrum expert, one safety experts and one representative for a company that routinely have to have their software products certified.

IEC 61508-1 walkthrough of chapter 5 "Documentation"
We have gone through the section 5.2 - Requirements on documentation - in IEC 61508, part 1. The documentation requirements in IEC 61508, part 3 is just a reference to part 1 of the standard. The result from the first iteration of the IEC 61508, part 1, section 5.2 walkthrough was that out of a total of 11 issues, we found that
- Five was "OK".
- One was "not OK" (5.2.3 below)
- Five needed further investigation – "?"

The second iteration focused on the following six issues:
- 5.2.1. The documentation shall contain sufficient information, for
  - each phase of the overall, E/E/PES and software safety lifecycles completed.
    These documents will fall in the class Reusable documents (see ch. 4 below)
  - necessary for effective performance of subsequent phases.
    SafeScrum is mainly performed as part of phase 10 Realisation.
  - verification activities.
    The verification process should use automatic testing tools – e.g., Cucumber (http://cukes.info/) or Fitnesse (http://fitnesse.org/). This will also enable a considerable amount of pragmatic reuse.

  The problem for Scrum is traceability. In order to handle this problem, we have added an extra activity to handle all traceability in SafeScrum.
- 5.2.3 The documentation shall contain sufficient information required for the implementation of a functional safety assessment, <u>together with</u>
  - the information and
  - results derived from any functional safety assessment.

  This problem is partly taken care of by the SafeScrum process but the assessor will need more information, which is not available from Scrum as it is practiced now. This means that SafeScrum needs to be complemented by normal functional safety assessment.
- 5.2.4 The information to be documented shall be as stated in the various clauses of this standard unless justified or shall be as specified in the product or application sector international standard relevant to the application
  We should be pragmatic when fulfilling this clause, since this opens up for a wide range of interpretations for what should be accepted as PoC. The most important thing here is,

*however, to discuss this with the assessor **before** the project starts in order to get an agreement on the information that will be needed.*

- 5.2.5 The availability of documentation shall be sufficient for the duties to be performed in respect of the clauses of this standard.
  In order to make all relevant documents available for the assessor we need first of all to register all relevant information. The simplest way to do this is to use a whiteboard and to take snap-shots. Theses snap-shots, together with the date and a list of participants should be accepted as process documentation. When the relevant documents are registered there exist several tools for sharing information like e.g. www.projectplace.com.
- 5.2.10. The documents or set of information shall be so structured as to make it possible to search for relevant information. It shall be possible to identify the latest revision (version) of a document or set of information.
  All relevant documents must be stored in a project database and indexed properly.
- 5.2.11. All relevant documents shall be revised, amended, reviewed and approved under the control of an appropriate document control scheme.
  The important question here is when – e.g., after each iteration, after some iterations or just when we have finished all development iterations. Using the methods suggested for section 5.2.5 it is easy to conform to the two first points – revised and amended – while the last two – reviewed and approved – might be problematic in the sense that it will bureaucratize and delay the Scrum process, thus reducing its effect. These review aspects are normally included in the contract between the manufacturer and the assessor.

  Two important things can be done:
  - *Move much of the necessary documents out of the Scrum iteration loop.*
  - *Get an agreement with the assessor as to which iterations need to be included in 5.2.11 and how this can be performed when using e.g. databases.*

IEC 61508 walkthrough of the normative Annex A "Guide to the selection of techniques and measures" of Part 3

Although annex A in IEC 61508, part 3 is not directly related to documents and PoC, it gives an overview of the needed activities and thus indirectly an overview of the necessary PoC. The 10 tables – A1 – A10 – contains a total of 70 requirements. In order to simplify a walkthrough of these tables we have decided to assume SIL 2 development, remove all issues related to maintenance and only consider the activities that are marked as HR – Highly Recommended (although, in practice, some R activities should be performed). This reduces the number of issues to 19. The two tables A3 and A4 are only concerned with pre-development activities. Three tables – A5, A6 and A7 – are only concerned with testing and the PoCs can be sufficiently covered by the automatically generated test logs. Table A2 is concerned with design activities. In our opinion, the PoC will in some cases be satisfied by white-board snapshots plus a list of participants. High level design – architecture – is decided before we enter SafeScrum. Using the whiteboard for detailed design has some pros and cons. Pro: quick, can document the design process, not only the final result. Con – may lack the formality achieved by a document.

The only challenge is table A9, which is concerned with static and dynamic analyses. When we check the more detailed tables – B2 and B8 – we see that the PoC for the requirements in B2 are covered by the test logs. The only remaining challenges are in B8, which requires analysis of control- and data flow. This document will have to be done separately (outside SafeScrum) but only when the system is finished and ready for certification.

4. Classification of the documentation

The relevant documents for Part 3 are presented in Table A.3[3] "Example of a documentation structure for information related to the software lifecycle" in Part 1 of IEC 61508.

Copy from Part 1:
Tables A.1, A.2 and A.3 provide an example documentation structure for structuring the information in order to meet the requirements specified in Clause 5. The tables indicate the safety lifecycle phase that is mainly associated with the documents (usually the phase in which they are developed). The names given to the documents in the tables are in accordance with the scheme outlined in A.1. In addition to the documents listed in Tables A.1, A.2 and A.3, there may be supplementary documents giving detailed additional information or information structured for a specific purpose, for example parts lists, signal lists, cable lists, wiring tables, loop diagrams and list of variables.

There are several levels of documentation in a software project. The documents at these levels have different sources, different costs but often the same roles, both in the project itself and when it comes to certification.

- **Reusable documents** – low extra costs. This is documents where large parts are reused as is, while small parts need to be adapted for each project and even for each sprint for some documents. If reuse is the goal right from the start, the changes between projects or iterations will be small. For further information about reuse see IEEE std 1517 [18].
- **Combined** - Identify documents that can be combined into one document
- **Automatically generated documents** – high initial costs but later low costs. This is documents that are generated for each new project or iteration by one or more tools. Examples are test results and test logs from Jira and requirements documents from Doors (www-03.ibm.com/software/products/en/ratidoor/).
- **New documents** – high costs. This is documents that have to be written more or less from scratch for each new project.

In the table below, we have classified the documents that are specified in table A.3 regarding software in Part 1 of IEC 61508.

IEC 61508-1, table A.3 for SW	Classification and comments
1. Specification (software safety requirements, comprising: software safety functions requirements and software safety integrity requirements)	Generated from e.g. a requirement management tool and/or backlog management tool and is reusable. For further information see IEEE Std 830-1998 [19] and IEEE Std 1233-1998 [20].
2. Plan (software safety validation)	Reusable. The document can be combined with document 26. For further information see IEEE Std 730-2002 [21].
3. Description (software architecture design)	Reusable. For further information, see ISO/IEC/IEEE Std 42010 [22], IEEE Std 1016 [23] and www.sysmlforum.com/ regarding SysML model management.
4. Specification (software architecture integration tests);	Reusable. The standard ISO/IEC/IEEE 29119-3:2013 [24] "Test Documentation" includes relevant information related to specification of tests.
5. Specification (programmable electronic hardware and software [4]integration tests);	Reusable

---

[3] Similar tables exists for Part 1 and Part 2
[4] Observe definition 3.8.1 in Part 4 related to integration tests

IEC 61508-1, table A.3 for SW	Classification and comments
6. Instruction (development tools and coding manual)	Reusable. New development tools have to have relevant instructions. See existing coding manuals/information issued by Exida for C/C++ [25] and a Guideline issued by MISRA[5] for C++ [26]. See www.misra-cpp.com/ for further information.
7. Description (software system design);	Reusable For further information, see IEEE Std 1016 [23].
8. Specification (software system integration tests)	Reusable. The document can be combined with documents 9 and 10.
9. Specification (software module design);	Reusable. The document can be combined with documents 8 and 10. For further information, see IEEE Std 1016 [23].
10. Specification (software module tests)	Reusable Can be combined with documents 8 and 9
11. List (source code);	Source code can easily be generated directly from the code management system. Also, there are many tools that may automatically produce code documentations. E.g. Doxygen (www.doxygen.org) and other similar tools.
12. SW module design: Report (software module tests);	Generated. Some of the tests are generated automatically, others are semi-automatic and some are manually.
13. Report (code review)	Combined. Doc 13, 14, 15, 16 and 17 can be one report. The documents can be developed gradually so. There exist several tools for static code analysis (e.g. http://cppcheck.sourceforge.net/ for for static C/C++ code analysis) and code review (e.g. www.parasoft.com/cpptest). See also IEEE 1028:2008, IEEE Standard for software reviews and audits [27]. This standard defines five types of software review and audits. In this edition of the standard there is a clear progression in informality from the most formal, audits, followed by management and technical review, to the less formal inspections, and finishing with the least formal inspection process - walkthroughs.
14. SW module testing: Report (software module tests)	Generated. Doc. 13, 14, 15, 16 and 17 can be one report Some of the tests are generated automatically, others are semi-automatic and some are manually.
15. Report (software module integration tests);	Generated. Doc 13, 14, 15, 16 and 17 can be one report. Some of the tests are generated automatically, others are semi-automatic and some are manually.
16. Report (software system integration tests);	Generated. Doc 13, 14, 15, 16 and 17 can be one report Some of the tests are generated automatically, others are semi-automatic and some are manually.
17. Report (software architecture integration tests)	Generated. Doc 13, 14, 15, 16 and 17 can be one report. Some of the tests are generated automatically, others are semi-automatic and some are manually.

---

[5] MISRA: The Motor Industry Software Reliability Association. www.misra.org.uk

IEC 61508-1, table A.3 for SW	Classification and comments
18. Report (programmable electronic hardware and software integration tests)	Generated. Some of the tests are generated automatically, others are semi-automatic and some are manually.
19. Instruction (user);	Reusable Can be combined with 20. For further information, see IEEE Std 1063 [28].
20. Instruction (operation and maintenance)	Reusable. Can be combined with document 19
21. Report (software safety validation)	Newly developed.
22. Instruction (software modification procedures);	Reusable
23. Request (software modification);	Newly developed. Can be combined with document/database 25.
24. Report (software modification impact analysis);	Newly developed. A template has been presented in [29].
25. Log (software modification)	Newly developed. Tools exist for software modifications like e.g. the open source tool bugzilla, www.bugzilla.org. Can be combined with document/database 23.
26. Plan (software safety);	Reusable The document can be combined with document 2. For further information, see IEEE Std 1228 [30].
27. Plan (software verification);	Reusable
28. Report (software verification);	Generated. Some of the tests are generated automatically, others are semi-automatic and some are manually.
29. Plan (software functional safety assessment);	Reusable.
30. Report (software functional safety assessment)	Reusable. Finished after the last test/verification/validation report
31. Safety manual for compliant items	Reusable. May have a few remaining parts after the last test/verification/validation report

Table 1: Table A.3 regarding SW documentation in Part 1 and corresponding classification

Overview of document types as presented in A.3 in Part 1 of IEC 61508:

Documents Nu as listed in Table 1 Above	Comments
11 reports (Nu 13, 14, 15, 16, 17, 18, 21, 24, 28 and 30).	The standard ISO/IEC/IEEE 29119-3:2013 includes procedures and templates for Test status report, Test completion report, Test data readiness report, Test environment readiness report, Test incident report, Test status report and Test completion report.
6 specifications (Nu 1, 4, 5, 8, 9 and 10. 4, 5, 8 and 10 are test specifications)	The standard ISO/IEC/IEEE 29119-3:2013 includes both agile and traditional procedures for specifications and examples regarding Test design, Test case and Test procedure.
four plans (Nu 2, 26, 27, 29)	Validation, safety (can be based on e.g. EN 50126 [31] or IEEE Std 1228 [30]), verification and functional safety assessment
four instructions (Nu 6, 19, 20 and 22)	Development tools and coding manuals User, operation and maintenance instructions Modification procedure

Documents Nu as listed in Table 1 Above	Comments
two descriptions (Nu 3 and 7)	SW architecture design and SW system design
a list (Nu 11)	List source code
a request (Nu 23)	Request SW modification. Tools exist for software modifications like e.g. the open source tool bugzilla, www.bugzilla.org. Can be combined with document/database 23.
a log (Nu 25)	SW modification
a manual (Nu 31)	Safety manual for compliant items

Table 2: Overview of Table A.3 SW documents

The main documents are the reports, specifications and plans. As seen from the overview above, these documents should be the focus when trying to reduce the documentation work.

Overview of the document classes is shown in the table 3 below.

Class	Document number	Comments
Reusable	16 documents: 2, 3, 4, 5, 6, 7, 8, 9, 10, 19, 20, 22, 25, 26, 29 and 30	Reusable documents should be made more generic by the manufacturer. For documents that shall be updated as part of several sprints, reuse solutions is very important. These documents could e.g. include tables or a point list that are easily updated. For more information, see IEEE std 1517:2010 [18].
Combined	2 documents: 2 and 26 3 documents: 8, 9 and 10 5 documents: 13, 14, 15, 16 and 17 2 documents: 19 and 20 2 documents/databases: 23 and 25	12 documents can be merged to four documents. References are simplified when combining documents. The general parts are often the same. The relation between activities etc, is more visible. However this, to some extent, depends on e.g. the size of the project.
Generated	9 documents: 1, 11, 12, 14, 15, 16, 17, 18 and 28	Several possibilities exist. This will be studied later in the project.
New documents	5 documents: 6 (new tools) 21 (SW safety validation), 23 (request: SW modification), 24 (SW modification impact analysis) and 25 (log: SW modification).	Discussions with the assessor: As part of the Scrum mindset it is important to reduce the amount of documentation and it is assumed that the assessor should be involved early in the project. What could be a minimum of documentation should therefore be discussed with the assessor before starting to develop any new document. Templates and examples: For some documents templates and examples has already been developed as part of research, standardization and organizational work. See e.g. [29], ISO/IEC/IEEE 29119-3:2013 and www.misra.org.uk.

Table 3: Classes of documents

5. Discussion and conclusion

The acceptance of a system that has safety critical components rests on three pillars – agreements with the assessor, trust in the developers and competent work. This holds, independent of standard and development methods applied. The pillars are, however, not constructed independently. In our experience, an agreement with the assessor must come first. This will enable us to settle important questions such as:

- Which parts of Scrum may pose problems later in the project?
- What is accepted as PoC for each activity?
- Which documents are needed, in which form and when?

When this is in place, we can start to build trust based on demonstration of competence and strict adherence to all agreements.

Our conclusion is simple – the requirement that we need to certify a system according to IEC 61508 cannot be used as an argument against using the Scrum development process. The problems that exist are not a consequence of formulations of the standard's requirements but are related to what the individual assessor will accept as PoC for an activity.

We have looked into the documents necessary for approval of the software and grouped them according to the opportunity for reuse, combination of several documents into one, documents generated automatically and new documents. Only five of the documents are new documents when doing recertification. In addition we suggest that new documents should initially be discussed with the assessor, having trust and Scrum philosophy in mind to ensure correct level of documentation.

As part of our ongoing work on safety critical systems development we will try out the described approach in an industrial environment. This will partly be done to see if the approach needs modifications and partly to see how the assessors can be involved so that we can get a more efficient cooperation. We will also study how we can build trust between developers and assessors. This will not remove the need for PoCs but it will allow the assessor to focus on the few, critical parts of his works and leave the rest to the developers.

References

[1] K. Schwaber, Beedle, M., *Agile Software Development with Scrum*. New Jersey: Prentice Hall, 2001.

[2] M. Müller, "Functional Safety, Automotive SPICE® and Agile Methodology at KUGLER MAAG CIE GmbH," presented at the 8th Automotive Software Workshop, 2011.

[3] C. Webster, N. Shi, and I. S. Smith, "Delivering Software into NASA's Mission Control Centre Using Agile Development Techniques," presented at the Aerospace Conference, Big Sky, USA, 2012.

[4] IEC, "61508:2010 Functional Safety of Electrical/Electronic/Programmable Electronic Safety-related Systems (E/E/PE, or E/E/PES)," ed.

[5] T. e. a. Wien, "Reducing Lifecycle Costs of Industrial Safety Products with CESAR " presented at the Emerging Technologies and Factory Automation (ETFA), Bilbao, Spain, 2010.

[6] T. Stålhane and G. K. Hanssen, "The application of ISO 9001 to agile software development," presented at the Product Focused Software Process Improvement (PROFES 2008), Frascati, Italy, 2008.

[7] T. Stålhane, T. Myklebust, and G. K. Hanssen, "The application of Scrum IEC 61508 certifiable software," presented at the ESREL, Helsinki, Finland, 2012.

[8] T. Stålhane, V. Katta, and T. Myklebust, "Scrum and IEC 60880," presented at the FFI seminar, Storefjell, Norway, 2013.

[9] ISO, "19011: Guidelines for auditing of management systems. Ed. 2," ed, 2011.

[10] ISO/IEC, "17000: Conformity assessment – Vocabulary and general principles. Ed.1," ed, 2004.

[11] ISO/IEC, "17011: Conformity assessment – General requirements for accreditation bodies accrediting conformity assessments bodies. Ed. 1," ed, 2004.

[12] ISO/IEC, "17020: General criteria for the operation of various types of bodies performing inspection. Ed. 2," ed, 2012.

[13] ISO/IEC, "ISO/IEC 17021 Conformity assessment – Requirements for bodies providing audit and certification of management systems," ed, 2011.
[14] ISO/IEC, "17024: General requirements for bodies operating certification of persons. Ed. 2," ed, 2012.
[15] ISO/IEC, "17025: General requirements for the competence of testing and calibration laboratories. Ed. 2," ed, 2005.
[16] ISO/IEC, "17065: Conformity assessment – Requirements for bodies certifying products, processes and services. Ed. 1," ed, 2012.
[17] ISO/IEC, "17067: Conformity assessment – Fundamentals of product certification and guidelines for product certification schemes," ed, 2013.
[18] IEEE, "1517 standard for information technology – System and software life cycle processes – Reuse processes. Ed. 2," ed, 2010.
[19] IEEE, "Std 830 Recommended Practice for Software Requirements Specifications," ed, 1998.
[20] IEEE, "Std 1233 Guide for Developing System Requirements Specifications," ed, 1998.
[21] IEEE, "Std 730 Standard for Software Quality Assurance Plans," ed, 2002.
[22] ISO/IEC/IEEE, "Std 42010 Systems and software engineering - Architecture description. ," ed, 2011.
[23] IEEE, "Std 1016 Recommended Practice for Software Design Descriptions," ed, 2009.
[24] ISO/IEC/IEEE, "29119-3. Software and systems engineering – software testing – Part 3: Test documentation. Ed. 1," ed, 2013.
[25] Exida, "C/C++ Coding Standard recommendations for IEC 61508," ed, 2011.
[26] MISRA, "Guidelines for the use of the C++ language in critical systems," ed, 2008.
[27] IEEE, "Standard for software reviews and audits. Ed. 2," ed, 2008.
[28] IEEE, "Std 1063 Standard for Software User Documentation," ed, 2001.
[29] T. Myklebust, T. Stålhane, G. K. Hanssen, and B. Haugset, "Change Impact Analysis as required by safety standards, what to do?," presented at the Probabilistic Safety Assessment & Management conference (PSAM12), Honolulu, USA, 2014.
[30] IEEE, "Std 1228 Standard for Software Safety Plans," 1994.
[31] EN, "50126 Railway applications - The specification and demonstration of Reliability, Availability, Maintainability and Safety (RAMS)," 1999.

A Software Package for the Assessment of Proliferation Resistance of Nuclear Energy Systems

Zachary Jankovsky*[1], Tunc Aldemir[1] Richard Denning[1] Lap-Yan Cheng[2] and Meng Yue[2]

[1]The Ohio State University, Columbus, Ohio, USA
[2]Brookhaven National Laboratory, Uptown, New York, USA

Abstract: In order to better safeguard nuclear material from diversion by a malicious actor, it is important to search the input parameter space to gauge the attractiveness of various strategies that could be employed by such an actor. The ability to create and cluster a large number of scenarios based on similarity allows for a more complete and faster investigation of this parameter space. The software tool PRCALC was developed by the Brookhaven National Laboratory to estimate the various measures for covert diversion of nuclear material from a hypothetical fuel cycle system. The software package OSUPR (Ohio State University Proliferation Resistance) was written to extend PRCALC's abilities to allow for the creation of many scenarios at a time, as well as to take advantage of multiple processing threads in the computation of proliferation resistance measures. OSUPR also allows for clustering of the outputs of PRCALC using three methods: mean-shift, *k*-means, and adaptive mean-shift. The clustered results can yield insights to vulnerable aspects of the fuel system.

Keywords: PRCALC, Proliferation Resistance & Physical Protection, Clustering, Mean Shift, *k*-means.

1. INTRODUCTION

Since the Treaty on the Non-Proliferation of Nuclear Weapons took effect in 1970, efforts have been made to better identify and secure vulnerable parts of the nuclear fuel cycle. The treaty is primarily enforced through safeguards – methods by which diversion of material may be deterred or detected. A cornerstone of international safeguards is nuclear materials accountancy, complemented by containment and surveillance. One aspect of safeguards accountancy is to detect material unaccounted for (MUF), by comparing the mass of certain materials entering and exiting a process to the changes expected for the process. Other safeguards focus on consistent record-keeping, evidence of tampering, and surveillance of sensitive areas.

Due to the cost associated with changing an already-built nuclear fuel system, it is desirable to implement in the design of the system features to enhance proliferation resistance & physical protection (PR&PP) before construction. An effort by Brookhaven National Laboratory to model PR&PP of a Generation IV reactor and reprocessing system [1] had led to the development of the software tool PRCALC [2] towards such a purpose. This paper describes an extension of PRCALC that can create and process multiple material diversion scenarios, as well as cluster the results with a goal of identifying vulnerable areas of the system.

2. BACKGROUND AND MOTIVATION

2.1. PRCALC

One of the systems for which data are available in PRCALC is a hypothetical Example Sodium-cooled Fast-spectrum Reactor (ESFR) [3] and a hypothetical reprocessing plant. The reprocessing plant accepts both ESFR and light water reactor (LWR) spent fuel as inputs, and outputs ESFR fuel assemblies and various waste products. It is assumed that the facility is under International Atomic Energy Agency (IAEA) safeguards represented by 4 categories:

*Author Contact: jankovsky.3@osu.edu

1. Audit of various nuclear material accounting records or reports
2. Material verification such as physical inventory verification (PIV) of all nuclear material in a nuclear energy system
3. Surveillance and monitoring of spent fuel pool, reactor and reprocessing areas
4. Containment seals on reactor, shipping casks, and safeguards equipment

PRCALC represents the diversion from the fuel system as a Markov model as shown in Fig. 1, with each stage and stage element listed in Table 1. Each stage element is a potential target for material diversion by some malicious actor. Each diversion attempt may be detected, fail for technical reasons, or be successful. Once enough material is diverted, the actor converts it to weapons-usable material in a clandestine facility which is not subject to safeguards. Scenarios are created one at a time by varying targets, diversion rates, and safeguards in place. A scenario is a single PRCALC input file, with a unique set of input parameters. Multiple scenarios may be batched together and run all at once, so that the user does not have to manually load each scenario.

Figure 1: PRCALC Markov Model

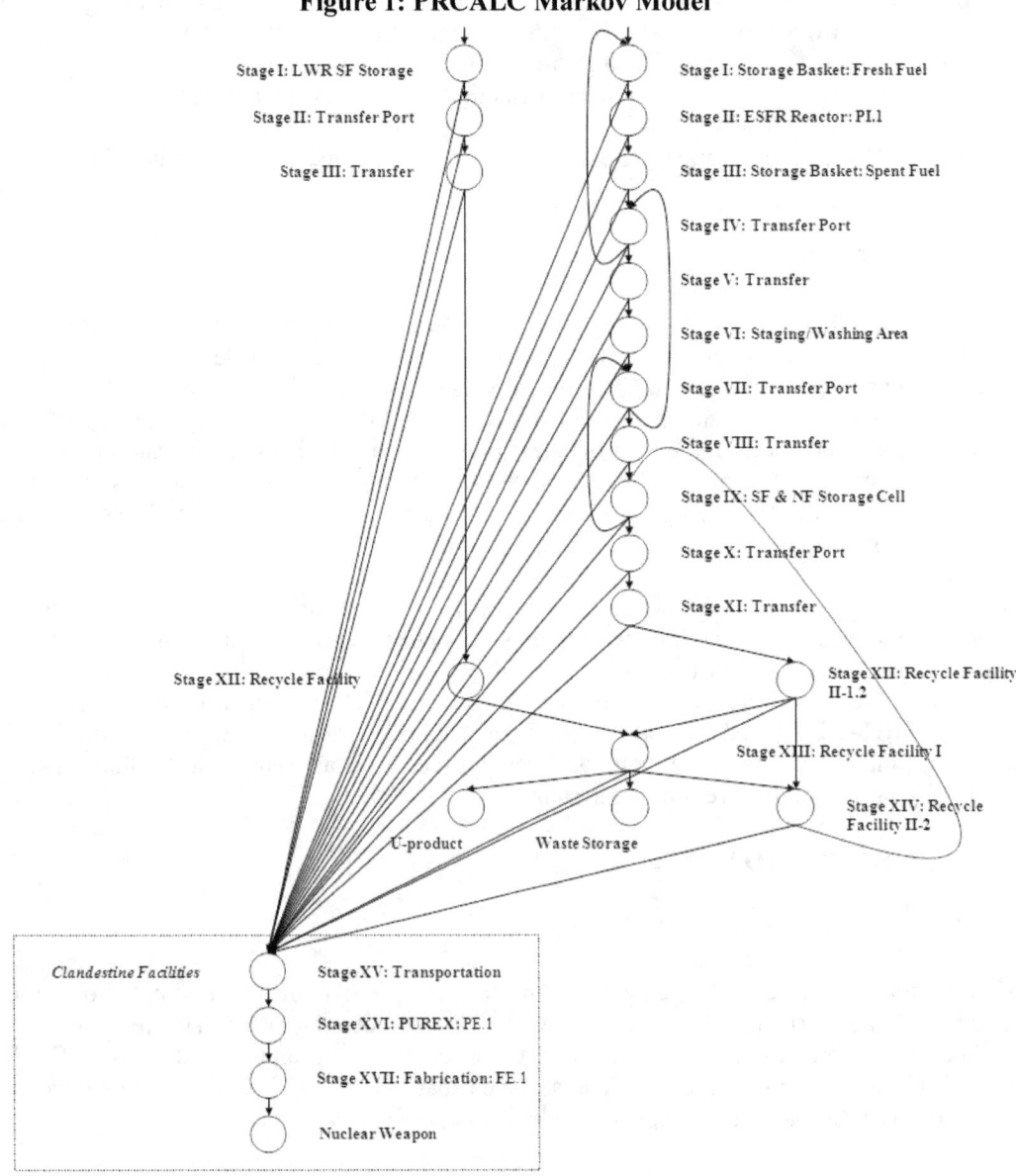

Table 1: PRCALC Stages & Elements

Stage I	LWR SF Storage
	Storage Basket: Fresh Fuel
Stage II	LWR Transfer Port
	ESFR Reactor
Stage III	LWR Transfer
	Storage Basket: Spent Fuel
Stage IV	ESFR Transfer Port
Stage V	ESFR Transfer
Stage VI	Staging/Washing Area
Stage VII	Staging/Washing Transfer Port
Stage VIII	Staging/Washing Transfer
Stage IX	SF & NF Storage Cell
Stage X	SF & NF Transfer Port
Stage XI	SF & NF Transfer
Stage XII	LWR SF Disassembly
	ESFR SF Disassembly
Stage XIII	Chopping
Stage XIV	Electro-refiner
Stage XV	U-product Processing
	TRU Extraction
Stage XVI	Product Preparation
Stage XVII	Pin Fabrication
Stage XVIII	Assembly
Stage XIX	Clandestine Transportation
Stage XX	Clandestine Chemical Conversion
Stage XXI	Clandestine Chemical Separation

Diversion rates are inputs to the PRCALC analysis. Thus, if the proliferators are able to overcome technical challenges and avoid detection, the time to achieve a goal quantity of material is determined by input. For each type of safeguards activity there is a characteristic time to detect an anomaly, and a time to confirm that the anomaly was caused by diversion and is not a false alarm. The detection rate is given by

$$r_i = \sum_{j=1}^{n} \frac{1}{T_{D(i,j)}} \qquad (1)$$

where r_i is the total detection rate for stage i, n is the number of safeguards approaches for stage i, and $T_{D(i,j)}$ is the total detection time for safeguard j on stage i. This greatly simplifies the Markov model, as opposed to including a detection state and transition rate associated with confirmation of detection for each safeguard approach at each stage of the fuel cycle. The total technical failure rate for each stage is handled in a similar fashion, being a combination of multiple technical failure mechanisms.

PRCALC provides five outputs relating to proliferation resistance. Probability of Detection (DP) is the likelihood that a diversion attempt will be detected by safeguards and subsequently halted by outside intervention. Probability of Failure (PF) is the likelihood that a proliferation attempt will fail either in obtaining material or in converting it to weapons-usable material because of failure to overcome technical barriers. For example, the material could be too hot, thermally or radioactively, for the actor to move it in a manner that would not be detected. The actor also may lack the expertise or equipment to reliably convert the material once obtained. Probability of Success (PS) is the likelihood that the proliferation attempt will succeed, and is equal to 1-(DP+PF). Proliferation Time (PT) is measured in weeks and is the time that would be required to obtain and convert the material without intervention or technical difficulties. Finally, the Material Type index (MT) reflects the average attractiveness of diverted material. This is based on the effort that would be required to convert the material to weapons-usable plutonium. For example, reactor-grade plutonium has an MT of 0.95 while LWR spent fuel has an MT of 0.50.

2.2. Motivation

The goal of this work is to develop a methodology for PR&PP analysis of proposed fuel systems that involves analyzing a large number of scenarios for common factors. Each scenario is a unique strategy that an actor might use to obtain nuclear material. The product is OSUPR (Ohio State University Proliferation Resistance), with a MATLAB graphical user interface (GUI). PRCALC was chosen as the basis of the methodology and used as a module of OSUPR due to its ease of use and expandability. Because PRCALC was originally written to create one scenario at a time, a method was developed to create a large number of scenarios with little input from the user. By using multiple processing threads it was found possible to greatly reduce the time required to process the large data sets that were produced.. Finally, clustering of a data set can be accomplished using different methods (Section 3), allowing the user to inspect the data in diverse ways.

3. OSUPR

OSUPR is shown schematically in Fig. 2. It comprises two interfaces that each accept user input and output data to MATLAB-compatible files. The Scenario Creation interface is shown as Fig. 3, and has 11 steps which are listed in Table 2. The number of scenarios created is estimated (Step 5) using Eq. (2), where r is the number of diversion rates (set in Step 2), N_{SG} is the number of safeguards conditions to be considered (Step 3), and N_{stage} is the number of targets selected (Step 4). In Step 6 the folder for the scenario creation data is chosen, and Step 7 creates the scenario files. Each PRCALC scenario is one file. To simplify data collection, a single file (referred to as data set in the rest of the paper) is kept for the set with all relevant information as noted in Table 3. It first includes scenario creation data, and is later grown to include PRCALC outputs and clustering results. Step 8 informs the user of the number of processing threads available. Step 9 selects the number of equal-sized batches to create, typically the same as the number of processing threads listed in Step 8. Step 10 creates the batch files. Finally, scenarios are run through PRCALC using Step 11. If a set of scenarios has been previously created, there is also an option to load the Information File to run through PRCALC.

$$N_{scen} = r^{N_{stage}} * N_{SG} \qquad (2)$$

Once a set of scenarios has been run through the PRCALC engine, the Information File is updated and may be examined immediately or clustered (see Fig.2). The general structure of the Information File is shown in Table 3. The BatchAssign, Batches, and ConfirmFile fields are used for multithread processing. PRResults contains both the raw output of PRCALC, and the outputs normalized so that each output varies from 0 to 1 over the set. Finally, each clustering algorithm information field contains the parameters used, the number of clusters, cluster centroid locations, and cluster assignment for each scenario. The cluster centroid for a certain dimension is the average value of the scenarios in that cluster along that dimension.

Figure 2: OSUPR Schematic

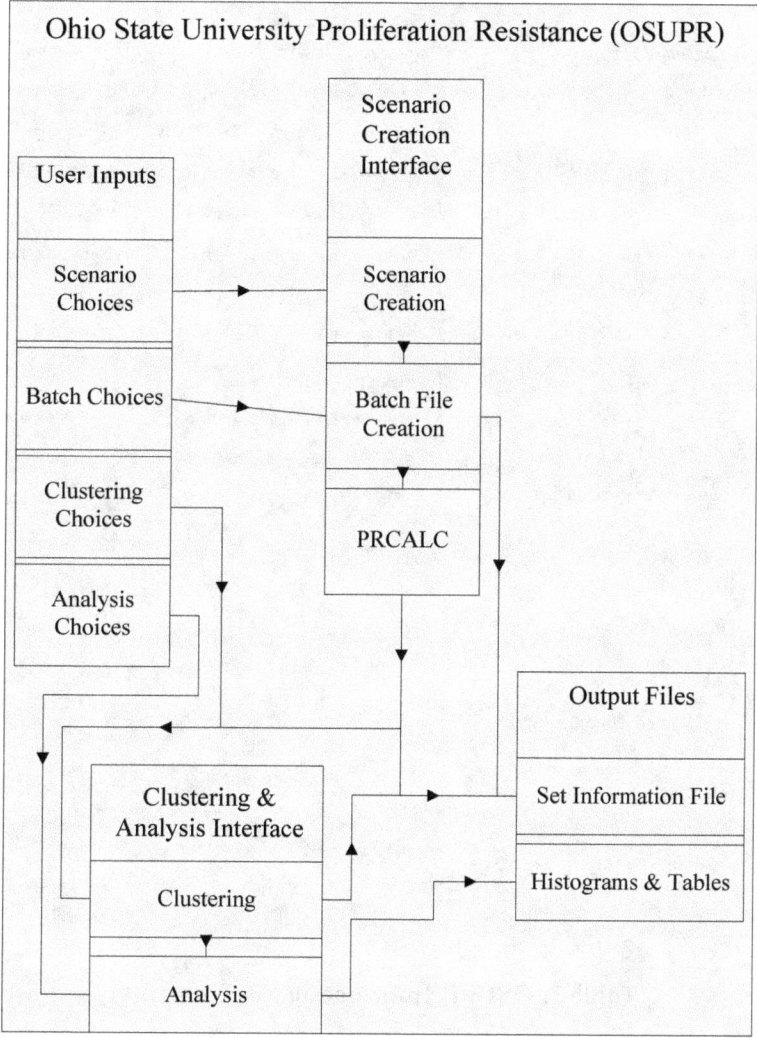

Table 2: OSUPR Scenario Creation Interface (see Fig.3) Steps

Step #	Description
1	Series Name
2	Diversion Rate Selection
3	Safeguards Condition Selection
4	Target Selection
5	Set Size Estimation
6	Folder Selection
7	Scenario Creation
8	Processing Core Availability
9	Batch Number Selection
10	Batch File Creation
11	PRCALC Operation

Figure 3: OSUPR Scenario Creation Interface

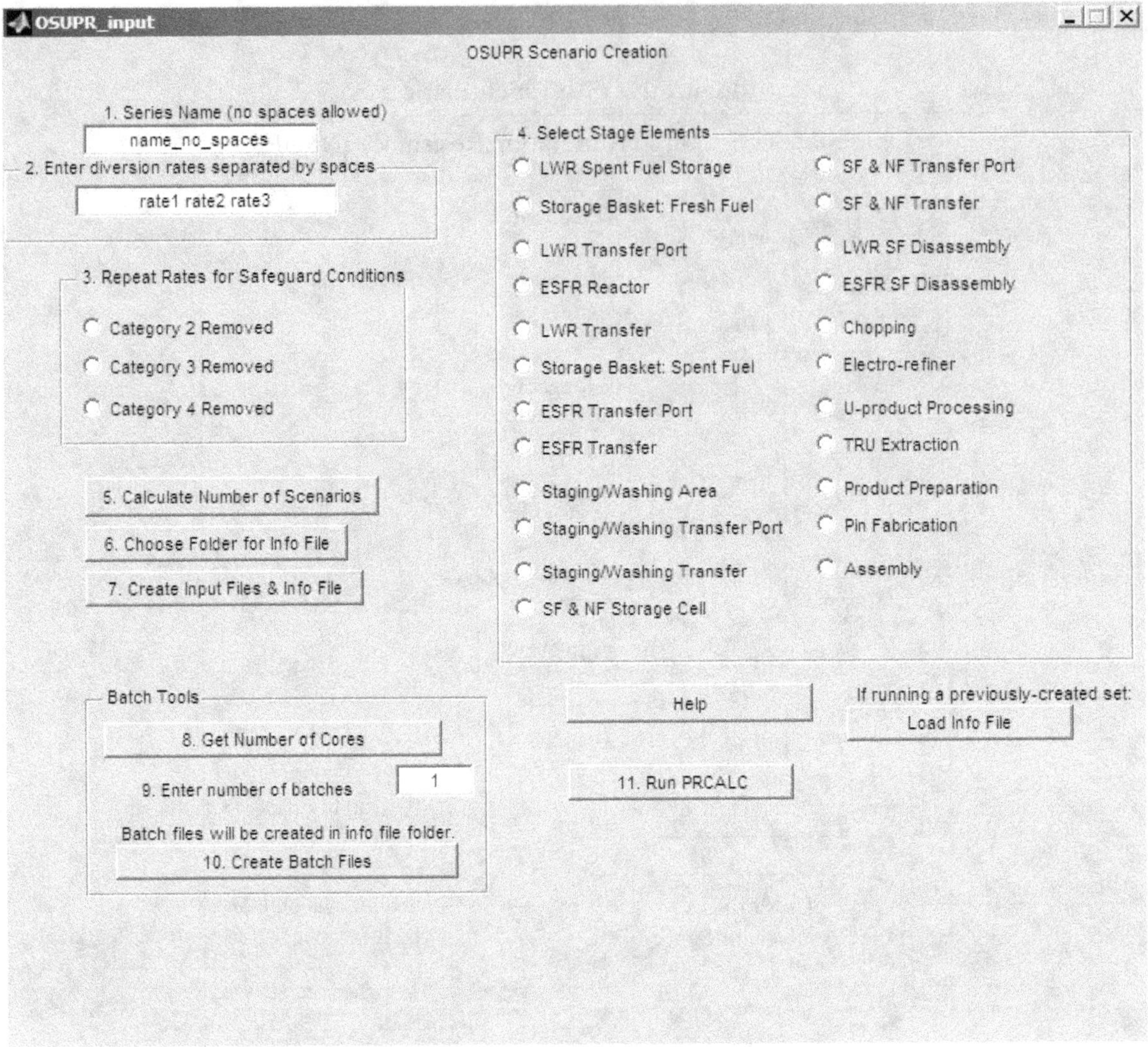

Table 3: OSUPR Information File Contents

Field	Description
Names	The name of each scenario file
Rates	Diversion rates associated with each scenario
BatchAssign	Assignment of each scenario to a batch
Batches	Batch file names and locations
ConfirmFile	The name of a small file used to confirm that a thread is finished
PRResults	PRCALC outputs, both raw and normalized
FOM	A rudimentary figure of merit for each scenario
KmeansInfo	Cluster assignment and statistics for k-means
MeanShiftInfo	Cluster assignment and statistics for mean-shift
FAMSInfo	Cluster assignment and statistics for adaptive mean-shift

Currently OSUPR runs on a HP xw6400 workstation with 2 quad-core processors, which appear as eight processing threads under Microsoft Windows 7. Using Step 11 each batch is run in a separate instance of OSUPR, and the operating system automatically allocates the work efficiently across the available threads. Table 4 shows the performance of OSUPR using multiple threads for a set of 6,558 scenarios. Time is kept from when scenarios are sent to PRCALC to when all batched outputs are assembled into a single file. It can be seen that the wall clock time required to run a set almost halves with each doubling of threads used. Perfect scaling would result in 8 threads requiring 12.5% of the time for 1 thread, while in the experiment it required 15.1%. This deviation is likely due to the overhead of re-uniting the outputs after all batches are finished running. It was expected that PRCALC would benefit from parallelization, as the scenario are independent of each other.

Table 4: OSUPR Multithreading Performance

Threads	Time (s)	vs previous	vs 1 thread
1	80460	N/A	100%
2	43848	55%	55%
4	22761	52%	28%
8	12111	53%	15%

In its default form, PRCALC displays a progress bar and a plot of outputs for each scenario. Some minor cosmetic changes were required for its use as a component of OSUPR. The progress bar in PRCALC, which tracks only the current scenario, was removed and replaced with a progress bar that tracks the entire set. The output plot was completely removed because opening a new MATLAB plot window for each scenario in a large set would tie up computer resources and overwhelm the user.

The Clustering & Analysis interface in Fig.2 is shown in more detail in Fig. 4, and its 7 steps are listed in Table 5. In Step 1 of the clustering process, an Information File (from the scenario creation interface) is loaded. Each scenario is a data point, to be clustered on the dimensions DP, PS, and PT. The user may choose between three algorithms (mean-shift, adaptive mean-shift and k-means), which are more fully described for this application in [4]. Mean-shift algorithm requires the user to specify a bandwidth, or neighborhood size, for each cluster. The algorithm iteratively searches for the highest data density within the neighborhood. If cluster centroids meet, they are merged into one cluster. Adaptive mean-shift is a variation on mean-shift in which the user does not need to specify a parameter. This is especially valuable if little is known about the nature of the data set. The bandwidth is varied for each point based on a pilot run including, by default, 10% of scenarios chosen at random. K-means requires the parameter k, which is the number of clusters desired. k cluster centers are randomly placed in the data space, and scenarios are assigned to the cluster with the nearest center. The center location of each cluster is then updated to the centroid of the data points assigned to it. This continues iteratively until the cluster assignment of points is stable. In Step 2 the user specifies a clustering parameter if necessary, and Step 3a, 3b, or 3c runs the chosen clustering algorithm.

Clustering algorithms may only be run one at a time, but each algorithm can be run and the results recorded to the Information File for a given set of scenarios. This allows the user to compare the outputs of each algorithm. The user is notified when clustering is complete, but no progress bar is available due to the nature of clustering. Based on experience a set of approximately 50,000 scenarios on the same computer requires roughly a second for k-means, an hour for adaptive mean-shift, and tens of hours for mean-shift. The adaptive mean-shift program currently used is in the form of a Windows executable file, and has not been tested with other operating systems.

Table 5: OSUPR Clustering & Analysis Interface (see Fig. 4) Steps

Step #	Description
1	Load Information File
2	Enter Clustering Parameter
3	Run Clustering Algorithm
4	Choose Algorithm for Analysis
5	View Cluster Centroids
6	Choose Cluster of Interest
7	Analyze Cluster of Interest

Figure 4: OSUPR Clustering & Analysis Interface

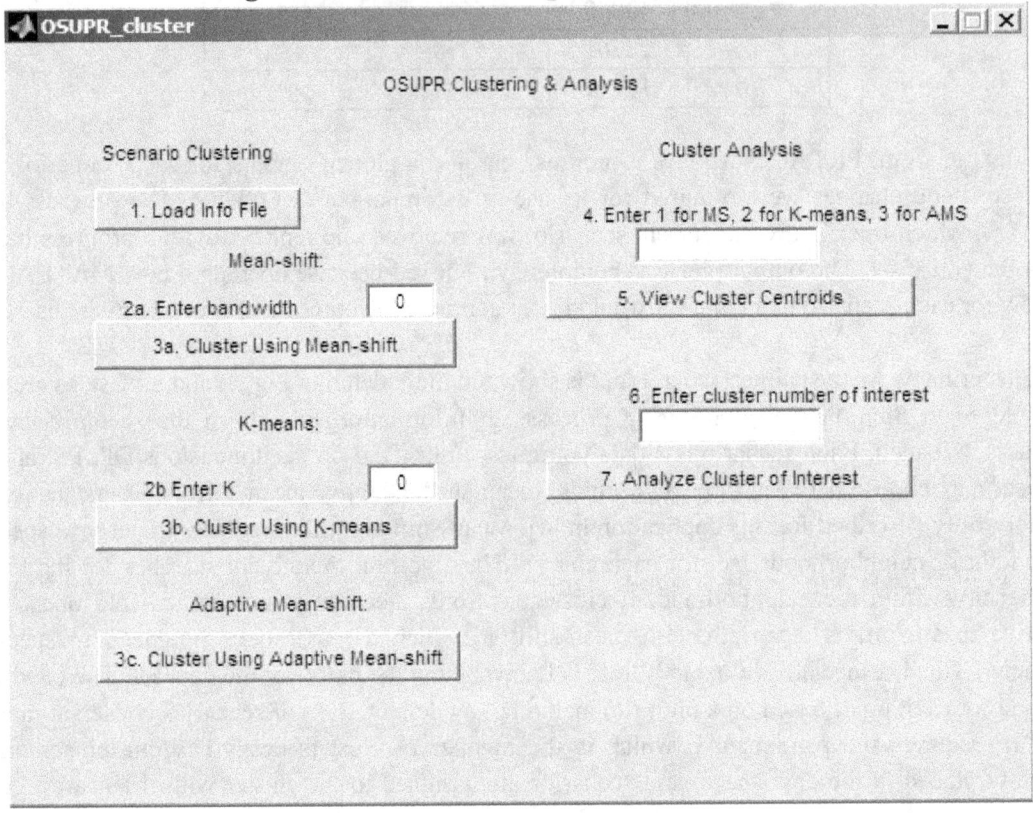

For the user to analyze the results of a particular clustering algorithm, the algorithm must be chosen in Step 4. The analysis of a single cluster is demonstrated more fully in [5]. Step 5 presents the cluster centroids, as demonstrated in Table 6. This set of scenarios was created using 4 diversion rates, 7 potential targets, and 4 safeguards conditions. The use of Eq. 2 returns 65,536 scenarios. Four scenarios were created with diversion rates of 0 at every target, and these were removed for a total of 65,532 scenarios. This allows the user to choose a cluster that is of interest for further examination. In this example the scenarios in Cluster 4 have on average the lowest probability of detection, the shortest time requirement, and the highest probability of success. These properties may cause these scenarios to be of interest to a would-be proliferator, and therefore also to a safeguards analyst. The cluster of interest is chosen in Step 6 of the clustering & analysis interface.

In Step 6, the cluster of interest is compared to the full set of scenarios. One type of comparison that is output is a histogram for each clustering dimension, as demonstrated for DP and PS in Fig. 5. The same set of data is used as in Table 6. A bimodal distribution is seen on both outputs across the set, with the cluster of interest containing the "best" scenarios from the point of view of a proliferator. The group of scenarios centered around a DP value of 0.5 for Fig.5(a) are nearly all in Cluster 4 in Fig.5(b). Similarly for PS, much of the group of scenarios seen in c centered around 0.65 are present in Fig.5(d), in Cluster 4. These histograms are output as image files in the same folder as the Information File, specified in Step 6 in Fig. 3.

Table 6: Sample Cluster Centroid Output
(DP: Detection Probability, PT: Proliferation Time, PS: Probability of Success)

Cluster	DP	PT (weeks)	PS
1	0.980	756.159	0.002
2	0.984	390.579	0.002
3	0.818	32.706	0.034
4	0.493	31.067	0.096

Figure 5: Sample Cluster of Interest Output Histograms
(DP: Detection Probability, PS: Probability of Success)

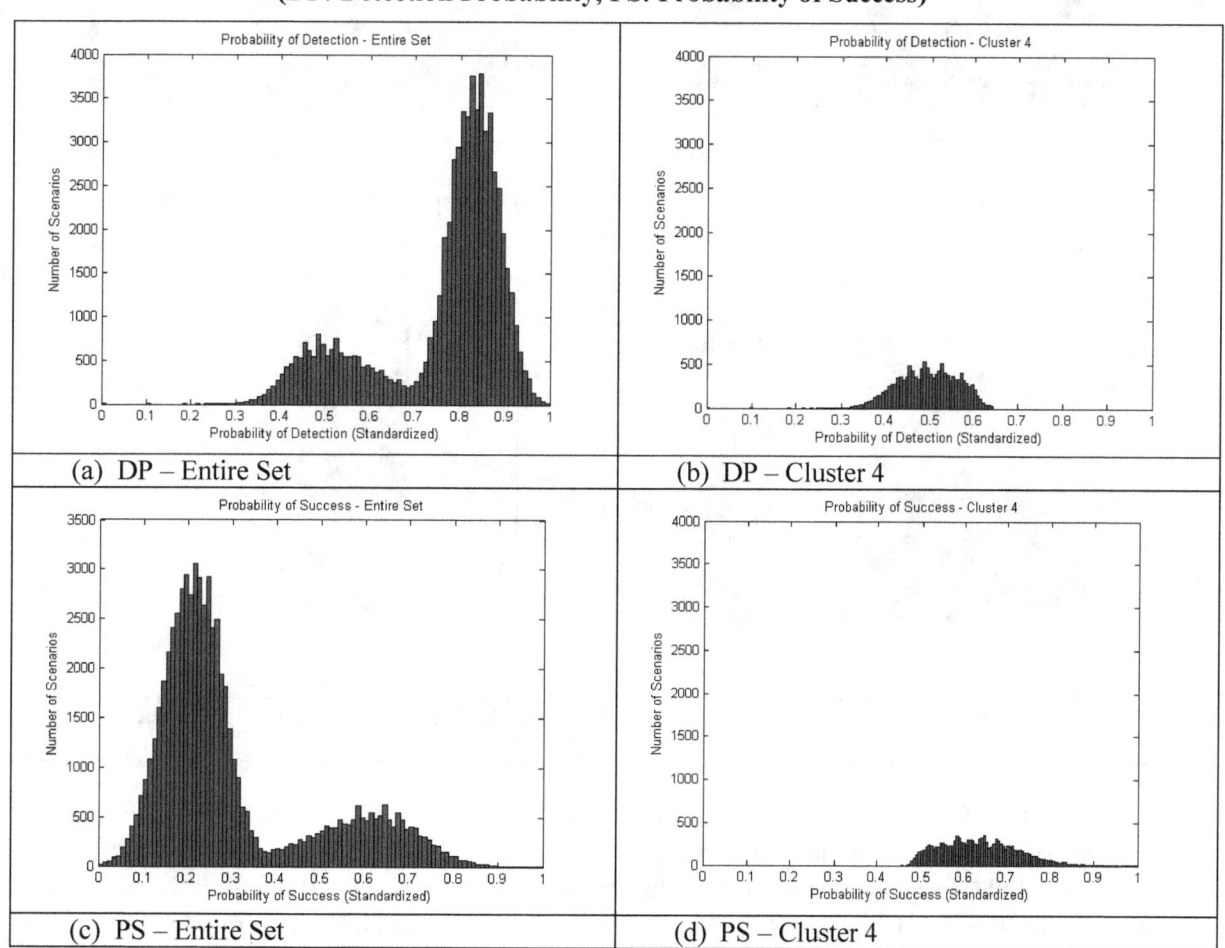

(a) DP – Entire Set   (b) DP – Cluster 4
(c) PS – Entire Set   (d) PS – Cluster 4

Another set of histograms is created to show the inputs that led to a scenario being in the cluster of interest. This is most closely tied to the goal of the work, which is to identify vulnerabilities in a fuel system. Sample outputs are shown in Fig. 6 for the same set as in Table 6. The diversion rates used in this set were 0, 1, 2, and 4. Safeguards Condition 1 has all default safeguards (see Section 2.1) in place. Conditions 2-4 have Categories 2, 3, or 4 removed. This represents a successful attempt by the proliferator to circumvent various safeguards in the interest of material diversion. Because scenarios are created using every combination of the input rates and targets with equal likelihood , the histograms shown in Fig. 6 would be flat across the entire set of scenarios. Thus, a deviation from level in one cluster may indicate that a strategy is more or less attractive to a proliferator. It must be noted that this analysis is for a hypothetical fuel system, and is only to demonstrate the methodology that has been developed.

Figure 6: Sample Cluster of Interest Input Histograms

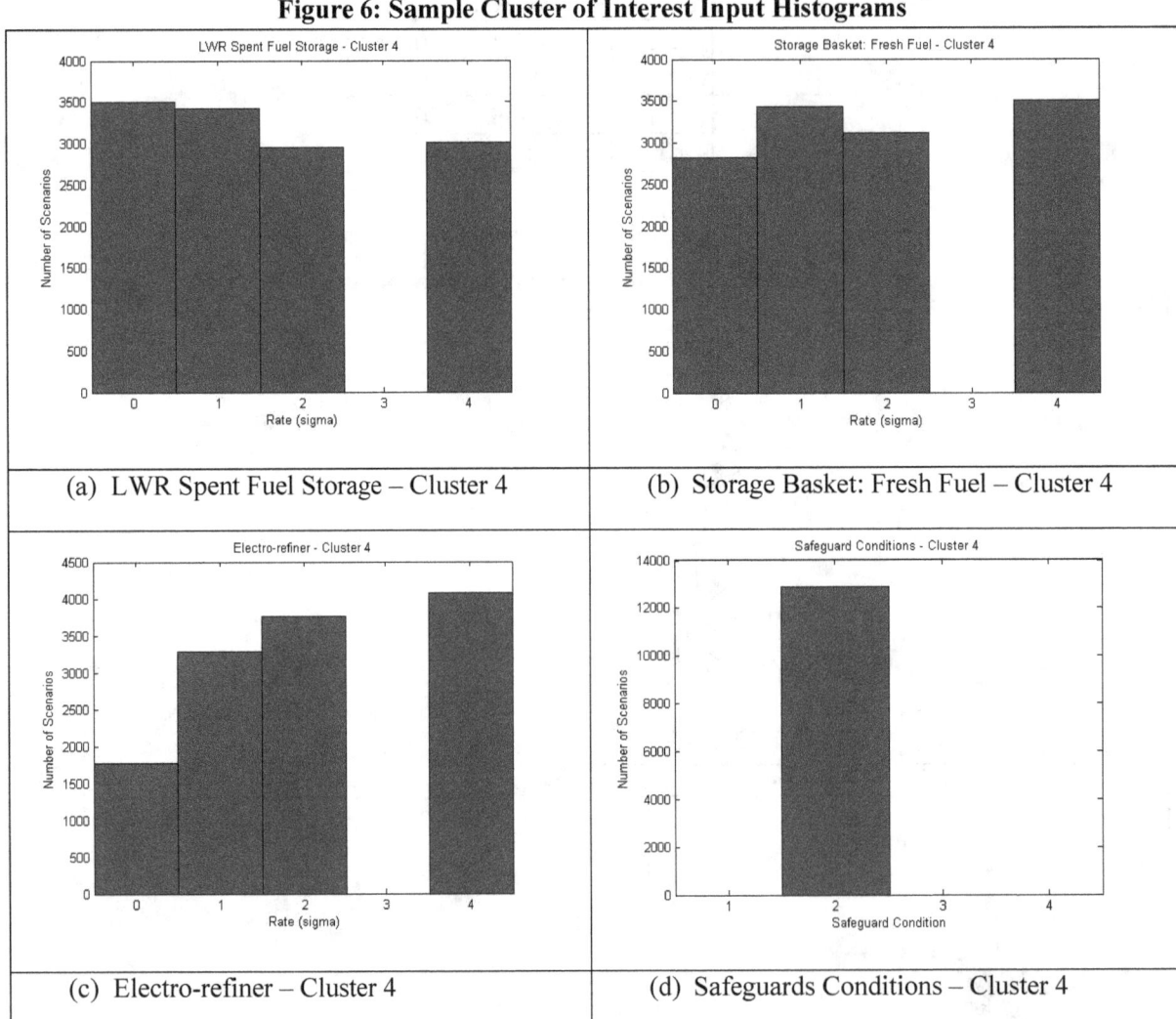

(a) LWR Spent Fuel Storage – Cluster 4
(b) Storage Basket: Fresh Fuel – Cluster 4
(c) Electro-refiner – Cluster 4
(d) Safeguards Conditions – Cluster 4

In Fig. 6(a), LWR Spent Fuel Storage, increased diversion seems to lead to a decrease in attractiveness. It may be that increased diversion at this target contributes highly to detectability while contributing little to speeding up diversion. Fig. 6(b) shows that for Storage Basket: Fresh Fuel, which represents the storage of new ESFR fuel, no advantage is evident for high or low diversion rates. For part Fig. 6(c), scenarios with a high diversion rate are well-represented in the cluster of interest. This indicates that an increase in diversion from the Electro-refiner would be attractive to a proliferator.

Finally, Fig. 6(d) shows the Safeguards Conditions present in Cluster 4. Only scenarios with Physical Inventory Verification safeguards removed are in the cluster of interest, indicating that the integrity of these safeguards is important in the fuel system under consideration.

4. CONCLUSIONS

The analysis of a large number of scenarios may yield insights into proliferation resistance for future and current nuclear fuel systems. A mechanized process of defining scenarios, estimating PR-related outputs, clustering the outputs and presenting meaningful results to a user is developed and built into an easy-to-use graphical user interface.

Acknowledgements

This research has been performed using funding received from the DOE Office of Nuclear Energy's Nuclear Energy University Programs.

References

[1] R. Nishimura, R. Bari, P. Peterson, J. Roglans-Ribas, D. Kalenchuk, "Development of a Methodology to Assess Proliferation Resistance and Physical Protection for Generation IV Systems," *America's Nuclear Energy Symposium – ANES 2004,* Miami Beach, FL, October 3-6 (2004).

[2] M. Yue, L. Cheng, R. Bari, "Quantitative Assessment of Probabilistic Measures for Proliferation Resistance", *American Nuclear Society 2005 Winter Meeting,* Washington, DC, November 13-17 (2005).

[3] "PR&PP Evaluation: ESFR Full System Case Study Final Report," http://www.gen-4.org/Technology/horizontal/documents/PRPP_CSReport_and_Appendices_2009_10-29.pdf (2009).

[4] Z. Jankovsky, D. Zamalieva, R. Denning, A. Yilmaz, T. Aldemir, "A Comparison of Various Clustering Schemes for Proliferation Resistance Measures," *2013 American Nuclear Society Winter Meeting,* Washington, DC, November 10-14 (2013).

[5] Z. Jankovsky, D. Zamalieva, A. Yilmaz, R. Denning, T. Aldemir, "A Clustering Analysis of Probabilistic Proliferation Resistance Measures in an Example Nuclear Fuel System," *2013 International Topical Meeting on Probabilistic Safety Analysis,* Columbia, SC, September 22-26 (2013).

Risk Estimation Methodology for Launch Accidents

Daniel J. Clayton[*][a], Ronald J. Lipinski[a], and Ryan D. Bechtel[b]
[a] Sandia National Laboratories, Albuquerque, NM, USA
[b] Office of Space and Defense Power Systems, NE-75,
U.S. Department of Energy, Germantown, MD, USA

Abstract: As compact and light weight power sources with reliable, long lives, Radioisotope Power Systems (RPSs) have made space missions to explore the solar system possible. Due to the hazardous material that can be released during a launch accident, the potential health risk of an accident must be quantified, so that appropriate launch approval decisions can be made. One part of the risk estimation involves modeling the response of the RPS to potential accident environments. Due to the complexity of modeling the full RPS response deterministically on dynamic variables, the evaluation is performed in a stochastic manner with a Monte Carlo simulation. The potential consequences can be determined by modeling the transport of the hazardous material in the environment and in human biological pathways. The consequence analysis results are summed and weighted by appropriate likelihood values to give a collection of probabilistic results for the estimation of the potential health risk. This information is used to guide RPS designs, spacecraft designs, mission architecture, or launch procedures to potentially reduce the risk, as well as to inform decision makers of the potential health risks resulting from the use of RPSs for space missions.

Keywords: Radioisotope Power Systems, launch safety, launch accident analysis, probabilistic risk assessment.

1. INTRODUCTION

For fifty years nuclear power sources have enabled exploration missions to deep space and locations where solar panels are impractical or inefficient [1]. The United States (U.S.) Department of Energy (DOE) provides space nuclear systems to the National Aeronautics and Space Administration (NASA) for use on civilian space missions with special requirements for spacecraft electrical power and thermal heating. These energy sources fall into two general classes, either Radioisotope Power Systems (RPSs) for electrical power or Radioisotope Heater Units (RHUs) for local component heating. RPSs are compact and light weight, have long lives, and are highly reliable. These qualities enable space missions with high power requirements. Figure 1 shows the General Purpose Heat Source (GPHS), which contains and protects the fuel pellet, and serves as the heat source in various RPS designs and the Multi-Mission Radioisotope Thermoelectric Generator (MMRTG) [2], which is the current-generation RPS.

Due to the radioactive nature of the RPS and RHU, use on a specific mission must be approved by the U.S. Executive Branch per Presidential Directive / National Security Council Memorandum 25 (PD/NSC-25). As part of the launch approval process, the DOE prepares a Safety Analysis Report (SAR) in order to quantify the potential health risks from a launch accident. Figure 2 shows the flow of information and calculations with the SAR code suite. The suite consists of several hundred thousand lines of code and scripts and has been developed under control of a detailed quality assurance program. The risk estimations compare the potential risks involved with the probability of occurrence. This SAR is provided to the Interagency Nuclear Safety Review Panel (INSRP), which performs an independent review and assessment of the potential risks posed by the mission. The INSRP in turn prepares a Safety Evaluation Report (SER) documenting the review. The results of the SAR and SER are submitted to the White House Office of Science and Technology Policy (OSTP) for

---

[*] djclayt@sandia.gov

approval to proceed with the mission. If the risk estimates are acceptable, OSTP recommends to the President that the mission be approved.

Figure 1: a) General Purpose Heat Source (GPHS), b) Multi-Mission Radioisotope Thermoelectric Generator (MMRTG)

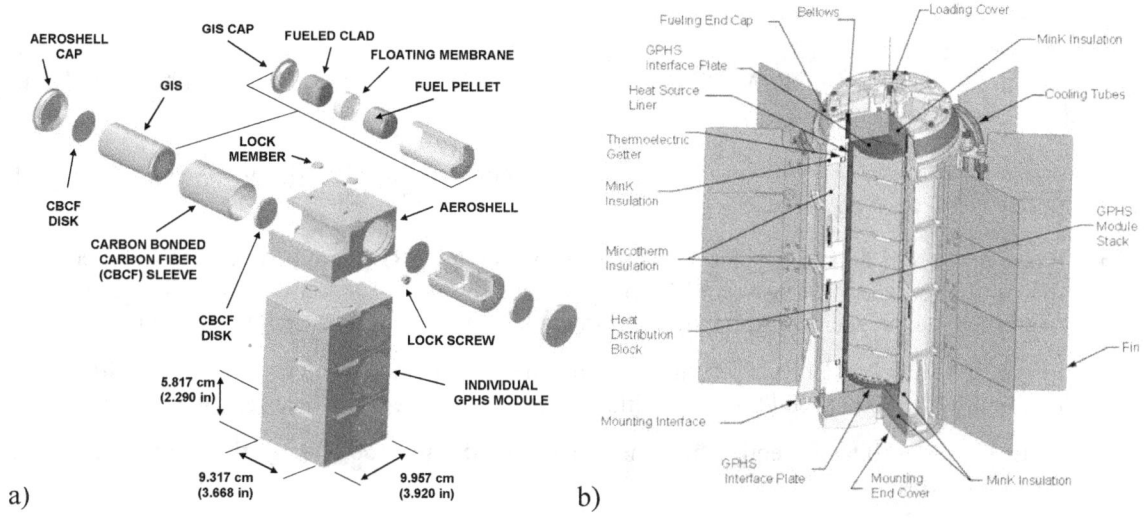

Figure 2: Code Suite Used in SAR Calculations

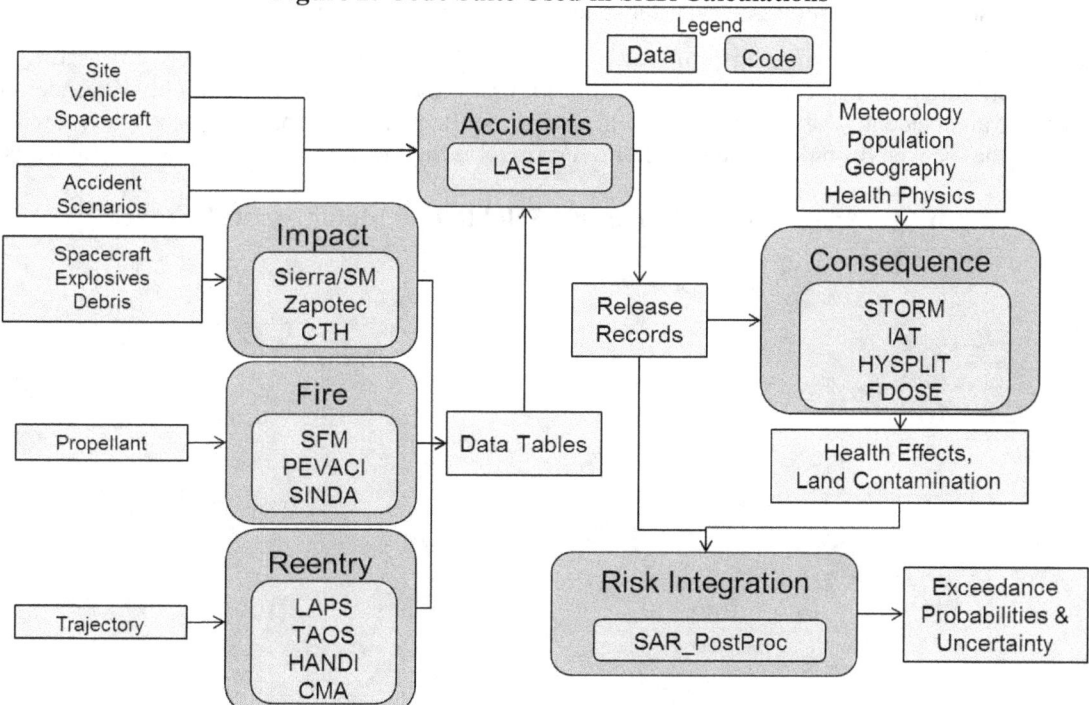

The safety analysis documented in the SAR includes an extensive series of analyses, primarily computer simulations of both a probabilistic and a deterministic nature. The analyses consider mechanistic and phenomenological models to simulate the progression of a launch accident, environments associated with such accidents, transport of radioactive material, and subsequent consequences to the public. Models are executed many times to accurately characterize the probabilistic nature of the event and in turn determine the potential risk. The methodology of this risk estimation is the focus of this paper. The risk estimation considers: 1) potential accidents associated with the launch, and their probabilities and accident environments; 2) the response of the radioisotope hardware to accident environments with respect to source terms (that portion of the release that becomes airborne) and their probabilities, and 3) the radiological consequences and mission risks associated with such releases.

2. REPRESENTATIVE ACCIDENT SCENARIOS

For the purpose of the risk analysis, the mission is usually divided into five mission phases on the basis of the mission elapsed time (MET, the time (T) relative to launch), reflecting principal events during the mission as follows:

- Phase 0: Pre-Launch, $T < t_1$, from installation of the RPS to just prior to start of the Stage 1 liquid rocket engines (LREs) at t_1.

- Phase 1: Early Launch, $t_1 \leq T < t_x$, from start of Stage 1 LRE(s), to just prior to t_x, where t_x is the time after which there would be no potential for debris or intact vehicle configurations resulting from an accident to impact land in the launch area, and water impact would occur.

- Phase 2: Late Launch, $t_x \leq T$, from t_x to when the launch vehicle reaches an altitude of nominally 30,480 m (100,000 ft), an altitude above which re-entry heating could occur.

- Phase 3: Suborbital Re-entry, from nominally 30,480 m (100,000 ft) altitude to the end of Stage 2 burn 1.

- Phase 4: Orbital Re-entry, from end of Stage 2 burn 1 to Stage 2 / spacecraft separation.

- Phase 5: Long-Term Re-entry, after spacecraft separation until no chance of Earth re-entry.

Figure 3 shows a typical mission profile for a solid rocket motor assisted launch, illustrating the various configurations that can occur. As seen in Figure 3, the RPS response is highly dependent on the time of an accident. The accident time will determine the possible impact surfaces and velocities, as well as the local environment, such as blast overpressure, fragment impacts and fire environments.

Figure 3: Typical Mission Profile of a Solid Rocket Motor Assisted Launch

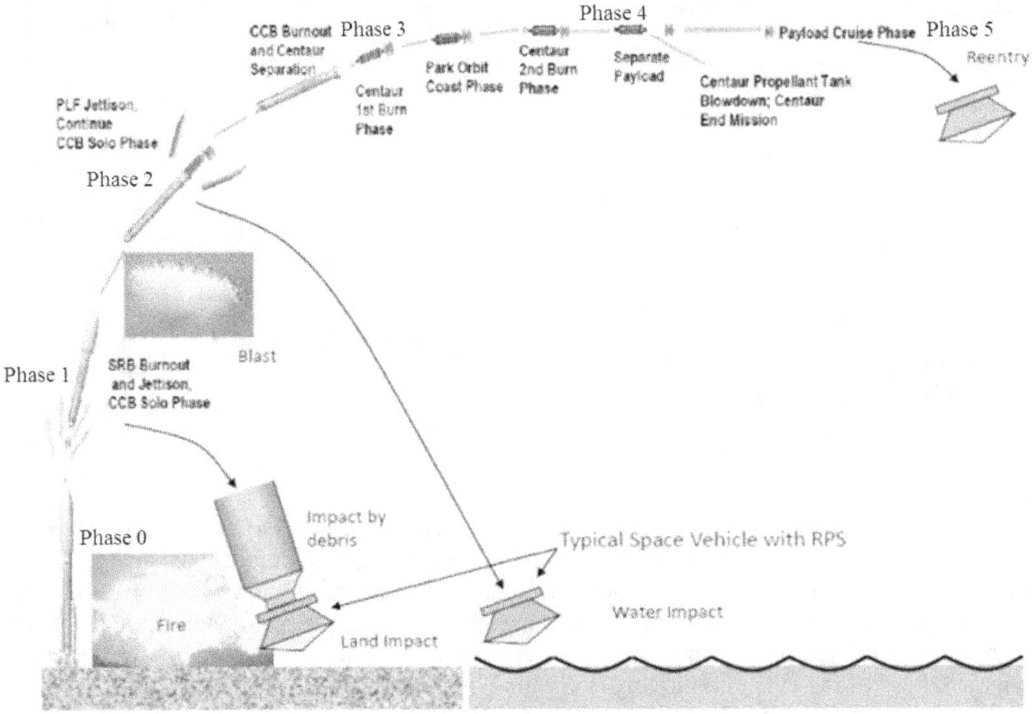

The various potential accident environments are grouped into Representative Accident Scenarios (RASs) based on similarities to the environment and sequence of events experienced by the RPS. Similar RASs can be found within the different mission phases, but are kept separate as to not preclude the ability to determine the risk for each phase independently. Results for each RAS are calculated and combined together to determine the risk for each phase, as well as the overall mission risk, based on the relative probability of each RAS.

3. ACCIDENT ENVIRONMENT MODELING

The simulation of the RPS response to the accident environments is embodied in a computer code entitled Launch Accident Sequence Evaluation Program (LASEP) [3]. The location and state of the RPS is simulated from the initial insult, generally occurring at altitude, through Earth impact and any subsequent thermal environments associated with the accident. The outcome of the simulation involves determining whether a release of hazardous material occurs and, if so, the characteristics of the release, which include the release's quantity, location, and particle size distribution.

The calculated response of the RPS to accident environments is based on physical principles, prior safety testing of RPSs and their components, along with modeling of the response of the RPS and its components to accident environments using computer codes. This information allows estimates to be made of the probability of release of material and the amount of the release for the range of accident scenarios and environments that could potentially occur during the mission. The protection provided by the RPS components minimizes the potential for release in accident environments. Potential responses of the RPS and its components in accident environments are summarized generally as follows:

- Explosion Overpressure and Fragments: Liquid propellant explosions from launch vehicle destruct and resulting fragments are estimated to result in some RPS damage but no release.

- Impact: Fracturing of the RPS and its components under mechanical impact conditions provide energy absorbing protection to the radioactive material. Some impacts of an intact RPS or GPHS modules on steel or concrete near the launch pad could result in small releases, depending on the impact velocity. Ground impact of an intact space vehicle (SV) for an early launch accident is expected. The combined effect of the SV hitting the ground and the RPS subsequently being hit by the SV components above it occasionally results in a release, depending on the impact velocity and orientation. Larger debris impacting the RPS could result in higher releases for certain orientations.

- Thermal: Exposure of released material to a liquid propellant fireball environment would be of short duration (nominally 20 s or less). Very minor vaporization of the exposed particulates would occur depending on the timing of the ground impact release and the fireball development. For the launch vehicles which include solid rocket motors, exposure of released material to the higher-temperature, longer-burning, solid propellant could lead to more substantial vaporization of exposed material.

- Re-entry: Most of these impacts occur in water with no release. Land impact can result in releases that are similar in nature to those from impact near the launch pad, but without the presence of solid propellant fires. Re-entry will result in some heating and ablation of the surface of the GPHS modules, but no containment failure or release into the air. When these separated components impact land, there is a potential for release from the GPHS module during impact on rock. No release is expected from a water impact or soil/sand impact.

4. SOURCE TERMS

Due to the complexity of modeling the full RPS response deterministically on dynamic variables, the evaluation is performed in a stochastic manner with Monte Carlo simulations. The variability in the time, relative position/orientation and strength of the impact, explosion or fires, along with the variability in the response of the RPS, add to the complexity. The models are executed many times to accurately characterize each RAS. Each execution cycle, which simulates the occurrence of a single accident, is termed a trial. Within a RAS, results for each trial are independent because a new set of random numbers are generated for each trial. A large number of the simulations can result in no release.

By gathering statistics on the outcomes for all trials, the probability of release and the distribution of releases for each RAS, according to geographic location, altitude, particle size, and total quantity can be estimated. Each trial is assumed to be equally probable. Probability distributions for the releases are constructed for each RAS. Probability distributions can be presented in several different ways. Complementary cumulative distribution functions (CCDFs), which give the probability that any given level is exceeded, are used in this analysis.

The release CCDFs (referred to as source terms) for each RAS are assembled together into source terms for each phase of the mission and the overall mission. Figure 4 shows an example of typical source terms for each phase, as well as for the overall mission. The source terms in Figure 4 are normalized by the total inventory for convenience. As seen in Figure 4, for this mission and configuration, the source term for Phase 1, Early Launch, dominates the total mission and completey overlaps the curve for the overall mission. The other phases contribute much less to the overall mission, with Phases 4 and 5 contributing the least. Comparing the relative source terms for the various phases gives an indication of which phenomena need detailed modeling and validation. Comparing the source terms for the individual RASs within the foremost phase can illuminate which phenomena or sequence of events dominate the response. Examination of the RAS source term can help determine what aspects of the RPS, space vehicle or mission could be modified to reduce the mission risk.

Figure 4: Typical Normalized Source Term CCDFs for Each Phase and Overall Mission

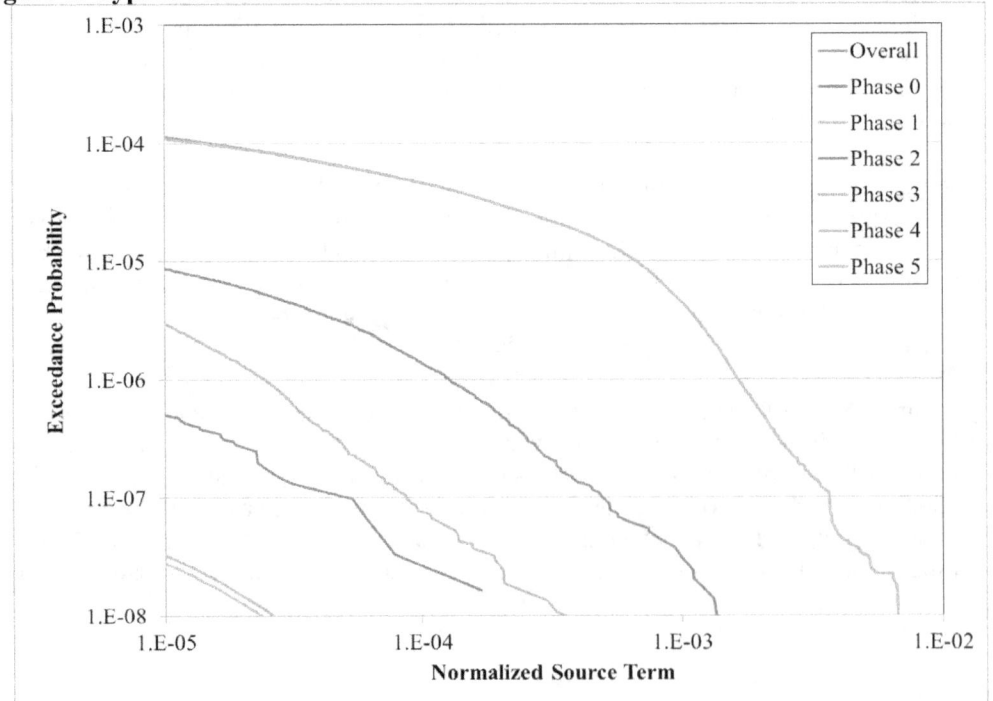

5. TRANSPORT MODELING

The source terms calculated from the accident modeling are composed of a wide range of particle sizes. Large particles tend to deposit rapidly near the point of release and produce a high contamination gradient in ground surface concentrations, while scarcely contributing to material inhalation. In contrast, small particles remain airborne for a longer time and, due to diffusion effects, develop a small gradient in ground surface concentrations, while largely contributing to material inhalation in the surrounding areas.

The source term particles can be elevated by thermal buoyancy effects from liquid propellant fireballs or from solid propellant fires during launch accidents. Meteorological conditions vary in space and time, which governs the transport and diffusion of the released material. These conditions include wind velocity components, relative humidity, atmospheric turbulence and pressure. The local meteorology strongly affects both the potential rise of the particles from the fire environments and the transport of the particles to the surrounding areas. These conditions can vary greatly day to day or even within a few hours. The majority of the uncertainty in the transport modeling arises from the large variations in meteorological conditions.

The transport of the hazardous material is calculated for a collection of independent observations with a sampled value of three main input variables: release trial result, meteorological date, and time of day of the accident. Each combination is termed an observation. Exhaustively considering all possible combinations would be computationally prohibitive. A quasi-Monte Carlo method (a Halton sequence) is used to generate combinations of source term, weather day and launch time. This method facilitates construction of consequence CCDFs and estimation of uncertainty in the CCDFs, but the resolution of the low-probability, high-consequence tail of the CCDF is necessarily limited by the sample size and the computational expense of consequence calculations. To provide additional resolution in the tail of the CCDF within computational constraints, a form of importance sampling is used.

Conceptually, the importance sampling technique used involves partitioning the probability space being sampled into two disjoint sets, A and B, with B representing the low-probability, high-consequence events. The sets A and B are sampled separately, applying a higher sampling density to B, and consequences are calculated for each sample element. In importance sampling, the portion of a parameter's range that is associated with high consequences is preferentially sampled; results obtained with this sampling technique are then appropriately weighted when combining the results to obtain a CCDF. Analysis of results has shown a strong correlation between the amount released and the resulting consequences to the public. The importance sampling technique is applied to the source term trials that have larger amounts of material released. This reduces the total amount of computational time and resources needed in the transport calculations.

6. CONSEQUENCES

The radiological consequences resulting from the given accident scenarios are calculated in terms of: 1) maximum individual dose, 2) collective dose, 3) health effects, and 4) land area contaminated at or above specified levels. The radiological consequences are based on atmospheric transport and settling simulations. Biological effects models, based on methods prescribed by the National Council on Radiation Protection and Measurements (NCRP) and the International Commission on Radiological Protection (ICRP), have been applied in past missions to predict the number of incremental latent cancer fatalities over 50 years (health effects) induced following a release and assuming no mitigation measures.

Multiple exposure pathways are considered in these types of analysis. Figure 5 illustrates the potential exposure pathways and their relationships. One pathway is direct inhalation of the released cloud, which could occur over a short duration (minutes to hours). The other exposure pathways result from deposition onto the ground and are calculated over a 50-yr exposure period. These pathways include groundshine, ingestion, and additional inhalation from resuspension. A 50-year committed dose period is assumed for the material that is inhaled or ingested.

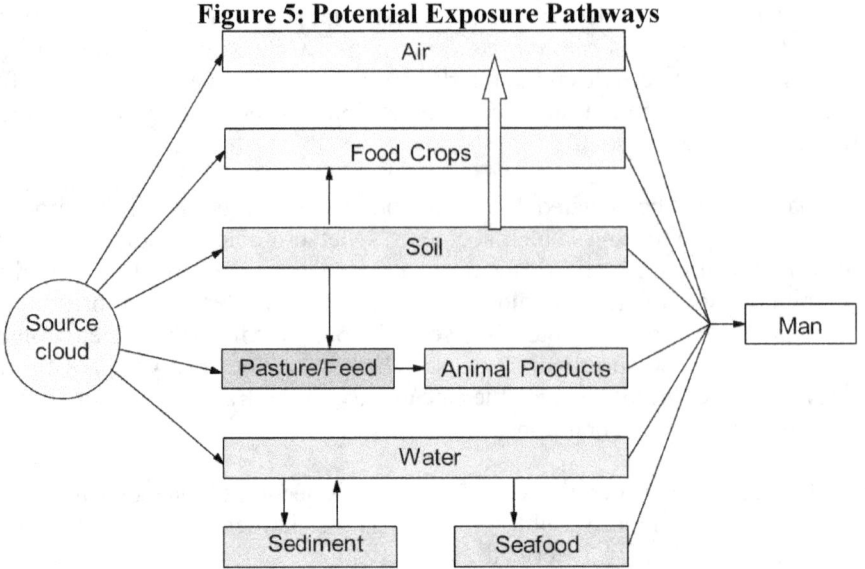

Figure 5: Potential Exposure Pathways

The maximum individual dose is the mean (for historical meteorological conditions) maximum (for location) dose delivered to a single individual for a given accident, considering the probability distribution over all release conditions. Collective dose is the sum of the radiation dose received by all individuals exposed to radiation from a given release in units of "person-rem." Internal doses are determined using particle-size dependent dose conversion factors based on ICRP-66/67 [4] [5] and ICRP-60 [6].

The health effects represent incremental cancer fatalities over 50 years induced by releases, determined using a health effect estimator for the general population based on recommendations by the Interagency Steering Committee on Radiation Standards (ISCORS) [7]. The health effects estimators are based on a linear, no-threshold model relating health effects and effective dose. This means that health effects scale linearly as the dose decreases to zero, rather than assuming a threshold dose below which there would be no health effects. To estimate the total health effects within the population, the probability of incurring a health effect is estimated for each individual in the exposed population, given a release, and then the probabilities are summed over that population.

Potential environmental contamination criteria for assessing contaminated land areas are 1) areas exceeding specified screening activity concentration levels and 2) dose-rate related criteria considered by the U.S. Environmental Protection Agency (EPA), the Nuclear Regulatory Commission (NRC), and the DOE in evaluating the need for land clean up following radioactive contamination [8]. The resuspension contribution to dose assumes that no mitigation measures are taken. The potential for crop contamination is based on the Derived Intervention Limit (DIL), as defined by the Food and Drug Administration (FDA) [9]. The DIL is converted to a cropland deposition threshold by considering the annual average uptake factor of deposited radionuclides and annual crop yields (kilogram of edible food per square meter of land). The number of square kilometers of cropland that exceeds this value for each crop type is determined from atmospheric transport calculations, cropland location maps, and the average fraction of each crop type in the area.

Risk is defined as the expectation of health effects in a statistical sense (i.e., the product of total probability times the health effects resulting from a release, and then summed over all conditions leading to a release). The risk is determined for each mission phase and the overall mission. Since the health effects resulting from a release equals the sum of the probability of a health effect for each individual in the exposed population, risk can also be interpreted as the total probability of one health effect given the mission (for risk much less than one).

7. CONCLUSIONS

Due to the hazardous material that can be released during a launch accident, the potential health risk of an accident must be quantified, so that appropriate launch approval decisions can be made. The risk is calculated by modeling the response of the RPS to potential accident scenarios and the subsequent transport of the hazardous material in the environment and in human biological pathways. Due to the complexity of modeling, the evaluation is performed in a stochastic manner with a Monte Carlo simulation. The results are summed and weighted by appropriate likelihood values to give a collection of probabilistic results for the estimation of the potential health risk. This information is used to guide RPS designs, spacecraft designs, mission architecture, or launch procedures to potentially reduce the risk, as well as to inform decision makers of the potential health risks resulting from the use of RPSs for space missions.

Acknowledgements

Sandia National Laboratories is a multi-program laboratory managed and operated by Sandia Corporation, a wholly owned subsidiary of Lockheed Martin Corporation, for the U.S. Department of Energy's National Nuclear Security Administration under contract DE-AC04-94AL85000. This work is supported by the DOE Office of Space and Defense Power Systems. SAND2014-1670C.

References

[1] L. L. Rutger, et al. "*Radioisotope Power Systems Launch Safety and Space Science*", 3rd IAAS Conference, Rome, Italy, Oct. 21-23, 2008, (2008).
[2] A. V. von Arx. "*MMRTG Heat Rejection Summary*", Space Technology and Applications International Forum, (STAIF) 2006, Albuquerque, NM, Feb 2006, pp. 743-750, (2008).
[3] T. E. Radel and D. G. Robinson. "*Launch Safety Analysis Code for Radioisotope Power Systems*," PSA 2008, American Nuclear Society, Knoxville, Tennessee, Sept. 7-11, 2008, (2008).
[4] ICRP 1994. "*International Commission on Radiological Protection, Human Respiratory Tract Model for Radiological Protection*", ICRP-66 (1994).
[5] ICRP 1993. "*International Commission on Radiological Protection, Age-dependent Doses to Members of the Public from Intake of Radionuclides: Part 2, Ingestion Dose Coefficients*", ICRP-67 (1993).
[6] ICRP 1990. "*International Commission on Radiological Protection, 1990 Recommendations of the International Commission on Radiological Protection*", ICRP-60 (1990).
[7] ISCORS 2002-02. "*Interagency Steering Committee on Radiation Standards, A Method for Estimating Radiation Risk from Total Effective Dose Equivalent (TEDE) Final Report*", ISCORS Technical Report 2002-02, ISCORS, Environmental Protection Agency, Washington, DC (2002).
[8] U.S. EPA, "*Radiation Site Cleanup Regulations: Technical Support Document for the Development of Radionuclide Cleanup Levels for Soil*", EPA 402-R-96-0111 A, Environmental Protection Agency, (1994).
[9] D. Thompson, "*Accidental Radioactive Contamination of Human Food and Animal Feeds: Recommendations for State and Local Agencies*", U.S. Department Of Health and Human Services Food and Drug Administration Center for Devices and Radiological Health Rockville, MD (1998).

Development of online reliability monitors software for component cooling water system in nuclear power plant

Yunli Deng[a], He Wang[a]*, Biao Guo[a]

[a] Fundamental Science on Nuclear Safety and Simulation Technology Laboratory, College of Nuclear Science and Technology, Harbin Engineering University, 145-1 Nantong Street, Nangang District, Heilongjiang, Harbin, 150001, P.R. China
276961539@qq.com, wangheboy@hotmail.com[a]*

Abstract: The online risk monitoring system(OLRS) of Nuclear power plants(NPP) composed of digital instrument and control system by the way of data unilateral transmission. It automatically obtains the actual status of system and components to determine the instantaneous risk on time and used by the plant staff in support of operational decisions.

During normal operation of NPP, the safety systems work continuously for the long period of time and also there are regular equipment alteration and maintenance activities. Therefore, the high reliability of equipment or safety system is important in the aspect of safe operation of NPP. The objectives of this paper are to introduce software in order to develop the online reliability monitor of NPP safety system. The Component Cooling Water System (CCWS) has been considered as an example of safety system in present study.

The online reliability monitor software of CCWS is developed by using the Visual Basic 6.0 as an application development platform and Microsoft SQL Server 2000 taken as database environment. In present study, it has been shown from the verification of Qinshan Master-Control Room simulator that the system achieves the function to monitor the reliability of critical components of CCWS.

Keywords: Online Reliability Monitor, Nuclear safety, Component Cooling Water System (CCWS)

1. INTRODUCTION

The recently research in nuclear industry has been more focused on safety of NPP after the Fukushima Daiichi accident in Japan happened in March 2011 which still plagues Fukushima citizen and Japanese society [1]. The many meetings have been organized worldwide by the international atomic energy agency (IAEA), nuclear regulatory commission (NRC), European nuclear commission (ENC) in order to confirm that how to enhance the safety of NPP after the big accidents.

In order to ensure the safety of nuclear power plant, comprehensive methodology of risk assessment called probabilistic risk assessment (PRA) or probabilistic safety assessment (PSA) was established in USA in 1974. The WASH-1400 report of probabilistic risk assessment (PRA) for nuclear power plant had been first published in 1975 by U.S. Nuclear Regulatory Commission. [2, 3]. PSA techniques have been used increasingly widely in many countries in the risk-informed decision making process in NPP design, operation, and licensing activities. Especially, for the purpose of maintenance, planning shutdown maintenance PSA was updated as a Living PSA first introduced in USA in 1989.

Living PSA can reflect the current design and operational features of the plant is used to evaluate the safety and to support the plant staff making operational decisions. The special tool to determine the instantaneous risk bases on the actual status of the systems and components. The tool is called safety monitor or risk monitor [4]. Since the Risk Monitor ESSM was first put into service in 1988 in England, and then ESOP1/LINKITT, EOOS were developed. And more than 140 Monitor Systems had been developed in the world at the year of 2005 [5].

The reliability monitor is the part of risk monitor system of NPP and the function of reliability monitor is to evaluate the reliability of individual safety subsystems or evaluate the numerical value of risk of whole NPP systems.

According to the equipment states of system, the reliability monitor software can compute the reliability and calculate the level of risk. When those equipment states happen to change, the reliability monitor can reflect this condition and it is known as the online reliability monitor. This paper will introduce a Reliability Monitor system of Component Cooling Water system (RMCCWS) in Qinshan NPP, which is using the Living PSA technology and realizing the function– monitoring the key components, computing the reliability of CCWS and giving the importance rank of MCSs –to support the plant staff making decisions.

2. Details on RMCCWS

2.1 Description of Component Cooling Water System (CCWS)

Qinshan NPP is the first NPP in china, The CCWS system in Qinshan NPP is selected for the software development. The CCWS belong to the three-level in the nuclear safety, and one -level in anti-seismic. The configuration of system contains three heat exchangers; three CCWS pumps and a CCWS ripple tank, and the relational valves, pipeline, instruments as shown in Fig.1. They constitute three independent trains. When the plant is on normal operation, CCWS system offers the cooling water for the reactor primary pumps, auxiliary components, spent fuel pit heat exchangers, etc. and carries the heat from exchangers or components to service water system, which can maintain the plant safe operation.

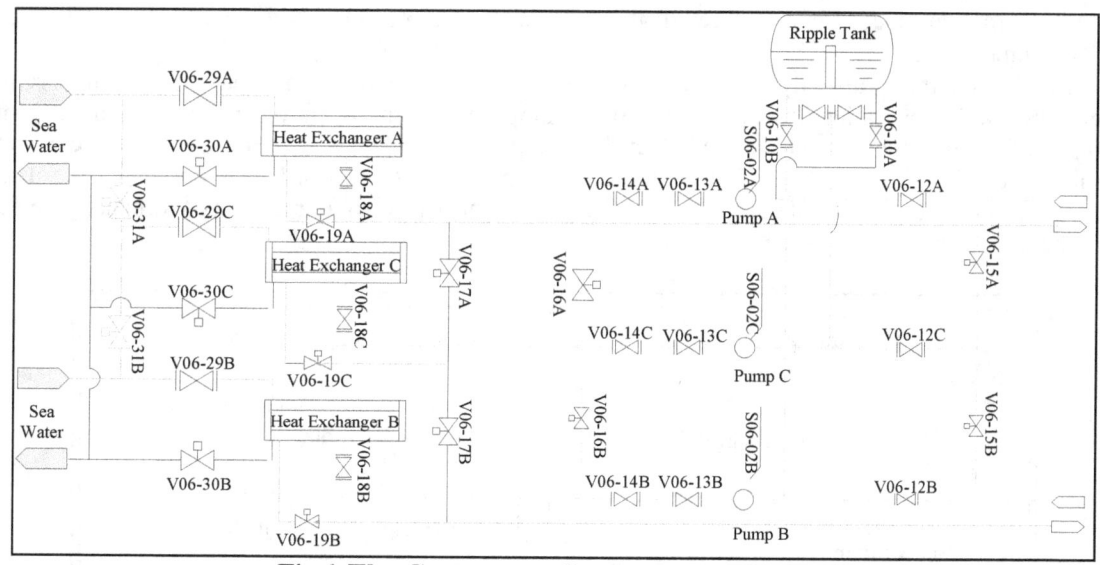

Fig.1.The Component Cooling Water System

Three trains generally are called A, B and C, and each one can provide the 100% ability to transfer the heat from exchangers or components to service water system. Normally A is on operation, B is standby, and C is being overhauled. Every month it will test pump A and pump B and change their states to keep their usability, but if B can't start, C is standby and begin to start. There is an assumption that the failure pump can be fixed in a month. The system can succeed to maintain the function until A, B and C fail at the same time. Usually system has seven states listed in Table 1 about the three trains during NPP operation [6].

Table 1:The states of three trains

Train	A	B	C
State 1	operation	standby	overhaul
State 2	standby	operation	overhaul
State 3	operation	operation	overhaul
State 4	operation	overhaul	overhaul
State 5	operation	overhaul	standby
State 6	overhaul	operation	overhaul
State 7	overhaul	operation	standby

The Qinshan NPP PSA report is the basis of RMCCWS, the reliable model is built according to the report and the reliable data of the components also collected from the report which base on the operational experience.

2.2 The Main Functions of RMCCWS

The RMCCWS is a real-time analysis tool used to evaluate the reliability (the system's failure probability) of the CCWS system based on the actual states of the systems and components [7]. When CCWS is at a specific state, the RMCCWS can compute the reliability of the CCWS system. With the passage of time, the states of system may happen to change and the RMCCWS evaluate the reliability with time. The series values of reliability results are indicated to operators through the Human-machine interface (HMI) with a reliability curve. Because the reliability results values are appear with the time and it is called dynamic reliability of system and RMCCWS is called online reliability monitor.

The dynamic reliability information is very significant for the operators in order to check the current state of system and operators can get Minimal Cut Sets (MCSs) important rank information of the CCWS from RMCCWS. The operators will also find the weak links or the problems in the system more easily with information of MCSs important rank. The RMCCWS has an ability to display the states information of components through the HMI such as motor operated valves, heat exchangers, CCWS pumps, etc.

MCSs important rank, dynamic reliability information, and other information can support the operators making operational decisions. For an example when the operators manually change some equipment's states, they can observe whether the change of reliability is in keeping with their expected result and the dynamic reliability is acceptable or equipment/system fails. If it is not, then there must be some problem in the CCWS system, so the operators must take some measures and make sure that dynamic reliability is in the acceptable range.

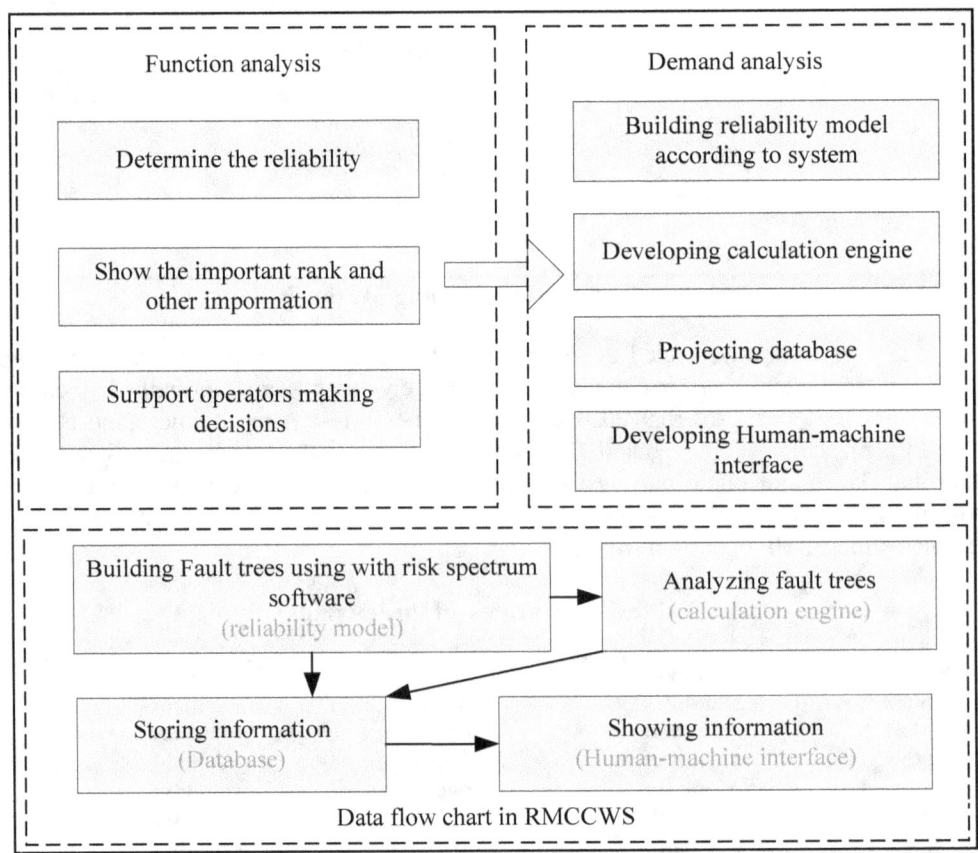

Fig.2 RMCCWS implementation

As shown in Fig.2, in order to confirm the main functions of RMCCWS, there are four different demands corresponding to four models. These models are reliability model, calculation engine, database and human-machine interface. The reliability model is built first and it needs to develop a calculation engine to compute the reliability and MCSs important rank of CCWS. The reliability model and its related parameters are collected to store in database. At the last, the information read from database are shown on HMI to operators.

2.3. Reliability Model

Fault trees analysis (FTA) is appropriate tool used to examine the reliability of CCWS. The fault trees model will change with the change happens in state of system. Therefore, in order to reflect all states of the system through fault trees model, seven fault trees have been built according to seven states of three trains of CCWS system. In every fault trees model, the top event is selected with failure state of train because of no cooling water. The component faults are thought to be basic events in the fault trees models. The basic events and their code are listed in Table 2 [6]. There are some other basic events such as unlocking of stop valves, the break of steel tube and so on, which aren't listed.

Table 2: the basic events list

No.	Event Description	Event Code	Component Code
1	Ripple tank S06-4 loss of function	RCCWS-0004SCLFF	RCCWS-0004
2	Leak of the heat exchanger S06-1A tube	RCCWS-001CSHTIL	RCCWS-001C
3	Power supply switch can't keep the position of pump A	RCCWS-002AFH3RP	RCCWS-002A
4	Power supply switch can't keep the position of pump B	RCCWS-002BFH3RP	RCCWS-002B
5	Power supply switch can't keep the position of pump C	RCCWS-002CFH3RP	RCCWS-002C
6	Pump A fails to operate	RCCWS-002APM5FR	RCCWS-002A
7	Pump B fails to operate	RCCWS-002BPM5FR	RCCWS-002B
8	Pump B fails to operate	RCCWS-002CPM5FR	RCCWS-002C

The fault trees are built with the Commercial software Risk Spectrum®, and the data of basic events' probability is input in the software. And the Risk Spectrum® has the whole fault trees information, and it generates an ASCII file which will be stored in the database of RMCCWS through a code interface.

2.4. Calculation Engine

The calculation engine of the RMCCWS is made up with two parts, one is qualitative analysis, and other is quantitative analysis. The qualitative analysis is the base of quantitative analysis, whose main function is to calculate the fault tree model to get the MCSs. The quantitative analysis calculates the reliability of CCWS and MCSs important rank. The detail description of qualitative analysis and quantitative analysis is given in subsequent sections below.

2.4.1 Qualitative Analysis

In the qualitative analysis, the Fussell method is utilized to conduct the fault tree analysis which is a basic and common algorithm [8]. The Fussell algorithm method read the fault trees from top to the bottom and to use sub events to replace the up events until all sub events are basic events. It will get the MCSs after simplifying and absorbing with Boolean rules. The flow diagram of the Fussell method to calculate the MCSs is shown in Fig.3.

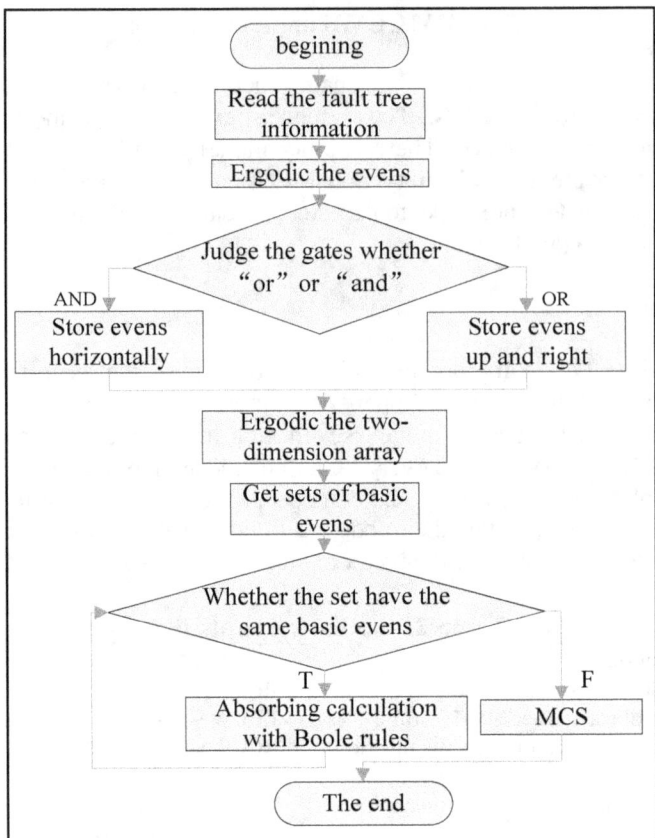

Fig.3 The Fussell flow chart

2.4.2 Quantitative Analysis

Quantitative analysis calculates the failure probability of CCWS system of Qinshan NPP and confirms the accidents which will cause the failure of CCWS. It is also calculate the contribution to system's failure, and in other words it needs to analyze the important degree. In the quantitative analysis, there are three steps of calculations: first, the evaluation of probability of each MCS, and then assessment of failure probability of the system, at last get the some important degree analysis with the results that have been calculated. These three steps of calculation are described below with the mathematic equations.

1) Probability of MCSs
Usually, the basic events are independent, so Eq. (1) can be used to calculate the probability of MCS of independent events.

$$P(K_i) = \prod P(B_i) \qquad (1)$$

Where, $\prod$ is the product of failure probability of basic events in the MCS.

2) Reliability of CCWS
Approximate calculation has been used to calculate the failure frequency of the top events in the software with a high precision. The computational formula:

$$\begin{aligned}F_s(t) &= P(t) \\ &= P(K_1 \cup K_2 \ldots \cup K_k) \\ &= \sum_{i=1}^n P(K_i) - \sum_{i<j=2}^n P(K_i K_j) + \sum_{i<j<l=3}^n P(K_i K_j K_l)\end{aligned} \qquad (2)$$

Where K_i stands for the $(MCS)_i$, i, j, k stands for the order number, and n stands for the number of the MCS, $P(K_i)$ or $P(K_iK_j)$ stands for the probability of the $(MCS)_i$ happened or both the $(MCS)_i$ and $(MCS)_j$ happened.

$$R(t) = 1 - P(t) \quad (3)$$

Where R(t) is the reliability of CCWS system.

3) Important degree analysis

The MCSs important calculation refers to the contribution of the probability of MCSs. And the basic events important calculation refers to contribution of the basic failure probability. The computational formula of the two important degree analyses:

$$I_i^{*FV} = \frac{P(M_i)}{P(U)} \quad (4)$$

Eq. (4) is the MCSs important degree formula where $P(M_i)$ stands for probability of the $(MCS)_i$ happening and $P(U)$ stands for the top events failure frequency.

$$I_i^{FV} = \frac{P(U_i)}{P(U)} = \frac{P(U) - P(U)_{(x_i=0)}}{P(U)} \quad (5)$$

In Eq. (5), the $P(U_i)$ stands for the frequency of basic event happening whereas, the $P(U)_{(x_i=0)}$ stands for the frequency of basic event not happening. The weakness of the system can be found through the analysis of important degree, and it will improve the safety of system to take some measures to improve the weakness.

2.5. Database

The RMCCWS needs a database to store the data, parameters and information, which are input parameters, data or information produced during the process of calculation. As shown in Fig.4, there are five parts of input data consist of reliable database [9]. The fault tree and basic parameter are input into database by developers, and the other three parts belong to data produced during the calculation process and the process of system operation. The five parts of reliable database are given below.

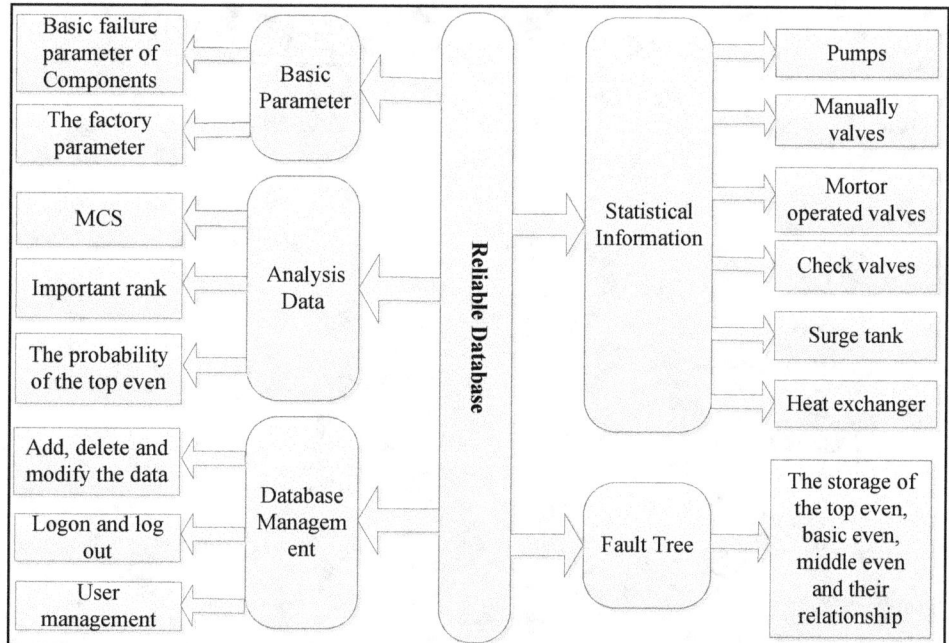

Fig.4 The structure of the Reliable Database

a) *Basic parameter*
    This part includes the failure parameter and the factory parameter of components, which will not change during the operation.
b) *Statistical information*
    It is important and meaningful to record the current state of the components, which not only can be used to support the operator for decisions making, but also can be useful data for the components designing and operation. The states information of pumps, ripple tank, heat exchangers and valves will be recorded in the RMCCWS.
c) *Analysis data*
    This part store the data produced from internal calculation progress, including MCS qualitative analysis results and failure probability, and important degree rank results of quantitative analysis.
d) *Fault trees information*
    It needs to store all fault trees models input by developers before the system to be operate. There are seven fault trees in the RMCCWS, and the system will select a fault tree depending on the state of CCWS.
e) *Database management*
    It stores the user information and user login logs. The manager can add, delete and modify the data in the database through setting the database management function.

The reliability of database concerns with the reliability of the system, so the database must be given regular maintenance and backing up data to improve the database reliability.

2.6. Human-machine Interface

RMCCWS has a personal and colourful human-machine interface (HMI), the operators can get the information certainly, which gives operators a good environment to work and make decisions. Fig.5 shows the main interface of RMCCWS, which has been divided mainly three parts such as menu bar, toolbar and information and information further subdivided into five different areas as the depicted in Fig.5 [10].

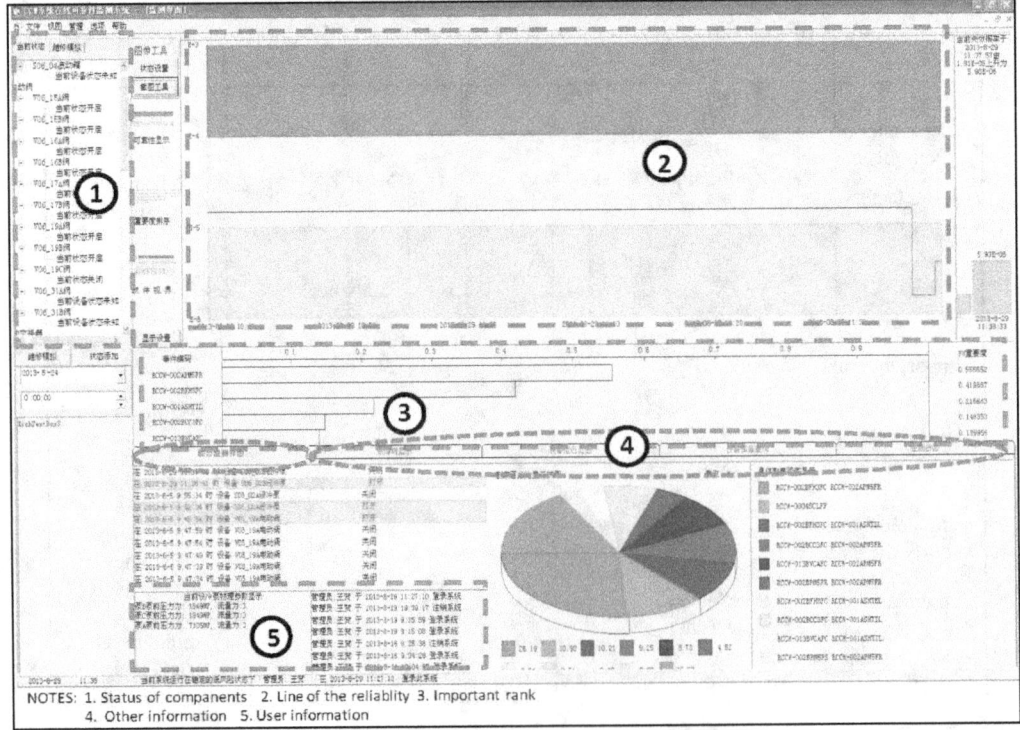

Fig.5 The main interface form of RMCCWS

a) Menu bar, it provides different forms for users to select, such as the login form, database management form and the main interface form.
b) Toolbar, there are some auxiliary functions such as inquiring history information and saving the pictures.
c) Information (see the Fig.6)
- Status of components in area 1, it reads the status of the component with Statistical information in database and it helps operators to know the current states of the system.
- System reliability in area 2, the system reliability is the most important information of the CCWS which reflects the security level. There are three different colors in this area which stand for different security levels, the green is concerned with high security and low system failure probability, the yellow is middle, and red stands for unsafe level. As the time going, three is a line on this area, which can reflect the current system reliability, and also it is allowed to inquire the history reliable information.
- Important degree rank is shown with histogram and pie graph in area 3, the operators can easily find out the contributors of system failure.
- User's login logs are shown in area 4, which is convenient for user's management.
- Area 5 is a menu bar, other information such as fault tree information, basic parameters of components and statistical information can be inquired in this area.

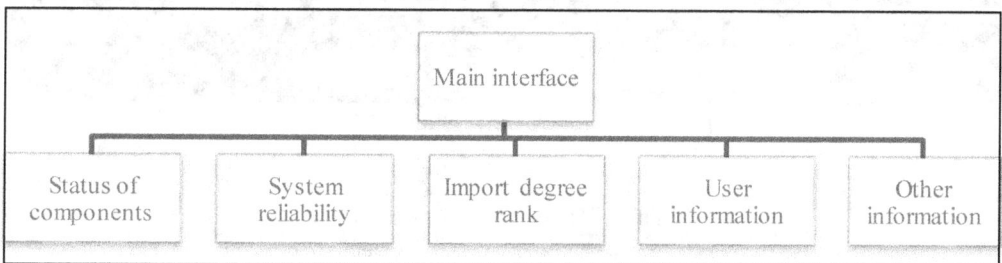

Fig.6 Information showing on the HMI

3. Function Testing

In order to guarantee the software's reliability, RMCCWS must be tested strictly before being used to NPP. The test place is the Fundamental Science on Nuclear Safety and Simulation Technology Laboratory in the Harbin Engineering University, Harbin, China and the test environment is the Qinshan NPP simulation machine. RMCCWS is installed on a computer of the simulation machine. As Fig.7 shows, this is the CCWS control system interface, and the states of the components can be changed manually. RMCCWS collects with simulation machine via the code interfaces so that the RMCCWS could be able to get the states information from control system of simulation machine [11].

The test progress is to change the states of pumps. Just as the trains' states, the three pumps also have seven states listed in Table 1. It can be thought to succeed that the database of RMCCWS changes accurately with the different states of pumps. During the test program, the states of pumps change from A operation, B standby, C overhaul to A standby, B operation, C overhaul. The valves and pumps' states are showing in Table 3. The user's operation is to open pump B firstly, when pump B is sure to start, and then close pump A. The operation records are listed in Table 4.

Table 3: The states of pumps and valves before test

valves	state	pumps	state
V06-19A	open	S06-02A	close
V06-19B	open	S06-02B	open
V06-19C	close	S06-02C	close

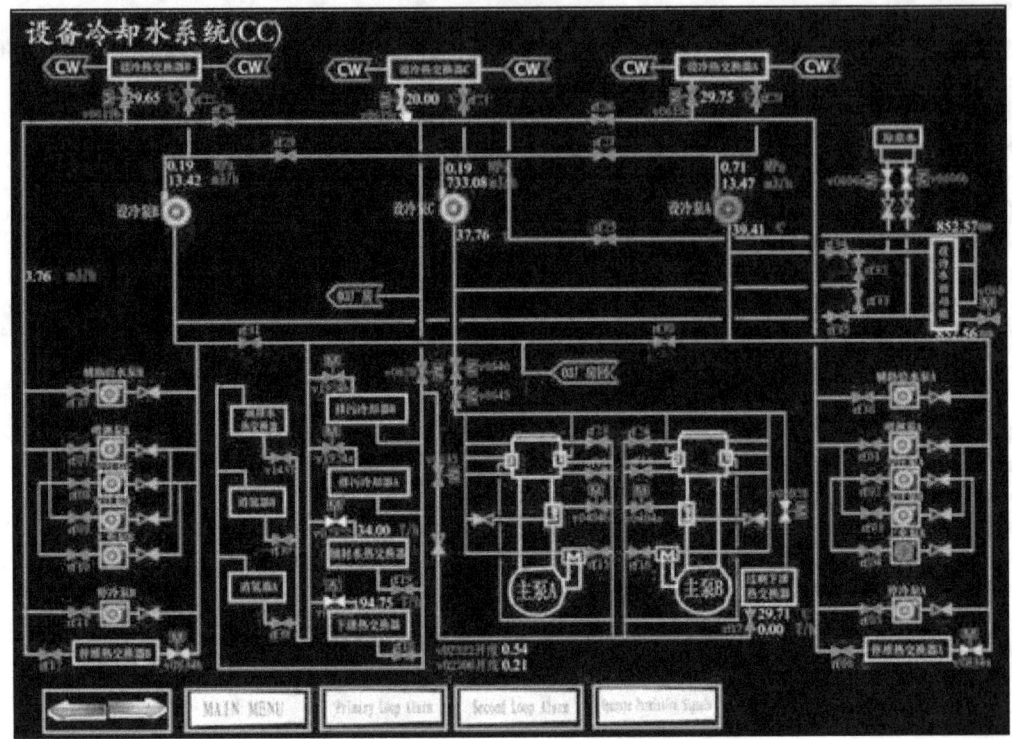

Fig.7 The control interface of CCWS

Table 4: operation records

Time	Operation
2013-12-30 14:51:19	Open S06B
2013-12-30 14:55:03	Close S06A

Fig.8 is the screenshot about the system failure probability during the test progress, the line's value in the picture is the failure probability of CCWS, it is clear that there are three values, because there are three different states of pumps during the test progress, as Table 5 shows.

Table 5: The probability of different periods

State	Value	Period
A operation, B standby, C overhaul	1.80E-5	14:51:44
A operation, B operation, C overhaul	5.93E-6	14:51:44—14:55:31
A standby, B operation, C overhaul	1.80E-5	14:55:31—

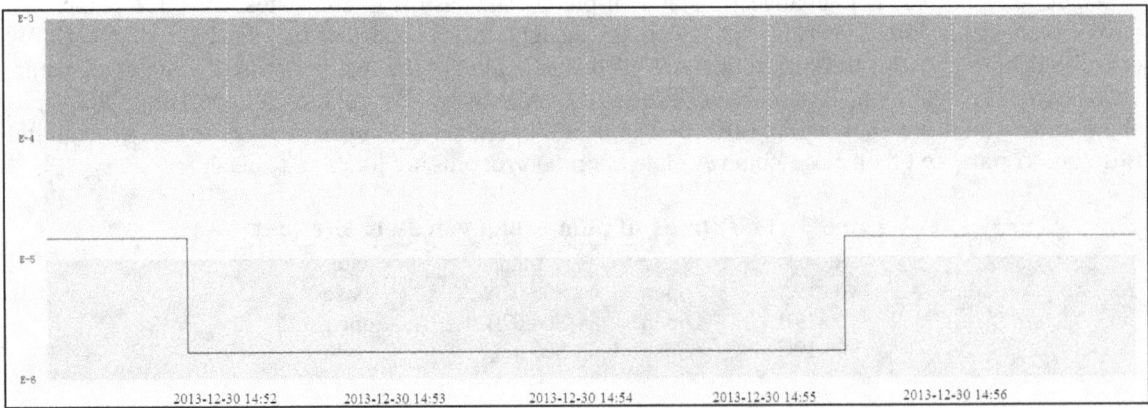

Fig.8 Failure probability of CCWS changing with time

Comparing Table 5 with Table 4, it concludes that the time of calculation engine is about 30seconds. There also is the important degree rank of MCS during different periods listed in Table 6.

Table 6: Important degree rank of test progress

Rank	Period 1	Period 2	Period 3
1	RCCWS-002APM5FR RCCWS-002CFH3FC	RCCWS-002BFH3FC RCCWS-002APM5FR	RCCWS-002BPM5FR RCCWS-002CFH3FC
2	RCCWS-001ASHTIL RCCWS-002CFH3FC	RCCWS-0004SCLFF	RCCWS-001BSHTIL RCCWS-002CFH3FC
3	RCCWS-0004SCLFF	RCCWS-002BFH3FC RCCWS-001ASHTIL	RCCWS-0004SCLFF
4	RCCWS-002APM5FR RCCWS-002CCC3FC	RCCWS-002BCC3FC RCCWS-002APM5FR	RCCWS-002BPM5FR RCCWS-002CCC3FC
5	RCCWS-002APM5FR RCCWS-013CVCAFC	RCCWS-002BCC3FC RCCWS-002APM5FR	RCCWS-002BPM5FR RCCWS-013CVCAFC

From the test progress, about 30 seconds after the operator started the pump B, the system failure probability dropped from 1.80E-5 to 5.93E-6. It indicated that the pump B succeeded to start. Then the operator was allowed to shutdown pump A. This progress helps operator change the system states safely.

If the system failure's value is in the red area in fig.8, it is an unacceptable situation. From the MCS important rank, the operator can easily find the problem and will take some measures to reduce the system failure probability and enhance the system reliability, or it needs to showdown the NPP in order to keep the safety. So the RMCCWS can support operator making decisions.

4. Conclusion

RMCCWS is designed to monitor the reliability of CCWS and have the ability to reflect the states of current system and components. According to the test results, it can be concluded that RMCCWS is able to show the reliability of current system and give the important degree rank. The RMCCWS also can support operators to make decision during operation.

Acknowledgement

This paper is funded by the international Exchange Program of Harbin Engineering University for Innovation-oriented Talents Cultivation, and this paper is supported by Chinese "863" plan, "Research on On-line Risk Monitor and Management Technology of Nuclear Power Plant"

Thanks for the help of Yang Li etc. on the development of the software.

References

[1] IAEA, *"Nuclear Safety Review for the Year 2012"*, Austria, IAEA-NSR-2012, (2012)
[2] NRC, *"Reactor safety study an assessment of accident risk in U.S. commercial Nuclear Power Plants* (WASH-1400)*"* , (1975)
[3] IAEA, *"Code of Conduct on the Safety of Research Reactors"* , IAEA-CODEOC-RR-2006 ,(2004)
[4] IAEA, *"Living probabilistic safety assessment (LPSA)"*, IAEA-TECDOC-1106, pp.6-10, (1999)
[5] NEA, *"Risk Monitors: The State of the Art in their Development and Use at Nuclear*

Power Plant", NEA/CSNI/R (2004)20, pp.35-45, (2004)
[6] Jun Tan, "*The development of Qinshan Nuclear Power Station CCWS System On-line Reliability Model*", pp.7-11, (2013)
[7] Yichan Wu, Lingqing Hu, Yazhou Li. "*The progress of the risk monitor of third Qinshan heavy water reactor*". Chinese Journal of Nuclear Science and Engineering, Vol.31, No.1, (2011)
[8] W. E. Vesely, F. F. Goldberg. "*Fault Tree Handbook*", NRC, 1981, Washington: U.S.
[9] Qiang Xu, "*The CCWS On-line Reliability Database Development of Qinshan Nuclear Power Plant*", pp.3-16, (2013)
[10] Yang Li, "*Design HMI For The On-Line Reliability Monitor System Of Component Cooling Water*", pp.4-6, (2013)
[11] Qinglong Xu, "*The Testing of CCWS's On-Line Reliability Monitor System of Qinshan NPP*", pp.20-28, (2013)

A Parallel Manipulation Method for Zero-suppressed Binary Decision Diagram

Jin Wang, Shanqi Chen, Liqin Hu*, Rongxiang Hu, Fang Wang, FDS Team
Institute of Nuclear Energy Safety Technology, Chinese Academy of Sciences, Hefei Anhui
China

liqin.hu@fds.org.cn

Abstract: A parallel algorithm for the manipulation of Zero-suppressed Binary Decision Diagrams (ZBDDs) on a shared memory multi-processor system was described. Theoretical analysis showed that parallel manipulation of ZBDD has a better time performance than the sequential operation of ZBDD. Since the parallel ZBDD algorithm uses much less time than the ordinary ZBDD algorithm, a real-time calculation can be done in the risk monitoring of a nuclear power plant, which would do a favor of accelerating emergency response and improving safety of nuclear power plants.

Key words: Parallel, Zero-suppressed Binary Decision Diagram, Reliability, Probabilistic Safety Assessment

1. INTRODUCTION

Fault tree analysis is one of the main methods used in Probabilistic Safety Assessment (PSA) of a nuclear power plant. For solving a large fault tree, many new methods were introduced in the past years, such as Binary Decision Diagram (BDD) [1-3], Zero-suppressed Binary Decision Diagram (ZBDD) [4-6], variable ordering [7-10] and truncation technology [11], functional decomposition [12]. BDD Algorithm made the large fault tree analysis based on computer possible, and ZBDD algorithm improved the calculation speed greatly. However, in some cases like risk monitor which needs a real-time analysis, traditional ZBDD algorithm still need further improvement to meet such demand.

ZBDDs are very efficient representations of a factorized structure of minimal cut sets (MCSs), and are widely used for solving a fault tree. The ZBDD algorithm is known as an efficient replacement of a cutset-based algorithm that is based on traditional Boolean algebra, since that logic operations on logic functions, such as AND and OR, are reduced to operations on ZBDDs with a set of new ZBDD operation formulae developed in 2004 by Woo Sik Jung. However, operations on ZBDDs are time-consuming in some cases, and a fast manipulation method is needed.

Algorithm implemented by serial program traditionally cannot be accelerated in Multi-core computer widely used nowadays, so hardware performance advantage was constrained. There are two approaches transforming serial program to parallel program, one of which is transformed automatically by complier and the other is re-write the code. The second one was adopted by the paper because it can adapt the hardware environment of Multi-core computer more appropriately. A parallel ZBDD manipulation method was proposed and its performance was theoretically analyzed. This method would be applied in the Reliability and Probabilistic Safety Analysis Program RiskA [13-21] in later work.

2. TRADITIONAL MANIPULATION

ZBDD was proposed by S. Minato, and a set of new operation formulas shown below based on ZBDD was developed by Woo Sik Jung. In the four formulas (1)~(4), $Y=ite(y,Y_1,Y_2)$ where the variable ordering of $y>x$. X_1 and Y_1 are the left sub-trees of ZBDD, and X_2 and Y_2 are the right sub-trees.

$$ite(x,X_1,X_2)ite(x,Y_1,Y_2)=ite(x,(X_1Y_1+X_1Y_2+X_2Y_1),X_2Y_2) \quad (1)$$
$$ite(x,X_1,X_2)+ite(x,Y_1,Y_2)=ite(x,X_1+Y_1,X_2+Y_2) \quad (2)$$
$$ite(x,X_1,X_2)ite(y,Y_1,Y_2)=ite(x,X_1Y,X_2Y) \quad (3)$$
$$ite(x,X_1,X_2)+ite(y,Y_1,Y_2)=ite(x,X_1,X_2+Y) \quad (4)$$

As with the manipulation, the construction of a ZBDD from a fault tree was focused on. Transformation from basic events in fault tree to variable in ZBDD was done one by one from bottom to up as post-order traversal in traditional implementation of ZBDD algorithm.

Take the fault tree illustrated in Fig.1 for example, all fault tree nodes form a sequence as D, E, G4, A, B, G2, C, F, G, G5, G3, G1 according to the post-order traversal. Based on this sequence, ZBDD manipulations were done from D to G1, and the construction of a ZBDD from the whole fault tree was completed when the last operation on G1 was made. Suppose the variable ordering was A<B<C<D<E<F<G, and the corresponding variable in ZBDD was identified as a, b, c, d, e, f, g. The operation sequence would be shown by a set of expressions below. The result shown in Fig.2 of the last operation on G1 was the corresponding ZBDD of fault tree in Fig.1.

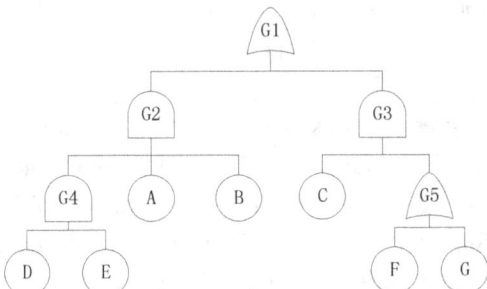

Fig.1 An example of a fault tree

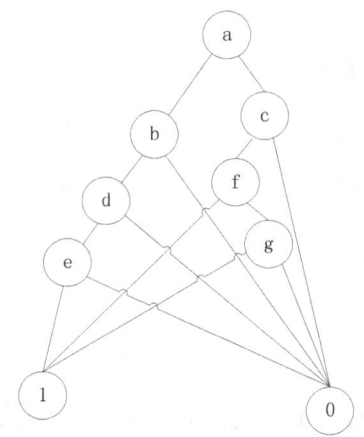

Fig. 2 ZBDD of the fault tree in fig. 1

Table 1: ZBDD operation sequence for the fault tree in fig.1

Order	Fault tree node	ZBDD operation
1	D	ite(d,1,0)
2	E	ite(e,1,0)
3	G4=DE	ite(d,1,0) ite(e,1,0)=ite(d,ite(e,1,0),0)
4	A	ite(a,1,0)
5	B	ite(b,1,0)
6	G2=G4AB	ite(d,ite(e,1,0),0) ite(a,1,0) ite(b,1,0)=ite(a,ite(b,ite(d,ite(e,1,0),0),0),0)
7	C	ite(c,1,0)
8	F	ite(f,1,0)
9	G	ite(g,1,0)
10	G5=F+G	ite(f,1,0)+ite(g,1,0)=ite(f,1,ite(g,1,0))
11	G3=CG5	ite(c,1,0) ite(f,1,ite(g,1,0))=ite(c,ite(f,1,ite(g,1,0)),0)
12	G1=G2+G3	*ite(a,ite(b,ite(d,ite(e,1,0),0),0),0)+ite(c,ite(f,1,ite(g,1,0)),0)=ite(a,ite(b, ite(d,ite(e,1,0),0),0),ite(c,ite(f,1,ite(g,1,0)),0))*

In this case, suppose each operation lasted t_i ($i = 1, 2..., 12$) seconds, so the whole process of ZBDD construction needed t seconds which can be obtained by the following formulae.

$$t = \sum_{i=1}^{12} t_i \quad (5)$$

3. PARALLEL MANIPULATION

As mentioned in section 2, we focus on the construction of a ZBDD from a fault tree, and devised a parallel algorithm for the construction. The construction was just done through a set of operations on ZBDDs. In the construction, there are many logic operations to be processed, and some of them can be processed in parallel. At first, we introduce an extraction method and a parallel-execution method for such parallelizable operations. This is the parallel execution method for an operation sequence (or a set of operations). To extract more parallelism, we introduce a dynamic expansion method of a logic operation. The dynamic expansion is a method to obtain sub-operations from ZBDD operation formulae. These sub-operations are executed in parallel and the results of these sub-operations are merged to obtain the result of the original operation.

The parallel execution method was to construct ZBDD from the fault tree layer by layer from bottom to top. We also take the fault tree in fig.1 for example. The whole tree could be divided into four layers which were {D, E, F, G}, {G4, A, B, C, G5}, {G2, G3} and {G1} shown in fig.3. So the construction could be done by four steps: in step 1, basic events D, E, F and G were transformed to ZBDD in parallel, which were done as operation 1, 2, 8 and 9 in table 1; in step 2, G4, A, B, C and G5 were processed to construct ZBDD in parallel, which were done as operation 3, 4, 5, 7 and 10 in table 1; in step 3, G2 and G3 were processed to construct ZBDD in parallel, which were done as operation 6 and 11; in the last step, G1 was processed to construct the whole ZBDD as shown in fig.2, which was corresponding to the last operation in table 1.

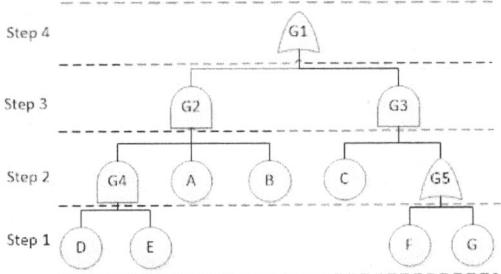

Fig. 3 Steps of the parallel algorithm

In each operation from fault tree to ZBDD, some sub-operations could be done in parallel too. In the four formulas (1)~(4), for example, in the first one, the sub-operation like X_1Y_1, X_1Y_2, X_2Y_1 and X_2Y_2 could be done in parallel.

If suppose each operation lasted t_i' (i = 1, 2..., 12) seconds, so $t_i' \leq t_i$ because of the parallel of the sub-operations of each operation. Then, step 1 would last $max\{t_1', t_2', t_8', t_9'\}$, while step 2 lasting $max\{t_3', t_4', t_5', t_7', t_{10}'\}$, step 3 lasting $max\{t_6', t_{11}'\}$, step 4 lasting t_{12}'. So the whole process of ZBDD construction needed t' seconds which can be obtained by the following formulae.

$$t' = max\{t_1', t_2', t_8', t_9'\} + max\{t_3', t_4', t_5', t_7', t_{10}'\} + max\{t_6', t_{11}'\} + t_{12}' \quad (6)$$

$$t' \leq max\{t_1, t_2, t_8, t_9\} + max\{t_3, t_4, t_5, t_7, t_{10}\} + max\{t_6, t_{11}\} + t_{12}$$
$$< \{t_1 + t_2 + t_8 + t_9\} + \{t_3 + t_4 + t_5 + t_7 + t_{10}\} + \{t_6 + t_{11}\} + t_{12} = t \quad (7)$$

That is, $t' < t$ was established.

4. SUMMARY

A parallel manipulation method for ZBDD was presented in the paper, and theoretical analysis demonstrated the higher efficiency of such method. Our parallel algorithm would be implemented in C++ on a shared memory multi-processor system, and its efficiency would be demonstrated quantificationally by performing benchmark tests in future work.

Acknowledgments

This work was supported by the Strategic Priority Research Program of Chinese Academy of Sciences (No. XDA03040000), the National Natural Science Foundation of China (No. 91026004), the Informatizational Special Projects of Chinese Academy of Sciences (No. XXH12504-1-09), and the Foundation of President of Hefei Institutes of Physical Science (No. YZJJ201327).

References

[1] C. Lee, "Representation of switching circuits by binary-decision programs," BELL SYSTEM TECHNICAL JOURNAL, **38**, 985-999 (1959).

[2] S. B. Akers, "Binary decision diagrams," IEEE Transactions on Computers, **c-27**, 509-516 (1978).

[3] A. Rauzy, "New Algorithms for Fault-Trees Analysis," Reliability Engineering & System Safety, **40**, 203-211 (1993).

[4] S. Minato, "Zero-suppressed BDDs for set manipulation in combinatorial problems," in DAC '93: 30th ACM/IEEE-CS Design Automation Conference, Dallas, TX, 272-277(1993).

[5] J. Woo Sik, H. Sang Hoon, and H. Jaejoo, "A fast BDD algorithm for large coherent fault trees analysis," Reliability Engineering & System Safety, **83**, 369-374 (2004).

[6] J. Woo Sik, "ZBDD algorithm features for an efficient Probabilistic Safety Assessment," Nuclear Engineering and Design, **239**, 2085-2092 (2009).

[7] P. Liu, Y. C. Wu, Y. Z. Li, et al., "An ordering scheme of the basic events based on zero-suppressed binary decision diagrams for the large-scale fault tree analysis," Chinese Journal of Nuclear Science and Engineering, **27**, 282-288, (2007).

[8] C. Ibanez-Liano, A. Rauzy, E. Melendez, et al., "Variable ordering techniques for the application of Binary Decision Diagrams on PSA linked Fault Tree models", Safety, Reliability and Risk Analysis: Theory, Methods and Applications, **1-4**, 2051-2059 (2009).

[9] Y. Z. Li, J. Wang, L. Q. Hu, et al., "A Variable Ordering Heuristic based on Zero-suppressed Binary Decision Diagrams," in Proceedings of International Conference on Nuclear Engineering, Xi'an, Shanxi, China, 691-695 (2010).

[10] J. Wang, F. Wang, J. Q. Wang, et al., " A variable ordering heuristic for risk monitors based on zero-suppressed binary decision diagram," Chinese Journal of Nuclear Science and Engineering, **30**, 360-364 (2010).

[11] W. S. Jung, S. H. Han, and J. E. Yang, "FAST BDD TRUNCATION METHOD FOR EFFICIENT TOP EVENT PROBABILITY CALCULATION," Nuclear Engineering and Technology, **40**, 571-580 (2008).

[12] S. Contini and V. Matuzas, "Analysis of large fault trees based on functional decomposition," Reliability Engineering & System Safety, **96**, 383-390 (2011).

[13] Y. C. Wu, P. Liu, L. Hu, et al., "Development of an integrated probabilistic safety

assessment program," Chinese Journal of Nuclear Science and Engineering, **27(3)**, 270-276 (2007).

[14] M. Nie, Y. C. Wu, X. J. Deng, et al., "Comparative analysis between RiskA and Risk Spectrum in fault tree calculation," Chinese Journal of Nuclear Science and Engineering, **26(4)**, 358-362 (2006).

[15] Y. C. Wu, Y. Z. Li, L. Q. Hu, et al., "Development of the Third Qinshan Nuclear Power Plant Risk Monitor," in Proceedings of International Conference on Nuclear Engineering, Xi'an, Shanxi, China, 697-701 (2010).

[16] F. WANG, Y. LI, J. Q. WANG, et al., "ARCHITECTURE AND DESIGN OF THIRD QINSHAN NUCLEAR POWER PLANT RISK MONITOR," in International Conference on the Physics of Reactors 2012, PHYSOR 2012: Advances in Reactor Physics, Knoxville, Tennessee, USA, 4591-4598(2012).

[17] J. Q. WANG, Y. Z. LI, F. WANG, et al., "DEVELOPMENT AND VALIDATION OF INSTANTANEOUS RISK MODEL IN NUCLEAR POWER PLANT'S RISK MONITOR," in International Conference on the Physics of Reactors 2012, PHYSOR 2012: Advances in Reactor Physics, Knoxville, Tennessee, USA, 4599-4606(2012).

[18] J. W. Xu, J. Wang, S. Q. Chen, et al., "Web-based Fault Tree Collaborative Modeling in RiskA," in ANS PSA 2013 International Topical Meeting on Probabilistic Safety Assessment and Analysis, Columbia, SC(2013).

[19] F. Wang, L. Q. Hu, J. Wang, et al., "A Nuclear Power Plant Risk Monitor Based On Cloud Computing," in 2013 International Conference on Quality, Reliability, Risk, Maintenance, and Safety Engineering, Emeishan, Sichuan, China, 194-197(2013).

[20] J. Wang, L. Q. Hu, S. Q. Chen, et al., "Verification Of RiskA Calculation Engine Based On Open-PSA Platform," in 2013 International Conference on Quality, Reliability, Risk, Maintenance, and Safety Engineering, Emeishan, Sichuan, China, 32-35(2013).

[21] J. Q. Wang, F. Wang, J. Wang, et al., "Application of calculation engine of RiskA to nuclear power plant's probabilistic safety assessment," Chinese Journal of Nuclear Science and Engineering, **31(1)**, 75-79 (2011).

Realistic Modelling of External Flooding Scenarios
A Multi-Disciplinary Approach

J. L. Brinkman[1]
NRG, Arnhem, The Netherlands

Abstract: Extreme phenomena, such as storm surges or high river water levels, may endanger the safety of nuclear power plants (NPPs) by inundation of the plant site with subsequent damage on safety-related buildings. Flooding may result in simultaneous failures of safety-related components, such as service water pumps and electrical equipment. In addition, the accessibility of the plant may be impeded due to flooding the plant environment. Therefore, (re)assessments of flood risk and flood protection measures should be based on accurate state-of-the-art methods.

The Dutch nuclear regulations require that a nuclear power plant shall withstand all external initiating events with a return period not exceeding one million years. For external flooding, this requirement is the basis of the so-called nuclear design level (Nucleair Ontwerp Peil, NOP) of the buildings, i.e. the water level at which a system – among others, the nuclear island and the ultimate heat sink – should still function properly. In determining the NOP, the mean water level, wave height and wave behaviour during storm surges are taken into account. This concept could also be used to implement external flooding in a PSA, by assuming that floods exceeding NOP levels directly lead to core damage. However, this straightforward modelling ignores some important aspects: the first is the mitigative effect of the external flood protection as dikes or dunes; the second aspect is that although water levels lower than NOP will not directly lead to core damage, they could do so indirectly as a result of combinations of system loss by flooding and random failure of required safety systems to bring the plant in a safe, stable state. A third aspect is time: failure mechanisms need time to develop and time (via duration of the flood) determines the amount of water on site.

This paper describes a PSA approach that takes the (structural) reliability of the external defences against flooding and timing of the events into account as basis for the development and screening of flooding scenarios.:

Keywords: PRA, Hazards, External Flooding, Modelling.

1. INTRODUCTION

Extreme phenomena, such as storm surges or high river water levels, may endanger the safety of nuclear power plants (NPPs) by inundation of the plant site with subsequent damage to safety-related buildings. Flooding may result in simultaneous failures of safety-related components, such as service water pumps and electrical equipment. In addition, the accessibility of the plant may be impeded due to flooding of the plant environment. These consequences are so severe that, (re)assessments of flood risk and flood protection measures should be based on accurate state-of-the-art methods.

Dutch nuclear regulations require that a nuclear power plant shall withstand all external initiating events with a return period lower than one million years. For external flooding, this requirement is the basis of the so-called *nuclear design level* (nucleair ontwerp peil, NOP) of the buildings for external flooding, i.e. the water level at which a system – among others, the nuclear island and the ultimate heat sink – should still function properly. In determining the NOP, the mean water level, wave height and

---

[1] *brinkman@nrg.eu*

wave behaviour during storm surges are taken into account. This concept could also be used to implement external flooding in a PSA, by assuming that floods exceeding NOP levels directly lead to core damage. However, this straightforward modelling ignores some important aspects: the first is the mitigating effect of the external flood protection as dikes or dunes; the second aspect is that although water levels lower than NOP will not directly lead to core damage, they could do so indirectly as a result of combinations of system loss by flooding and random failure of required safety systems that have to bring the plant in a safe, stable state. Time is a third ignored aspect: failure mechanisms need time to develop and time (via duration of the flood) determines the amount of water on site.

This paper describes a PSA approach that takes the (structural) reliability of the external defences against flooding and timing of the events into account as basis for the development, screening and quantification of flooding scenarios.

2. SITES IN THE NETHERLANDS

In the Netherlands there are four sites where nuclear reactors were or are located. Figure 1 gives their locations. The first nuclear power plant built in the Netherlands was a 50 MW_e BWR (GKN, a pre MK I with two suppression tanks). This plant – shut down since 1997 - was located in the floodplains of the river Waal. The second power plant is located close at the North Sea coast in the Westerschelde estuary: KCB 500 MW_e PWR. The third and fourth reactors are pool type research reactors built in the early 60-ties of the last century. The smallest one (HOR: 3 MW_{th}) is located several meters below sea level in a polder area near the city of Delft and the other (HFR: 45 MW_{th}) is located in the dunes in the North West part of the Netherlands.

Given their location, it will be clear that all 4 plants needed to consider external flooding as part of the design basis and later in their PSA. The four site locations illustrate the fact that external flooding is site specific. River floods differ in height and duration from sea floods, river dikes fail differently compared to sea dikes, dunes in their turn fail in a different way compared to dikes. In case of sea flooding the impact of waves has to be assessed. In river flooding waves play a minor rule.

Figure 1: Sites of Nuclear Reactors.

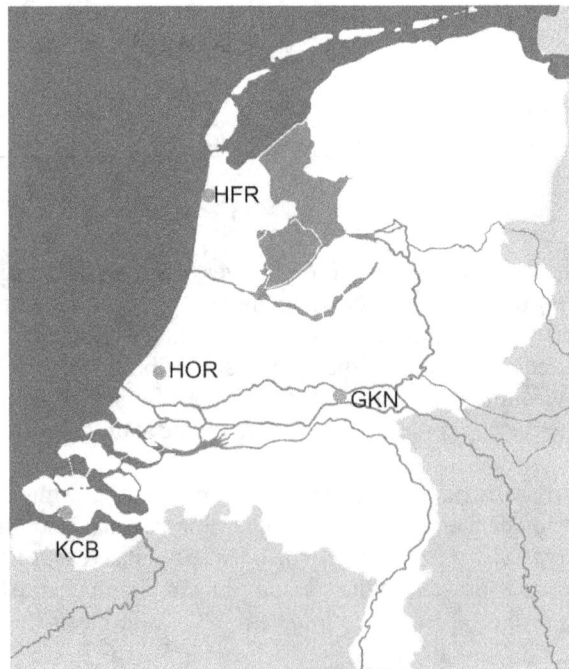

3. DETERMINISTIC DESIGN

3.1 Nuclear Base level

In 1980, the Nuclear Base Level (in Dutch: Nucleair BasisPeil, NBP) and the Nuclear Design Level (Nucleair Ontwerp Peil, NOP) were introduced. The NBP results from the requirement that a nuclear power plant should be protected against external hazards in such a way that the probability of an accident with serious consequences caused by external events - in this case floods-, will be small compared to the risk of serious accidents originating from causes within the plant itself. This requirement is met if the safety measures are such that an external event with a return period of 1 million years (frequency of 1E-6 per year) or more can be withstood.

Basis of the NBP assessment is the official Water Level Exceedence Frequency line as used by the authorities in the design of the Dutch flooding defences. Figure 2 gives an example for a sea location at the west coast.

Figure 2: Water Level Exceedence Curve

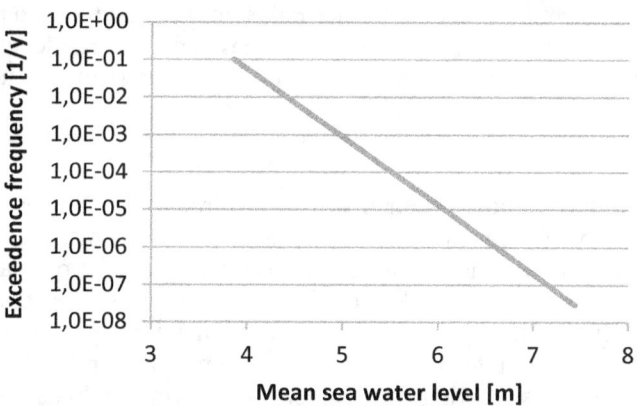

3.2. Nuclear Design Level

The next step is to add various surcharges to the NBP, as defined in the regulations of the IAEA. The resulting level is the calculated nuclear design level (calculated NOP). Examples of surcharges to take into account are:
- Effects of showers;
- Compensation for rising sea level and decreasing soil level;
- Building settlement;
- Wave height.

Because of the dynamic effects of the water (waves), the calculated NOP can be distinguished in:
- **Static NOP** The level at which a constant water load acts on the walls of the buildings in which the safety-related systems and components are housed. This water level is used in the stress - strength calculations for the building design, to withstand the water pressure.

- **Dynamic NOP** This level takes the wave action into account and is used to determine the minimum elevation at which systems have to be placed or to which height buildings should be water tight.

The expected life time of the plant has to be taken into account when calculating the surcharges for Building settlement and rising sea water level. Regarding of safety functions, the calculated dynamic NOP is decisive.

4. PSA

4.1. PSA and NBP

This NOP concept, as it has a frequency base, could also be used to implement external flooding in a PSA, by assuming that floods exceeding NOP levels directly lead to core damage. However, as mentioned earlier, this straightforward modelling ignores three important aspects: the first is the mitigating effect of the external flood defences protecting the plant; the second aspect is that although water levels lower than NOP will not directly lead to core damage, they could do so indirectly as a result of combinations of system loss by flooding and random failure of required safety systems to bring the plant in a safe, stable state, and thirdly, the time aspect is ignored in two ways: 1) failure mechanisms need time to develop and 2) time (via duration of the flood) determines the amount of water on site. Consequently, a more sophisticated approach is needed. In the development of this approach, use is made of the work of the Netherlands' Department of Water Management (Rijkswaterstaat), which applies a comparable probabilistic method for evaluating the designs of new and existing dikes and dunes.

From the three aspects mentioned above, it is clear that the change in approach is not so much in the flooding scenario development and modelling, but rather in the way the initiating event: the relation between water levels outside the external flooding defences and the water levels on site or in the plant buildings. This relationship is as well physical (water level) as numerical (frequency).

4.2 Flooding scenario's

The development of external flooding scenarios in event trees starts with establishing which water levels will impact the safety relevant structures, systems and components, e.g. what on site water level causes loss of off-site power, what level loss of the secondary plant. Loss can simply be caused by inundation of components or by collapse of a building. In the first case the static water level inside the building is determining. In the latter case not only direct (dynamic) forces from the water on the walls of the buildings have to be taken into account, but also - depending on the distance between building and the point where the water is entering the plant site – undermining phenomena of the foundations need attention. With respect to waves, one should bear in mind that the wave height after the breach is far less than for instance at sea, as long as the water is flowing fast through the breach. The plant internal design features against external flooding play a dominant role.

Once the discrete water levels are established, the scenario development is – as with all hazards – in principle straight forward. The basis of the event trees describing the flooding scenarios is the PSA internal events model. In general the event trees for a normal plant trip, loss of off-site power and loss off ultimate heat sink are used. These trees are pruned or modified to account for (part of) systems lost as result of the flooding level.

Before any external flooding scenario (event tree) can be developed, the relationship between water level outside the defences against flooding and the water level and thus consequences inside the plant should be clear. In fact the reasoning starts backwards as compared to the scenario description given by the event tree: what are critical flooding levels inside or onsite around the plant that impact safety relevant structures, systems and components and how can those levels be related to water levels in the river or at sea. In general this will not be a one to one relationship.

4.3. Flooding frequencies

Generally less straight forward is determining the initiating event frequencies for floods that should be taken into account in the PSA model. This requires some sort of translation from the water levels off-site to the critical water levels on site. Two issues influence this translation:
   1   The conditional failure probability of the external flood defence.

2   The duration of the flood in combination with the flood height, the way the flood defence fails and the site characteristics: a) the height of the site as compared to the sea and to its surrounding area and b) the area that can flooded. These parameters determine the water level that is reached behind the failed flood defence.

Both issues lead to a reduction of the initiating frequency. The first issue results in a reduction factor on the initiating frequency at a given water level. The second issue makes that a higher water level (with a lower frequency) is needed off-site to obtain a certain water level on site. The next paragraphs will elaborate this.

4.3.1. Failure of dikes and dunes

Flood defences can fail in different ways. Although it looks like the most obvious mechanism overtopping is not the only and also not per definition the dominant failure mechanism of a flooding defence. Figure 3 gives an overview of the main failure mechanisms of dikes and dunes:

- **Overtopping**
  In this case the dike fails because large amounts of water overrun the dike; the dike is simply not high enough;
- **Macro-stability**
  The dike becomes unstable by water penetrating and saturating the core of the dike. As a consequence the inside slope of the dike starts sliding under the sea or river side water pressure;
- **Sea side erosion**
  The top layer (grass plus clay, stone, tarmac) is damaged by wave attack. Once this protective top layer is gone the main dike structures are eroded away.
- **Piping**
  The water pressure forces water under the clay layer that covers the main structure of the dike or under the clay layer that forms its foundation. So called pipes form and the sand in or under the dike is washed away causing the dike to collapse. Piping also plays a major role where for instance the pipework of the ultimate heat sink penetrates the dike and no design precautions e.g. in the form of addition screens, are taken to counteract this mechanism.
- **Erosion of dunes**
  Dunes fail in general simply by the wave action of the sea. Every wave reaching the dune row erodes the dune by removing sand. The erosion speed is influenced by the length and slope of the beach in front of the dunes.

From the description of the possible failure mechanisms it will be clear that flood defences can and will fail at water levels below their maximum height; e.g. before overtopping becomes the dominant failure mechanism.

When trying to quantify the probability of failure a definition of what a failed defence is, is necessary. In all cases failure is defined as the condition that the amount of water passing the flood defence exceeds a predefined amount. Before this amount is reached the water that passes the flood defence will not lead to problems behind the defence. For a dike for instance it signifies the starting point of the development of a breach. From this point on it will take time to develop a full size breach.

To obtain the (conditional) failure probability the structural reliability of the flood defence is calculated by evaluating the resistance of the flooding defence against the possible failure mechanisms (being the strength of the flood defence) initiated by the high tide (being the stress on the flood defence). Interactions between the different failure modes are taken into account. Parameters influencing the strength of the flooding defence are the dimensions (e.g. width, height, the inside and outside slope of dike), the material used for the underground, the core, and top layer (clay) and cover (grass, tarmac, cobbles, stone), density and grain size distribution of the sand and clay, permeability, subsoil type etc. For dunes and sea dikes the slope of the sea bottom and the width of the beach play an important role. Mean water level, wave height, wave frequency and wave direction are factors that determine the stress.

Figure 3: Major Failure Mechanisms for Dikes and Dunes [1].

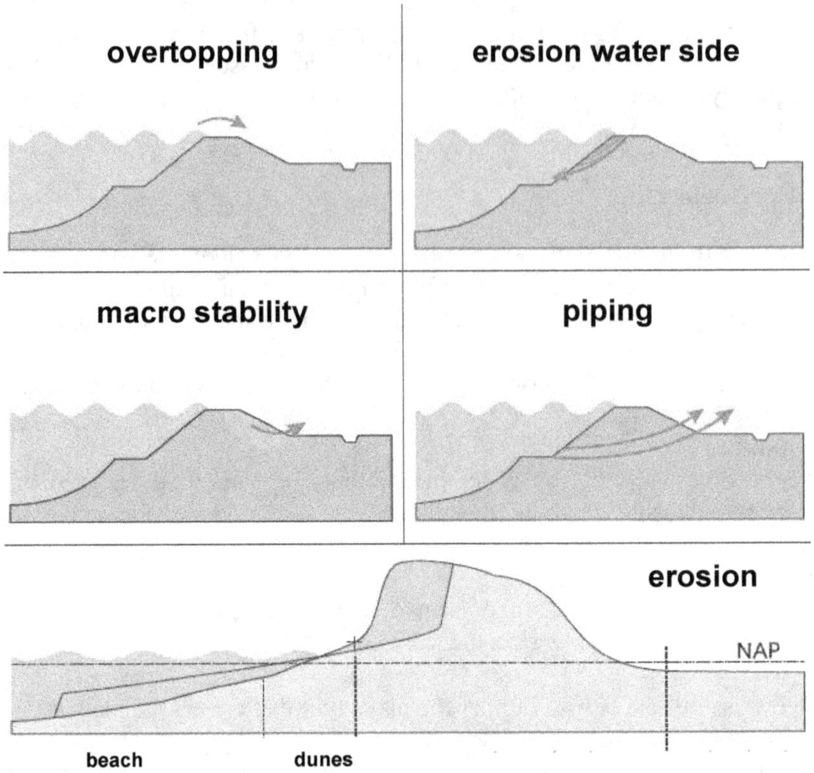

In table 1 an example of the output of the calculation for a section of a sea dike at a given storm surge level is presented. It shows that erosion of the outer slope at the locations with a grass cover dominate the probability of failure. Overtopping is not a major concern. Which of the mechanisms is dominant, changes with the water level. It will be clear that overtopping will become more and more dominant when the water level comes nearer to the height of the dike. Also the type of flooding influences the dominant failure mechanism. In case of river dikes the stability of the dikes is a major concern, Piping and macro-instability are in general the dominating failure mechanisms. There will in general be less dynamic attack by waves, but the much longer time water will stand against the dike, as compared to high water levels at sea, can cause saturation of the core of the dike and thus instability and the one sided water pressure promotes piping..

Table 1: Example of a conditional failure probability, total and per failure mechanism, for a flooding height of 2.9 m.

Failure mechanism	Failure Prob.	Combined Failure Prob.
Overtopping	2.9E-08	
Sea side erosion: stone cover	8.6E-10	
Sea side erosion: grass cover	9.4E-07	9.9E-07
Piping	1.2E-08	
Macro stability	1.3E-08	

Figure 4 gives a result of a complete set of stress strength evaluations of a dike section over a range of water levels for an example river dike. As expected the conditional failure probability is very low for normal water levels between 0 and 2m above the local reference level. It approaches unity when the water level tends towards the maximum height of the dike (6.3m).

Figure 4: Conditional failure probability of a dike as function of flood level [m above reference level]

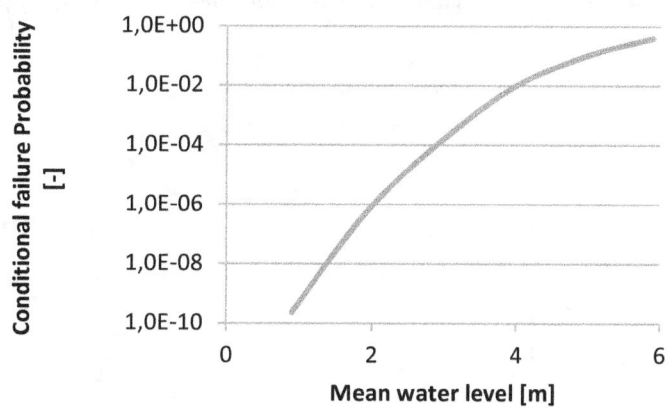

4.3.2 Water level on site

The water level on site is determined by two factors: the amount of water that can enter the site through the breach and the amount of water that is needed to reach a certain water level on site.

Breach calculations
The amount of water that can enter the site is depending on the duration of the high water level, and the size of the breach. High water levels in a river caused by for instance melting snow or heavy or prolonged rain can last for a long time (several days to over a week), while high flood levels on sea are mostly limited by the duration of the storm and the normal tide (12 - 48 hours). Also the breach size and thus the amount of water that can enter the site is a function of time. Time is needed for the process of developing a breach and for the growth process of a breach.

Erosion starts - for instance, depending on the dominant failure mechanism - at the inner slope by the small amounts of water that are flowing down. The inner slope will erode until the crown of the dike is reached. The amount of water entering the site will remain small and constant until the crown of the dike is completely eroded away and the height of the dike starts dropping and the breach starts growing in width. This growth will stop when the flow rate of water through the breach is so low that no further erosion is possible [2].

As this process takes time and the speed it develops increases with increasing water level, it is imaginable that - certainly at lower flood levels at sea - the breach has no time to develop fully before the flooding level at sea drops. This means that although the flooding defence has failed no water will enter the site.

Basin calculations
If a full breach develops, the next step is to evaluate the resulting water level on site taking into account the surroundings of the site. Factors to consider are the size of the area that is open to flooding, its elevation with respect to the normal mean sea level, secondary flood defences, and the height differences within the flood threatened area. Also in this case it is possible that flooding levels will be very limited, as the amount of water available could limited in relation to the available area.

An example result of such an evaluation (from breach and basin calculations) is given in figure 5. For instance a flood level outside of the flood defences (blue line) of 4 m corresponds with a water level on site of approximately 2.8 m (red line). The corresponding conditional probability of the flood defence failing at these levels is 1E-4. Outside flood levels below approximately 2.1 m do not result in significant amounts of water on site, because although the flood defence fails, this relatively low water level has no potential to form a breach of any significance.

Figure 5: Relation between water level on site (red line), and the flood level (blue line)

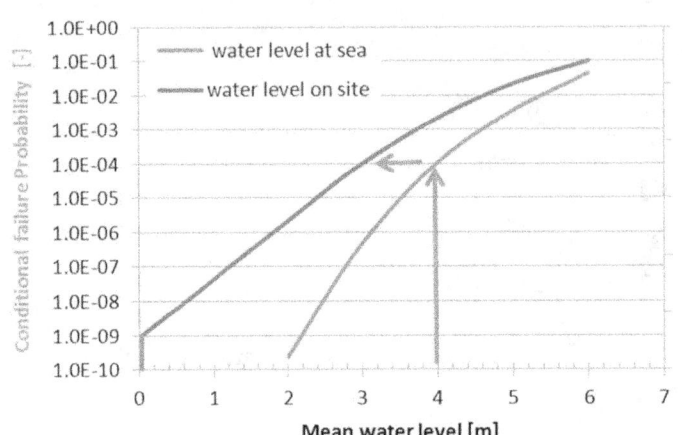

4.3.3. Initiating event calculation

The last step in the process is to obtain the initiating event frequencies for identified threatening water levels on site (plant flooding scenarios). This is done by combining the conditional failure probability given a certain water level on site from figure 5 with the exceedence frequency from figure 2.

The process is illustrated in the two figures below. Suppose the following flooding scenario: off-site power is lost at a water level of 3m on-site (red arrows in figure 6) and that additional systems fail at 4.4m on-site (green arrows in figure 6). The loss off-site power situation then exists between off site water levels of 4 and 5.1 m with a conditional probability of failure of the dike varying between approximately 1E-4 and 7E-3. The accompanying exceedence frequencies lie roughly between 5E-2 and 5E-4 (red and green arrows in figure 7).

Figure 6: Relation water level on site and the flood level: red arrows: start of flooding scenario, green arrows end of scenario

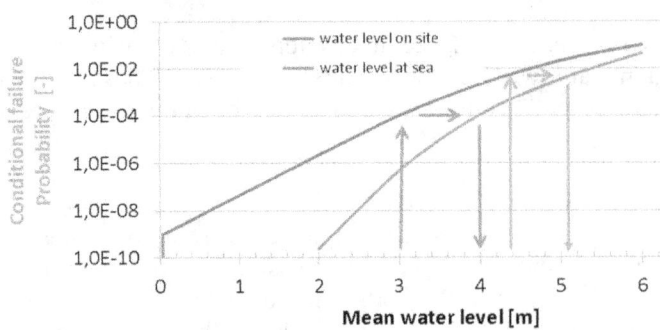

Figure 7: Exceedence frequency: red arrows: start of scenario, green arrows end of scenario

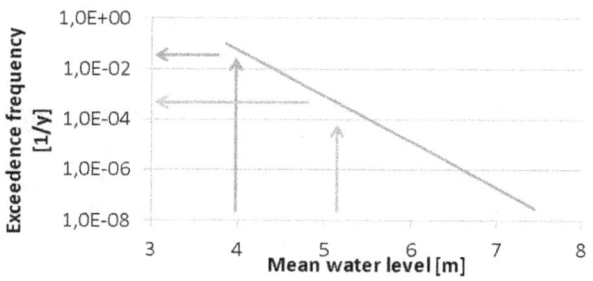

The resulting initiating frequency for loss of off-site power due to flooding is approximately 2.3E-5 per year. This value is calculated by discretising the exceedence curve between 4m and 4.8m resulting in an approximated frequency per water level, multiplying these frequencies with their the corresponding conditional failure probabilities and summing the results. This process is illustrated in table 2.

Table 2: Initiating frequency of LOSP scenario caused by external flooding

Event	Water level in plant [m]	Water level at sea [m]	Exceedence frequency [1/y]	Frequency [1/y]	Cond. prob. of dike failure [-]	Initiating frequency [1/y]
start of LOSP	3	4	0,0524	0,0179	0,0001	1,9E-06
		4,1	0,0345	0,0118	0,0002	2,0E-06
		4,2	0,0227	0,0078	0,0003	2,1E-06
		4,3	0,0149	0,0051	0,0004	2,1E-06
		4,4	0,0098	0,0034	0,0006	2,1E-06
		4,5	0,0064	0,0022	0,0010	2,2E-06
		4,6	0,0042	0,0015	0,0015	2,1E-06
		4,7	0,0028	0,0010	0,0022	2,1E-06
		4,8	0,0018	0,0006	0,0033	2,1E-06
		4,9	0,0012	0,0004	0,0048	2,0E-06
		5	0,0008	0,0003	0,0071	1,9E-06
additional failures	4,4	5,1	0,0005			
				Initiating frequency LOSP scenario due to flooding		**2,3E-05**

6. PRACTICAL EXPERIENCE

The method has been applied in updating an existing external flooding analysis and in the development of a new analysis. For the existing analysis, the frequency of identified flooding scenarios of the plant turned out to be significantly lower than in the previous study. The decrease has to main reasons. The main reason is the calculated difference between the water level at sea and the water level on site. The former model assumed the same water level off-site and on-site in case of a breach of the flood defences. An additional insight gained from the breach and basin calculations made for this study was that the wave height on site was much lower than originally assumed. The effect of the lower waves is that a higher water level on site is needed to cause a specific scenario to happen. The higher water on site results in a required higher water level on sea with a corresponding lower frequency.

In case of the new study for the second site, the results of the structural reliability analyses of the flood defences show that the rows of dunes in front of the plant have a failure frequency that is below 1E-8 per year. Two weak spots with a much higher frequency have been identified north (approximate distance 3 km) and south (1km) of the plant. The water might then reach the plant through the valleys between the dunes. Preliminary flow path analysis based on detailed contour maps of the area that are publicly available [3] show that the water will probably not reach the plant, because the water will be diverted to the hinterland through low spots in the last dune row. These spots have a height that is lower that the minimum plant elevation.

7. CONCLUSIONS

Realistic modelling of external flooding scenarios in a PSA requires a multi-disciplinary approach. Next to being thoroughly familiar with the design features of the plant against flooding, like its critical elevations for safety (related) equipment and the strength and stability of buildings, additional knowledge is necessary on design of flood protection measures as dikes and dunes, their failure behaviour and the modelling of this failure behaviour.

The approach does not change the basic flooding scenarios – the event tree structure – itself, but impacts the initiating event of the specific flooding scenarios and results more realistic and better underpinned initiating event frequencies.

References

[1] http://www.helpdeskwater.nl/onderwerpen/waterveiligheid/programma%27-projecten/veiligheid-nederland/publicaties/illustratiemiddelen/
[2] P.J. Visser, Breach growth in sand-dikes, doctoral dissertation, Delft University of Technology, 1998, Delft."
[3] www.ahn.nl

Insights from the Analyses of Other External Hazards for Nuclear Power Plants

James C. Lin*
ABSG Consulting Inc., Irvine, California, United States

Abstract: Because the probable maximum events selected in FSAR for nuclear plants may not be the maximum possible events, they could possibly be exceeded by more severe events in the future. As such, there is a need to re-evaluate the other external hazards, especially those associated with the natural phenomena. To ensure that the maximum possible intensities of the natural phenomenon hazards are identified and analyzed, one has to be able to identify the physical limits of the parameters that define the intensities of the hazards. However, in some cases, it is truly difficult to identify the absolute, physical limits of parameters associated with selected natural hazards. One way to address the issue of exceeding the probable maximum event is to evaluate the quantitative risk in terms of core damage and large early release frequencies resulting from the specific hazard of concern. This will require the estimation of the hazard frequency. While it may be possible to assess the occurrence frequencies of selected natural phenomena of limited intensity, the uncertainty in the assessed frequencies of events with magnitude beyond the range of historical occurrences may be uncomfortably high. Furthermore, some of the external hazards may not lend themselves to an easy assessment of their occurrence frequencies. As such, deterministic criteria will still need to be used for the risk evaluation of selected hazard events. This paper groups the entire set of other external hazards into a number of categories and discusses the characteristics, PRA evaluation methods, and other aspects of each of these groups.

Keywords: Other External Hazards, Internal Events PRA, Core Damage Risk, Nuclear Power Plant, SSC

1. INTRODUCTION

Many of the other external hazards, including natural phenomena, have typically been evaluated in the Final Safety Analysis Report (FSAR) as part of the nuclear plant licensing process. In general, the approach used in the FSAR is to specify a probable maximum event (i.e., maximum credible event) substantiated by the historical records, and demonstrate that the plant design can withstand the effects of the probable maximum event selected. Since the Fukushima Daiichi accident, it is thought that, because these probable maximum events selected may not be the maximum possible events, there is still the possibility that these probable maximum events be exceeded by more severe events in the future. As such, the need to re-examine the other external hazards especially those associated with the natural phenomena has been reinvigorated.

To ensure that the maximum possible intensities of the other external hazards are identified and used in the design analysis of nuclear power plants, one has to be able to identify the physical limits of the parameters that define the intensities of the hazards. However, it is, in some cases, truly difficult to identify the absolute, physical limits of parameters associated with selected natural hazards. Yet, we cannot afford to establish an unreasonably high intensity such that the plants cannot be economically designed to withstand these super intensity hazards.

One way to address the issue of exceeding the probable maximum event is to evaluate the quantitative risk in terms of core damage or large early release frequencies resulting from the specific hazard of concern. This will require the estimation of the hazard frequency. While it may be possible to assess the occurrence frequencies of selected natural phenomena of limited intensity, the uncertainty in the

---

* jlin@absconsulting.com

assessed frequencies of events with magnitude beyond the range of historical occurrences may be uncomfortably high.

Furthermore, some of the external hazards may not be amenable to an easy assessment of their occurrence frequencies. For example, some hazards will only occur at locations with specific geologic or soil conditions. Therefore, the hazard frequency cannot be readily determined because no applicable data is available for the occurrence of the hazard at that specific location. As such, deterministic criteria will still need to be used for the risk evaluation of selected hazard events.

In view of the preceding considerations, this paper groups the entire set of other external hazards into a number of categories and discusses the characteristics, PRA evaluation methods, and other aspects of each of these groups. In addition, this paper will also discuss the analyses of hazards that are not included in Appendix 6-A of the ASME/ANS Probabilistic Risk Assessment (PRA) standard.

2. CATEGORIES OF OTHER EXTERNAL HAZARDS

The major categories of the other external hazards for nuclear power plants may include: weather related events, events resulting from specific soil or geologic conditions, external flooding, high winds, extraterrestrial events, aircraft, marine, and ground (including rail and truck) transportation accidents, onsite chemical storage and nearby facility hazards, other man-made hazards, etc. Addendum B of the ASME/ANS PRA standard grouped the other external hazards into the following categories: biological events, external fires, extraterrestrial events, extreme temperature, ground shifts, heat-sink effects, heavy load drop, high winds, industrial accidents, lightning, site flooding, snow, transportation accidents, turbine-generated missiles, and volcanic activity. From the characteristics of the hazards and the methods that can be used for evaluation, the other external hazards are defined in this paper by the following categories: Aircraft Impacts, Biological Events, External Fires, Extraterrestrial Events, Extreme Temperatures and Selected Atmospheric/Weather Conditions, Geologic Conditions or Soil Related Events, Heavy-Load Drop, Industrial Accidents, Lightning, Marine and Ground Transportation Accidents, Site Flooding, Turbine-Generated Missiles, Ultimate Heat Sink (UHS) Degradation/Loss, Extreme Winds, etc.

3. CHARACRTERISTICS AND EVALUATION OF OTHER EXTERNAL HAZARDS

3.1. Aircraft Impacts

Aircraft impacts are considered separately from the rest of the transportation accidents because the evaluation of the frequency and consequences of aircraft crashes are different from the other types of transportation accidents. Aircrafts are generally defined in terms of the following categories: commercial, military, and general aviation. Commercial aviation involves air carriers and air taxi. Military aircrafts include large bomber, cargo, and tanker planes, as well as smaller fighters, attack aircrafts, and trainers. General aviation aircrafts include fixed wing single engine (reciprocating), fixed wing multiengine (reciprocating), fixed wing turboprop, fixed wind turbojet, and helicopters.

The frequency of aircraft crashes is typically calculated by a formula that considers the frequency of airplane flights, crash rate per unit distance traveled or per operation, flight distance nearby the plant within which an aircraft crash can impact the plant area, potential impact area of an aircraft crash, the effective target area, and the likelihood of the location of crashes with respect to the airport or air corridor. In the U.S., the aircraft crash rate is usually derived from the aviation accident data compiled by National Transportation Safety Administration and from data on airport operations and airway flights maintained by Federal Aviation Administration. Aircraft crashes can result from airport operations (including takeoff and landing) and enroute flights on low and high altitude airways. Data on airport operations can be obtained relatively easy. However, the derivation of the frequency of airway flights is more difficult because most of the flights today (including commercial airlines) do not strictly follow the airways. Flight routes are primarily determined by the shortest routes between the origin and destination navigated by the Global Positioning System (GPS). Instead, a rule of 5 miles apart and 1,000 feet vertical separation is used. Aircraft crash rates are dependent on the types of the aircrafts.

Because an airplane can typically glide for some distance following failure or loss of its propulsion, the flight distance nearby the plant within which an aircraft crash can impact the plant area is determined by flight altitude, the shortest distance between the plant and the airway, and the gliding ratio for the type of the aircraft. The potential impact area of an aircraft crash is dependent on the flight altitude and the gliding ratio.

The most important differences between the accidents of aircrafts and other modes of transportation are the consequences of the accidents. For aircraft crashes, only damages to the Systems, Structures, and Components (SSCs) caused by the crash impact are considered. The effective target area includes the critical plant area, the shadow (fly-in) area, and the skidding area. The shadow area is attributed to the height of the plant structures; i.e., impacts resulting from airplanes flying directly into the structures on its descent. It is based on the projection of the vertical face of the structures onto the horizontal surface and is dependent on the aircraft impact angle and the height of the structures. The skidding area is simply the product of the skidding distance and the length of the side of the structures exposed to the aircraft. It is dependent on the flight phase (takeoff, landing, in-flight), dimension of the structures, aircraft's wingspan, impact angle, direction of approach relative to the heading of the structures, and the length of the skid. The aircraft wingspan, impact angle (which is a function of flight altitude), and skid distance are dependent on the type of the airplanes. The average skidding distance is also different between crashes that occurred during takeoff and landing. Finally, the extent of damage that can be produced by the aircraft impact on a building is dependent on the airplane crash speed.

3.2. Biological Events

The types of biological growth or intrusion may include macro-organisms/macro invertebrate (mussels and clams), algae, micro-organisms, and silt, etc. These events are applicable to nuclear plant sites that use once-through water systems drawing water from rivers, lakes, ponds, or the ocean. Collection of silt can occur for plants taking suction of water from rivers. Algae growth often occurs for plants taking suction from lakes or ponds. Agricultural nutrient runoff can contribute to the algae growth. In addition, light penetration provides excellent incubation conditions for algae present in the lake waters entering the raw water systems. The potential impacts of biological events include fouling or plugging of service water or circulating water systems, silt/sediment deposition and buildup, Microbiologically Influenced Corrosion (MIC)/under deposit corrosion (e.g., leading to pitting, pin-hole leaks in raw water system piping and heat exchangers), etc. The specific types of biological events that are applicable to a plant depend on the environment (i.e., ultimate heat sink) that the plant is exposed to.

The typical chemical treatments include the use of dispersant to prevent silt deposition, corrosion inhibitor (e.g., inorganic phosphate by cathodic corrosion and scale inhibition), oxidizing biocide, and non-oxidizing biocide. Examples of oxidizing biocide are chlorine and bleach/bromine (which uses bleach to activate bromine). Due to the environmental pollution concerns, there is only limited-duration application of chlorine at nuclear plants. Oxidizing biocide is typically not compatible with stainless steel material used for tanks, piping, etc. Oxidizing biocide is generally ineffective for clams, but effective for most other biological growth. Chlorination by the use of Sodium Bromide in combination with Sodium Hypochlorite solution can be used for algae control and serves to prevent slime and algae growth in the raw water systems. Nuclear plants have been using Sodium Hypochlorite to replace chlorine for biological growth. However, it is less than effective in controlling microbiological fouling/corrosion. An example of non-oxidizing biocide is quaternary amine (clamicide) which can be used in addition to oxidizing biocide to treat clams and mussels. Non-oxidizing biocide has also been used to control aerobic slime forming bacteria and anaerobic corrosive bacteria.

The nature of this hazard does not really lend itself for easy calculation of the frequency of occurrence. As such, the evaluation of biological events is typically based on whether the chemical treatment performed at the plant can effectively control the biological/microbiological growth or intrusion. The biological event hazard can generally be screened out from detailed evaluation because it occurs slowly and can be monitored, controlled, and mitigated.

3.3. External Fires

The category of external fires includes forest fires, grass fires, and nonsafety building fires. Forest and grass fires that occur outside the plant site boundary. Potential impacts from forest fires include loss or degraded offsite power, plant ventilation impacts (clogging of filters or isolation), degraded conditions (visibility and air quality) for plant personnel, or even failure of plant systems. Most nuclear plants are physically separated from fire hazards in the adjacent areas; i.e., the site is sufficiently cleared in adjacent areas so that forest or brush fires pose no safety hazards. In addition to the mitigating effect of the separation distances of potential fires from the site, the control room is equipped with smoke detectors and manually operated intake dampers to identify and isolate outside smoke, respectively.

The vegetative cover of land surrounding the nuclear plant sites is usually not significant. Also, the plant site itself typically has limited vegetation because it is mostly paved and the area in proximity of plant buildings is cleared to preclude the possibility of external fires damaging equipment or impacting control room operations. From time to time, Security may also require some vegetation removal outside the plant fence line. Therefore, because of the sparse vegetation and because the site is clear, fire cannot propagate to the site. In addition, the plant design and fire-protection provisions are adequate to mitigate the effects.

The main concern of fires in the non-safety buildings include fire damages to accident initiation and mitigation equipment contained in these buildings and propagation of fires from non-safety buildings to safety-related or Category I buildings causing damage to PRA equipment contained in the safety-related buildings. The fire protection program features typically do not differentiate between safety and non-safety buildings. The fire protection design, protection, and considerations are applied equally to both non-safety buildings and safety-related buildings. Reviews of the nuclear plant operating experiences indicate that the detection, response, and suppression for fire incidents in non-safety buildings are generally conducted in the same manner as the safety buildings.

3.4. Extraterrestrial Events

This category includes primarily the meteorite and satellite impact hazards. Both of these hazards have the potential to damage structures and outdoor equipment when they fall to the earth. Almost all meteorites would fragment during entry into the earth's atmosphere because of heating and aerodynamic forces and the fragments quickly decelerate to around 150 miles per hour. The frequency of meteorite impacts can be estimated by evaluating the fraction of meteorite impacting the earth with weights in excess of a specified value sufficient to cause damage to structures or outdoor equipment. Most of the man-made objects in earth orbit that are large enough to cause damage would burn up in the atmosphere before reaching the earth, but some survive reentry and fall to the earth. The frequency of satellite impacts may also be estimated by the frequency of man-made objects in excess of certain weight impacting structures of specific sizes. Both the meteorite and satellite impact hazards can be screened out based on low frequency of occurrence.

3.5. Extreme Temperatures and Selected Atmospheric and Weather Conditions

Fog, frost, hail, high summer temperature, ice cover, low winter temperature, and snow are included in this category. Impacts from fog include poor visibility and humid conditions. Due to the installation of security barriers around the plant facilities, any increased traffic accidents due to heavy fog are not expected to lead to vehicle impact with critical plant structures. Nevertheless, the impacts of fog on traffic accidents are accounted for in the general vehicle accident rates.

Compared to the snow and ice hazards, frost has less occurrence and its impacts on the plant equipment are less severe. For plants located in cold weather regions, cold weather operation is usually assured in that the yard piping is buried below the frost line and outdoor piping serving transformers is of the dry pipe design. Most systems are generally located in heated areas and are not subject to freezing conditions.

The main concern for hail is damage to structures or outdoor equipment from impact. However, most nuclear plant buildings are designed to withstand impacts and loadings greater than those caused by hail. Outside electrical equipment and offsite power could be affected by hail, but given the low frequency of severe hailstorms, their effects are covered under equipment and offsite power failure rates in the Internal Events PRA. In addition to the outdoor transformers and switchyard equipment, there are additional pieces of PRA equipment also exposed to the inclement weather. These include tanks, tank level transmitters, manual valves, check valves, outdoor diesel generator radiator, outdoor diesel generator radiator cooling fans and air flow switches, and expansion joints. In general, the function of the outdoor tanks as a water source would not be affected by hailstone impact because only the top of the tank could be hit by hailstones and any possible damage to the tank top would not lead to the loss of the tank water. For the manual and check valves, no functional damage could result due to the robust design of the valve body. For the tank level transmitters, there are usually enclosures that are sufficiently robust to prevent hail damage. Since most of the outdoor diesel generator radiator cooling fans and air flow switches are located underneath the outdoor diesel generator radiator, they are not directly exposed to the hail impact. With respect to the expansion joints, they are often protected by enclosures which should be sufficient to prevent hail damage.

As such, the only piece of outdoor PRA equipment that could potentially be damaged by hailstone impact is the outdoor diesel generator radiator cooling coils. However, the top of the outdoor diesel generator radiator cooling coils is typically covered by a metal mesh structure which is judged to be built with sufficient strength to withstand the impact of the largest hailstone expected. The impact with the metal mesh is expected to cause the disintegration of the hailstone thus preventing any subsequent damaging impact with the radiator cooling coils down below. However, even in the event that the falling hailstone penetrates through the openings in the metal mesh structure, the amount of energy and fall speed would decrease substantially after contact with the metal mesh structure. Besides, the size of the hailstone that can actually hit the radiator cooling coils down below would be reduced to no greater than the size of the holes in the metal mesh structure. As such, the hailstones would most likely be either stopped by the metal mesh structure above the radiator cooling coils or reduced to a smaller size with substantially lower speed and impact energy after penetrating through the holes in the metal mesh structure.

High summer temperatures can potentially impact air-conditioning, heating, and ventilation (HVAC) system efficiency, the UHS, offsite power reliability, or the electrical system. Typically, the indoor heat loads resulting from the highest temperature ever reached are still within the design conditions for the HVAC systems. For the consideration of indoor temperature based on the air conditioning loads resulting from the indoor heat gains due to solar radiation and convection through the concrete walls, the site design base temperature may still be conservative, even if the actual outdoor temperature is higher. The calculated hourly indoor heat gains are usually greater for every hour using a constant design outdoor air temperature than they are using the actual outdoor air temperature experienced. Therefore, even though the outdoor temperature may occasionally exceed design temperature for a short time period, the actual heat loads experienced indoors are still below and bounded by the design conditions. The maximum water temperature in the UHS must also be shown to be below the UHS design temperature.

The most important effects of ice cover are ice jam flooding and blockage causing impacts on the cooling water intake. Normally, ice jams form at obstructions and irregularities such as bridge piers, islands, sharp bends, and at the upstream edge of a reach of solid ice. The water level behind the ice jam increases rapidly until the head and/or more ice flow destroys the plug. The characteristics of the

UHS usually contribute to a very low possibility of an ice jam forming. Heavy water traffic and the mitigating effect of warm water discharges from industry upstream can also reduce the possibility of ice jam formation. Blockage of the intake and thus the inability to supply with sufficient water could occur by means of ice floes plugging the front of the structure or by formation of ice on the trash racks or traveling water screens. Typically, intake withdraws water from several feet below the water surface; it is unlikely that ice floes could pile up in such a way as to block a significant portion of the intake opening. In addition, the cleaning mechanism for the trash racks should remove ice just as it removes leaves, branches, and other debris should the broken ice floes pass through the intake opening and block the trash racks. As such, blockage of the intake by accumulation of floating ice on the racks and screens is not expected to occur. Also, due to the size of the ice blocks that may be formed, the physical impact of ice on the intake structure does not typically present a hazard to the safe operation of the plant. Therefore, the formation of an ice jam that would cause a significant rise in the water elevation or that would physically block the intake structure is unlikely to occur.

Low winter temperatures can result in freezing of water in pipes, tanks, or reservoirs. Plant design usually protects against freezing and accounts for the possibility of water in pipes freezing. Typically, outside pipes are either installed underground below frost line or have insulation and heat tracing provided to prevent freezing of the pipes in unheated spaces. For example, freezing in the Refueling Water Storage Tank (RWST) during cold weather periods is prevented by the RWST insulation and by maintaining the RWST temperature above the minimum temperature specified in the Technical Specifications. Heat tracing is provided on all RWST connecting lines exposed to the weather. Cold weather operation for the fire protection system is assured in that yard piping is buried below the frost line and outdoor piping serving transformers is of the dry pipe design. All other portions of the fire protection systems are located in heated areas and are not subject to freezing conditions. Although rarely, the UHS water temperature does occasionally during the extremely cold condition decrease to below the freezing point for a brief period of time. However, because of the turbulent nature of the water flow, the UHS water in the intake structure suction bays will usually not freeze because the duration of below freezing temperatures is short. The most significant impact of the low winter temperatures is freezing of the accident initiation and mitigation equipment/functions. However, because of the short durations of below freezing water temperatures, the freeze protection design of systems and equipment, and the slowly developing impacts of this type of hazard, low winter temperatures can usually be screened out.

Excessive snow can result in additional loading on roofs, impacts on onsite and offsite power, and flooding during melting. Protection from rain, ice, snow, and lightning is typically inherent in both plant building and electrical system design. Buildings are usually designed to withstand impacts and loadings greater than those caused by snow. For containment structure, the roof is designed to support at least 30 lb/ft^2 of snow loading on projected area. For buildings except containment, the roofs are likely designed with a live load (including snow load) of at least 30 to 40 lb/ft^2. The piping which is exposed to atmospheric conditions is also protected from clogging due to ice and snow. Because the design snow load typically exceeds the highest snowfall that can be realistically expected and the roofs over Category I components are constructed such that they can safely hold a minimum depth of rainfall, the snow hazard can usually be screened out.

Historic data does exist for these weather conditions. However, the actual occurrences of extreme weather conditions to the extent of beyond the plant design basis are still rare. As such, the screening of these hazards is still based primarily on the capability of the plant design to preclude damages that may result from these hazards.

3.6. Geologic Conditions or Soil Related Events

This category of hazards includes avalanche, coastal erosion, landslide, sinkholes, soil shrink-swell, and volcanic activity.

Avalanche is strongly dependent on the plant site and surrounding topography. This hazard can typically be screened out if there are no mountains near the site, or snow and ice do not accumulate around the site because of mild weather. Coastal erosion involves the erosion of the coastal properties caused by such weather phenomena as hurricanes or other severe storms. However, individual storms may not be sufficient to cause damage to plant structures. The extent of impact by this hazard can be monitored by shoreline survey for shoreline movement, field measurements, map measurements, aerial photograph comparisons, etc. This hazard can be screened out if there is a long spatial separation between the shoreline and plant structures, or if the plant is not located near a coast. The landslide hazard can be screened out if the topography of the areas surrounding the site indicates no immediate adjacent hills or mountains that are susceptible to landslide and could impact the plant structures and critical equipment (i.e., the areas that are most susceptible to sliding are sufficiently removed from the main plant areas to present significant risk in the event of a landslide).

Sink hole are generally caused by karst processes. Sinkholes are common where the rock below the land surface is limestone, carbonate rock, salt beds, or rocks that can naturally be dissolved by ground water circulating through them. The mechanisms of formation involve natural processes of erosion or gradual removal of slightly soluble bedrock (such as limestone) by percolating water, the collapse of a cave roof, or a lowering of the water table. Minor sink holes may also be created by leaking underground pipes. In general, however, the types of soils at the plant site and the groundwater level determine if this hazard would occur. During plant excavation and construction, any soils exposed that may cause this hazard would be removed and backfilled with concrete or soils that would not impact the integrity of the foundation of plant structures.

With respect to soil shrink-swell, due to their physical and chemical properties, some clays may swell (expand) when water is absorbed (i.e., wet), and shrink (contract) when the water dries up (i.e., dry). Significant expansion or contraction due to changes in moisture content can damage the foundations of the plant buildings/structures. Typically, during construction, the looser near surface soils and clay soils would be removed and replaced with densely compacted soil (e.g., granular fill), and the safety-related buildings and structures would be founded on bedrock or on compacted granular fill and medium dense to dense in situ granular soils. As such, these buildings/structures are not expected to be affected by any subsurface soil shrink/swell consolidation that may occur.

There are two types of volcanic activities, explosive and quiet flows of lava. Either could impact site structures if they occur nearby. The ash from an explosive eruption could threaten the UHS and also lead to additional roof loading. The only active areas of the contiguous U.S. for volcanic activities are in the Pacific Northwest.

This group of hazards are strongly dependent on the local geologic and soil conditions in the vicinity of the plant. As such, these hazards do not easily lend themselves to frequency assessment. The evaluation of these hazards focuses primarily on the applicability of the conditions that cause these hazards to the plant proximity.

3.7. Heavy-Load Drop

Failures of cranes during movement of heavy loads (e.g., equipment, structures, etc.) could cause impact damage to risk-significant equipment causing loss or degradation of accident mitigation functions. This hazard is typically managed by station procedure for heavy load lifting. Prior to raising any boom or lifting device, usually, the entire work area must be surveyed to ensure adequate clearance exists between systems, structures, components, etc. Generally, a second person must also be used for verification during all boom or lifting activities. The evaluation of this hazard can be performed by examining the procedures intended to preclude the hazard and postulating scenarios that may result from the combinations of failure to observe the procedure and malfunction of the lifting device or other relevant equipment.

3.8. Industrial Accidents

This category of hazard events includes industrial or military facility accident, pipeline accident, release of chemicals from onsite storage, and toxic gas. The main concerns of these events are release of toxic or asphyxiant gas that may cause control room habitability problem, overpressure resulting from flammable gas or chemical explosion damaging critical plant structures and outdoor equipment, explosion-generated missiles impacting critical plant structures and outdoor equipment, and thermal radiation from fires caused by ignition of the flammable materials released.

The key in the evaluation of all of the hazard events in this group is to identify the specific hazardous chemicals that may be released within the hazard impact range from the critical plant area. For pool fires, jet fires, flash fires, and Boiling Liquid Expanding Vapor Explosion (BLEVE) fire balls, the impact range of thermal radiation from these fire hazards is usually shorter than the distance between the critical plant area and the closest offsite facility if the flammable/combustibles are ignited offsite.

For the release of flammable gases which are lighter than air (e.g., natural gas), in general, they will quickly dissipates in the air to concentrations below the lower flammability limit. As such, the risk from the release of lighter-than-air flammable gas is usually not significant unless it is close to the critical plant area. For heavier-than-air flammable gases (e.g., propane), however, it takes much longer distance to disperse to concentrations below the lower flammability limit. They tend to remain close to the ground and drift with the wind until it is encountered by an ignition source or dispersed to concentrations below the lower flammability limit. Ignition of a flammable gas can result in a flash fire (which could flash all the way back to the release source) or vapor cloud explosion (VCE) which could cause overpressure damage or explosion-generated missile impact to the critical SSCs in the plant. The criterion used for the consideration of damage to SSCs due to explosion overpressure is typically 1 psi. In terms of explosion-generated missiles, SSCs could be considered damaged if the explosion fragments can fly sufficient distance to reach the SSCs considered. Of course, a more detailed analysis can also consider if the missile that reach the plant has sufficient energy to penetrate the building exterior walls.

Flammable gases may be stored at nearby industrial or military facilities, transported in nearby pipelines, or kept in onsite storage locations. The distance from the release point at which the flammable gas has dispersed to concentrations below the lower flammability limit can be calculated using dispersion analysis software program. For the kind of flammable gases that may be present onsite, transported in the nearby pipeline, or stored in nearby industrial or military facilities, VCE in the form of deflagration can only occur in confined or congested spaces. For such explosive charge as dynamite, detonation could occur and would produce blast pressure and impact range significantly greater than that generated by deflagration. The other type of explosion that may also occur is BLEVE resulting from excessive external heating of the storage vessels containing selected flammable liquids (e.g., propane) due to a pool fire or a jet fire.

For the evaluation of toxic gas, the analysis determines if, after dispersion from the release point to the control room ventilation intake, the concentration of the toxic gas is above the toxicity limit at the control room ventilation intake or inside the control room. Although not toxic, some gases stored onsite or offsite (e.g., propane, nitrogen, CO_2) may still be hazardous to the control room operator because they can cause asphyxiation by displacing sufficient oxygen in the air. The dispersion and concentrations of toxic gas or asphyxiant as a function of distance from the release point can also be calculated using a dispersion analysis software program which would use such input atmospheric data as wind direction, wind speed, and atmospheric stability. Typically, releases of toxic or asphyxiant gases from offsite locations would not result in control room habitability issue due to the sufficient dispersion distance.

The screening of chemicals that may potentially be hazardous may use criteria such as solid chemicals; chemicals with all hazard ratings less than 3; chemicals with vapor pressure at ambient conditions less than 10 torr; chemicals with flash point in excess of 100°F (or flammability rating < 3), but no asphyxiation and toxicity concern; chemicals of small quantities (e.g., with storage weights less than 100 lbs, individual storage container no greater than 5 gallons), etc.

3.9. Lightning

Lightning strikes can damage onsite electrical equipment and can impact the availability of offsite power. Protection from lightning is generally inherent in both plant building and electrical system design. Lightning grounds have typically been provided where necessary to prevent lightning from adversely affecting the plant. Lightning arrestors are usually provided for station service transformers, main transformers, and various buses, which are located in the switchyard.

Partial or complete loss of offsite power due to lightning or other causes have been examined as part of the Internal Events PRA, and other effects of lightning on nuclear power plants are generally insignificant. The approach used with respect to the evaluation of lightning is to review all the lightning events that have occurred at the plant site and determine the impact of the lightning events on the plant. If the impact of lightning is no more severe than the loss of offsite power with respect to plant safety, then no further analysis is necessary because they have already been accounted for in the Internal Events PRA. Otherwise, a probabilistic analysis to determine the lightning strike frequency and the lightning-induced core damage frequency would be needed.

3.10. Marine and Ground Transportation Accidents

The group of marine and ground transportation accidents includes such hazards as ship impact, vehicle impact, and vehicle or ship explosion. Generally speaking, toxic gas (which was already discussed previously under the group of industrial accidents) can also result from these transportation accidents. Ground transportation includes both trucks and railways. Due to the terrorist protection measures, the risk of vehicle impact at nuclear plants has reduced substantially. All U.S. plants have very strict vehicle control programs. In addition, vehicle barrier systems and other physical barriers are installed in every part of the critical plant areas. Also, due to the separation distance between the critical plant areas and the adjacent main roads or highways, truck accidents offsite causing physical impact to the onsite SSCs are extremely unlikely. For waterway traffic, the only credible ship impact is with the intake structure. For many nuclear plants, some kind of structures (e.g., breakwater, reef) may be present that can keep strayed or floating marine vessels from colliding with the intake structure. Nevertheless, the evaluation of ship impact can be based on the determination whether the intake structure can withstand the impact of the largest marine vessels allowed in the nearby waterways. The effects of vehicle or ship explosion are similar to those discussed for the industrial accidents. The evaluation needs to determine the kinds of chemicals in addition to gasoline that may be carried by trucks, railroad cars, and marine vessels that may use the nearby waterways.

3.11. Site Flooding

The category of site flooding includes such hazards as external flooding, high tide, intense precipitation, seiche, storm surge, tsunami, and waves. External flooding can result from a number of different sources of water; e.g., intense precipitation, snow melt, river flooding, dam failures, hurricane in conjunction with storm surge and waves, high tide in combination with waves, intense precipitation in combination with snowmelt, intense precipitation in conjunction with dam failure, intense precipitation in conjunction with snow melt and river flooding, etc. The primary analysis for external flooding is to evaluate the water surface elevations and determine if the maximum water surface elevation would exceed the grade level causing intrusion of the flood water into the critical plant buildings. The estimation of the site water surface elevations involves the analysis of the rate of water addition (e.g., the hydrological analysis) and the hydraulic analysis to determine onsite water flow and runoff depth.

Seiche may be generated by meteorological effects, wind, seismic activity, or tsunamis. It is only applicable to plants located near a large body of water where surge or seiche flooding could be a credible source of flooding. By itself, high tide is usually protected by the plant design. As such, for site flooding evaluation, high tide is considered in combination with other meteorological and/or hydrologic conditions that may occur in the plant area. As listed previously, intense precipitation may also occur in combination with snow melt, dam failure, high tide, storm surge, and waves to result in the maximum water surface elevations possible at the plant site.

Both the low pressure weather system (e.g., during a hurricane or typhoon) and wind can cause surges in the water level. With sufficient speed and duration, the wind could move the surface waters from one position to another, causing an increase in the water level. The magnitude of the wind induced surges and waves depend primarily on the wind speed, the distance over the water the wind blows, and the depth of the water. As such, a low pressure system associated with a hurricane/typhoon in combination with the wind generated by a hurricane/typhoon can induce a significant storm surge and waves. In addition, the precipitation that accompanies a hurricane/typhoon will further increase the water level. The evaluation of the increase in water level due to these natural phenomena is to some extent determined by the physical parameters used to model these natural phenomena. The maximum possible values of these parameters are, sometimes, difficult to determine because they may be affected, in some cases, by many other factors.

Tsunami can result from underwater earthquake, landslide, and volcanic eruption. The most severe tsunami is caused by underwater seismic activity occurring in the subduction zones off the coast where the tectonic plates join. Flooding induced by tsunami caused by landslide is more limited to its local area; i.e., has smaller impact area. Most nuclear plants are sited in locations far away from the subduction zone, landslide, and volcanic areas.

In general, the evaluation of the site flooding hazards is still deterministic in nature. Assessment of the frequency of occurrence of most of the natural phenomena in this group involves very large uncertainties because the data for beyond-design-basis extreme events can only be extrapolated from the historical records.

3.12. Turbine-Generated Missiles

The wheel capable of producing the largest missile is the last stage wheel of the low pressure turbine. Compared to low pressure turbine with shrunk-on disks, it is much less likely to generate turbine missiles for rotors consisting of the shaft with the turbine wheels as one forging. For some of the one-piece design, the speed capability of the rotors is higher than the maximum attainable speed of the turbine and the probability of missiles being generated from this rotor is not present. Turbine-generated missiles are thus not credible for the design. The turbine manufacturers will typically perform an analysis of the turbine reliability, which considers known and likely failure mechanisms (i.e., use data for crack initiation and growth) and expresses such failure probability in terms of the intervals between inservice inspection and test. The results of this analysis can be used for the evaluation of this hazard.

3.13. Ultimate Heat Sink Degradation/Loss

The ultimate heat sink provides the cooling water supply to the plant. The hazards that can affect the functionality of the ultimate heat sink include drought, low lake/river water level, and river channel diversion. When evaporation greatly exceeds precipitation for prolonged periods during a drought, a condition of low water level in the UHS may occur. In addition, severe wind could also cause tilted water level and possibly a low water level at the location of the plant cooling water intake. However, nuclear plants are typically designed with a large volume of water impounded to meet the cooling water requirements for an extended period of time even in the worst low UHS water level condition.

Another possible cause of low lake water level is the loss of lake water into a salt mine due to oil and gas drilling. In the 1980 salt mine drilling accident on Lake Peigneur, the Texaco's oil rig drilled directly into the Crystal Diamond salt mine instead of under the lake (caused by a miscalculation) which resulted in draining of the lake water and along with it 11 barges plus a tug boat.

Upstream river diversion cannot occur if the river valley is deeply entrenched in bedrock of sandstones and shales.

3.14. Extreme Winds

The wind hazards include extreme straight winds and tornadoes, hurricanes/typhoons, and sandstorms. Since the safety-related buildings at nuclear plants are designed as a minimum to withstand tornado wind pressure, these structures are generally more than capable to withstand the wind loading from the extreme straight winds. Compared to the wind loading for a hurricane/typhoon, the wind loading from a design basis tornado is usually more limiting.

The risk-sinficant impacts of tornadoes include the tornado wind loading and tornado-missile impact. Because Category I buildings are designed for earthquakes which represent significantly greater loading than tornado wind, they can withstand the strongest tornado that can be expected. As such, only the non-Category I buildings (e.g., turbine building) are likely to be damaged by tornado wind loading. In addition, outdoor equipment (e.g., switchyard, yard transformers, outdoor diesel generator radiator, etc.) may also be damaged. Therefore, the main impacts of tornadoes include the potential failures of the PRA equipment located outdoors and inside the non-Category I buildings (e.g., by building debris) as well as the potential failures of the operator actions that must pass through or are performed in the non-Category I buildings or on the yard. Nevertheless, the frame and concrete floors of the turbine building are expected to be largely undamaged in the event of a tornado. However, equipment in the turbine building at or above ground level could fail due to the effects of wind and debris.

To protect against tornado missiles, the exterior structure elements of safety related buildings are typically designed with missile barriers (e.g., greater thickness to prevent missile penetration). To evaluate the likelihood of SSC damage by tornado missiles, Monte Carlo simulation techniques may be used.

The concerns of sand and dust storms involve blockage of HVAC systems, impacts on the UHS, and effects on onsite and offsite electrical equipment. This is only applicable to plants situated at a location susceptible to this hazard (e.g., at or nearby a desert).

4. PLANT UNIQUE EXTERNAL HAZARDS

For each nuclear plant, there may also be a few unique or additional external hazards that are not listed in the ASME/ANS PRA standard. This may include such hazards as cottonwood debris, frazil ice, industrial sabotage, mayfly activity, military action, and solar flares. Industrial sabotage and military action are generally not evaluated in the PRA analysis.

For plants with raw water intake, frazil ice may, in general, form on the equipment/structure surfaces at the intake structure or pumphouse/forebay. Generically, for plants with intake structure located offshore at the bottom of the water, the most likely location of frazil ice formation is the intake surfaces at the offshore intake structure. For these plants, although frazil ice may also form on the trash rack and traveling screens at the pumphouse/forebay, it is usually less likely due to the various design protection (e.g., natural earth heating of the intake water in the underground intake tunnel, recirculation of the warm discharge water for suction during winter, and backwash of the traveling screen). For plants with intake structure serving both the intake and the pumphouse functions, the metal trash racks and traveling screens at the intake structure may, in general, get plugged with frazil ice, resulting in significant reduction of the available head, the blockage of the ultimate heat sink, and the freezing open of gates. However, the frazil ice phenomenon is a relatively slow developing event. The timing of the event is multiple hours before the phenomenon is obvious with a significant number of additional hours before sufficient flow blockage could occur and the unit would need to shut down. Also, the frazil ice conditions are short lived and usually pass after 12 hours.

Infestation of mayflies could affect offsite power and the associated outdoor electrical equipment. Mayfly activity may also cause blockage of roof drains. In addition, mayflies can potentially cause clogging of safety-related and non-safety related ventilation intakes and outside heat exchanger fans. An important plant measures to minimize the impact of mayflies involves inspections of areas where mayflies may collect. Reduction in station lightning prior to dusk keeps the mayflies from entering the plant. Frequent mayfly infestation checks during operator rounds and clearing of the ventilation intakes and mayfly fouling in the outdoors main transformer coolers and other heat exchangers help maintain the ventilation systems and electrical component cooling equipment functional. Checking and clearing mayflies from plant structures and roofs as well as flushing drain spouts to ensure no blockage have also been effective.

In addition to mayflies, cottonwood debris can also potentially cause similar impact on the HVAC equipment (e.g., clogging of ventilation intakes). To eliminate this problem, seasonal plant checklist needs to include clear instructions on the actions for protecting plant equipment from cottonwood debris. As part of the operator round checks which inspect equipment for mayfly fouling, operators should also examine clogging/blockage that may be caused by cottonwood debris.

A massive solar flare could potentially disable large portions of the U.S. electrical grid for an extended period of time (i.e., long-term, widespread power outages). As such, the primary impact of solar flares on the risk of nuclear plants is the potential to cause the loss of offsite power.

5. CONCLUSION

Since some of the other external hazards do not lend themselves to quantitative frequency analysis, the evaluation of other external hazards must be performed using a combination of deterministic and probabilistic methods. Quantitative frequency analysis of selected other external hazards may involve substantial uncertainties because the historic data may need to be extrapolated to derive the frequency of extreme events; e.g., selected natural phenomena.

Combinations of natural phenomena can realistically occur to produce the most severe impacts challenging the safe operations of the nuclear plants. Therefore, evaluations of the combination events are also essential part of the evaluation of other external hazards.

References

[1] The American Society of Mechanical Engineers and American Nuclear Society, "*Addenda to ASME/ANS RA-S-2008 – Standard for Level 1/Large Early Release Frequency Probabilistic Risk Assessment for Nuclear Power Plant Applications*", ASME/ANS RA-Sb-2013, The American Society of Mechanical Engineers and American Nuclear Society, 2013, New York.

The next-generation risk assessment method about the effect of a slope and foundation ground on a facility in a nuclear power plant

Susumu Nakamura[a], Ikumasa Yoshida[b], Masahiro Shinoda[c], Tadasi Kawai[d], Hidetaka Nakamura[e] and Masaaki Murata[f]

[a] Dept. of Civil & Environmental Eng., College of Engineering, Nihon University, Koriyama, Japan
[b] Tokyo City University, Tokyo, Japan
[c] Railway technical research institute, Kunitachi, Japan
[d] Tohoku University, Sendai, Japan
[e] Japan nuclear regulation authority, Tokyo, Japan
[f] Mitsubishi heavy industry, Takasago, Japan

Abstract: From the background of the accident of the nuclear power plant caused by The 2011 off the Pacific coast of Tohoku Earthquake, the view about the effect of ground such as a slope and a foundation on the nuclear power plant in not only the regulatory guidance for seismic design but also the standard about seismic probabilistic safety assessment was also revised remarkably in JAPAN. A view of the limit state to evaluate the fragility curve about the effect of a slope failure on the facilities described in the latter standard was improved by geotechnical approach such as considering the dynamic behavior of geomaterials after collapse. The view should be called the next-generation assessment about slope stability. The limit state regarding on the slope failure on the facility was specified based on an experimental consideration. Here, the view is reported with experimental results obtained from shaking table tests and its numerical analysis. The experimental examples are also described to verify the effect of countermeasure against the seismic action exceeding the limit state. As a examples to evaluate the movement of rock block induced by slope collapse, the numerical method and its example of application were also described:

Keywords: Slope, Limit state, Shaking table test, Countermeasure, Numerical analysis

1. INTRODUCTION

The regulatory guidance for seismic design of nuclear power plant in JAPAN was revised in 2006. As a phenomenon accompanying an earthquake, the consideration about the effect of a slope failure around a reactor building on the safe performance of a nuclear power plant was newly specified. Then, the standard about seismic probabilistic safety assessment was published by Atomic Energy Society of Japan in 2007, and the revision was carried out in 2013. Although the standard in the 2007 was also considered about the effect of a slope failure on nuclear power plant, the view about the effect was revised remarkably based on the loss accident of the external power source by collapse of the power transmission steel tower by slope failure of the embanked ground which produced within the Fukushima Daiichi nuclear power plant by The 2011 off the Pacific coast of Tohoku Earthquake. Furthermore, the range of a slope for which consideration of AM is required was expanded greatly from the natural slopes around facilities to the slope of embanked ground around a passageway.

An important matter in the revision is a view of the limit state to evaluate the fragility curve about the effect of a slope failure on facilities in nuclear power plant. Based on the procedure described in the 2007 standard about the slope stability assessment around a reactor building in nuclear power plant, the effect of a slope failure on the facility has been estimated by considering a possibility to cause a slope failure as a limit state in the safe side. As for the revision, the state that rock mass reached at a facility after collapse occurred was considered as a limit state. The movement of collapsed rock and soil are used as index to evaluate the limit state. These limit states consider the ground behavior as either the structural damage or the functional damage, and is used to evaluate the damage probability of facilities indirectly. On the other hand, in order to evaluate directly either the structural damage or the functional damage to facilities after collapse of slope, it was specified that an action such as an impulse force at the time of a rock mass reaching to a facility was evaluated as a hazard which acted to a

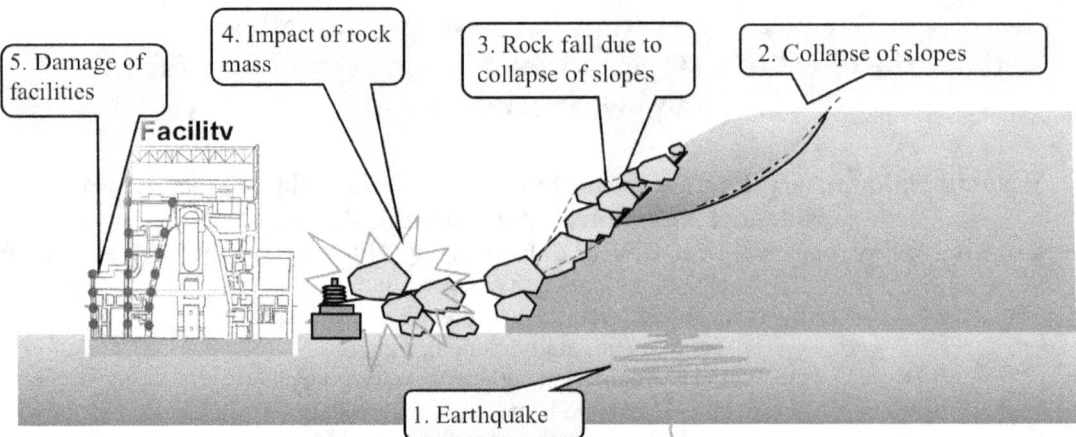

Figure 1 Image of behavior of rock block after collapse of slope

structure. While the previous limit state is indirect evaluation of the effect of a slope failure on a facility, it is the difference among both that the latter limit state is direct evaluation of the effect. The latter limit state was specified based on the experimental results described in Annex of the new standard

This revision associated with the effect of slope failure was carried out by the outcomes of not only experimental study about the slope failure mechanism but also numerical study about slope failure behavior ground response deformation in slope based on many shaking table tests of slope models. These studies had been conducted for the development of the slope stability assessment technology by Japan nuclear energy safety organization. The outcome of Japan nuclear energy safety organization was published as "a guideline of the design and risk evaluation against the seismic stability of the ground foundation and the slope, 2013". In this paper, the fundamental concept about the effect of slope failure on nuclear power plant is described by using not only some experimental examples but also some verification examples of numerical method to evaluate the collapsed behavior of slope based on the experimental results. The important feature of the guideline is to have used the numerical method to be able to evaluate seamlessly the behavior from a seismic response to collapse.

2. THE RISK SCENARIO AND LIMIT STATE CAUSED BY SLOPE FAILURE

Many of the nuclear power plants in Japan are located in a coastal area because of the necessity to acquire a lot of cooling water. According to the geographical condition around the seashore, a natural slope may exist near some of the nuclear power plants. In the plant, we are afraid of about the secondary damage caused by not only the collision of the rock masses to facilities but also slope failure against strong earthquake ground motion, and it becomes important to take into consideration the accident scenario by slope collapse.

The outline of a accident scenario in a nuclear power plant caused by the slope failure against strong earthquake ground motion is shown in Figure 1. As one of accident scenarios, the following scenario is thought about. First of all, a slope failure occurs due to strong earthquake motion. Then, roll of some rocks in a slope occurs, and collides with a reactor building, outdoor important apparatus, or equipment. And it is assumed that apparatus in a building and outdoor apparatus are damaged and that the functional safety of facility in the plant is lost. Moreover, as another scenarios due to slope failure, a collapse of a power transmission steel tower, interruption of passageway for the AM, the effect on AM apparatus such as a water supply car and the power supply car on the slope are also assumed.

In the risk assessment to the effect of a slope in the plant, the fragility characteristics of each facility is evaluated by using the relationship between the limit value and response value. The both values are specified by the physically meaningful index associated with the limit state of some facilities having the important functional safety of the nuclear power plant, and are obtained by the realistic values modeled by characteristic value whose uncertainty is considered adequately.

The following two limit states can be considered for evaluating the effect of unstable behavior of slope on the facilities.
i) The state that a collapse of slope occurs.

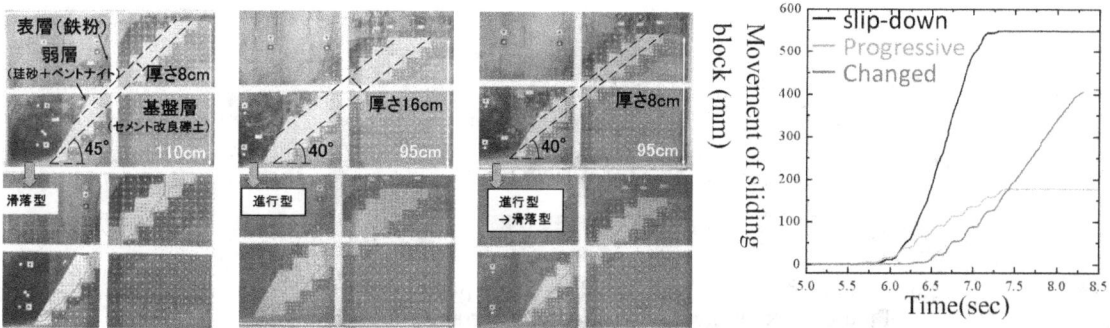

(a) Slip-down type (b) Progressive collapse type (c) A type changed from (b) to (a)

Figure2 Collapse modes for three small-scale slope models by shaking table tests

Figure 3 Time histories of slide block for each collapse mode

ii) The state that the damage of facilities in the plant due to the collapse of slope causes the loss of the functional safety.

Here, the limit state described in i) specifies the effect of slope unstability on the facilities in the plant indirectly by the possibility of collapse of the slope. The safety factor of slip (resistance / acting shear force on a slip surface) has so far been used as an evaluation index. However, even if the safety factor of slip is less than 1.0, the slope failure behavior during earthquake differs significantly depending on not only the strength and deformation characteristics but also the ground structure. The method to evaluate a possibility of occurrence of slope failure by using the movement of rock mass obtained numerically based on the results of the shaking table tests was proposed by authors in order to evaluate a actual failure behavior.

On the other hand, the limit state described in ii) is the state that the damage of the facilities due to the movement and impulse force of the collapsed ground which reached the reactor building, the switching station, and the condensate tank causes the loss of the functional safety directly. The state may estimate not only as a hazard about the action to an facility like impulse force but also as the movement of the rock mass which is indirectly equivalent to the damage of the facilities to cause the loss of the functional safety. These were evaluated by using the numerical methods which were verified based on the comparison of the experimental result about the collapse of slope as described in Chapter 3.

3. FAILURE MECHANISM OF A SLOPE AND EFFECT OF COUNTERMEASURE BY EXPERIMENT

3.1. Failure mode and the limit state of a slope

The collapse mode of a rock slope can be classified into collapse of slide down, slide collapse, toppling collapse, and buckling collapse. A slide type was selected as a target collapse, collapse which generates in the slope with the decreased strength by the surface weathering, Slide collapse which generates within the almost horizontal weak layer of a slope, collapse which generates in weak layers, such as seam were used as the experimental slope model. Based on the experiment, the behavior after the collapse was also evaluated in accordance with the relationship between the failure mode and the slope characteristics. Here, a failure mode corresponds to the slope state of changing to the unstability from stability. The characteristics are specified by the ground structure, strength and deformation.

In order to evaluate those characteristics, the shaking table tests by using some small-scale models of rock slope, the medium scale models and the large scale models have been carried out. Here, based on some results obtained by the shaking table tests [1] [2] of the small-scale models of rock slope, the relationship between failure mode and a limit state are described. The typical failure modes are shown in Figure 2. The slope models consisted of a base layer, a weak layer, and a surface layer. And inclination and thickness of the weak layer were changed as an experimental parameter. The base layer was made by using stability treated particle size adjustment rubble with the cement to regard the layer as the stable rock layer. Furthermore, the layer was completely fixed with container by an anchor to control slide during shaking. The weak layer was made by using the materials which mixed bentonite with quartz sand 6 at 1% of weight ratios. The surface layer was made by using the materials which mixed bentonite at 10% of weight ratios to iron powder for keeping sufficient inertia force. In addition, stepping was used as the structural model around the layer boundary to prevent the sliding in the layer boundary.

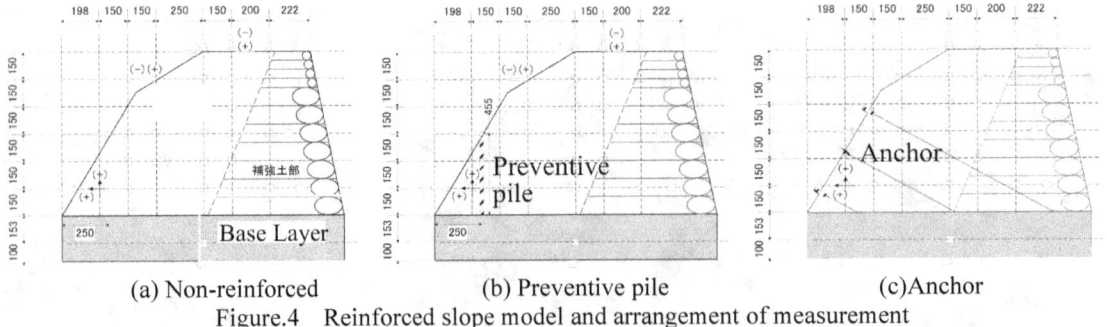

(a) Non-reinforced   (b) Preventive pile   (c) Anchor

Figure.4 Reinforced slope model and arrangement of measurement

In order to measure the collapse behavior of the slope model during shaking every moment, marked points for image analysis were installed in the side face of the model. By using a high-speed camera, displacements in the two-dimensional plane were measured during shaking. The input waveform was assumed ten cycle of sine wave which have a period of 5Hz. The amplitude of the waveform increased gradually by 100Gal from 100Gal. And the shaking table test was finished at the stage which reached collapse. Comparison of time history of the movement of the surface layers after the failure was shown in Figure.3. After a tension crack occurred near the shoulder of slope, surface layer collapsed as a block along slip surface which occurred in a weak layer. At first, the behavior of the slip-down type that a surface layer on the slip surface suddenly slid down after collapse was shown in Figure.3a). The slip-down collapse occurred suddenly at the shaking at input acceleration 400gal. After failure occurred due to the formation of the slip surface, the behavior of the progressive collapse mode which surface layer on the slip surface moved gradually was shown in Figure.3b). The behavior occurred at the shaking at input acceleration 500gal. The behavior changed to the slipping down type after a progressive collapse occurred as shown in figure 3b) was shown in Figure.3c). Although the behavior according to a progressive collapse mode occurred during the shaking at input acceleration 500gal, slip-down collapse occurred suddenly after collapse at toe of slope.

As the behavior of the progressive collapse mode that rock block slides gradually during shaking, a sliding block stops when slide movement after shaking is not large. For this reason, the slope stability can be evaluated reasonably by specifying a standard value in a safe side. However, since collapse of either toe of slope or surface layer may be induced during the behavior of the progressive collapse mode, collapse mode may change to slip-down type from progressive collapse mode. Therefore, it is important to make sure of a collapse behavior. On the other hand, since collapse mode of slip-down type is a phenomenon that the whole slide block slips down to the toe of slope in an instant, the influence on the facility in a nuclear power plant near the slope is serious. Thus, by judging the collapse mode appropriately according to a ground structure, a strength characteristic and a deformation characteristic, it becomes possible to set up a suitable limit state.

3.2. Effect of countermeasure

When a collapse of a slope affects on the facilities in a nuclear power plant, it is necessary to take a adequate countermeasure. Although the anchor and preventive pile which have been used generally as countermeasure will be executed even in a nuclear power plant, the seismic design method to be able to consider the influences has not established so far. In order to establish the method to evaluate the stability of the slope with reinforcement against strong earthquake motion, not only the dynamic response characteristics of the slope with reinforcement but also the effect of reinforcement are made clear based on the experimental results obtained by the shaking table tests of small-scale reinforced slope model. The experimental results are described here.

There are three kinds of slope models, an unreinforced slope, the slope reinforced by anchor, and the slope reinforced by preventive pile. Shape of slope model and arrangement of measuring instruments are shown in Figure 4. The shapes of slope models are equal each other. The height and the width are 1.15 m and 1.5 m respectively. This model consists of a base part, a general part, and a reinforced part, The material of a base part is improved gravel mixed with cement and the general part is imitating the weathering layer. The improved gravel was made of a gravel, cement, and water which are 100 vs 7 vs 4 as ratio of weight. The general part and the reinforced part were considered to satisfy not only the condition that they are stable when making model but also the condition that they collapse due to the strong shaking. The general part was made by using silica sand 6 grade 100, bentonite 1 and water 10 as the ratio of weight. In reinforced part, geo-net with the low strength was laid every 10cm. The vibration condition of a shaking table is the same with that mentioned in 3.1.

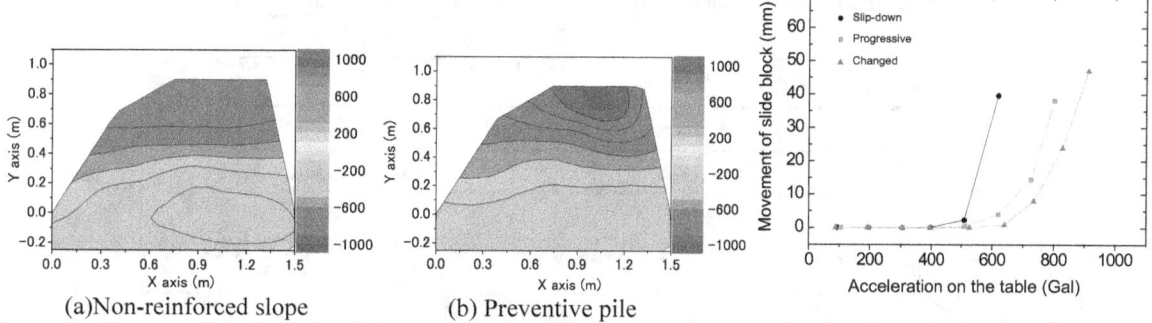

(a) Non-reinforced slope   (b) Preventive pile

Figure 5 Contour of Horizontal acceleration when the maximum acceleration generates at the shoulder of slope

Figure 6 Relationship between acceleration at shaking table test and settlement at the shoulder of slope

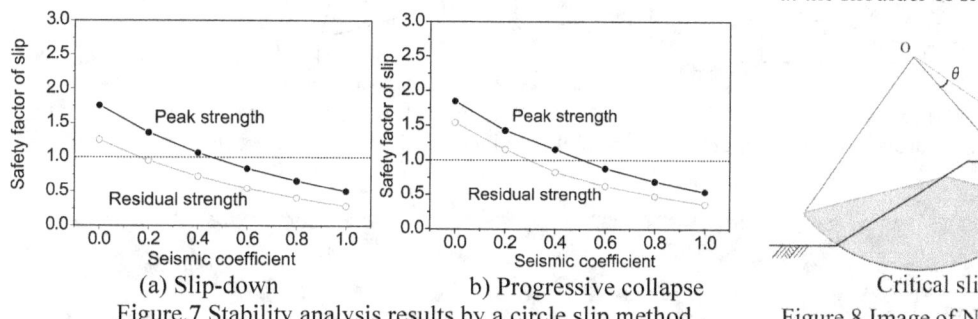

(a) Slip-down   b) Progressive collapse

Figure.7 Stability analysis results by a circle slip method

Figure.8 Image of New mark method

Contour lines of a horizontal acceleration at the time that the horizontal acceleration at the shoulder of slope becomes minimum are shown in Figure 5. A case of non-reinforced slope and the case of the slope reinforced with preventive pile are shown. This figure indicates at the state that the inertia force to the direction of the slope front becomes maximum. Accelerations on the table are 800gal for non-reinforced slope and 600gal for reinforced slope. It is found that the response at the shoulder of slope is amplified greatly against the bottom of slope. Especially, as for the reinforced slope, it is found that the amplification becomes larger due to the increase of the stability of the slope according to the effect of countermeasure. The relationship between the amount of shoulder subsidence and the acceleration on the table is shown in Figure 6. The relationship between the amount of shoulder subsidence and acceleration on the table is shown in Figure 6. Although the non-reinforced model collapsed at 600gal, the model reinforced by a preventive pile and the model reinforced by an anchor collapsed at 800gal and at 900gal, respectively. Based on these results, the effect of countermeasure was verified.

4. EVALUATION METHOD OF THE LIMIT STATE AND THE VERIFICATION

4.1. Examination of the limit state about the stability by the sliding safety factor

Stability analysis of the slope model in which different collapse mode occurred as shown in figure.2 was carried out by the circle slip method using seismic coefficient as a seismic action. As the different collapse modes, slipped-down type, progressive collapse type and type changed slipped-down type from progressive collapse type were selected. The analysis results of a slipped-down type and a progressive collapse type are shown in Figure 7 as a representative case. Using peak strength and residual strength as strength characteristics, The safety factors of those slope models were calculated against the horizontal seismic coefficient changed at every 0.2 from 1.0 to 0.0. First of all, as a result of shaking table test against each collapse model, accelerations which collapses generated, became slipped-down type 400Gal, progressive type 500Gal, and progressive/slipping-down type 500Gal. Next, the validity of stability analysis verifies by checking that the slide safety factor becomes about 1.0 when the horizontal seismic coefficient 0.4, 0.5, and 0.5 acted to each model. As a result of stability analysis, the safety factor in each collapse mode became slipped-down type 1.065, a progressive type 0.987, and a progressive/slipping-down type 1.065. Stability analysis was verified to be appropriate based on the results that the safety factor of slip for each collapse mode was almost 1.0.

4.2. Estimate of movement of rock block for progressive collapse type by New Mark method

Table 1 Comparison of experimental results with numerical results

Collapse mode	Experimental results	Numerical results
Slip-down	560mm	Not applicable
Progressive	195mm	182mm
Changed type	70mm (progressive)	40mm (Progressive)

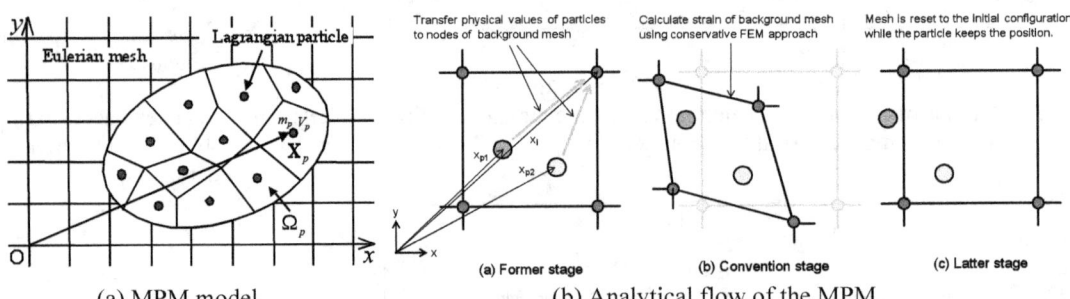

(a) MPM model   (b) Analytical flow of the MPM

Figure 8 Schematic figures of the model and the analytical flow of the MPM

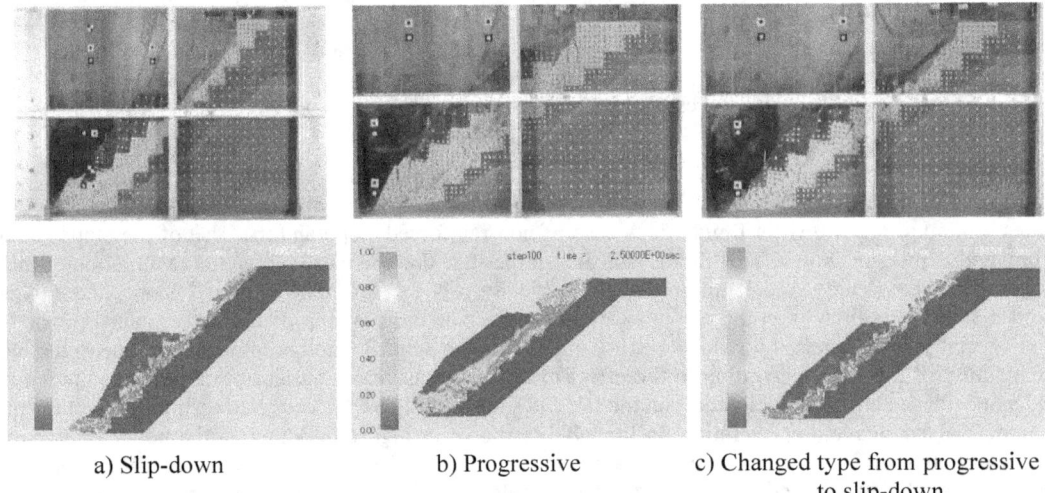

a) Slip-down   b) Progressive   c) Changed type from progressive to slip-down

Figure 9 Comparison of deformation obtained by MPM with experimental behaviors

In order to make sure of the deformation of the rock slope after collapse, deformation analysis was carried out by the Newmark method as shown in Figure.8 using the material properties obtained from the laboratory tests. Newmark method is the method to evaluate a sliding displacement on the sliding surface by integrating acceleration as a inertia force to sliding block after generating a sliding surface (critical slip surface) when the safety factor on the sliding surface calculated by the circle slide method becomes equal to 1.0. Furthermore, in the case that the safety factor is more than 1.0, the deformation analysis is carried out by using a cohesion and a internal frictional angle at a peak strength. And in the case that the safety factor is less than 1.0, the deformation analysis is carried out by using a cohesion and a internal frictional angle at a residual strength. Based on these processes, more realistic displacement is possible to be estimated.

An analysis result is shown in table 1. Calculation of a slipped-down type by Newmark method was not completed since a yield seismic coefficient was negative value at the time to calculate a displacement. The calculated value for slope model with an progressive collapse type is good agreement with the experimental value. Then it is found that Newmark method has a good applicability. Moreover, for a slope model with collapse mode which changes from a progressive collapse type to a slipped-down type, the experimental value was good agreement with the calculated value at the time when the collapse mode changed from progressive collapse type in early stages of shaking. As mentioned above, it is thought that the Newmark method has a good applicability to evaluate the movement of slide block in the mode which slope collapses gradually.

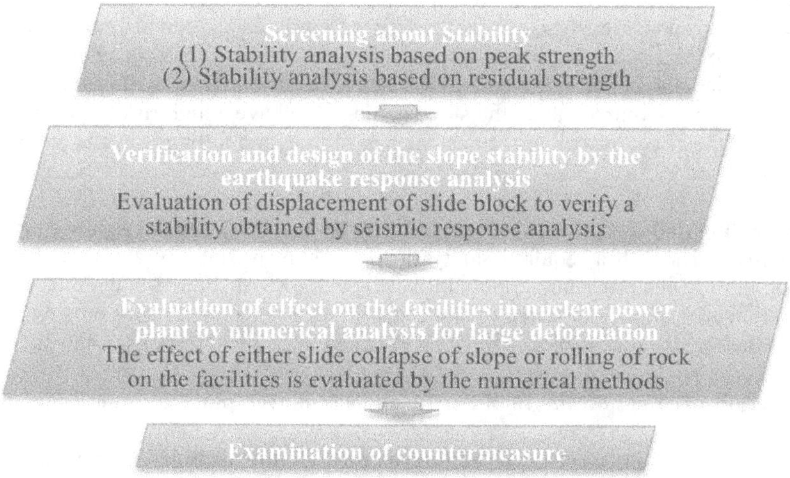

Screening about Stability
(1) Stability analysis based on peak strength
(2) Stability analysis based on residual strength

Verification and design of the slope stability by the earthquake response analysis
Evaluation of displacement of slide block to verify a stability obtained by seismic response analysis

Evaluation of effect on the facilities in nuclear power plant by numerical analysis for large deformation
The effect of either slide collapse of slope or rolling of rock on the facilities is evaluated by the numerical methods

Examination of countermeasure

Figure 10 Design procedure about the slope stability

4.3. Estimation of the movement of a rock block to each collapse mode by a particle method

In order to establish the numerical method which can evaluate the behavior before and after collapse of the rock slope, a particle method among some methods about large deformation analysis was applied to evaluate the experimental results. The material properties obtained from the laboratory tests were used to make a slope model. The particle method is a method called MPM (Material Point Method), and is a kind of a particle method called PIC (Particle in Cell) which calculates an advective term by particles and calculates other clauses with a lattice. This method uses the technique to calculate an advection by the perfect Lagrange method using particles, therefore has following characteristics; One is hard to generate numerical diffusion. The other is that the boundary where particles can move is possible to be easily set up in a lattice. Moreover, since MPM performs the formation of a weak form type and discretization using the interpolation function to the lattice like FEM as shown in Figure 8, it can utilize the numerical-analysis technology of FEM accumulated until now.

As for modeling the slope model mentioned in 3.1, perfect-plasticity model is used as a constitutive relationship in weak layer. The base layer was modeled as an elastic body. Although material properties were obtained from the triaxial compression test, a cohesion at residual strength was set to 1.0kPa for the slope model except the progressive collapse type. The value was set as a strength at the time when a safety factor by a circle slip method becomes 1.0 at a horizontal seismic coefficient 0.0. With the slip-down type, the idea about setting this value is based on the evidence that horizontal acceleration in surface layer becomes near the zero, after a slip surface occurs in a weak layer.

Not only the contour lines about the maximum shear strain but also the collapse situation of the slope model by a shaking table test is shown in Figure 9. In all cases, it is found that the behavior of collapse obtained by MPM analysis is good agreement with an experimental result. Furthermore, different collapse behaviors according to the collapse modes in which a slip-down type collapses suddenly and a progressive collapse type collapses gradually have reappeared in MPM analysis. In addition, although generation of the tension crack in the collapse behavior was recognized clearly by a shaking table test, the behavior was not able to imitate completely in numerical analysis by MPM. Hereafter, it is necessary to develop in consideration of the tension characteristic of the ground under low confined pressure.

5. VIEW OF THE DESIGN ABOUT SLOPE STABILITY

Based on not only the knowledge about the collapse mode and limit state of a slope mentioned above but also the verification about the numerical method to evaluate the index associated with limit state, the design procedure about the slope stability which will be required for risk assessment is shown in Figure 10.
The outline is as follows.

First step : Screening about stability

Slope stability is evaluated based on the safety factor of slip using the dynamic response characteristic of slope obtained by the seismic response analysis against reference earthquake ground motion. Here, for the slope judged as a safety factor of slip becoming below a required value, stability analysis against a slope's own weight is carried out using a residual strength. For a case that a safety factor obtained by the stability analysis against a slope's own weight is below a required value, the slope is judged to have a high risk to generate the collapse of a slipped-down type. And the effect of rock movement such as sliding and falling is evaluated quantitatively based on the following procedures.

Second step : Verification and design of the slope stability by the earthquake response analysis
For the slope which satisfies with a required safety factor of slip by stability analysis against a slope's own weight as mentioned in previous procedure, a possibility that collapse of slipped-down type occurs is considered to be small. Therefore, verification and design of slope stability are performed based on a response displacement of slope calculated from the dynamic response of slope obtained by seismic response analysis.

Third step : Evaluation of effect on the facilities in nuclear power plant by numerical analysis for large deformation
The effect of either slide collapse of slope or rolling of rock on the facilities is evaluated by the numerical methods which can consider a collapse behavior of slope. First of all, it is evaluated whether a rock block generated by collapse of slope arrives at a facility. Next, impulse force is evaluated when judged with reaching to the facilities. Finally, an adequate countermeasure is installed when it is thought that the effect is serious.

Fourth step : Examination of countermeasure
As countermeasures installed on slope around the facilities in a nuclear power plant, there is an earth removal work, an anchor work, a preventive pile work. The design method of countermeasure is based on the design method of a non-reinforced slope. By considering two or more slip surfaces which have the high possibility about the loss of slope stability, slope stability is evaluated based on the safety factor of slip obtained by the dynamic response of the slope.

6. CONCLUSION

In order to establish the stability assessment procedure of a next generation to evaluate the effect of the collapse of the slope on the functional safety of a nuclear power plant, the developed view in a geotechnical aspect based on the experiment and numerical approach was described. The limit states which depend on the collapse modes were specified based on the experiment. Furthermore, the applicability of numerical method and the effect of countermeasure were made clear. The numerical method was used to evaluate movement of rock block after a collapse of slope, and is able to evaluate seamlessly from a dynamic response to a collapse behavior. The major results in this report are described as follows.

1) The following two limit states were specified for evaluating the effect of unstable behavior of slope on the facilities having the important functional safety of the nuclear power plant.
i) The state that a collapse of slope occurs.
ii) The state that the damage of facilities in the plant due to the collapse of slope causes the loss of the functional safety.

2) Based on the experimental results obtained by the shaking table tests using three scale models of slope, it was made clear that there are three modes of collapse. The first type was a slipped-down type. Second type was a progressive collapse type. The third type was a type which changed from a progressive collapse type to a slipped-down type. A progressive collapse mode is that slide block stands it still when the movement of the block is not large after shaking. For the mode, considering the failure induced at a toe of slope and a surface, it was showed that the slope stability is able to evaluate rationally by using the appropriate value in a safety side. Moreover, the acceleration when collapse occurred at the slope reinforced by not only a preventive pile work but also an anchor work was larger in comparison with the non-reinforced slope, and the effect of countermeasures was verified.

3) As a result of stability analysis by a circle slip surface method, stability analysis was verified to be appropriate based on the results that the safety factor of slip for each collapse mode was almost 1.0.

4) Newmark method has a good applicability for the mode which slope collapses gradually.

5) The collapse situation of the slope model by a shaking table test is good agreement with the collapse behavior obtained by MPM analysis. Furthermore, different collapse behaviors according to the collapse modes in which a slipped-down type collapses suddenly and a progressive collapse type collapses gradually was able to reappear in MPM analysis. In addition, generation of the tension crack in the collapse behavior was not able to imitate completely in numerical analysis by MPM.

6) Based on not only the knowledge about the collapse mode and limit state of a slope as mentioned above but also the verification about the numerical method to evaluate the index related with limit state, the design procedure about the slope stability which will be required for risk assessment was established.

References

[1]   Shinoda M. Nakajima S. Nakamura H. Kawai T and Nakamura S., *Shaking table test of large-scaled slope model subject to horizontal and vertical seismic loading using E-Defense*, Proc. of 18th international Conference on soil mechanics and Geotechnical Engineering, pp.1603-1606, (2013.9)

[2]   Abe K. Izawa J. Nakamura H. Kawai T and Nakamura S., *Analytical study of seismic slope behavior in a large-scale shaking table model test using FEM and MPM*, Proc. of 18th international Conference on soil mechanics and Geotechnical Engineering, pp.1407-1410, (2013.9)

Probabilistic Tsunami Hazard Analysis for Nuclear Power Plants on the East Coast of Korean Peninsula

In-Kil Choi[a], Min Kyu Kim[a] Hyun-Me Rhee[a]
[a] Korea Atomic Energy Research Institute, Daejeon, Korea

Abstract: On March 11, 2011, there was a tremendous earthquake and tsunami on the east coast of Japan. The earthquake and tsunami caused a severe accident at the Fukushima I NPP. Before the 2011 event, a tsunami was one of the many external events for a NPP, but after the Fukushima accident, a tsunami has become a very important external hazard that should be considered for the safety on NPP. After the Fukushima accident, many countries have attempted to develop a tsunami safety assessment method for nuclear power plants. To perform a tsunami safety assessment for a NPP, deterministic and probabilistic approaches can be applied. In this study, a probabilistic tsunami hazard analysis was performed for the east coast of Korea. There are three NPP sites located on the east coast of Korea. An empirical analysis and a numerical analysis were performed for an assessment of a tsunami hazard.

Keywords: Tsunami, Hazard, Probabilistic Approach, Empirical Method, Numerical Method

1. INTRODUCTION

Extreme external events are emerged as a significant risk contributor to a nuclear power plant after the Fukushima accident. There are many kinds of extreme external events which threaten the safe operation of nuclear power plants. It is impossible to cope with all of the extreme external events. To secure the safety of a nuclear power plant, it is necessary to identify the extreme external events that can potentially threaten the safety of nuclear power plants and to estimate the frequency and intensity level of the identified extreme external events. The design level of an external event, such as earthquake and tsunami, has been determined by a deterministic and/or a probabilistic hazard analysis. After the Fukushima accident, the safety of a nuclear power plant for a beyond design level became important due to the possibility that exceed the initially determined level at a design stage. Even though the design level was determined based on the best estimated results at that time, it should be reevaluated periodically to maintain the safe operation of a nuclear power plant by reflecting the up-to-date and accurate information of external event hazard.

All of the Korean nuclear power plants are located in the coastal area, 3 sites in the east coast and 1 site in the west coast. So the Korean nuclear power plants can hardly be free from tsunami attack. It can certainly be guessed from that the Korean peninsula has historically experienced tsunami several times. The design level of tsunami wave for the Korean nuclear power plants has been determined by the deterministic hazard analysis method. For the realistic consideration of a tsunami risk for a nuclear power plant, it is necessary to perform a probabilistic tsunami hazard analysis.

The first is a tsunami hazard assessment that determines a tsunami return period for a target nuclear power plant site. The second is a tsunami fragility assessment that evaluates a failure probability of safety-related equipment and structures caused by the force and inundation height of a tsunami wave. The last part is a system analysis that calculates the risk caused by a tsunami using event trees and fault trees. This study focused on a probabilistic tsunami hazard assessment for nuclear power plants located on the east coast of Korea. The probabilistic tsunami hazard assessment (PTHA) is based on the methodology of probabilistic seismic hazard assessment (PSHA). The PTHA can be performed using an empirical and numerical method. In this study, both empirical and numerical methods are applied to develop the tsunami hazard curves and are compared.

2. EMPRICAL METHOD

2.1. Tsunami Return Period Assessment Method

For an evaluation of tsunami hazard curves for the east coast of Korea, a tsunami propagation analysis should be performed from the seismic source. However, a tsunami propagation analysis needs a lot of effort and has many uncertainties because of the lack of seismic source information. Therefore, in this study, both an empirical method and a numerical method were applied for an evaluation of a tsunami hazard curve. For a regression of the return period of a tsunami on the east coast of Korea, the power law, upper-truncated power law, and exponential function were considered, but finally, the power law and general exponential function were used. The equations for a power law and upper-truncated power law are shown in equations (1) and (2), respectively [1,2].

$$\dot{N}(r) = Cr^{-\alpha} \qquad (1)$$
$$\dot{N}_T(r) = C(r^{-\alpha} - r_T^{-\alpha}) \qquad (2)$$

2.2. Development of Tsunami Catalogue

For the development of a tsunami catalogue, instrumental records after 1900 were considered. After 1900, there were four tsunamis that occurred on the east coast of Korea. The most vulnerable tsunami event occurred in 1983. In 1983, the Akita earthquake occurred on the west side of Japan. At this time, a maximum wave height was recorded at about 4.2m in the Imwon harbor in Korea. One person was killed and two persons went missing. Hundreds of boats and houses were destroyed and damaged. All tsunami events after 1900, including the 1983 event, are summarized in Table 1.

Table 1: Tsunami events on the east coast of Korea after 1900

	Earthquake	Damage in Korea	Max. Wave Run up
1940. 8. 2.	Hokkaido Magnitude 7.0	No damage recorded	Mukho: 1.2m
			Najin: 0.5m
1964. 6. 16.	Niigata earthquake Magnitude 7.5	No damage recorded	Busan: 0.32m
			Ulsan: 0.39m
1983. 5. 26.	Akita earthquake Magnitude 7.7	Death: 1 Missing: 2 Ships: 81 Buildings: 100	Sokcho: 1.56m
			Mukho: 3.9m
			Imwon: 4.2m
1993. 7. 12	Hokkaido Magnitude 7.8	Ships: 35 Fishing implements: 3000	Sokcho: 2.76m
			Mukho: 2.03m
			Pohang: 0.93m

For an assessment of tsunami events before 1900, the historical records were determined. "The annals of the Chosun dynasty" were referred for an evaluation of the tsunami catalogue. Through the historical records assessment, five tsunami events on the east coast of Korea were found. All tsunami records in the 'The annals of the Chosun dynasty' are summarized in Table 2 [3].

Finally, the tsunami catalogue was developed using a combination of historical and instrumental record as shown in Figure 1. This catalogue covers from 1392 to 2009, which is a 618 years period. However, as shown in figure 1, it can be recognized that tsunami events were recorded only in a limited period. From 1392 to 1642, and from 1741 to 1939, there were no tsunami events recorded. However, from 1940 to 2009, four tsunamis occurred. This unequal occurrence of tsunami events indicates that this tsunami catalogue has many uncertainties.

Table 2: Tsunami events at the east coast of Korea before 1900

Date	Location	Damage
1643.6. 21.	Ulsan	Big waves reach to a 12 steps from a seashore
1668. 7. 25	Cheolsan	Waves were very high and an earthquake happened
1681. 6. 24	Yangyang	Sea water drawdown to 100 steps from a seashore
1702. 11. 28.	Gangwondo	Tsunami run up at the east coast of Korea, so many houses were inundated
1741. 7. 19	East coast	The sea level increased and inundated to the nine villages of east coast of Korea. Many houses and fishing boats were destroyed.

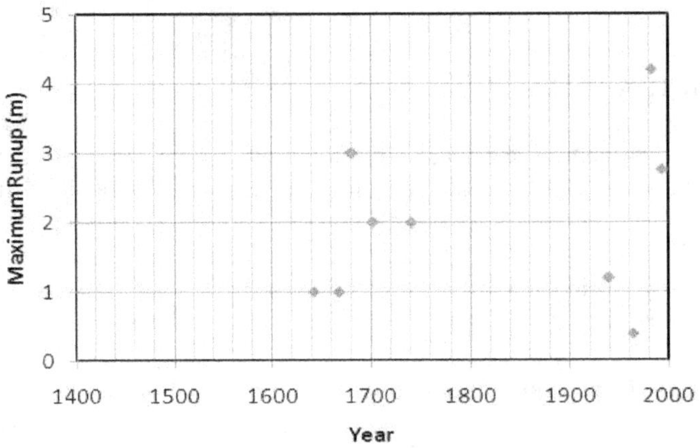

Figure 1: A tsunami catalogue of the east coast of Korea

2.3. Development of tsunami hazard curve

The return period of tsunami events was determined using a power law and exponential function as shown in Figure 2. As shown in Figure 2, the exponential function matches the tsunami return period better than that of the power law. The exponential function was more appropriate for an estimation of tsunami return period.

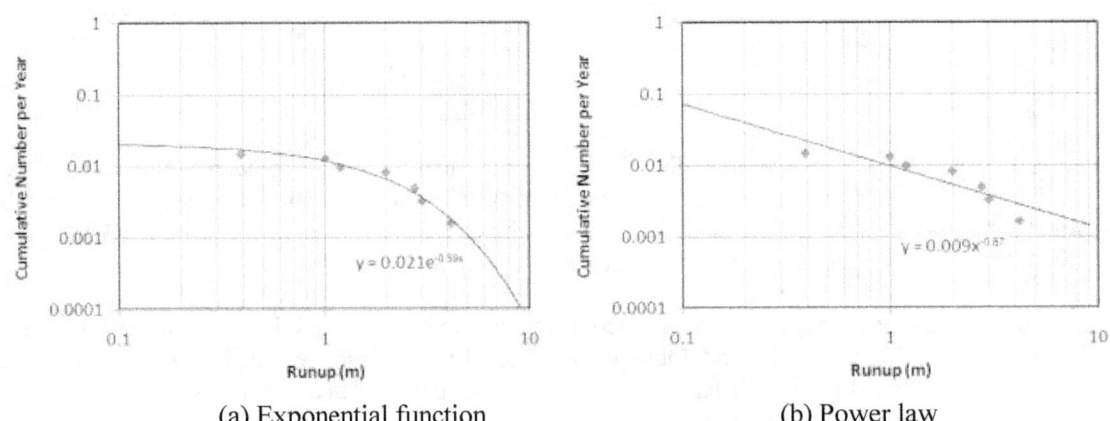

(a) Exponential function   (b) Power law
Figure 2: Tsunamis return period evaluation by using empirical method

However, as shown in figure 1, there was only one tsunami event where the maximum wave height was below 1 meter. That is because small tsunami events were not recorded in the historical record. A small tsunami event makes the tsunami return period become overestimated. For a decrease in the uncertainty of the tsunami return period, a 1940 tsunami event where the maximum wave height was recorded as 0.39m was deleted. Through this method, the tsunami return period was re-evaluated as shown in Figure 3. As shown in figure 3, the tsunami return period was decreased compared to figure 2.

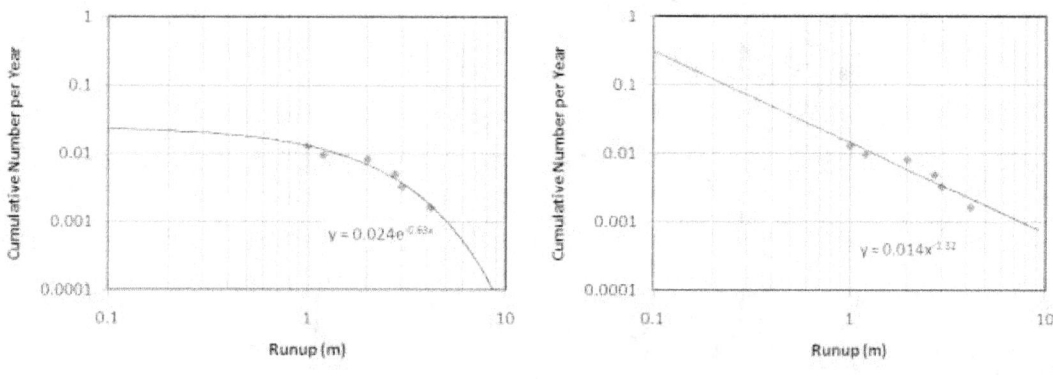

(a) Exponential function (b) Power law

Figure 3. Tsunamis return period evaluation using empirical method in the case of exclusion of a 0.39m event

Finally, the tsunami return periods were summarized according to the 0.39 m tsunami event in Table 3. As shown in Table 3, the return period of tsunami run up events were slightly changed according to the 0.39 m tsunami event. In the case of the 10 m maximum run-up height caused by a tsunami event, the return period was 17,383 and 22,690 years, respectively. The meaning of a 10 m maximum run up height is the ground level of the Ulchin NPP site.

Table 3: The return period of maximum run-up height caused by a tsunami event in the east coast of Korea

	Include 0.39m		Exclude 0.39m	
Max Runup	Prob.	Return Period	Prob.	Return Period
1	1.16E-02	86	1.28E-02	78
5	1.10E-03	910	1.03E-03	972
10	5.75E-05	17383	4.41E-05	22690
15	3.01E-06	332114	1.89E-06	529507

3. NUMERICAL METHOD

3.1. Determination of tsunami source

For an evaluation of the tsunami hazard of Korea, tsunami sources should be determined. Five tsunami source areas were considered for a tsunami hazard analysis, as shown in Figure 4. The historical and instrumental tsunami records in Korea are summarized in Table 4 according to the hypocenter. As shown in Table 4, five tsunamis were caused by earthquakes from area A, and two tsunamis were caused by earthquakes from area C. The source areas of another two tsunamis have not yet been identified. Therefore, areas A and C were selected for the tsunami sources of the Korean peninsula. However, in this study, only the tsunami sources in area A were considered for the tsunami propagation analysis. The detailed location of the tsunami sources in areas A and C are shown in

Figure 5. The fault parameters for the tsunami simulation were used in the JSCE method [4,5] for the case of area A.

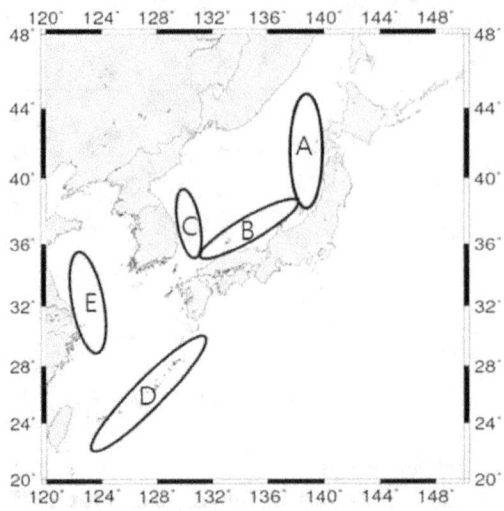

Figure 4: Selected tsunami source areas for Korean peninsular

Table 4: Tsunami catalogue according to the source area

Date	M	Hypocenter	Area
1643-7-24	6.5	Ulsan	?
1681-6-12	6.8	Yangyang	C
1810-2-19	6.5	Cheong-jin	?
1702-11-28	?	Gangwon	C
1741-7-19	?	Peonghae	A
1940-08-02	7.5	West part of Hokkaido	A
1964-06-16	7.5	North part of Niigata	A
1983-05-26	7.7	West part of Aomori	A
1993-07-12	7.8	South west part of Hokkaido	A

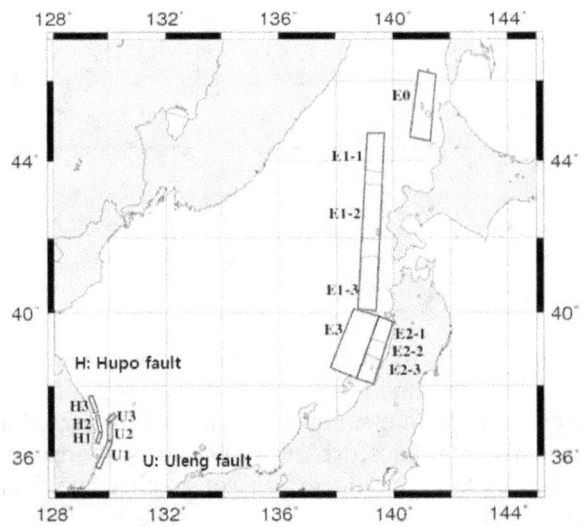

Figure 5: Selected tsunami source areas for Korean peninsular

3.2. Establishing a logic tree

For the tsunami hazard analysis, we should consider various kinds of uncertainties of tsunami sources. To consider the uncertainties of tsunami fault parameter, a logic tree method was applied. The tsunami sources, magnitude distribution, recurrence intervals, and tsunami height estimations were considered for the uncertainties. A sample logic tree for a tsunami hazard analysis is shown in figure 6.

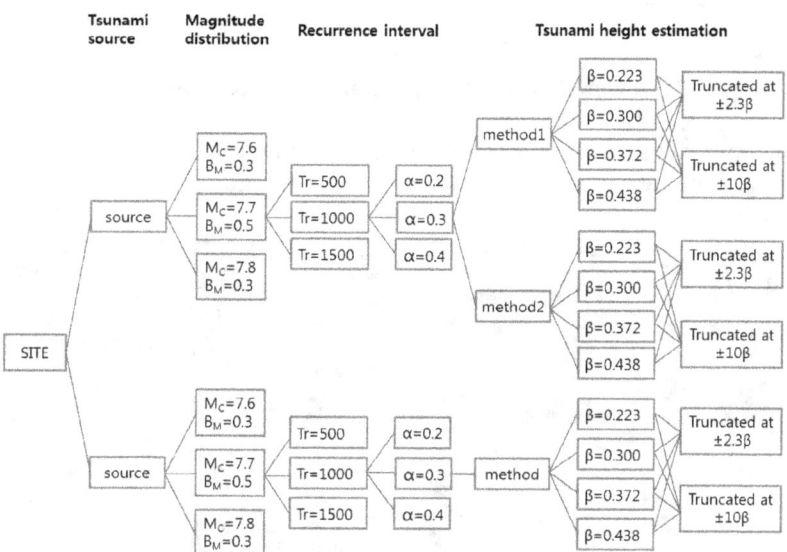

Figure 6: Sample Logic tree for Tsunami Hazard Analysis

3.3. Tsunami Simulation

For the tsunami propagation analysis, TSUNAMI_ver 1.0 [6] developed by JNES for using the IAEA international collaboration research program was used. Before a tsunami simulation for determining a tsunami hazard, a verification analysis was performed. In the case the Akita earthquake in Japan in 1983, a tsunami run-up occurred on the east coast of Korea. In the 1983 tsunami, the Imwon harbor in Korea was severely damaged and inundated. There are some researches on the 1983 tsunami because the 1983 tsunami was a very good example that can be used for the verification of a tsunami simulation code [7, 8]. One of research about the 1983 Akita earthquake and tsunami calculated the wave run-up of the Ulchin NPP site. There was no nuclear power plant in the Ulchin area in 1983, and thus this analysis calculated the artificial wave run-up if the same earthquake and tsunami were to occur in the same area. The fault parameters of the Akita earthquake were verified by several researchers [8]. The verification analyses were performed using the verified fault parameter for the Ulchin NPP site, as shown in Table 5. The analysis results are shown in Figure 7 according to the simulation method. As shown in Figure 7, a tsunami wave arrives almost 110 minutes after an earthquake occurs on the west coast of Japan. The arriving times are almost similar to the numerical results, and the time history wave run-up height is also similar.

Table 5: The fault parameters for tsunami simulation for 1983 Akita earthquake

No.	θ	δ	λ	D	L	W	S
1	22	40	90	2	40	40	7.6
2	355	25	80	3	60	40	3.0

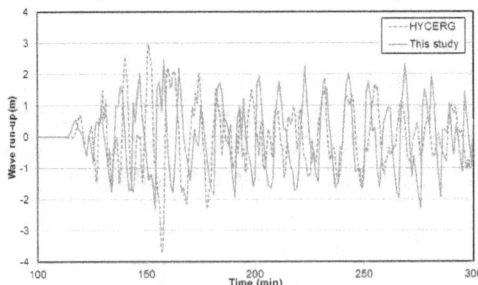

Figure 7: Selected tsunami source areas for Korean peninsular

Eighty numerical simulations were performed for determining a tsunami hazard. The maximum and minimum wave heights of the calculation results are shown in Figure 8.

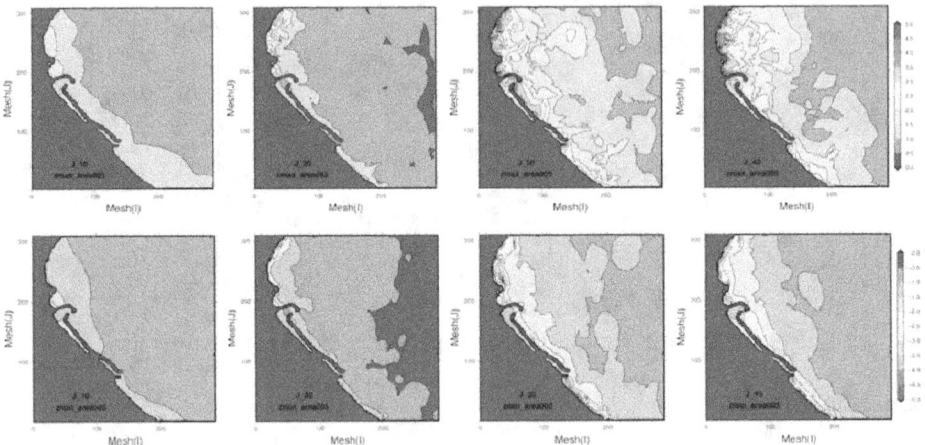

Fig. 8 The distributions of maximum and minimum wave height (up: maximum wave, low: minimum wave)

3.4. Tsunami Hazard

A tsunami hazard analysis was performed according to a branch of the logic tree. Although a round-robin algorithm and the Monte Carlo simulation should be considered for determining a fractile curve of a tsunami hazard, and only the Monte Carlo simulation was performed in this study. Temporary tsunami hazard results for the Ulchin NPP site are shown in Figure 9.

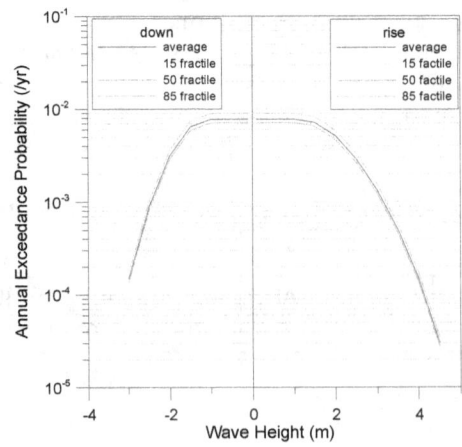

Figure 9: Tsunami hazard analysis results

4. CONCLUSION

In this study, a tsunami hazard curve was determined for a probabilistic safety assessment (PSA) induced tsunami event at a Nuclear Power Plant site. Empirical and numerical methods were also applied for the tsunami hazard analysis. In the case of the empirical method, a tsunami catalogue was developed using previous tsunami records. For an evaluation of the return period of the tsunami run-up height, the power-law and exponential function were considered. In the case of a numerical analysis for a tsunami hazard assessment, TSUNAMI_ver1.0 was used. The logic tree method was applied for considering uncertainties in a tsunami hazard. Through this study, the return period of the maximum and minimum tsunami run-up was evaluated using the empirical and numerical methods temporarily, but a more accurate tsunami hazard analysis is needed for a more accurate tsunami hazard curve.

Acknowledgements

This work was supported by Nuclear Research & Development Program of the National Research Foundation (NRF) grant funded by the Korean government (MEST).

References

[1] Burroughs, S.M. and Tebbens, S.F., Upper-truncated Power-law in Natural Systems, Pure applied geophysics, 158, pp.331-342, 2001.
[2] Burroughs, S.M. and Tebbens, S.F., Power-law Scaling and Probabilistic Forecasting of Tsunami Runup Heights, Pure applied geophysics, 162, pp.331-342, 2005.
[3] Min Kyu Kim, In-Kil Choi, A Tsunami PSA methodology and application for NPP site in Korea, Nuclear Engineering and Design 244(2012) 92-99
[4] JSCE (Japan Society of Civil Engineering), The Nuclear Civil Engineering Committee, The Tsunami Evaluation Subcommittee, Probabilistic Tsunami Hazard Assessment Method, September 2011 (in Japanese)
[5] Atomic Energy Society of Japan. Implementation standard concerning the tsunami probabilistic risk assessment of nuclear power plant: 2011, AESJ-SC-RK004E:2011, NISSEI EBLO INC., Tokyo, Japan, 2013.
[6] Japan Nuclear Energy Safety Organization. Tsunami simulation code "TSUNAMI". Japan, 2008.
[7] Korea Hydro Nuclear Power. A study on the probabilistic safety assessment method for tsunami. 08SF07, Daejeo, Korea, 2010.
[8] Korea Energy Power Research Institute. Numerical simulation of tsunami on the coastal area of the Korean Peninsula. Daejeon, Korea, 2007.

External Events PSA for the spent fuel pool of the Paks NPP

Attila Bareith[a], Jozsef Elter[b], Zoltan Karsa[a], Tamas Siklossy[a*]
[a] NUBIKI Nuclear Safety Research Institute, Budapest, Hungary
[b] Paks Nuclear Power Plant Ltd., Paks, Hungary

Abstract: Originally, probabilistic safety assessment of external events was limited to the analysis of earthquakes for the Paks Nuclear Power Plant in Hungary. The level 1 PSA for external events other than earthquakes was completed in 2012 showing a significant contribution of wind and snow related failures to core damage risk. On the basis of the external events PSA for the reactor, a similar assessment was subsequently performed for a selected spent fuel pool of the Paks plant in 2013. The analysis proved to be a significant challenge due to scarcity of data, lack of knowledge, as well as limitations of existing PSA methodologies. This paper presents an overview of the external events PSA performed for the spent fuel pool of the Paks NPP. Important methodological aspects are summarized, which are relevant to the spent fuel pool external hazard PSA. Although some important challenges had already been experienced during the reactor PSA – that initiated follow-on analyses and developmental efforts –, the most important lessons and analysis areas that need further elaboration are summarized and highlighted in the example of the spent fuel pool PSA to ensure completeness in discussing key analysis findings and unresolved issues.

Keywords: PSA, External Events, Extreme Weather Conditions, Spent Fuel Pool, Fragility Analysis.

1. BACKGROUND

The Hungarian Nuclear Safety Codes [1] list the most important internal and external hazards which shall be taken into consideration during the justification of the design and safety. In particular, the Codes highlight that severe weather conditions and seismic events shall be addressed in the PSA. Originally, probabilistic safety assessment of external events was limited to the analysis of earthquakes for the Paks Nuclear Power Plant in Hungary. The level 1 seismic PSA for the reactor was completed in 2002. Although other external events of natural origin had previously been screened out from detailed plant PSA mostly on the basis of event frequencies, a review of recent experience on extreme weather phenomena made during the periodic safety review of the plant led to the initiation of PSA for external events other than earthquakes in 2009. Hungarian nuclear safety regulations prescribe that the design basis for loads from natural external hazards shall be set at 10^{-4}/a hazard frequency for operating nuclear power plants. According to the regulations, the risk from natural external hazards beyond the design basis shall be assessed at least in the range of $10^{-7} \div 10^{-4}$/a hazard frequency. Therefore probabilistic safety assessment of external hazards has to be performed unless it can be shown that the design basis of the plant ensures that the plant can withstand the loads induced by a hazard with 10^{-7}/a frequency. In addition to these requirements, the accident of the Fukushima Dai-ichi Nuclear Power Plant and the Targeted Safety Reassessment of the nuclear power plants located in the European Union confirmed further the importance of risk analysis for external hazards. The level 1 PSA for external events other than earthquakes was completed in 2012 showing a significant contribution of wind and snow related failures to core damage risk.

2. OBJECTIVES

On the basis of the external events PSA for the reactor, a similar assessment was subsequently performed for a selected spent fuel pool of the Paks nuclear power plant in 2013. Among others, the objectives of the assessment were to quantify, to the extent possible, the level of risk induced by natural external hazards and to identify the main risk contributors relevant to the spent fuel pool. It was foreseeable from the beginning of the assessment that all the risk contributors from the various hazards could not be determined and quantified adequately on the basis of the available background analyses. Therefore a main further objective was to identify analysis areas that would need to be further dealt with in order to develop a full scope external event PSA for the spent fuel pool, as well as to reduce uncertainties and conservatism where necessary. Consolidated proposals on safety enhancement can only be made after resolving these analysis issues, although an objective was to identify apparently important safety concerns in this assessment phase. Hereby we remark, that lessons learned from the external events PSA for the reactor enabled the initiation of some obviously relevant follow-on analyses and developmental efforts for the spent fuel pool based on the similarities of the two analyses.

As to the scope of the analysis, potential hazard induced accidents in all feasible combinations of the plant operational states related to the reactor and the spent fuel pool (hereafter: all plant operational states) had to be dealt with. Concerning low power and shutdown states of the reactor, the plant operational states of a typical refuelling outage were looked at.

3. MAJOR ANALYSIS STEPS

The analysis proved to be a significant challenge due to scarcity of data, lack of knowledge, as well as limitations of existing PSA methodologies. Although some important challenges had already been experienced during the reactor PSA, hereby important methodological aspects are summarized in the example of the spent fuel pool PSA to ensure completeness of an overview on every major analysis step.

The external event PSA for the spent fuel pool of the Paks plant followed the commonly known steps: selection and screening of external hazards, hazard assessment for screened-in external events, analysis of plant response and fragility, PSA model development, and risk quantification and interpretation of results.

3.1. Selection and Screening of External Hazards

During the first step of identifying external hazards that required detailed analysis, we made an attempt to develop a comprehensive list of potential site specific external hazards. At first we performed a review of regulatory requirements nationally and internationally. Relevant requirements of the Hungarian Nuclear Safety Codes [1] and WENRA reference levels [2] enabled to determine the vast majority of potential external hazards. In addition, use was made of the following documents to identify the initial list of potential external hazards:
- the stand-alone volume of the joint ANS-ASME PRA standard that sets forth probabilistic safety assessment methodology for external hazards [3,4],
- a guidance document of the Swedish nuclear safety authority that builds upon the Finnish and Swedish external hazard assessment experience [5],
- the Specific Safety Guide of the International Atomic Energy Agency on level 1 PSA [6].

We applied a successive approach with combined deterministic and partially probabilistic screening of all the potential external hazards to identify the risk significant ones that needed detailed analysis to quantify the risk of the spent fuel pool. During this screening it was found that available hazard analyses did not enable to decide if tornados and blockage of the water intake filters could be screened out or not. Additional hazard assessment has been proposed to clarify these questions.

After screening the following natural external hazards were subject to detailed analysis:
- extreme wind,
- extreme rainfall,
- extreme snow,
- extremely high and low air temperature,
- lightning,
- extreme frost and ice formation.

3.2. Hazard Assessment

The objective of hazard assessment was to determine event frequencies for different magnitudes of the parameter which represents best the load induced by an external hazard. Hazard assessment was based on the data collected by the Hungarian Meteorological Service at station Paks during the past few decades. The following observations were taken into consideration:
- maximum gust of wind [m/s],
- instantaneous and daily average maximum and minimum air temperature [°C],
- maximum 10, 20, 60 minute and daily precipitation intensity [mm/min],
- maximum thickness of snow [cm],
- maximum load of frost and icing [g/mm].

The main difficulty in determining the occurrence frequency of extreme weather conditions is the lack of observations for those events whose probability should be estimated, since data samples from experience are available for short durations only. The results include significant uncertainties irrespectively of the computational method applied. In accordance with the international practice of climatological applications, we made use of the extreme value theory to characterize and quantify each external hazard. Hazard curves were established by fitting Gumbel distribution to the annual extreme values of the most up to date site specific meteorological data. Hypothesis testing was conducted to justify that the Gumbel distribution was an appropriate approximation of the hazard curves. It is noted that lightning as an external hazard required a different analysis approach because several physical properties of lightning had to be assessed in order to be able to characterise the vulnerability of plant structures and equipment.

Extreme weather conditions were estimated at different confidence levels (5, 15, 30, 50, 70, 85 and 95%) for 1 to 10^{-7} 1/a frequency of exceedance. The results of hazard analyses are not discussed hereby for every single hazard, but Figure 1 demonstrates the hazard curves for extreme snow as an example. The results of the analysis show – among others – the plant design basis value for the occurrence frequency of 10^{-4} /a at 50% confidence level (107 cm) and the lower limit of the safety assessment which has the occurrence frequency of 10^{-7} /a (e.g. 175 cm at 50% confidence level). The hazard curves also demonstrate the uncertainty limits of the Gumbel approximation, e.g. the expected thickness of snow for occurrence frequency of 10^{-5} /a is 104 cm at 5% confidence level, while it is 166 cm at 95% confidence level.

Figure 1: Hazard curves for extreme snow

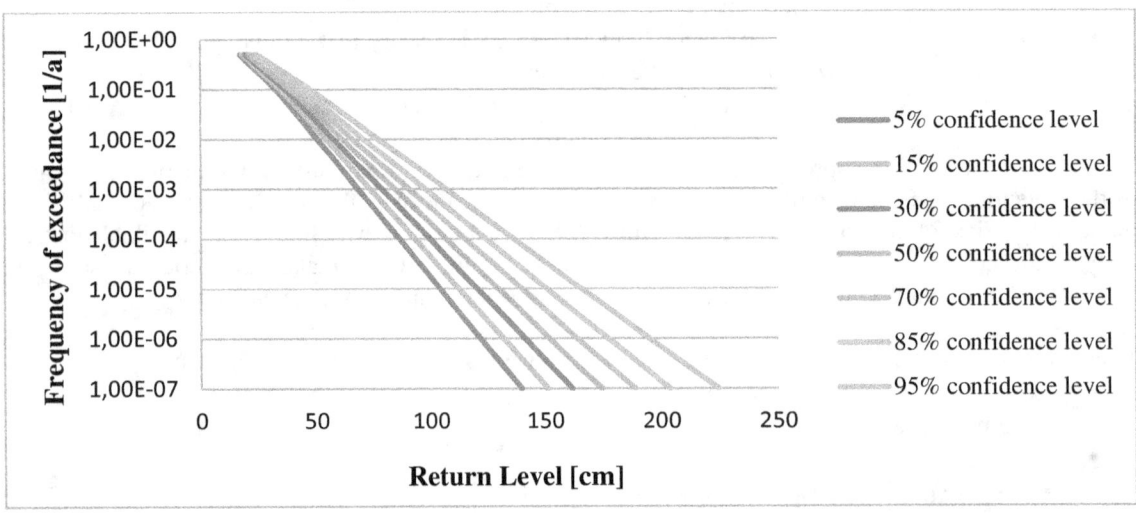

3.3. Plant Response and Fragility Analysis

In the analysis of plant response to external hazards we characterized the loads induced by each external hazard on safety related systems, structures and components (SSCs) relevant to the spent fuel pool in such a form that was appropriate for use in probabilistic safety assessment. We determined the probability of loss of essential safety functions and spurious actuations for different levels of load by means of fragility curves. The methods applied to describe fragility varied among characteristic groups of external hazards.

The effects of loads from wind and snow on structures and outdoor facilities were analysed in detail for the purposes of plant response analysis. Vulnerability of power transmission lines to extreme frost and ice formation (hereafter: frost) was also taken into consideration during plant response analysis. Wind, snow and frost related fragility curves, as an outcome of the corresponding fragility analysis, were established by using a closed mathematical expression for different confidence levels. Design data were reviewed, safety margins ensured by the standards applied during structural design were assessed, and use was made of a recent large scale structural re-analysis for the plant to determine fragility. Figure 2 demonstrates the wind related fragility curves for the reactor hall as an example.

Figure 2: Wind related fragility curves for the reactor hall

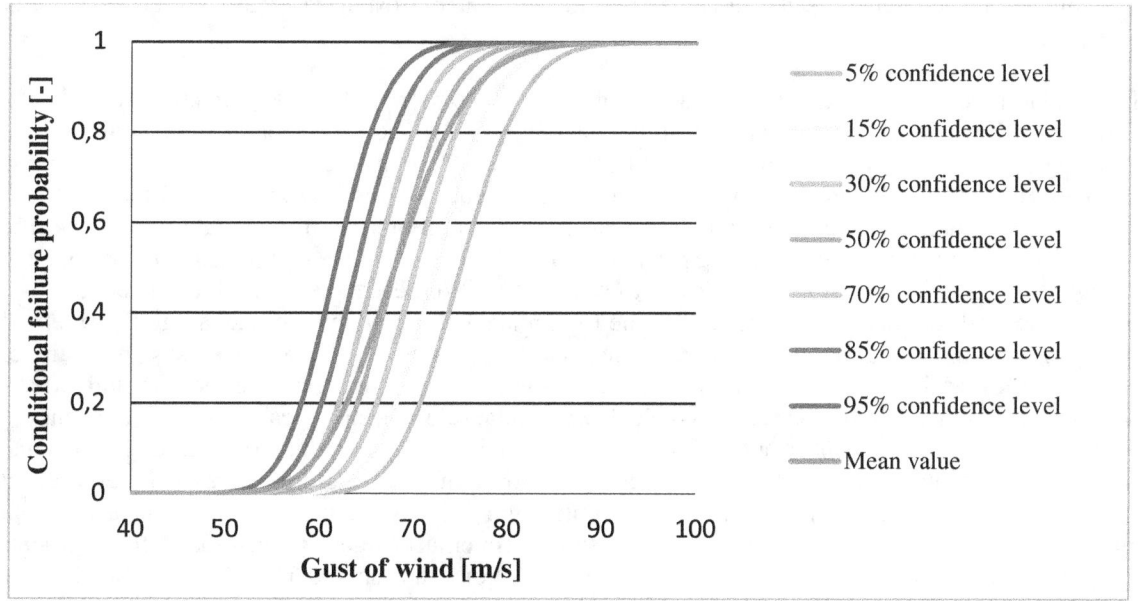

Primarily hydraulic load assessment for the canalization system helped to evaluate how external flooding caused by extreme precipitation would impact the operability of safety related SSCs. The plant response evaluation of lightning strikes required a different methodology than the analysis of other meteorological events, because lightning could cause various failure modes depending on lightning properties that cannot be characterised by a single parameter. Accordingly, lightning related fragility was described by examining the fulfilment of the design requirements prescribed in the applicable lightning protection standards and thus by evaluating the effectiveness of the lightning protection system at the plant. Primary and secondary hazardous effects of a lightning strike were taken into consideration in this evaluation. To determine the plant response to extreme temperatures, we compared the manufacturers' data on temperature resistance of each safety related component relevant to the spent fuel pool to the expected environmental temperature at the location of the component in different plant operational states with considerations to the applicable operational strategies in such extreme conditions and to the capacity of the connected HVAC (heating, ventilation, air conditioning) systems.

The plant response analysis proved to be the most challenging task in the PSA for external events mainly due to the lack of supporting analyses as well as data on component capacity that could be usefully and sufficiently applied in fragility assessment for PSA. Therefore, high priority was given to assemble an expert panel that could support the PSA with knowledge and experience about plant design, operation and safety analyses in relation to external hazards. Staff members of the plant had the most important role in that expert panel.

3.4. PSA Model Development

Based on the findings of hazard assessment and plant response analysis, the risk of fuel damage in the spent fuel pool induced by extreme precipitation and lightning was found to be insignificant. However, some follow-on analyses were proposed and safety enhancement measures were conceptualised to fully underpin this conclusion. Due to lack of appropriate data and supporting analysis on the capacity of spent fuel pool systems and components no PSA model has been developed yet for extreme temperatures. At present efforts are being made to enable risk quantification in relation to extreme temperatures. Consequently, PSA models have been developed for extreme wind, snow and frost hazards at this stage of the analysis. The RiskSpectrum PSA Professional software was applied for modelling purposes, utilizing to the extent possible the spent fuel pool PSA model for internal events and the reactor PSA models for external hazards developed earlier for full power as well as low power and shutdown states. Models developed for wind, snow and frost hazards are discussed in brief hereafter.

The initiating event of each PSA model is the relevant external hazard (wind, snow or frost) characterized by hazard curves (as demonstrated in Section 3.2). The loads from a wind, snow or frost initiating event might cause damage to structures or outdoor facilities identified during plant response analysis. Hazard induced damage and failure forms were put into fragility groups. All the structures and equipment that were found virtually identical from the point of view of vulnerability to a specific hazard were grouped together, assuming fully correlated failures of all the components in a group, and a single set of fragility curves was assigned to each group. We determined six wind related and also six snow related fragility groups, as well as one frost related group which are relevant to the safety of the spent fuel pool. Hazard induced transient initiating failures and additional system, train or component level failures and degradations were identified by a thorough examination of failure effects within each fragility group. The impact of block wall collapse on electrical cables was also taken into consideration during the identification of hazard induced failures. During this examination failures that could be caused by the simultaneous occurrences of different group failures were also identified. It was found that the plant responses to and the mitigation process for the identified single transient initiating failures were virtually the same for random (internal) initiating events and for transients induced by external hazards. The scope of safety functions that should be fulfilled following the occurrence of multiple transient initiating failures is assumed to be a union of the safety functions modeled for single transient initiating failures, taking into account the external hazard induced failures of the mitigation systems.

A so-called generic event tree was built up for each hazard in every plant operational state to identify hazard-induced fuel damage sequences. This event tree models both single and multiple hazard-induced transients together with the associated consequences on plant and human responses. On one hand each potential hazard induced transient is represented by a single dedicated event tree header in the generic event tree, on the other hand one separate header demonstrates the mitigation function of establishing plant operation in island-mode in case of loss of off-site power. Moreover the last header in the tree corresponds to the heat removal function of the spent fuel pool. A simple reading of the event tree is that upper branches represent (as usual) the success of the given event tree header (e.g. the associated transient initiating failure does not occur), while lower branches represent failure of the given event tree header (e.g. occurrence of the given transient initiating failure). By setting the appropriate boundary condition sets on each event sequence, the last header represents all the mitigation functions and systems for the transients modelled in the corresponding event sequence.

Some failure modes considered in the spent fuel pool PSA for internal events can be induced by an external hazard, too. As a first modelling step the failure modes that were found sensitive to the effects of external hazards were listed. Thus a failure mode included in this list can occur as a consequence of an external hazard or due to random, non-hazard related effects. For these failure modes the basic events of the PSA model for internal events were transferred into an OR gate that defined the connection logic between the two types of failure causes (i.e. hazard and non-hazard related ones).

Pre-initiator (type A) human actions considered in the spent fuel pool PSA for internal events are included in the external event PSA without any modification because these actions are independent of the nature of the initiator. Initiator (type B) human actions that contribute to the development of a spent fuel pool transient are generally not considered in the external event PSA where the external hazard is the only (common cause) initiator, although the occurrence of spent fuel pool transients initiated by snow load can be prevented if snow is removed from some designated areas in a timely manner. To model this effect failure to remove snow from the roofs of some technological buildings and other facilities in time was taken into account as a contributor to the development of snow related transients. Most post-initiator (type C) actions considered in the spent fuel pool PSA for internal events are identically included in the external event PSA. However, in the external event PSA no credit is given to a type C action, if major structural or equipment failures incapacitate the personnel to successfully interact either in the control room or by means of local actions.

During data assessment for PSA quantification the hazard potential was characterised by a family of continuous hazard curves, while hazard-induced equipment and structural failures were described by continuous fragility curves within the hazard levels of interest. This approach was preferred to defining discrete hazard ranges. The reliability data for random equipment failures were taken from the PSA for internal events.

3.5. Risk Quantification and Interpretation of Results

As stated above, for risk quantification purposes we used a family of continuous hazard and fragility curves, rather than using discrete values for different hazard magnitude ranges. The occurrence frequency of a minimal cutset induced by a specific external hazard *(f(MCS))* was determined partly by convoluting the input hazard curves with the relevant family of fragility curves, as well as by taking into account the probability of random equipment failures using the following formula of approximation:

$$f(MCS) = FP(NEBE_1) \cdot \ldots \cdot FP(NEBE_{NE}) \cdot \sum_{i=1}^{160} (FF_i(EBE_1) \cdot \ldots \cdot FF_i(EBE_E) \cdot h_i) \quad (1)$$

where:

$NEBE_j$ denotes basic events for random failures in the minimal cutset, i.e. failures that occur independently of the external hazard (j = 1, 2, ... NE);

$FP(NEBE_j)$ is the probability of a random failure in the minimal cutset;

EBE_k is a basic event in the minimal cutset representing a failure due to an external event (k = 1, 2, ... E);

$FF_i(EBE_k)$ is the mean conditional fragility probability for external hazard range "i" of a basic event in the minimal cutset representing a failure due to an external event;

h_i is the mean occurrence frequency of the external hazard range "i".

The conditional probability of fuel damage *(CFDP(MCS))* in relation to a minimal cutset can be assessed as:

$$CFDP(MCS) = \frac{f(MCS)}{\sum_{i=1}^{160} h_i} \quad (2)$$

The frequency of fuel damage induced by an external hazard *(FDF)* is determined as follows:

$$FDF = \left(1 - \prod_{n=1}^{N_{MCS}} (1 - CFDP(MCS_n))\right) \cdot \sum_{i=1}^{160} h_i \qquad (3)$$

The dominant fuel damage minimal cutsets of failures induced by external hazards were determined in the first place by using the RiskSpectrum PSA Professional software applied generally to model development and quantification in the Paks PSA. Since RiskSpectrum cannot be used to perform the numerical approximation of the convolution integral, following the generation of minimal cutsets, separate, stand-alone computer codes were applied to determine cutset frequencies, calculate the overall fuel damage frequency, and perform uncertainty and sensitivity analyses.

For risk characterisation, the point estimate of fuel damage frequency and the annual fuel damage probability were determined for the different external hazards in each plant operational state. By summing up the fuel damage probabilities for the various plant operational states, we calculated the cumulative spent fuel pool risk (annual fuel damage probability) induced by the different external hazards. We used qualitative analysis to identify and explain the minimal cutsets that were found dominant contributors to the cumulative spent fuel pool risk.

Importance and sensitivity analyses were used to calculate the following measures for each fragility group in relation to the cumulative spent fuel pool risk:
- Fussel-Vesely importance (fractional contribution - FC);
- Risk reduction worth (risk decrease factor - RDF);
- Sensitivity measures (S_U, S_L, $S_{U/L}$).

Sensitivity measures for each fragility group were determined by assuming a higher and a lower value of HCLPF[*] for the group. These higher and lower values were selected so that they represented one order of magnitude change in the hazard occurrence frequency. Moreover, we assessed the expected decrease in the cumulative annual fuel damage probability if the HCLPF of those fragility groups that have lower resistance than the design basis of the plant was increased up to the design basis value. The results of these analyses enabled the characterisation of expected risk reduction if certain safety improvements were made.

The complete set of the hazard curves for an external event and the full range of fragility distributions for each structure and component representing different confidence levels were combined through a convolution integral to develop true uncertainty distributions for external hazard induced failure frequencies. Also, uncertainties in hazard induced failures were combined with uncertainties in human error rates and non-hazard related equipment failures using Monte Carlo simulation. As a result the probability density function and the cumulative probability distribution function of the fuel damage frequency were obtained. Quantification was done by using a spreadsheet application developed earlier in support of the seismic PSA.

4. FINDINGS

The development of external events PSA for a selected spent fuel pool of the Paks NPP was completed by the end of 2013. Hereby we summarize the quantified fuel damage risk induced by natural external hazards and the identified main risk contributors. In addition, we highlight some of the most important analysis areas that need to be further dealt with in order to develop a full scope external event PSA for the spent fuel pool, as well as to reduce uncertainties and conservatism where necessary.

---

[*] High Confidence on Low Probability of Failure

4.1. Fuel Damage Risk

A detailed logic model was developed for extreme wind, snow and frost hazards and therefore fuel damage risk was only quantified for these hazards due to the following reasons:
- Risk induced by extreme rainfall and lightning was found insignificant on the basis of design characteristics and corrective actions that the plant management has already made commitment to in order to enhance safety.
- The assessment for extremely high air temperature was limited to an initial and rough estimation of the conditional fuel damage probability if loss of off-site power was assumed in hot weather conditions. Among others, this limitation is attributable mostly to the uncertainties in the operational strategy to be followed under harsh weather conditions and to the uncertainty in assessing the impact of high temperature on the off-site power system.
- Currently no solid assessment of fuel damage risk due to extremely low air temperature could be made. This is in the first place due to the uncertainties in the operational strategy to be followed under harsh weather conditions and uncertainties in hazard assessment. Moreover there is a need for performing further analyses to enable an appropriate quantification of the temperature related fragility of some systems and components.

Based on the results of PSA model quantification, the point estimate approximation of the annual fuel damage probability for the spent fuel pool induced by external hazards is:
- $2,04 \cdot 10^{-5}$ from extreme wind;
- $7,63 \cdot 10^{-6}$ from extreme snow;
- $6,69 \cdot 10^{-6}$ from extreme frost.

These figures include the contributions of all the plant operational states analysed. The results show that the risk from extreme weather phenomena is important in comparison to the risk originated from other types of initiating events analysed in the spent fuel pool PSA for the Paks plant.

Some results of the uncertainty analysis are indicated in Table 1 below. The figures witness large uncertainties in the risk estimates.

Table 1: Uncertainties in annual fuel damage probability estimates for different external hazards

	5 %	Median	95 %
extreme wind:	$1,46 \cdot 10^{-7}$	$5,87 \cdot 10^{-6}$	$2,36 \cdot 10^{-4}$
extreme snow:	$6,41 \cdot 10^{-8}$	$1,66 \cdot 10^{-6}$	$4,27 \cdot 10^{-5}$
extreme frost:	$9,24 \cdot 10^{-7}$	$4,48 \cdot 10^{-6}$	$2,17 \cdot 10^{-5}$

The main contributors to fuel damage risk from extreme wind were found to be the structural failure of the longitudinal electrical gallery (part of the main building complex), failure in the power lines of the off-site power system and the human failure event to establish plant operation in island-mode. Regarding extreme snow the main risk contributors are failure to remove snow from the roofs of safety related buildings, structural failure of the reactor hall and the turbine building and structural failure of the on-site substation control building located at the switchyard. The ultimate contributor to frost induced risk is the failure of power lines in the off-site power system and in the switchyard.

4.2. Unresolved issues

We proposed numerous follow-on efforts and corrective actions based on lessons learned from the different PSA analysis steps, as well as on the results of risk quantification and the associated sensitivity studies. These proposals can be grouped into the following major categories:
- Those that can reassure the adequacy of the technical basis to screen out hazards considered negligible from risk point of view (e.g. tornado, blockage of water intake system, extreme rainfall, lightning);

- Those that can enable risk assessment for hazards not characterized quantitatively yet (e.g. extreme air temperatures, hazards currently considered insignificant);
- Those that can, by means of reducing uncertainties, establishing a better technical basis of the applied analytical assumptions, or decreasing unnecessarily high conservatism, enable a more accurate assessment of risk from hazards already quantified (extreme wind, snow, or frost).

Some of the proposals belong to more than one of the above-mentioned categories. Based on the results of the current study, competent members of the plant management have defined their position as follows:
1. Safety enhancement measures already in preparation and follow-on analyses in order to ensure a refined and more complete risk assessment have to be completed first.
2. If the refined assessment shows an unacceptable level of fuel damage risk, then, among other risk reduction measures, it might be necessary to set-up a detailed operational and transient mitigation strategy to follow in case of extreme meteorological conditions, similarly to the seismic safety concept elaborated earlier at the plant.

On the basis of the current analysis, it has already been pointed out that the detailed strategy referred to in item 2 above could significantly lower the risk from external hazards and the probability of human errors in severe weather conditions.

The most important area of follow-on analyses regarding extreme wind is the need to review the available structural analyses of the plant more thoroughly in order to better assess structural fragilities and subsequently reduce assumed conservatism in risk assessment to the extent possible. Moreover the reliability of establishing plant operation in island-mode in case of loss of off-site power could be enhanced since the failure of the power grid proved to be a significant risk contributor due to its less stringent design criteria.

With respect to extreme snow, the potential snow induced blockage of air intake systems to the diesel generators needs to be further studied. Also, modification of the relevant plant procedure on removal of snow deposits from building roofs has been proposed together with identification and allocation of human and equipment resources to enhance the effectiveness of actions aiming at the prevention of transient initiating failures and thus to lower fuel damage risk. Furthermore a more detailed review of the available structural analyses of the plant has also been proposed in order to better assess structural fragilities and subsequently reduce assumed conservatism in risk assessment to the extent possible. Regarding extreme frost, complementary assessments are needed to decrease conservatism by assessing the safety margin of relevant components and power lines at the switchyard beyond the design basis.

To fully justify that spent fuel pool risk imposed by extreme rainfall and lightning is insignificant some unresolved issues need to be clarified. A reassessment of the response of the canalisation system to hydraulic loads is needed with modified boundary conditions in comparison to the existing analyses. It may become necessary to establish controlled flooding of the diesel generator building as a result of this reassessment. Although controlled flooding cannot prevent the rooms inside the building from flooding, it can ensure the functionality of all safety related components if a few components are installed at higher elevation. In addition, it is seen necessary to examine whether extreme rainfall could lead to the damage of safety related components due to flooding through underground structures (e.g. cable tunnels).

Concerning the risk from lightning, protection of safety related components against lightning is currently subject to a review at the whole plant with focus on the secondary effects of lightning in particular. Protection of spent fuel pool components will be improved where necessary.

The risk assessment for extremely high and low air temperatures proved to be the most challenging task. Therefore it requires the most significant follow-on efforts. Detailed analysis is needed to evaluate the effectiveness and reliability of the plant HVAC systems during harsh weather conditions. Temperature limits for the safe operation of all spent fuel pool components with considerations to the

9

actuation of temperature related protection need to be determined in order to assess the sequence of equipment trips during harsh temperature conditions. Temperature resistance of electrical, control and instrumentation components located outside of the plant buildings should be assessed in detail to determine the safety margin beyond design basis and to underpin fragility analysis. The vulnerability of mechanical components to extreme temperatures needs to be reviewed. Fragility assessment regarding extreme temperatures needs to be conducted for the off-site power system to quantify fuel damage risk in an appropriate manner. It should be analysed whether the safe stable conditions of the spent fuel pool can be ensured by using power supply from the emergency diesel generators in lack of off-site power during extremely high and low air temperature conditions.

4. CONCLUSION

Development of external events PSA for a selected spent fuel pool of the Paks NPP was completed by the end of 2013. The analysis followed the commonly known steps: selection and screening of external hazards, hazard assessment for screened-in external events, analysis of plant response and fragility, PSA model development, and risk quantification and interpretation of results. The risk of fuel damage induced by natural external hazards was quantified to the extent seen feasible. In addition to risk quantification, unresolved issues and necessary follow-on analyses were identified and proposed.

Fuel damage risk has been assessed quantitatively for wind, snow and frost hazards. Detailed importance, sensitivity and uncertainty analyses were conducted. Moreover the main risk contributors induced by these external events were also identified. Additional follow-on analyses were proposed to enable an improved risk quantification by means of reducing uncertainties, establishing a better technical basis for the applied analytical assumptions, or decreasing unnecessarily high conservatism.

Based on the findings of hazard assessment and plant response analysis, the fuel damage risk induced by extreme rainfall and lightning was found to be insignificant. However, some follow-on analyses were proposed and safety enhancement measures were conceptualised to fully underpin this conclusion. Due to lack of appropriate data and supporting analysis on the capacity of spent fuel pool systems and components no PSA model has been developed yet for extreme temperatures. Follow-on analyses necessary for quantifying the risk of fuel damage induced by extreme temperatures have been identified.

A plan of follow-on actions has been set up based on the analysis findings. Follow-on analyses have been started in accordance with this action plan.

References

[1] *Hungarian Nuclear Safety Code; Volume 3, Design Requirements for Nuclear Power Plants.* Hungarian Atomic Energy Authority, Budapest, Hungary, 2011.
[2] *WENRA Reactor Safety Reference Levels* (www.wenra.org), January 2008.
[3] *American National Standard for Level 1 / Large Early Release Frequency Probabilistic Risk Assessment for Nuclear Power Plant Applications*, ANSI/ASME/ANS RA-S-2008 (revision of ANSI/ASME/ANS RA-S-2002), ASME, New York, USA, 2008.
[4] *American National Standard for External Event PRA Methodology* ANSI/ANS-58.21-2007 (revision of ANS-58.21-2003), ANS La Grange Park, USA, 2007.
[5] M. Knochenhauer and P. Louko: *Guidance for External Events Analysis*, SKI Research Report 02:27, Swedish Nuclear Inspectorate, Stockholm, Sweden, 2003.
[6] *Development and Application of Level 1 Probabilistic Safety Assessment for Nuclear Power Plants*, Specific Safety Guide No. SSG-3, IAEA, Vienna, Austria, 2010.

Ramifications of Modeling Impact On Regulatory Decision-making - A Practical Example

Ching Guey
Tennessee Valley Authority, Chattanooga, TN, U.S.A.

Abstract: PRA models have been used for nuclear power plants in several areas including Maintenance Rule (MR) a(4), Reactor Oversight Process (ROP) and Mitigating System Performance Index (MSPI). As a part of the living PRA program, the PRA model has been updated to reflect operating experience, feedback from applications, and more recent PRA data and methodology.

PRA modeling detail which can affect the regulatory decision-making of Emergency Diesel Generator (EDG) MSPI (highlighting key PRA assumptions and design basis requirement under multi-unit accidents) and ROP process are discussed. Risk insights on the key assumptions in both deterministic and PRA modeling which may affect the MSPI program and ROP process are presented.

Areas of improvement to manage more effectively the living PRA program for regulatory decision-making as a result of the lessons learned from a practical example with several regulatory ramifications are summarized. These include the need of ralistic modeling of design features of interest, realistic success criteria for multi-unit accident scenarios and dependency treatment of human reliability analysis (HRA).

Keywords: PRA, Multi-unit accidents, MSPI, Parallel Function Of EDG, HRA.

1. INTRODUCTION

The plant analyzed in this paper consists of three units. Eight diesel generators, (four for Units 1 and 2, and four for Unit 3) are provided as a standby power supply to be used on loss of the Normal Auxiliary Power System. Each of the diesel generators is assigned primarily to one 4.16-kV shutdown board. It is possible, through manual action of breaker ties to the shutdown buses, to make any diesel generator available to any 4.16-kV shutdown board. Four (4) diesel generators are in standby and aligned to automatically start when degraded voltage or under-voltage is sensed on the associated Unit 1 and 2 4-kV shutdown boards. Additionally, when a Loss of Coolant Accident (LOCA) occurs, the diesel generators are automatically started and remain in standby with their output circuit breakers open. When the transient is a loss of offsite power, the diesel generators will start and supply power to their associated 4-kV shutdown boards. Loads on the 4-kV shutdown boards will be shed and sequenced back on the boards as necessary by sequencing relays.

A test of the paralleling feature of the EDGs was performed. During the test, the Unit 1&2 A-train EDG was running and providing power to the safety-related board. The Unit 3A EDG was started and was placed in parallel operation with the running EDG. If the system had performed as designed, the 3A EDG would have picked up load and the 1&2 A EDG would have relieved load until they reached equilibrium and would have continued to run in parallel supplying the safety related load(s). During the test, the 3A EDG continued to pick up load and the 1&2A EDG continued to shed load until the 1&2A EDG tripped on a "reverse power" signal and power was lost to the associated safety-related board. A half- scram was created for Unit 1 due to the loss of a Reactor Protection System (RPS) power. An operator error (instead of restoring power to the de-energized RPS, the operator disabled the energized RPS train) caused a loss of both RPS trains, which resulted in a full scram and Main Steam Isolation Valve (MSIV) closure. No other mitigation systems failed. The MSIV was opened approximately 3 hours after the reactor scram. The test and subsequent plant response had two regulatory ramifications: MSPI of EDG failure and safety significance of the reactor scram with MSIV closure.

The 1&2A EDG tripped approximately 1 hour and 9 minutes after the closure of its output breaker (connecting it to the safety-related board) but less than an hour after the paralleling switch was manipulated to bring the 3A EDG into parallel operation. The root cause of the test failure indicated that the cause was faulty wiring in a transfer switch. The transfer switch in question is not within the boundary of the EDG, presumably because generic scope of EDG does not include this unique capability.

2. MSPI RAMIFICATIONS OF PARALLEL FUNCTION OF EDGs FOR MUTI-UNIT ACCIDENTS

The current safety analysis (current Licensing Basis) credits 2 residual heat removal (RHR) pumps for post design-basis RCS heat removal. One EDG cannot provide sufficient power to drive 2 RHR pumps, so two EDGs must run in parallel mode of operation to provide sufficient power.

Four (4) additional diesel generators power the Unit 3 4kV shutdown boards. Hardware capability exists to cross-tie the Unit 3 4kV shutdown boards to the Unit 1/2 shutdown boards.

However, the PRA model only credits this capability when Unit 1/2 diesels have failed independently or by common cause that is not applicable to Unit 3. The cross-tie capability is not credited for multi-unit initiators.

During a design basis loss of coolant accident (LOCA) on Unit 1 (2) concurrent with a LOOP on all three units, given a single active failure affecting a Unit 1/2 EDG or 4160V shutdown board, there could be insufficient power for two RHR pumps to provide suppression pool cooling on Unit 2 (1). Suppression pool cooling on the non-LOCA units will be required to maintain suppression pool temperature within acceptable limits during RCIC and/or HPCI operation during a LOOP.

The function monitored for the EDG system for MSPI is the ability to provide AC power to the class 1E boards following a loss of offsite power, per NEI 99-02 Appendix F (Reference 1). This function is listed in the Table 1:

Table 1 EDG MSPI Function Matrix		
Scope	Risk Significant Function	System/Components
When the transient is a loss of offsite power, the diesel generators will start and supply power to their associated 4-kV shutdown boards.	Yes	Diesel Generators A, B, C and D

It is noted that the risk significant function of the EDGs based on the PRA model does not include the paralleling capability of the EDGs. However, the MSPI basis document states that the PRA success criteria are the same as the design basis success criteria, which requires two EDGs in parallel. Since the parallel feature requires the switch, failure of the switch would be included in the boundary of the EDG and count as a failure. However, upon further scrutiny, the PRA model implicitly assumes that the condition requiring the paralleling of the EDGs is highly unlikely and is not included in the PRA model. Under the existing PRA model, the switch is not required for the PRA function and would therefore be outside the scope of the EDG boundary and would NOT count as a failure.

The following questions were raised with respect to the unique feature for paralleling two EDGs for accidents involving one unit in LOCA while there is a loss of offsite power to all three units:

- Is Emergency Diesel Generator (EDG) paralleling modeled or taken credit for in the PRA?

The PRA model used for MSPI basis document does not include the component (key control switch for the parallel function) explicitly. In order to "estimate" the importance of this parallel function, the surrogate "EDG failure to start" or "EDG failure to run" is used, which overstates the importance of the unique design basis feature of EDG.

- Does the PRA require one or two RHR heat exchangers for successful decay heat removal?

The PRA model has a different success criterion for different accident conditions. For general transient scenarios, which include loss of offsite power initiators, 1 RHR pump and 1 RHR heat exchanger is required for successful decay heat removal. For ATWS and IORV (inadvertently open relief valve) scenarios, 2 RHR pumps and 2 RHR heat exchangers are required for success.

- What is the effect of including the failure to parallel the EDGs in the PRA model?

There are several ways to model the failure to parallel EDGs, varying from the explicit inclusion of the control circuits and switches to a higher level for function or a more coarse approximation by a "surrogate" basic event in the model. A more realistic modeling included more detailed scenario and timing, but require significant resource. A balanced approach was used to provide risk insights without unnecessary burden of creating excessively large number of cutsets involving subcomponents.

3. PRA LEVEL OF DETAIL FOR REGULATORY DECISION-MAKING

The following technical considerations were included to ensure a comprehensive coverage of factors which may facilitate regulatory decision-making. These were developed based on the review of current PRA model and ongoing effort including NFPA 805 and Fukushima NTTF responses.

3.1. Spatial Aspect of DG Parallel Feature

For internal events, the impact of the DG parallel feature is relevant only when two RHR pumps are required, mostly for LOCAs in which RHR service water cooling is required. The likelihood of having a loss of offsite power (as an initiator or consequential) in conjunction with failure of alternate decay heat removal is very small. For fire, the impact is heavily dependent on the data and assumptions of fire ignition frequency, the cable failure mode and its likelihood and the relative location and status of the alternative mitigation function; however, scoping estimate was performed and indicates that it is not significant if the same proposed modifications to improve the margin for fire-induced station blackout events are credited.

3.2. Temporal Aspect of EDG Parallel Feature

The paralleling feature of the EDGs is predominantly required for late (~ one hour or later after the accident) heat removal. For loss of offsite power only and small LOCAs caused by SRV cycling, the success criterion for the RHR system is less stringent both in capacity and timing than that for the design basis LOCA conditions. In addition, several alternate heat removal mechanisms are available (e.g., hardened wet well venting, continuous condensate injection, and drywell spray etc). The parallel function is not automatic and can be initiated at the discretion of the operator based on the plant conditions at the time. By properly including the applicable scenarios, the risk impact of including the failure of the parallel feature of EDGs was estimated to be negligible.

3.3. Risk Insights

The design basis function of the EDG parallel capability is placed in a holistic perspective with the detailed accident sequence scenario analysis by considering both the timing and relevant success criteria with additional procedural enhancement. Original design basis requirement does not consider the temporal aspects of the conditions and the likelihood of all mitigating factors such as the success criteria for the cooling capability (worst case scenarios require more RHR service water pumps for conservative design purposes), the operator action, and alternate decay heat removal capability. The PRA model provides a more realistic characterization of the importance of the parallel function. The MSPI importance of the function is explicitly estimated by the more detailed PRA model. The importance of the function is at least an order-of-magnitude lower than that of the EDG failure to start or failure to run due to the much lower likelihood of the need of the EDG parallel function.

3.4. Lessons Learned

Several areas of improvement for the PRA model to support regulatory decision-making are identified through the in-depth consideration of the design basis features for the EDGs. The likelihood of the hypothetical multi-unit event scenarios of interest is in general low. Improvement of the PRA modeling including scenario-based parallel function and associated success criteria and improved procedures yields a lower safety significance of the unique EDG parallel feature.

Another lesson learned is that there is a continuing interest in the PRA community for a more critical review of the current Risk Oversight Program dealing with a performance deficiency involving a reactor scram. The testing of the parallel feature of EDGs discussed in this paper involved an unexpected reactor scram due to a loose connection of a switch controlling the paralleling (which caused a half-scram) and an operator error (which created a full scram). Significant effort was involved in the Significant Determination Process associated with the reactor scram involving MSIV closure. Reference 2 provides an interim guidance for risk significance determination for events involving a reactor scram. The guidance does provide credit for all mitigating systems if available, However, PRA modeling with respect to HRA includes treatment of minimum failure probability of a single human action to be 1E-5 and 1E-6 for multiple human actions. Modeling of dependence of human actions is especially critical for accident scenarios in which many mitigation systems are available but require operator actions for which the level of the dependence and floor values of failure probability can vary by orders of magnitude due to different judgment assigned to performance shaping factors.

4. CONCLUSION

An integrated approach by realistically considering time-dependent accident scenarios (including success criterion), plant mitigating systems and procedural enhancement resulted in a more realistic risk significance of the EDG parallel function, which was deterministically based with significant conservatism. Additional details in the area of success criteria and operator actions which can be easily incorporated into the PRA model have provided insights which reduce the perceived risk significance of an unlikely demand for a required design basis function, while providing a defense-in-depth to enhance safety. The synergistic interaction of design engineers with PRA analysts indicates that PRA models can continue to provide risk insights by including explicitly components which may otherwise considered negligible until such circumstances when they become a focused issue with regulatory ramifications. Due to a high degree of redundancy and diversity of the mitigation systems for certain benign events, the SDP can be dominated by uncertainties in the treatment of HRA (both minimum value of a single human action and minimum dependency for multiple human actions).

References

[1] NEI 99-02, Revision 7, Regulatory Assessment Performance Indicator Guideline
[2] Interim Staff Guidance to Supplement RASP, (Revision 2.0), Volume 1, Section 8

A Fresh Look at Barriers from Alternative Perspectives on Risk

Xue Yang*, Stein Haugen

Norwegian University of Science and Technology, Trondheim, Norway

Abstract: This paper takes a fresh look at alternative perspectives on major accident causation theories to highlight the fact that these perspectives can supplement and improve the energy barrier perspective. The paper starts from a literature study of energy barrier perspective, Man-Made Disaster theory (MMD), Conflicting Objective Perspective (COP), Normal Accident Theory (NAT), High Reliability Organization theory (HRO), Resilience Engineering (RE), and System-Theoretic Accident Model and Processes (STAMP) model to find out main concepts and identify critical factors. A further study of safety barrier perspective is carried out using STAMP methodology to understand how barrier functions can fail. It was found that alternative perspectives can supplement the barrier perspective by structurally analyzing possible failure causes for barrier function (STAMP, MMD), looking for driving forces for unsafe decisions and unsafe actions when human interacts or be part of barrier systems (COP, HRO), and emphasizing possible complex interactions and tight coupling within barrier functions (NAT, RE). Furthermore, suggestions to barrier management based on best practices from these perspectives are presented, which will be developed into concrete risk reduction measures, such as checklists, audits schemes, or indicators to help decision-makers better comprehend and maintain the performance of barrier functions in further work.

Keywords: Accident causation, Barriers, STAMP, HRO, Resilience Engineering, Safety management

1. INTRODUCTION

Over the past decades, safety gains increasing interests among industries to prevent loss of lives, economical loss and adverse consequences to the environment due to major accidents. Parts of the reasons are high-visibility accidents that resulted in tragedies and significant environmental damage all over the world, such as Three Mile Island accident, Ocean Ranger sinking, Chernobyl disaster, Piper Alpha disaster, Texas city refinery explosion, Deepwater horizon oil spill, etc.

Motivated by the desire to understand deeply what causes accident and how to prevent major accidents, various accident causation theories have been developed. Each accident theory has its own characteristics based on causal factors it highlights [1]. The energy-barrier perspective [2, 3] emphases on energy flow control and mitigation of consequences caused by release of energy based on a defense-in-depth principle. Man-Made disasters theory [4-6] highlights lack of information flow and misperception among individuals and groups during an incubation period. Conflicting objectives perspective [7] looks into driving forces for unsafe decisions that push systems towards safety boundary. Normal Accident Theory [8] is a rather pessimistic perspective stating that major accidents are inevitable in complex systems due to *"interactive complexity"* and *"tight coupling"*. System-Theoretic Accident Model and Processes (STAMP) [9] perceives accident causation from a systemic viewpoint, indicating that accidents arise from inadequately enforced safety constraints, flawed control process and inconsistent, incomplete or incorrect process model. High Reliability Organization (HRO) [10, 11] and Resilience Engineering (RE) [12-14] perspectives, which aim at building up a robust organization, focus on a series of properties of organization that can contribute to avoid major accidents. Strictly speaking, HRO and RE are not accident causation models. However, due to their important implications to accident prevention, they are also covered in this paper.

Among these seemingly competing perspectives, the energy-barrier perspective is the most popularly applied accident causation theory in Norwegian Oil and Gas industry. This is mainly due to huge

---

* Tel.: +47 73 59 71 05; fax: +47 73 59 28 96.
E-mail address: xue.yang@ntnu.no

amount of energy that is handled in the industry and disastrous consequences of major accidents to human lives, environment, and economical losses. Subscription of one perspective doesn't mean denying others. Rosness, Grøtan [15] compared these perspectives (except STAMP) and concluded that they are complementary, rather than contradictory to each other. After all, most of these perspectives are conceptualizations of common characteristics of past accidents that have significant implications for future major accident prevention. The purpose of this paper is to take a fresh look at these perspectives, to highlight the fact that these alternative perspectives can supplement and improve the energy-barrier perspective from different angles. The following research questions are discussed in detail in the rest of the paper.

1. What are the main concepts and principles of these perspectives?
2. How can they contribute to supplement and improve the energy-barrier perspective?
3. What are the implications for safety management?

This paper is mainly based on a study of above alternative perspectives of major accident causation theories, with a special focus on barrier perspective. The remainder of this paper is organized as follows. In section 2, main principles of alternative perspectives are summarized with a focus on critical causal factors that they emphasize. In section 3, differentiation between *barrier function* and *barrier system*, analysis of possible flaws in *barrier function* are carried out as necessary steps before utilizing essences from alternative perspectives to improve the barrier perspective. In section 4, implications to safety management are discussed and section 5 concludes the work.

2. MAIN PRINCIPLES OF DIFFERENT PERSPECTIVES

2.1. Energy-Barrier Perspective

Energy-barrier perspective is widely applied in Norwegian oil and gas industry. Barrier perspective origins from energy model that was introduced by Gibson [3] and further popularized by Haddon [2] with ten strategies for accident prevention. The basic idea is that accidents occur when control of dangerous energy is lost and there are no effective barriers between the energy source and vulnerable assets. [2]. This is the classical interpretation of barrier. The hazard control strategies are commonly referred as defense-in-depth principle. This was further developed by the "Swiss cheese model" which shows how an accident emerges due to holes in multiple barriers [16] caused by *active failures* and *latent conditions*. The concept of "barrier" is further extended into process model, as a means to prevent transitions between accident developing phases. The extended barriers are not only related to energy anymore, but also radically interpreted as "a physical and/or nonphysical means planned to prevent, control or mitigate undesired events or accidents [17]". Energy-barrier perspective is further discussed in Section 3.

2.2. Man-Made Disaster Theory (MMD)

Man-Made disaster theory suggested that disasters can be systematically analyzed rather than thinking them as "acts of god" or chance events that have nothing in common [5]. The theory shifted the focus from engineering calculation of reliability to soft factors that lead to failures. Turner's essential conclusions based upon a systematic qualitative analysis of 84 British accident inquiry reports over ten years were:

- Accidents or disasters develop through a long chain which is called the *incubation period*. The *incubation period* is characterized by the "accumulation of an unnoticed set of events which are at odds with the accepted beliefs about hazards and the norms for their avoidance [5]".
- Accidents arise from an interaction between human and organizational arrangements of the socio-technical systems set up to manage complex risk problems
- The build-up of latent errors and unnoticed events is accompanied by a collective failure of organizational cognition and intelligence

Dekker [18] pointed out that this *incubation period* is the most fascinating time where "drifts" are happening and accumulated that end up with a surprise of failure. Why a set of drifts or accumulated drifts are not noticed is because of rigidities of belief, misperception of danger signals, or simply events are unnoticed or are misunderstood. The root causes of the *incubation period* are generalized as *lack of information flow* and *misperception among individuals and groups*.

2.3. Conflicting Objective Perspective (COP)

The conflicting objectives perspective (COP) explains the driving forces behind "bad" safety related decisions by pointing out that accidents are caused by a systematic migration of organizational behavior under the influence of pressure toward cost-effectiveness in an aggressive, competitive environment [7]. The danger is that safety may gradually be sacrificed to economic and workload pressures, consciously or unconsciously. The closeness to the acceptable risk boundary determines the degree of proneness to accident (Figure 1).

Figure 1 Boundaries of Safe Operation. Adapted from Migration Model [7]

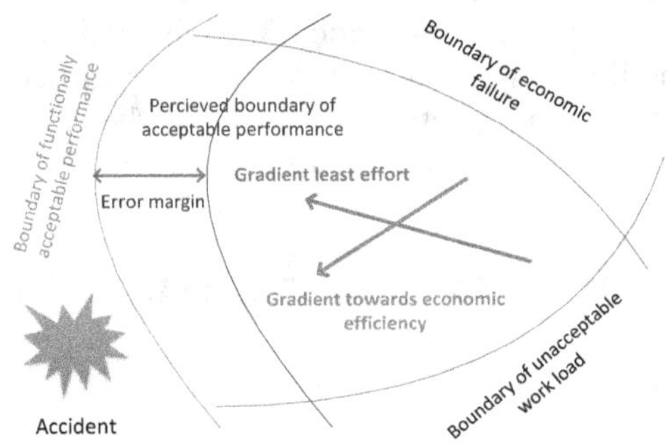

Rasmussen [7]'s migration model raises the need for identification of boundaries of safe operation to better control risk. To handle conflicting objectives, it is crucial to make boundaries visible and touchable, and to develop concrete coping skills at the boundaries. One way is to increase awareness of the boundary using instructions and motivation campaigns to create a counter gradient to the cost-effectiveness gradient to maintain the margin [7]. The biggest challenges are then: 1) how to identify where the boundaries are, and 2) how to make them visible to decision makers?

2.4. Normal Accident Theory (NAT)

The key idea suggested by NAT is that "major accidents are inevitable due to *"interactive complexity"* and *"tight coupling"* in complex systems [8]. Perrow [8] defined *complex interactions* as "those in which one component can interact with one or more other components outside of the normal production sequence, either by design or not by design". He further explained the definition of *Interactive complexity* in the preface of his new book in an more understandable way as [19]: "*Interactive complexity* is not simply many parts; it means that many of the parts can interact in ways no designer anticipated and no operator can understand. Since everything is subject to failure, the more complex the system the more opportunities for unexpected interactions of failures." *Tight coupling* means "there is no slack or buffer or give between two items" [8]. The *tight coupling* can happen to space, schedule, or resource. The tightness of coupling indicates how fast cause and effect can propagate through the system.

2.5. High Reliability Organization (HRO)

HRO research was initiated about 20 years ago and identified several characteristics that maintained the safety of the studied organizations [11, 20, 21]:
- Deference to expertise during emergencies

- Management by exception: managers monitor decisions but do not interfere unless there is a clear unplanned deviation in a course of action
- Climate of continuous training
- Several channels are used to communicate safety critical information
- In-built redundancy include back-up systems, internal cross-checks and continuous monitoring of safety critical activities

There has been much debate in HRO theories regarding whether to define and identify a HRO based on accident statistics or on the processes that it uses to successfully manage the risks [11, 20, 22, 23]. The focus of HRO research has changed to the types of processes and practices that enable certain organizations to achieve a safe state. One representative work is from Karl Weick, who conceptualized HROs as "mindful" organizations which highlights what an organization needs to do to achieve a continuous safe state [24]. *Mindfulness* is more about inquiry and interpretation grounded in capabilities for action [25]. The key messages of Weick and Sutcliffe [26] is to create a *mindful infrastructure* that continuously maintain HRO principles: preoccupation with failure; reluctance to simplify; sensitivity to operations; commitment to resilience; and deference to expertise (Figure 2).

Figure 2 HRO Principles Summarized from Weick and Sutcliffe [26]

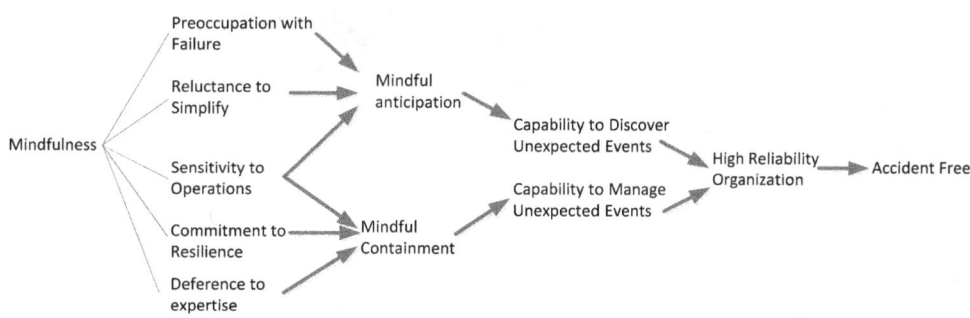

2.6. Resilience Engineering (RE)

A key concept in Resilience Engineering is that safety is not "freedom from unacceptable risk" anymore, but the "ability to succeed under varying conditions" [12]. Resilience is a family of related ideas, instead of one single thing [27]. Hollnagel first defines resilience as "the ability of a system or an organization to react and recover from disturbances at an early stage, with minimal effect on the dynamic stability" in the first volume of Resilience Engineering Perspectives series [14]. Woods [14] considers resilience as a wider capability more than adaptability. This definition emphasizes the *robustness* of the system. This means resilience is concerned with unanticipated perturbations, which arise because of the incomplete, limited or wrong competence envelope, or environmental changes so that new demands/pressures/vulnerabilities arise that undermine the effectiveness of the competence measure in play. Therefore, Woods argues that resilience engineering must monitor the boundary conditions and adjust or expand the model to accommodate changes. These boundaries are called textbook competence envelope, which is relative to unanticipated perturbations. Hollnagel provided a more elaborate working definition in the second volume to address other than adaptability [13], which was repeated in the third book [12] as "the intrinsic ability of a system to adjust its functioning prior to, during, or following changes and disturbances, so that it can sustain required operations under both expected and unexpected conditions". This definition expanded resilient reactions to changes, in addition to disturbances. Meanwhile, *robustness* under unexpected conditions is further emphasized. But somehow the "ability to recover" is diluted. This simplifies resilience into ability to dynamically steering activities, under both expected and unexpected conditions. Mattila, Hyttinen [28] defining '*managerial resilience*' based on findings from oil industry, interviews with offshore managers who had faced serious emergencies, showed that their trade-off decisions were key to maintaining the safety of installation. This is exactly what conflicting objective perspective is talking about. So Resilience, as a family of related ideas [27], seems to be a collection of at least barrier perspective, conflicting objective perspective, and HRO theories [29], with the aim of achieving safe state by *dynamically and wisely steering activities*.

2.7. System-Theoretic Accident Model and Processes (STAMP)

Leveson [30] conceives safety as a control problem and accidents arise from flawed processes; interactions among people, societal, and organizational structures, engineering activities, and physical system component. This is in line with Rasmussen's framework for risk management that addresses Structural hierarchy and System dynamics [7]. STAMP consists of three basic constructs: *safety constraints*, *hierarchical control structure*, and *process models*. Correspondingly, accidents can be studied by identifying which safety constraints were missing or violated or inadequately enforced; how inadequate control happened; and whether process model is inconsistent, incomplete or incorrect (Figure 3).

Figure 3 Possible Flaws in Control Loop that May Lead to Hazards [9]

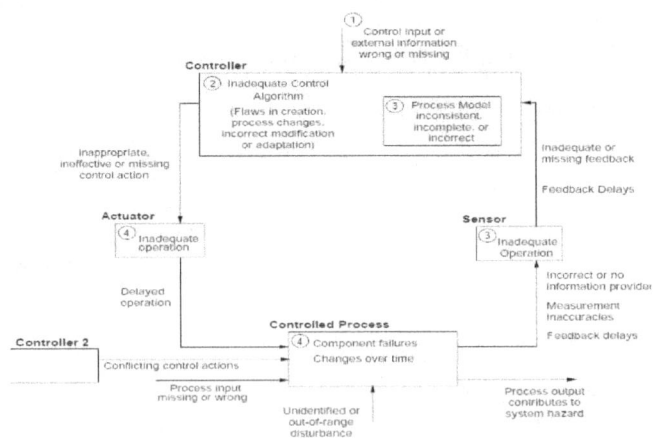

3. ENERGY-BARRIER PERSPECTIVES AND ALTERNATIVE PERSPECTIVES

The energy-barrier perspective is popularly applied in offshore oil production platforms, as a result of huge amount of energy involved. The scenarios of release of hydrocarbons and defense-in-depth barriers are modelled in sequence. The underlying assumption is that accidents happen because of absence or breach of these barriers. Subscription of one perspective does however not mean others are not applicable. The Snorre A blowout accident that had been analyzed from alternative perspectives (except STAMP), shows that each perspective tells a part of the story [15]. In order to see how other perspectives can supplement and improve barrier perspective in a systematic way, we need to see how barriers are working to prevent, control and mitigate unwanted outcomes first.

3.1. Barrier Function and Barrier System

Under energy-barrier perspective, it is useful to distinguish between *barrier function* and *barrier system*. *Barrier function* is "a function planned to prevent, control, or mitigate undesired events or accidents", while *barrier system* is "a system that has been designed and implemented to perform one or more *barrier functions*" [17]. Therefore, the *barrier function* is realized or executed by one or multiple *barrier systems*. These *barrier systems* are maintained or modified to maintain the desired *barrier function* during operation. The main focus in traditional offshore quantitative risk analysis (QRA) is on technical safety systems. However, the performance of *barrier systems* that are modelled in QRA may be far away from the real performance of *barrier function* during operation. BOP is one of the examples. OLF 070 [31], a widely followed guideline in the Norwegian oil and gas industry, requires that the Probability of Failure on Demand (PFD) of Blowout Preventer (BOP) as a *barrier system* should be between 10^{-3} and 10^{-2}. However, in practice, it was found only the 6 out of 11 cases on deep-water rigs that pushed the activation button of the BOP actually brought the well under control [32]. This means that BOP used by deep-water rigs as a *barrier function* had a "failure" rate of 45%, instead of between 0.1% and 1%.

3.2. Types of Barrier Systems and Alternative Perspectives

Barrier system may be classified according to several dimensions depending on the purpose of classification, as discussed in Sklet [17]. Our purpose is to identify different working mechanisms of *barrier systems* to see how other perspectives can influence their functions. Therefore the dimension from Hollnagel [33] that divide barrier systems into *physical (or material), functional (active or dynamic), symbolic* and *incorporeal* is selected for further discussion.

3.2.1. Physical (or material) barrier system

Physical barrier systems passively stand after installation to withstand forces up to a certain maximum beyond which it is no longer effective. Physical barrier systems are normally simple, passive systems. Some examples are fire walls, cages, explosion-proof container, and so on. The performance of these systems during operational phase is rather good and stable. Possible failures of the *barrier systems* could be: aging failures/damages, design deficiencies, installation errors, and excessive stress that beyond design limits [34]. When we dig further to find causes from alternative perspectives, STAMP provides possible explanations for design deficiencies: safety constraints are missing, violated or inadequately enforced throughout the hierarchical control structure; unmatched process model between designers and real situation. Installation failures, such as wrong location and wrong type of materials, are not uncommon in the field. This may happen due to insufficient information flow among designers and installation personnel. Aging failures, which can most possibly be avoided by scheduled proactive maintenances, maybe traced back to insufficient mindful anticipation that miss out early symptoms of degraded systems at early stage. Excessive stress, which means operational environment exceeds textbook competence envelope of the systems, is one type of *unanticipated perturbations* in RE which need more robust design to conquer. Conflicting objectives can be a reason behind aging failure, design deficiencies, and installation failure due to cost or schedule pressure. Above failure causes from alternative perspectives are summarized and structured in Figure 4.

Figure 4 Failure Causes of Physical Barrier System and Alternative Perspectives

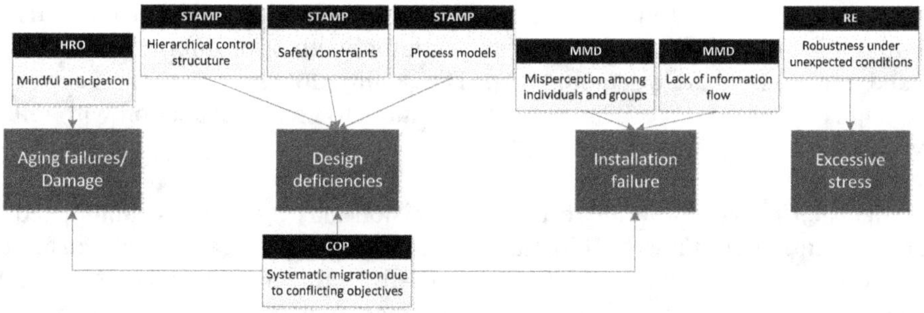

3.2.2. Functional barrier system

Functional barrier systems are active and can be activated when one or more pre-set conditions are met. Pre-designed actions will be carried out after a decision-making process. These systems vary from simple systems (e.g. interlock) to complex system (e.g. Safety Instrumented Systems), from technical systems to human operations. The three elements that are involved in the functional barrier system are sensor, decision making process and actuator (Figure 5). The performance of the systems becomes unreliable when human beings play the three roles

Figure 5 Key Elements in Functional Barrier System Adapted from [35]

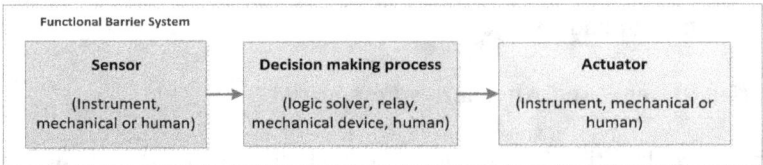

Since there is generally a process involved in the systems, the STAMP control framework (Figure 3) is adopted to identify possible flaws in the *barrier function*. We can see from Figure 6 that in addition to technical failures that may be caused by factors illustrated in Figure 4, more possible causes for failed barrier functions can be identified: information flow in between; inconsistent, incomplete, or incorrect process model (algorithm); inadequate/no operation; and unidentified out-of-range disturbance. When humans are involved in the *barrier function*, which mean humans are the ones to detect, make decision, or act upon, the failure causes become more complicated. After all, interaction failure between humans and machines has been realized as the largest contributor to the probability of system failure [36].

Figure 6 Possible Flaws in Controlled Barrier Function Executed by Functional Barrier System

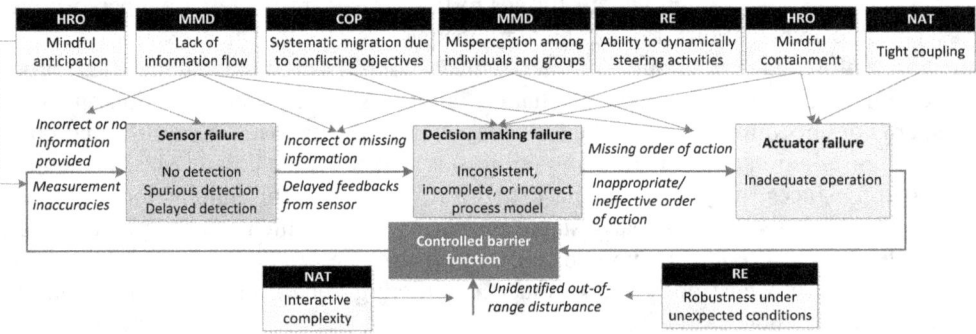

Human as sensors - Human has more flexibility than technical sensors. This has both positive and negative effects. When the pre-set condition is satisfied, there may be no output from the "sensor" due to: 1) we intentionally refuse to acknowledge that we know, is even by sub-consciousness. 2) Prior information is noted but not fully appreciated 3) Prior information is not correctly assembled 4) Relevant information is available, but when it is in conflict with prior information, rules or values, it is neglected and not taken into discussion [4]. The ability to "sense" is the essence of "*mindful anticipation*" in HRO perspective. This requires operators to be preoccupation with failure, reluctance to simplify, and sensitivity to operations.

Human as decision-makers - With respect to decision-making process, automated systems have basically static control algorithms, with periodic updates when necessary. In contrast, humans employ rather dynamic control algorithms, which can be easily influenced by other factors. Many process safety barriers need human's action like pressing Emergency Shutdown button to activate *barrier function*. Normally, it is a call about when to push the button. "Too late" activation of last defense barrier has appeared in several accident investigation reports [37]. Conflicting objectives due to high workload or cost pressure may prevent operators from making sound and timely decisions, consciously or unconsciously. The gap between operator's process model and real condition is a major cause for wrong decisions. In the Deepwater Horizon blowout accident, when Well Integrity Test was carried out to test cement casing, operators believed that as long as the pressure in the kill pipe is 0, based on U-tube effect principle, the integrity of the well can be verified. This process model is wrong. The fact is that the kill pipe was not in communication with the water-filled cavity below so that U-tube was not actually established [37]. Being aware of flaws in the process model, *dynamically* *steering activities* from RE and *mindful containment* from HRO contribute significantly to avoid failure of barrier function.

Human as actuators - Human is not a pump that can do a controlled start or stop following command. Commission failures, including violations, mistakes, slip and lapses, can happen when human act as an actuator [38]. The reasons behind might be *tight coupling* in terms of schedule, space or resource, or lack of *mindful containment*.

Unidentified out-of-range disturbance - Out-of-range disturbance is not part of barrier system or barrier function. The ideal design is supposed to cover all the possible situations the barrier system may face. We have to acknowledge it is impossible for designers to foresee everything. Handling disruptions and variations that fall outside the base mechanisms is addressed in NAT in terms of interactive complexity and Resilience Engineering in terms of unanticipated perturbations. Imaginations, learning from experience, early detection, close monitoring, and dynamically steering activities are proposed by RE to reduce the influence from out-of-range disturbance to the least degree [14, 39].

3.2.3. Symbolic barrier system

Road signaling system, signs, procedures, work permit, belong to this category. Symbolic barrier systems themselves cannot complete *barrier function*. Instruction, procedure, work permit cannot prevent any unwanted outcome from happening. It is the interpretation and action that complete the barrier function (Figure 7). Obviously, a road sign of speed limit of 50 only works when drivers notice them and actually slow down the speed under the limit. Symbolic barrier system is sometimes considered at the same level of efficiency, strength and robustness as functional barrier system or physical barrier system, which is generally not the case in practice [40].

Figure 7 Possible Flaws in Controlled Barrier Function Executed by Symbolic Barrier System

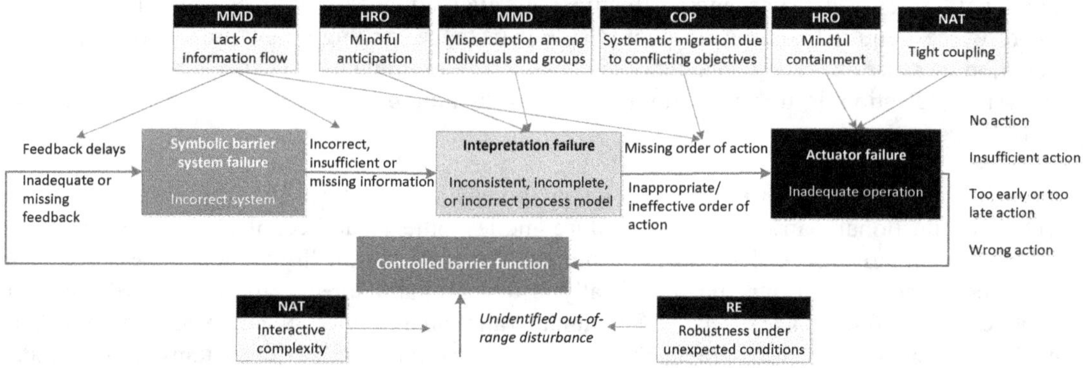

Comparing to functional barrier systems, the feedback loop that is addressed in STAMP and HRO deserves more attention to keep a symbolic barrier system reliable and effective. This is a weak link during operation. Vicente, Mumaw [41] found out that in nuclear power plants, where tasks and procedures are strictly prescribed, violations of instructions or skipping of steps have been repeatedly observed. On one hand, the behavior of operators appears to be quite rational given the actual high workload and time constraints. On the other hand, this reflected the deficiencies in the procedures themselves that need to be reported and improved. This is the same to warnings and signals. In spite of the great efforts putting into developing early warnings, many warnings are ignored and eventually accidents happened. A shift supervisor who worked in control room for Three Mile Island 2 nuclear power plant testified that there had never been less than 52 alarms lit in the control room and it had been a habit to ignore most of the alarms [42]. This is mainly due to the unreasonable design. User experience of these procedures and warning systems need to be reported and continuous improvement need to be implemented to bridge the gap between symbolic barrier systems and the practice.

3.2.4. Incorporeal barrier system

These barrier systems are largely synonymous with so called organizational barriers, i.e. rules that are imposed by the organization rather than being physically, functionally or symbolically present in the system, laws, safety culture, knowledge and skill [43]. Safety culture [16, 44] as a representative incorporeal barrier system, aims at building a foundation so that the designers and operators' can have 'good' safety beliefs, attitudes. Their behaviors are expected to act as additional accident barriers [45]. This is basically what HRO is talking about under *mindful infrastructure*. Another type of incorporeal barrier system is rules, laws and restrictions. This is emphasized in STAMP under concept of *safety constraints*, and Rasmussen's socio-technical framework. Enforcement of safety constraints sufficiently and adequately to lower levels in a hierarchical safety control structure, and finally implement to actions are the keys to utilize incorporeal barrier system.

4. IMPLICATIONS FOR SAFETY MANAGEMENT

Under a radical interpretation of barriers, all kinds of functions, elements and systems that are associated with safety are given the label "barrier" [40]. This created confusions and illusion that as long as we have a sufficient number of barriers in place, we are safe. The fact is that different types of barrier systems have different levels of adequacy, response time, effectiveness, specificity, reliability, robustness and independence. Discussions in section 3 illustrated that alternative perspectives can supplement the energy barrier perspective by structurally analyzing possible failure causes for barrier functions (STAMP, MMD), looking for driving forces for unsafe decisions and unsafe actions (COP, HRO), and emphasizing possible complex interactions and tight coupling among barrier systems (NAT, RE). In this section, the implications and suggestions to safety management of barriers are made to better manage *barrier functions* other than technical failures.

1. *Make sure the selected barrier system is optimal – No cost or workload pressure*
Realization of barrier function starts from selection of *barrier systems*. Generally speaking, physical and functional barrier systems are more effective and reliable than symbolic and incorporeal barriers, whereas resources required are much higher [43]. For instance, writing a new procedure is a common risk reduction measure as it is a quick and inexpensive way to implement [15]. Therefore, whether the barrier system selection is under economical or production pressure needs to be checked when designing the barrier systems.

2. *Periodically test and maintain information flow in technical functional barrier system*
For technical functional barrier systems, failure modes other than technical failures need to be identified and countermeasures have to be designed. The information flow among sensors, decision-making process, and actuator must be periodically tested and maintained. Barrier system doesn't equal to complete barrier function, especially for human functional barrier system and symbolic barrier system. This means further efforts are needed when these two types of barrier systems are selected.

3. *Check and enhance capability of human sensors*
When humans interact within functional barrier systems as sensors, we should enhance the capability of discovering unexpected events, require operators to be preoccupied with failures, avoid simplifications, and improve sensitivity to operations. Detecting deviations is not as straightforward as it looks. Make clear definitions and increase awareness of the deviations and abnormalities among operators; encourage raising of doubts and questions; treat all unexpected events as important information; develop sceptics; speak up [26]; all these needs organization to build up a culture fertilizer so that mindful behaviors and conscious inquiry [27] can grow.

4. *Facilitate sound decisions by efficient information management, accurate process model, relevant indicators and "safety first" mind*
When humans interact as decision-makers, make sure that the safety margin is not squeezed by the decision before taking any further action is crucial. "Deference to expertise" is necessary, but at the same time, we have to remember also centralization plays an important role to understand the big

picture. Sound decision depends upon sufficient, high quality, timely information. Systematic efforts are needed to build up efficient information management system. To facilitate decision-making process, we would need to identify indicators of the developing incubation period. Accurate process model needs continuous feedbacks, learning from past experience, training and knowledge sharing [26].

5. Reduce chances of human errors when they act as actuators

When human interact as an actuator, the key is to make sure the person fully acknowledges and understood the "mission", and reduce the chances for human error due to competence, disposable work descriptions, governing documents, technical documentation, design, Human Machine Interface (HMI), communication, supervision, time pressure, workload, work motivation, and attitude failures [38]. Organizational redundancy can be implemented as another barrier to detect the deviations of action from the "actuator" if necessary.

6. Emphasis on interpretation, action and feedback channel while using symbolic barrier systems

Effective symbolic barrier systems heavily rely on action and feedback channel. The gap between procedure and practice needs to be bridged by continuous feedbacks and updating. Design of signals and warnings has to be reasonable and practicable. Otherwise they could be counter-productive and speed up the development of accidents. These require robust reporting systems, comprehensive safety information systems, and rapid response [9].

7. Build up strong safety culture and sufficiently enforce safety constraints to reinforce incorporeal barrier systems

Incorporeal barrier systems are not physically present. Safety culture is the shared cognitions and administrative structure rather than individual attitudes to safety that deserves to be studied for development of organizational understanding regarding to risk and danger [4]. Four facets promoted by 'good' safety culture are: senior management commitment to safety; shared care and concern for hazards and solicitude over their impacts upon people; realistic and flexible norms and rules about hazards; and continual reflection upon practice through monitoring, analysis and feedback systems (organizational learning). For the other type of incorporeal barrier systems in terms of laws, restrictions, the enforcement of these safety constraints needs "vertical" alignment across the levels as indicated in social-technical framework [7] and Leveson's sociotechnical control model [30].

8. Keep in mind of "anticipated perturbations" and "interactive complexity"

There is textbook competence envelope for every barrier system. Anticipated perturbations and interactive complexity are unavoidable during operation. Monitoring boundary conditions and dynamically steering activities under both expected and unexpected conditions are required to accommodate changes.

5. CONCLUSION AND FURTHER WORK

Safety management on barriers has been focused on technical failures of physical and functional barrier systems. Along with extension from energy-oriented barriers to unwanted outcome-oriented barriers, barrier's performance during operational phase becomes unreliable due to human's interaction. The paper improves the understanding of how barrier functions can fail in terms of different types of barrier system. The possible failure modes of barrier functions identified in section 3.2 can be used as a guide to design and follow the performance of barrier systems. It was found that alternative perspectives on major accident provide explanations for possible flaws that exist in barrier functions and countermeasures from design to operation. This result testified the conclusion from Rosness, Grøtan [15]: these perspectives are complementing each other instead of competing. Furthermore, STAMP and barrier perspectives are conceived to be two totally different ways of risk modelling approach since STAMP is based on safety constraints emphasizing dynamic control, while barrier perspective is based on events (i.e. barrier failures) [30] which is rather static. Using STAMP methodology to model barrier function executed by functional barrier system and symbolic barrier system revealed its potential to systematically analyze how barrier functions can fail, such as how unsafe decision can be made and where information flow deficiency can happen. The applicability of possible flaws in controlled barrier function frameworks (Figure 6 and 7) need to be tested in case studies. However, this still opens the door to further research on how to manage barriers dynamically during operational phase.

Suggestions to better management of barriers are made from best practices of reviewed perspectives. Some suggestions are still quite conceptual and general that need to be further developed into concrete risk reduction measures, such as checklists, audits schemes, or indicators that can help decision-makers better comprehend and maintain the performance of barrier functions.

References

[1] Kjellén, U., *Prevention of accidents through experience feedback.* 2002: CRC Press.
[2] Haddon, W., *The basic strategies for reducing damage from hazards of all kinds.* Hazard Prevention, 1980.
[3] Gibson, J.J., *The contribution of experimental psychology to the formulation of the problem of safety-a brief for basic research.* 1961, New York, Association for the Aid of Crippled Children.
[4] Pidgeon, N. and M. O'Leary, *Man-made disasters: why technology and organizations (sometimes) fail.* Safety Science, 2000. **34**(1–3): p. 15-30.
[5] Turner, B.A. and N.F. Pidgeon, *Man-Made Disasters.* 1997: Butterworth-Heinemann Limited.
[6] Turner, B.A., *Causes of disaster: sloppy management.* British Journal of Management, 1994. **5**(3): p. 215-219.
[7] Rasmussen, J., *Risk management in a dynamic society: a modelling problem.* Safety Science, 1997. **27**(2–3): p. 183-213.
[8] Perrow, C., *Normal accidents: Living with high risk technologies.* 1999: Princeton University Press.
[9] Leveson, N., *Engineering a Safer World: Systems Thinking Applied to Safety (Engineering Systems).* 2012: The MIT Press.
[10] Laporte, T.R. and P.M. Consolini, *Working in practice but not in theory: Theoretical challenges of "high-reliability organizations".* Journal of Public Administration Research and Theory, 1991. **1**(1): p. 19-48.
[11] Roberts, K.H., *Some characteristics of one type of high reliability organization.* Organization Science, 1990. **1**(2): p. 160-176.
[12] Hollnagel, E., et al., *Resilience engineering in practice: A guidebook.* 2011: Ashgate Publishing, Ltd.
[13] Hollnagel, E., C.P. Nemeth, and S. Dekker, *Resilience engineering perspectives: remaining sensitive to the possibility of failure.* Vol. 1. 2008: Ashgate Publishing, Ltd.
[14] Hollnagel, E., D.D. Woods, and N. Leveson, *Resilience Engineering (Ebk) Concepts and Precepts.* 2006: Ashgate Publishing.
[15] Rosness, R., et al., *Organizational accidents and resilient organizations: six perspectives.* 2010, SINTEF Technology and Society.
[16] Reason, J.T. and J.T. Reason, *Managing the risks of organizational accidents.* Vol. 6. 1997: Ashgate Aldershot.
[17] Sklet, S., *Safety barriers: Definition, classification, and performance.* Journal of Loss Prevention in the Process Industries, 2006. **19**(5): p. 494-506.
[18] Dekker, S., *Drift Into Failure: From Hunting Broken Components to Understanding Complex Systems.* 2011: Ashgate Publishing Company.
[19] Perrow, C., *The Next Catastrophe: Reducing Our Vulnerabilities to Natural, Industrial, and Terrorist Disasters (New in Paper).* 2011: Princeton University Press.
[20] Roberts, K.H., *Cultural characteristics of reliability enhancing organizations.* Journal of Managerial Issues, 1993: p. 165-181.
[21] Rochlin, G.I., *Defining "high Reliability" Organizations in Practice: A Taxonomic Prologue,* in *New challenges to understanding organizations,* K.H. Roberts, Editor. 1993: New York: Macmillan. p. pp. 11-32.

[22] Hopkins, A., *The problem of defining high reliability organisations*. National Research Center for Occupational Safety and Health Regulation. January, 2007.

[23] La Porte, T.R., *High reliability organizations: unlikely, demanding and at risk*. Journal of Contingencies and Crisis Management, 1996. **4**(2): p. 60-71.

[24] Weick, K.E. and K.M. Sutcliffe, *Managing the unexpected: assuring high performance in an age of complexity*. 2001: Jossey-Bass.

[25] Eede, G., W. Muhren, and B. Walle, *Organizational learning for the incident management process: Lessons from high reliability organizations*. Journal of Information System Security, 2009. **4**(3): p. 3-23.

[26] Weick, K.E. and K.M. Sutcliffe, *Managing the unexpected: Resilient performance in an age of uncertainty*. 2nd ed. 2007: John Wiley & Sons.

[27] Westrum, R., *A typology of resilience situations*, in *Resilience engineering : concepts and precepts*, E. Hollnagel, D.D. Woods, and N. Leveson, Editors. 2006, Ashgate Publishing Limited. p. 35-41.

[28] Mattila, M., M. Hyttinen, and E. Rantanen, *Effective supervisory behaviour and safety at the building site*. International Journal of Industrial Ergonomics, 1994. **13**(2): p. 85-93.

[29] Hopkins, A., *Issues in safety science*. Safety Science, 2013(0).

[30] Leveson, N., *A new accident model for engineering safer systems*. Safety Science, 2004. **42**(4): p. 237-270.

[31] OLF, *OLF 070 - Norwegian oil and gas application of IEC 61508 and IEC 61511 in the Norwegian petroleum industry*. 2004.

[32] DNV, *Energy Report Beaufort Sea Drilling Risk Study*. 2009, Transocean Offshore Deepwater Drilling Inc.

[33] Hollnagel, E., *Barriers and Accident Prevention*. 2004: Ashgate.

[34] SINTEF, *Reliability Prediction Method for Safety Instrumented Systems - PDS method handbook 2010 edition*. 2010.

[35] Wei, C., W.J. Rogers, and M.S. Mannan, *Layer of protection analysis for reactive chemical risk assessment*. Journal of hazardous materials, 2008. **159**(1): p. 19-24.

[36] Kirwan, B., *A Guide To Practical Human Reliability Assessment*. 1994: Taylor & Francis.

[37] CCR, *Macondo The Gulf Oil Disaster. Chief Concels's Report*. 2011.

[38] Vinnem, J.E., et al., *Risk modelling of maintenance work on major process equipment on offshore petroleum installations*. Journal of Loss Prevention in the Process Industries, 2012. **25**(2): p. 274-292.

[39] Dinh, L.T.T., et al., *Resilience engineering of industrial processes: Principles and contributing factors*. Journal of Loss Prevention in the Process Industries, 2012. **25**(2): p. 233-241.

[40] Rollenhagen, C., *Event investigations at nuclear power plants in Sweden: Reflections about a method and some associated practices*. Safety Science, 2011. **49**(1): p. 21-26.

[41] Vicente, K.J., R.J. Mumaw, and E.M. Roth, *Operator monitoring in a complex dynamic work environment: a qualitative cognitive model based on field observations*. Theoretical Issues in Ergonomics Science, 2004. **5**(5): p. 359-384.

[42] Kemeny, J.G., *The need for change, the legacy of TMI: report of the President's Commission on the Accident at Three Mile Island*. 1979: The Commission.

[43] Hollnagel, E., *Risk + barriers = safety?* Safety Science, 2008. **46**: p. 221-229.

[44] Peters, G.A. and B.J. Peters, *Human error: Causes and control*. 2006: CRC Press.

[45] Taylor, J.B., *Safety Culture: Assessing and Changing the Behaviour of Organisations*. 2010: Gower.

Monitoring major accident risk in offshore oil and gas activities by leading indicators

Helene Kjær Thorsen[a] and Ove Njå[b]
[a] Safetec Nordic AS, Oslo, Norway[*]
[b] University of Stavanger, Stavanger, Norway

Abstract: In recent years, there has been a growing awareness that major accident risks should be monitored using risk indicators. We distinguish between leading and lagging indicators. The reason is that major accidents are rare events and the underlying causes are often fragmented and difficult to measure. However, it is a demanding task to develop appropriate leading indicators, because accident theories are disputed both in research literature and by practitioners. This paper presents the results from a study of a major oil and gas company's risk management processes and its use of indicators related to offshore installations. The work is based on analyses of accident reports, a literature review and interviews with offshore installation managers and platform integrity personnel.

We revealed major differences in attitudes among significant decision makers in relation to the use of risk indicators, spanning from skepticism and no use to in depth registration and analysis. However, all the offshore installation managers addressed the importance of a holistic view on risk and safety. Based on our findings we have developed an indicator set consisting of 16 leading indicators, covering technical, operational and organizational factors influencing major accident risk on offshore installations.

Keywords: Leading indicators, major accident risk, RIF, indicator criteria

1. INTRODUCTION

Major accidents in the oil and gas industry are greatly feared. Events such as the capsizing of Alexander Kielland (1980), the Piper Alpha explosion (1988) and the Norne helicopter accident (1997) are close to the minds of people working in the industry on the Norwegian continental shelf. These kind of accidents are very rare and are often perceived to occur as a completely surprise (today the concept of black swans is often discussed). However, are these events unpredictable? Accident investigations often reveal that there have been misjudgments in the organization, implying that early warnings, cues and signs of something serious and critical for the safety margins have been neglected. These issues have been noticed for a long time [1], addressing need for systemic and holistic approaches to monitor and reflect upon the performance of the high risk systems run by the organizations. Quantitative Risk Analyses (QRAs) have been carried out, identifying and implementing barriers in accordance with High Reliability Organization theory [2-4]. A further development has been to identify risk indicators in order to monitor barriers' performances and thus the total risk level at the various facilities, see e.g. [5-8]. The research on risk indicators and provisions of promising sets of indicator systems has been substantial, based on its normative ideological sense of governance.

Traditionally, the oil and gas industry has focused on so-called *lagging indicators* for monitoring major accident risk. These indicators are reactive as they measure "after the fact"-information, such as number of reported accidents/incidents last month. Examples of such indicators are Total Recordable Incident Frequency (TRIF) and Lost – Time Incident Frequency (LTIF). For a long time it was assumed that such indicators could reflect an installation's major accident risk [9]. This is in line with Heinrich accident triangle (iceberg theory) introduced in 1931, where the main principle was to focus on reducing the minor injuries and incidents [10]. This theory has been rejected by many researchers, who conclude; *relying on personal injury statistics will not reduce the major accident risk* [11-14].

---
[*] helene.kjar.thorsen@safetec.no

Several major accidents have further validated this conclusion, for example the BP Texas City refinery disaster in 2005, the Longford refinery accident in 1998 and the Deepwater Horizon blow out in 2010. The organizations all had excellent safety records with regard to personnel injuries and lost–time incidents before the accidents [9, 15-17]. Thorsen [18] analyzed accident statistics from some of the world's largest oil and gas companies over the years 2008-2011. The aim was to find out if the companies with the lowest TRIF–values also had the lowest FAR–values. No correlation was found. The lagging indicators such as TRIF and LTIF do not predict major accident risk.

Major accidents are rare events and the underlying causes are often fragmented and difficult to measure. Thus there is a need to observe features that might be related to the "production" of major accidents. However, it is a demanding task to develop appropriate leading indicators, because accident theories (explanations and causal relations) are disputed both in the research literature and by practitioners. Furthermore, risk conceptualizations and risk modelling are also highly disputed [19-22], which adds another challenge into the understanding of the risk picture. Risk management based upon sets of leading risk indicators might provide valuable information about changes in risk levels and aid the process of implementing effective risk reducing measures. The major issue of the study presented in this article is: *Which leading indicators have a potential to predict major accident risk in the operational phase of offshore oil and gas installations?*

Potential to predict major accident risk is an important, though difficult concept that must be taken into consideration. It is evident that there is a need to look for indicators that can provide a valid picture of major accident risks at offshore oil and gas installations. We emphasize that our foundational issue upon risk is purely epistemic and we claim that risk has no ontology [23]. Thus, potential to predict influences the involved and responsible parties' uncertainties regarding major accidents. We restrict major accident events to events due to hydrocarbon leakages in the operational phase of an offshore installation.

Our study object was a major worldwide offshore operator company in the oil & gas industry that operates up- and downstream facilities. To monitor and trend the company's risk level, an installation specific indicator set have had our primary focus. It is assumed that the indicators are reviewed by the offshore installation managers (OIMs) on a regular basis. The purpose of the indicators is to restrict attention to the areas that are considered especially important to ensure safe operations, including major accident risk. Examples of existing indicators are: Serious Incident Frequency (SIF); Total Recordable Incident Frequency (TRIF); Number of hydrocarbon leaks (> 0,1 kg/s); Falling Object Frequency (FOF); and Number of hours backlog in maintenance on safety critical equipment (both preventive and corrective maintenance). In addition to the indicator sets, the Company has implemented verification activities, covering both technical and operational barriers, and a technical barrier panel for continuous monitoring and follow-up of the technical integrity. In this study, only the latter is covered, as the platform integrity personnel (PIPs) are responsible for the monitoring tool and they perform evaluations of the technical integrity based on information from several indicators. Our development of leading indicators is to be seen as a supplement to major lagging indicators and verification activities that an offshore oil and gas company needs in its safety management system.

2. THE RECOMMENDED SET OF LEADING INDICATORS

Below we present our recommended indicator set, which covers technical, operational and organizational risk influencing factors that are assumed to influence major accident risks. Monitoring major accident risk requires indicators directed at underlying causes and latent conditions, in order for decision-makers to act upon early warnings before a major accident occurs. The indicator set is presented in table 1.

The indicator set is based on information from interviews, analyses of accident reports and a study of the research literature. None of the identified indicators fully satisfy the criteria that an indicator should meet to fulfill the intention to provide information to predict major accident risk. This implies that they are to be assessed holistically, i.e. as part of the key personnel's (such as OIM and PIP) continued risk image assessment. It has also been a premise for the development of indicators to meet a criterion of being specific to the facility considered. In this concrete case the lagging indicators: Number of hydrocarbon leaks (> 0,1 kg/s); Serious Incident Frequency (SIF), are by the OIMs considered important for major accident risk and should be seen in close relation to the recommended set in table 1.

Table 1: Recommended indicator set

Risk influencing factor (RIF)	Leading indicators	Measurement frequency	Indicator type (org, op, tech)
Monitoring technical barriers	Number of hours backlog in maintenance on safety critical equipment (both PM and CM)	Monthly	Tech/op/org
	Number of failures on safety critical equipment during testing	Monthly	Tech/op
	Status/condition of technical barriers (Number of red traffic lights in the system for barrier control)	Quarterly	Tech
Planning of activities	Number of plans sent onshore for reassessment and improvement.	Quarterly	Org
	Total number of work permits in one specific area (process area)	Monthly	Op
	Total number of work permits for hot work class A and B	Monthly	Op
	Maximum number of simultaneous activities last month	Monthly	Op
Dispensations (DISP)	Number of dispensations on HC – systems	Monthly	Org
Follow-up of and closing of actions and findings	Number of open findings from barrier verifications	Quarterly	Org/tech
	Number of overdue actions in Synergi with respect to HC-leaks	Monthly	Org
Competence and training (offshore and onshore)	Average number of years of experience with the specific systems for personnel	Quarterly	Org
	Average number of years of experience on the specific installation for personnel	Quarterly	Org
	Fraction of operating personnel that have received system training last 3 months	Quarterly	Org
	Number of workers in each personnel category whose training/courses are overdue	Monthly	Org
	Turnover of personnel during last 6 months	6 monthly	Org
Information about risk	Number of SJA operating personnel have attended during last 3 months	Quarterly	Org

The recommended indicator set is developed for use at the offshore installation by management both in offshore and onshore organizations. We claim that the indicator set of 16 individual indicators represents a manageable task to enable daily reflections upon major accident risk. It was deemed important to select indicators that the users could consider relevant, important and meaningful. Having indicators with face-validity [24] is a prerequisite for continuous reflection and learning.

During the time of study we obtained an understanding of plausible measurement frequencies of the indicators, but this recommendation is to be understood as a preliminary guideline. Practical adaptations are necessary. However, since the majority of indicators are organizational it is a demanding task for responsible key personnel to assess, understand, recognize and react within the organization. Below we discuss the rationale for choosing these indicators and the relationships between the indicators and major accident risks.

3. THEORETICAL AND EMPIRICAL PREMISES

3.1 Relations between major accidents, risk influencing factors, indicators and risk – a model

Figure 1 illustrates the relations between major accidents, leading and lagging indicators and risk in a risk management perspective. Major accident risk is seen as a combination of events (A), the consequences (C) of these events, and the associated uncertainties (U) [25]. In line with this risk perspective, we define a risk indicator as *"a measurable quantity which may provide information about risk factors influencing major accident risk on an offshore installation"*. Risk influencing factors are all conditions that either solely or in combination are assumed to influence the potential of a major accident occurrence. Often it is considered meaningful to categorize such conditions into technological, operational or organizational factors. Indicators are then the tools provided to operationalize the RIFs into a system for managing safety. They are observable. The status or condition of a RIF may be measured by the use of one or more indicators, depending on the complexity and nature of the factor [26]. Through the use of risk indicators, managers and decision-makers may increase their knowledge of important RIFs and hence reduce their uncertainty regarding future potential of major accidents. According to Hale [27], risk indicators have three main purposes: they monitor the level of safety in a system, provide the necessary information for decision-makers of where and how to act, and motivate action.

Figure 1: Relations between major accidents, RIF, indicators and risk

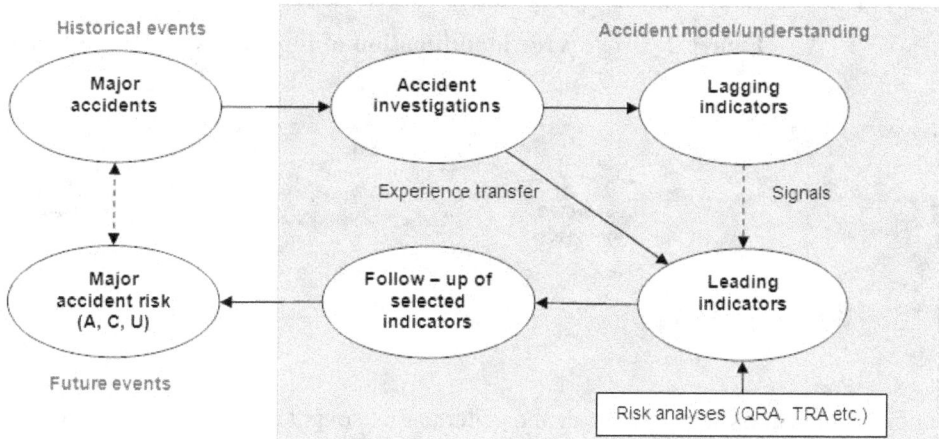

Investigations of historical accidents and precursors provide casual models of how accidents occur, which may provide valuable information of contributing and underlying causes, i.e. broken or defective barriers or RIFs. The knowledge of the casual chains, either derived from accident investigations or modeling of risk factors, may be used to establish indicators for the identified critical barriers, as seen in e.g. [5-7]. However, our understanding of accidents will affect what we look for in accident investigations and risk analysis, and thus what indicators we select for monitoring major accident risk. In addition, the quality of the risk analysis (completeness, level of detail, the goodness of models etc.) might restrict which RIFs to be identified. If the purpose of indicators is to increase our knowledge of major accident RIFs, lagging indicators might be important for prevention of new accidents by giving signals of where to place focus and guide the implementation of risk reducing measures. Hence, lagging indicators is a source of information for establishment and follow-up of leading indicators.

The process is to be seen as a learning process, based on the personnel's ability to assess the context, content of the system and organizational activities which they are committed to. Learning is understood as the personnel and organizations ability to change the systems, their ability to confirm and comprehend practices and activities at the offshore installations considered [28].

According to Kjellen [29] a leading indicator is *"an indicator that changes before the actual risk level has changed"*. This implies that leading indicators may provide valuable information of changes in risk levels before the occurrence of a major accident. Thus, leading indicators are preferred over lagging, due to their proactive value. Focusing on leading indicators support a proactive approach to risk management, as the focus is placed on reporting performance of preventive measures, compared to performance in the sense of occurrence of incidents and near-misses [30]. Risk management based on indicators may be problematic in the sense that there might be too much focus and effort on improving the indicator value and too little attention on whether the measure actually contributes to reducing the risk in a sustainable way [31]. Indicators are inputs to the risk management and decision-making processes; they bring attention to specific risk factors and shadow others. The ultimate goal is to provide an efficient set of safety considerations that enable key safety personnel (or all involved) to critically reflect upon their risk images [32] at the installation and support dialectical debates of safety[33].

3.2 Criteria for analyzing and developing the set of indicators

A prerequisite for the study of indicators was to use investigations performed and key personnel in the case organization and combine the data gathered with other literature on risk influencing factors (RIF). Based on existing indicator sets, literature review and governing documentation in the organization, we selected a large set of potential indicators. In order to identify useful indicators, we assessed the indicators against five criteria that the indicators should meet in order to fulfill the intention to provide information to predict major accident risk. The study process is illustrated in figure 2.

Figure 2: Process for identification of indicators

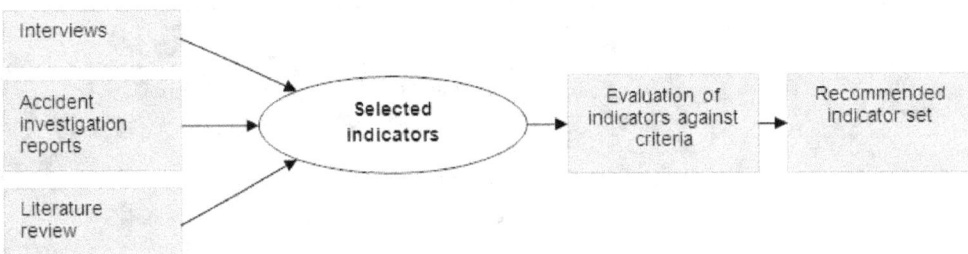

A combination of indicators that covers all the criteria was sought in accordance with Herrera [31]. Based on a holistic view on risk indicators we assessed each indicator against the following criteria:

1. It must be possible to *observe* and *measure* the identified RIFs. We must be able to see whether the results represent a deviation from a norm or not [34]. Further, it must be possible to express the status of the indicator in a way that can be recorded and compared with previous and future results.

2. The indicator must be *reliable*. The data which the indicators are based on need to have a high degree of consistency and accuracy [35]. Having reliable data is a prerequisite for meaningful analyses and establishment of risk reducing measures. The indicators also need to give the same measurement/result when used by different people on the same situation [27].

3. *Sensitive to changes*: The indicator must allow for early warnings by identifying changes that have an impact on the major accident risk. Herrera [31] argues that this criterion is especially important as leading indicators should provide a clear indication of changes over a reasonable time.

4. The indicator must be *intuitive* and *meaningful*. The meaning of the indicator must be self-evident and the measurement must be assessed as important for the prevention of major accidents. The indicator must be comprehendible and easy to use in order to be effective.

5. The indicator must be *robust to manipulation*. The indicator must not allow the organization to "look good" by e.g. change the reporting routines, rather than making the necessary changes to reduce the major accident risk[30].

3.3 Interviews with key personnel

We interviewed key personnel (7 OIMs and 5 PIPs) to get insight into the practical use of risk indicators and how indicators fit into the context of managing major accident risks. Focusing on major accident risk in day-to-day operations seems to be challenging. One of the OIMs expressed: *"It's a tendency to focus on minor injuries, which might overshadow what we really are afraid of – the major accidents"*. Another remarked: *"Operating personnel have inadequate understandings of major accidents"*. The same challenge was also revealed in the interviews with the PIPs. One of them said: *"A major accident have a huge potential, but very low probability which makes it hard to keep focus"*. However, several of the respondents claimed that there was much focus on avoiding major accidents in daily operations.

We found large variations between the interviewed OIMs regarding what indicators they perceived as important for monitoring major accident risk. About half of the OIMs highlighted the formally established indicators "SIF" and "number of HC-leaks", as these indicators are meant to measure serious accidents, incidents and near-misses, which was considered to hold major accident potential. Respondents had a clear understanding that these indicators were reactive, but it was stated that these indicators were important for implementing risk reducing measures. They were therefore considered to have a proactive value. On the other hand, some of the OIMs did not find any indicators suitable for monitoring major accident risk. It was stated that the statistical data was too limited for establishing reliable trends and it was deemed challenging to measure the effect of risk reducing measures. This specifically applied to the lagging indicators. However, it was stated that the indicator "Number of hours backlog on maintenance for safety critical equipment" could indicate weaknesses in the quality of technical barriers or a high workload. This could impact major accident risk.

Several of the OIMs pointed out that most of the established indicators were best suited for aggregation to company level, and that they were not applicable for their specific installation. The indicators were also used for benchmarking between installations and companies. This was considered to create a risk of overshadowing the installations specific risk factors. About half of the OIMs expressed that their focus on indicators in day-to-day operations was limited. Some quotes:

"I feel that it's wrong to spend lots of time on indicators, as these will not make us better. What makes us better is to avoid accidents and this is not reflected in the indicators".

"I strive to see which indicators that are related to major accident risk. I don't feel that TRIF and SIF are indicators that may provide information about major accident risk".

The PIPs had a different attitude towards the use of risk indicators. They are responsible for evaluation of the installations technical integrity, i.a. by collecting and analyzing information from several indicators. The majority of the PIPs addressed the importance of indicators, as several of the indicators measured important aspects of the safety critical systems and equipment. However, one of the PIPs explained that *"Indicators do not reflect major accident risk. The competence of the people assessing the indicators is much more important"*. When evaluating the barrier status, the PIPs claimed that they reviewed a selection of indicators and verified the quality and reliability of the data input. Offshore personnel are responsible for registration of failures of safety critical equipment during testing, which is critical for the quality of the safety information system. Hence, the PIPs' installation specific knowledge and experience plays an important part in understanding the indicator values. In addition, they valued the importance of having a holistic approach, where they assessed the entire indicator set with respect to major accident risk. The PIPs highlighted indicators like "Number of failures on safety critical equipment", "Backlog of maintenance on safety critical equipment" and "Number of open findings from technical barrier verifications" as especially important for managing major accident risk.

The respondents were divided with respect to management of major accidents through the use of risk indicators. Most of the OIMs perceived indicators as being lagging or reactive, only measuring previous incidents and accidents. In the daily risk management the OIMs emphasized being proactive, ensuring safe execution of work operations through proper activity planning, both offshore and onshore, field management, and having highly competent and trained personnel. Focus on such factors, was considered to constitute the most vital part of the proactive safety work. However, this was generally not seen in the context of indicators. The indicator term was mainly used in the context of the performance management system, which does not cover the organizational and operational factors which the OIMs deem as important for managing major hazards. The PIPs also recognized the importance of organizational and operational factors, as technical barriers must be maintained and tested in order to function as intended. This requires manual intervention on safety critical systems.

3.4 Accident investigations

Njå [36] problematized the selection of incidents that are investigated. Before anyone is asked to investigate an accident, there must be an event recognized by someone and the event must enforce an action. It is a "blink and wink"-situation, where the blink represents the events occurring in the company. While the blinks continue to occur, the winks are the sudden considerations – "what happened" – and the time is stopped. One initial question is thus: When does a blink become a wink. Which criteria should govern the winks? In the company studied there was a formal system for launching accident investigations which could be assessed as a systematic approach to responding to events. However, there might be biases in the systems, which we did not further assess, but rather selected amongst available investigations in the company's safety information system (Synergi).

We selected 6 random accident investigations in Synergi, within the most serious classification level; incidents with leak rates above 1 kg/s. The purpose was to analyze and assess whether existing indicators in the company could have provided early warnings before the leaks occurred. Two of the accidents were investigated by the Petroleum Safety Authority (PSA). They found that both accidents could have resulted in major accidents under slightly different circumstances. However, the internal investigation reports argued that neither of the accidents was likely to escalate to a major accident due to a low ignition probability. The reason for this claim was that the technical barriers installed to prevent escalation functioned as intended. The indicators "Backlog in maintenance for safety critical equipment" and "Number of failures on safety critical equipment" had positive values; not much backlog in maintenance and few failures of safety critical equipment. This was seen as contributing to preventing leaks form escalating into major accidents

However, all investigations revealed that none of the existing indicators in the company could have captured changes in the risk level with respect to the causes of the accidents. The underlying causes were to a large degree linked to organizational and operational factors, which are not reflected in the company's indicator sets. A short summery of the common root causes are given in the following:

- *Inadequate activity planning* with respect to risk assessments, coordination, and the time set out to perform the work operations, both onshore and offshore. In two investigations, high activity level was seen as a probable contributing cause.
- *Competence and training* was highlighted as contributing causes in all investigations. To have personnel with adequate installation specific competence, combined with experience, both offshore and onshore, was deemed especially important. One investigation stated that the turnover rate had been high, both in the offshore and onshore organization, which might had affected the overall installation specific competence in the organization. Lack of training and non-compliance with internal competence requirements was also noted.
- *Lack of experience transfer* was found to be a contributing cause in several of the investigations. PSA claimed that the Company had not sufficiently ensured that information from previous accidents had been used as learning basis for continuous improvement.

4 THE RATIONALE BEHIND THE SELECTED SET OF INDICATORS

4.1 Evaluation process

Through the interviews we developed insights into decision-makers' understanding of major accident risk. We identified several organizational and operational risk influencing factors, which were not covered by the Company's set of existing indicators. The investigations supported these findings. Through the literature review we identified and analyzed a large number of indicators [5-9, 37-42]. All identified indicators were then evaluated against the criteria described in section 3.2. Few documents/papers discuss the reasoning behind the selection of criteria and we have not seen any papers from the indicator research discussing how the recommended indicators are assessed against any criteria. According to Herrera [31], indicators are often selected because they are simple rather than inherently meaningful. In total we ended up with 31 indicators for further assessment.

The criteria were weighted according to their perceived importance, as shown in table 2, on the basis of the literature review and interviews. The criteria "Sensitive to change" was assigned the highest weight, as it is crucial to have indicators that may provide early warnings about changes in the risk levels [30, 31].

Table 2: Indicator criteria with assigned weight

Nr.	Indicator criteria	Weight
1	Observable and measurable	1
2	Reliable	2
3	Sensitive to change	3
4	Intuitive and meaningful	2
5	Robust to manipulation	1

All indicators were assigned a grade based on how they were considered to meet the various criteria. A letter grading system from B-F was chosen, with corresponding numerical values (B=3, C=2, D=1, E= -1 and F= -3). Grade B implies that the indicator is judged to satisfy the criteria, while F implies that the indicator is highly incongruent with the criteria. This approach was adapted from various assessment systems within the company in order to reflect their view. However, our concern was a generic approach for assessing the indicators' congruence with the intention of the criteria. Appendix A provides the total score for all identified indicators

4.2 Results

Of the 31 indicators included in the sample, our analysis gave 16 leading indicators covering technical, organizational and operational factors. The selection was based on the total score of each indicator, in addition to a subjective assessment to reduce the size of the indicator set, in order to have a manageable set. Initially, all indicators with a total score less than 10 were excluded. Several of these indicators could have provided information of the organization's ability to manage and understand the risk of major accidents (e.g. nr. 26, 29 and 30 in Appendix A) However, it is challenging to obtain reliable measurements. Some of the indicators had a total score above 10, but were excluded as they were assessed to have limited additional value in relation to the RIFs.

The RIFs that govern the set of indicators all implies that the organization deviates from norms, artifacts, assumptions deemed important. Below we argue for the rationale behind each indicator as a subset of the RIF category.

Monitoring technical barriers. The indicator *"Number of hours backlog in maintenance for safety critical equipment"* was considered to be an important leading indicator by the respondents. The same was found for the indicator *"Number of failures on safety critical equipment during testing"*. Poor or inadequate maintenance is often found to be a contributing cause to major accidents [43]. It is important to have a continuous focus on maintenance activities, as backlog in maintenance increases the risk of systems and equipment not functioning as intended. If this indicator shows a negative trend it implies a need to increase the maintenance activity. *"Number of failures on safety critical equipment* *during testing"* also provides early warnings. Further analysis of the indicator and test result are needed in order to identify proper actions. The test results may indicate weaknesses in the maintenance strategy, possible design weaknesses, incorrect use of components or equipment, or a need to adjust the test intervals. The status and condition of the individual barriers provides valuable information addressing need for follow-up actions. In the research literature, a more detailed follow-up of backlog and testing of particular equipment is suggested, e.g. for safety instrumented systems and alarms [39, 40]. This responsibility is assumed to lie within the platform integrity unit as part of the technical barrier panel.

Planning of activities. A high activity level may increase uncertainty with respect to the occurrence of accidents, and especially during simultaneous activities [37]. *Number of work permits* may reflect the activity level on the installation and thereby provide information about the risk level. The OIMs highlighted the Work Permit system (WP) and the interaction with the onshore organization as particularly important for managing major accidents. They were concerned with having control of number of WPs and types of WPs, and ensuring that all activities were based on risk assessments and that they were understood by the operating personnel. A challenge was revealed, regarding the interaction with the onshore organization. The onshore organization is responsible for the operations plan, and shall ensure that all plans are risk assessed and hold high quality before they are sent offshore. Several OIMs claimed that the operations plans often were inadequate, with lack of quality and risk assessments, which they considered to increase the risk of accidents.

Indicators related to activity level are frequently suggested in research literature, e.g. [6, 37, 40, 44]. Amongst our four recommended activity indicators one contain interaction with the onshore organization. If the indicator *"Number of plans sent onshore for reassessment and improvement"* is increasing this might indicate a risk of e.g. critical maintenance activities being postponed. In addition, it can provide insight into the organizations quality assessment and management processes.

Dispensations (DISP). Several OIMs addressed the importance of keeping control of number of dispensations (DISP). We have included the indicator *"Number of dispensations on HC – systems"*, as this may indicate a risk of major accidents due to non-compliance with regulatory requirements with respect to a highly critical system. In addition, all approved DISPs requires implementation of compensating arrangements. Some OIMs highlighted a challenge with maintaining adequate overview of all arrangements, especially with respect to the operational ones. A negative development of this indicator may thus provide early warning and indicate a need for deeper analysis to see how the DISPs and arrangements affect the major accident risk level.

Follow-up of and closing of actions and findings. A high number of overdue actions/risk reducing measures might increase the risk of major accidents and also reduce the performance of consequence reducing barriers [44]. The indicator *"Number of open findings from technical barrier verification"* might indicate weaknesses in the technical barriers, as long as the findings are not closed. The PIPs are responsible for closing findings. Through the interviews the PIPs complained about their ability to close findings due to the sheer volume of these. In one of the investigations it was highlighted that lack of resources within Platform Integrity created a situation where they did not have time to implement measures, in addition to evaluation and follow-up the technical barriers. We have also included the indicator *"Number of overdue actions in Synergi with respect to hydrocarbon leaks"*. For this indicator to be useful, the actions effect on the major accident risk must be analyzed and understood. The safety information system must allow for a categorization of severity level for the actions, in order to extract the data.

Competence and training. We selected five indicators covering several aspects of this RIF. In the research literature, indicators related to competence and training is frequently suggested, e.g. [7, 8, 37, 42]. The formal competence of personnel and/or experience from similar operations affects the ability to perform the work operations with high standards and in accordance with procedures and requirements. Competence and experience also plays an important part with respect to identification of potential danger/risk at an offshore installation [45]. All investigations highlighted the importance of having installation specific competence and experience, as knowledge of components, systems, barriers etc. is assumed to be better for personnel with more experience. The indicator *"Number of workers in each personnel category whose training/courses are overdue"* is assumed to reflect how updated the competence is for e.g. operators. However, training/courses needs to be specified with defined deadlines and the quality of the training will be an important aspect. The indicators should also cover the competence onshore, e.g. the competence of PIPs who are responsible for evaluating the status of technical barriers. In addition, the indicators should be divided for the different personnel group, e.g. operators, maintenance crew, inspection crew etc. *Turnover of personnel* was identified in the investigations as an important aspect to control, as high degree of turnover may increase the risk associated with inadequate experience and system specific competence.

Risk perception/Understanding. According to the OIMs, offshore personnel have a diverse understanding of the installation specific risks, especially major accident risk. It is not straight forward to establish indicators measuring this RIF. Risk perception is a diffuse concept and difficult to measure directly. According to Øien et al. [37], understanding of risk may be enhanced through basic knowledge of the risk concept and through specific knowledge of installation specific risks found in e.g. the TRA. They suggest the use of indicators measuring the proportion of personnel taking risk courses, proportion who are informed about risk analyses and attendance at SJA-meetings. However, the quality and content of the training/courses that is offered will also affect how this can contribute to increased risk understanding [46, 47]. We have selected the indicator *"Number of SJA operating personnel have attended during last month"* as increased attendance at SJA meetings is assumed to increase competence among offshore personnel regarding safety critical operations and associated risk factors.

5. DISCUSSION

We fully agree with the OIMs who claimed that their major aim was to obtain a holistic understanding of the safety level at the installation, in which major accident risk was included. However this must be more than gut feelings. We recognize that expert knowledge is characterized by an expert who generally knows what needs to be done based on mature and practiced understanding. An expert's skill has become so much a part of him that he needs to be more aware of it than he/she is of his/her own body. When things are proceeding normally at the installation, experts do not solve problems and do not make decisions, they do what normally works. While most expert performance is ongoing and non-reflective, when time permits and outcomes are crucial, an expert will deliberate before acting [48]. However the complex systems that make up offshore installations require more than individual expertise, and there is a need to specify concerns important for hazards developing into major accidents. Leading indicators is part of the risk image process, and we propose some important features for the set of indicators addressing comprehensiveness, applicability and manageability in the oil and gas industry:

- The indicators must all together be representative for and cover the major risks the oil and gas industry wants to be protected against.
- The indicators must be sufficiently detailed so that the variety and range of concerns are illustrated. It is important that the indicators provide information on how a situation or accident can develop in many directions, and that it is possible for the safety and emergency preparedness system and the personnel to influence on the development.
- The indicators must be realistic. Realistic could refer to logical and reasonable sequences of events that might occur.
- The indicators must be simple and easy to understand. A complicated indicator may lead to unnecessary disputes and may require that much time and many resources must be used to explain and discuss the indicators.
- The indicators should be logically consistent, in a way that it is possible to pinpoint connections between the indicator and characteristics of accident scenarios.

- The indicators must be dynamic and easy to change so that experiences from accidents or from training and exercises can be added and visualized. Such experience transfer is important for the organization in order to learn from own and others experiences.

The concept of leading and lagging indicators is disputed among researchers and practitioners. There are no agreed definitions of what "lead" and what "lag" are, and where along the casual chain one should make a distinction. Hopkins [49, 50] argues that the distinction is not fruitful, while Dyreborg [51] and Hale [27] claim that it has implications for organizational learning and that the distinction is essential when the indicators purpose is to provide information of where and how to act. The evaluation of all identified indicators clearly demonstrates that no single indicator can give adequate information about changes in the risk level with respect to major accidents. Therefore, one needs to look at the total indicator set to determine if changes in a single indicator are critical to the total major accident risk. If the lagging indicators show a negative trend, the leading indicators need to be analyzed and evaluated in order to see if these indicators have captured the changes. Such experience transfer may result in an alteration of the recommended indicator set. In addition, the indicator set might provide information about the quality of the management and organization, through the ability of planning work operations and activities, follow up of competence level and follow-up and closure of actions and finding. Lord Cullen's [52] report on Piper Alpha noted that the system was comprised of degraded systems with numerous indications of major accident risk. Without key personnel's attention, recognition and action prone behavior, major accident risk indicators might be dangerous[53].

OIMs claimed that consciousness about major accident risk in the day-to-day operations was challenging. Maintaining motivation and awareness that the indicators predict changes to the risk level is crucial [30]. Most hydrocarbon leaks occurring in offshore environments have been found to have operational and organizational causes [54]. The company's existing indicators were to a large degree lagging and we revealed a gap between the company's established indicator set and the actual causes of the investigated accidents. Further, several of the OIMs expressed a negative attitude towards risk indicators and misconceptions regarding the purpose of indicators. Hale [27] claims that indicators should motivate action. Hence the indicators must communicate this significance to key personnel. Our indicator set has been developed to reflect underlying factors and latent conditions that key personnel found to be important. We observed a general lack of trust amongst the OIMs regarding whether indicators might contribute to increased understanding of major accident risk. There seems to be a lack of understanding of the proactive and predicative value of indicators, especially with respect to measuring organizational and operational risk factors. This aspect of risk indicators needs to be communicated and utilized in the daily risk management process.

6. CONCLUSION

Through our study of major risk indicators we have found a comprehensive research literature that provides numerous suggestions about what indicators to use. The vast amount of normative well intended indicator sets stands in contrast to the practical approaches used by key personnel responsible for safety on the offshore installations.

Our recommended set of leading indicators have been developed on the basis of a literature review, existing systems in an offshore oil and gas operator company, six of its accident investigation reports and interviews with key personnel. Comprehensiveness, applicability and manageability formed basic prerequisites in our work. However, we do not think that there exists a universal set of indicators. We also think that our research needs to be challenged. We conclude that the need for further development and understanding of the use of leading indicators for monitoring the risk level on an offshore installation is important.

References
1. Njå, O., *Approach for assessing the performance of emergency response arrangements* 1998, Stavanger, Norway: Aalborg University/Stavanger University College. 1 b. (flere pag.).
2. Reason, J., *Managing the Risks of Organizational Accidents* 1997, Aldershot: Ashgate.

3. Weick, K.E., K.M. Sutcliffe, and D. Obstfeld, *Organizing for high reliability: Processes of collective mindfulness*, in Research in Organizational Behavior, R.I. Sutton and B.M. Staw, Editors. 1999. p. 81-123.
4. Rasmussen, J. and I. Svedung, *Proactive risk management in a dynamic society*2000, Karlstad: Swedish Rescue Services Agency. 196 s.
5. Øien, K., *Development of Early Warning Indicators Based on Incident Investigation*, in International Probabilistic Safety Assessment and Management Conference 2008: Hong Kong, China.
6. Øien, K. and S. Sklet, *Risikoindikatorer for overvåking av risikonivået på Statfjord A*, 1999, SINTEF: Trondheim.
7. Haugen, S., et al., *Major Accident Indicators for Monitoring and Predicting Risk Levels*, in SPE Health, Safety and Environmental Conference in Oil and Gas Exploration and Production2011: Vienna, Austria.
8. Øien, K. and S. Sklet, *Organisatoriske risikoindikatorer: Pilotstudie Statfjord A*, 2001, SINTEF: Trondheim.
9. Skogdalen, J.E., I.B. Utne, and J.E. Vinnem, *Developing safety indicators for preventing offshore oil and gas deepwater drilling blowouts.* Safety Science, 2011. **49**: p. 118-1199.
10. Heinrich, H.W., *Industrial Accident Prevention – A Scientific Approach*1931, New York: McGraw-Hill.
11. Groeneweg, J., *Small statistics study*, in Offshore European Conference2005: Aberdeen.
12. Apking, T.D. and D.K. Martin, *New Insight into the Prevention of Serious Injuries and Fatalities*, in SPE/APPEA Internation Conference on Health, Safety and Environment in Oil and Gas Exploration and Production2012: Perth, Australia.
13. Anderson, M. and M. Denkl, *The Heinrich Accident Triangle - Too Simplistic a Model for HSE Management in the 21st Century?*, in SPE International Conference on Health, Safety and Environment in Oil and Gas Exploration and Production2010: Rio de Janeiro, Brazil.
14. Teakle, J., et al., *Stepping Out of the Triangle and into the Field.*, in SPE/ATTEA International Conference on Health, Safety and Environment in Oil and Gas Exploration and Production2012: Perth, Australia.
15. Baker, J.S., et al. *The Report of the BP U.S Refineries Independendt Safety Review Panel.* 2007; Available from: www.bp.com/bakerpanelreport.
16. Hopkins, A., *Lessons from Longford – The Esso Gas Plant Explosion*2000, Australia: CCH Australia Limited.
17. Hopkins, A., *Failure to learn: the BP Texas City refinery disaster*2010, Sydney: CCH Australia Limited.
18. Thorsen, H.K., *Monitoring major accident risk in oil and gas operations – a research study on leading indicators*, 2013, University of Stavanger: Stavanger.
19. Aven, T., *On Some Recent Definitions and Analysis Frameworks for Risk, Vulnerability, and Resilience.* Risk Analysis: An International Journal, 2011. **31**(4): p. 515-522.
20. Aven, T., *Misconceptions of risk*2010, Chichester: Wiley. VIII, 240 s.
21. Haimes, Y.Y., *On the Complex Definition of Risk: A Systems-Based Approach.* Risk Analysis: An International Journal, 2009. **29**(12): p. 1647-1654.
22. Haimes, Y.Y., *Responses to Terje Aven's Paper: On Some Recent Definitions and Analysis Frameworks for Risk, Vulnerability, and Resilience.* Risk Analysis, 2011. **31**(5): p. 689-692.
23. Solberg, Ø. and O. Njå, *Reflections on the ontological status of risk.* Journal of Risk Research, 2012. **15**(9): p. 1201-1215.
24. Øien, K., et al., *Building Safety Inicators: Part 2 - Apllication, practices and results.* Safety Science, 2011. **49**: p. 162-171.
25. Flage, R. and T. Aven, *Expressing and communicating uncertainty in relation to quantitative risk analysis.* Reliability & Risk Analysis: Theory & Applications, 2009. **2**.
26. Haugen, S., et al., *A generic method for identifying major risk indicators*, in International Probabilistic Safety Assessment and Management Conference and the Annual European Safety and Reliability Conference2012: Helsinki, Finland.
27. Hale, A., *Why Safety Performance Indicators?* Safety Science, 2009. **47**: p. 479-480.

28. Braut, G.S. and O. Njå, *Learning from accidents (incidents). Theoretical perspectives on investigation reports as educational tools.*, in *Reliability, Risk and Safety. Theory and Applications*2010, Taylor & Francis Group: London:. p. 9-16.
29. Kjellen, U., *Why Safety Performance Indicators?* Safety Science, 2009. **47**: p. 486-489.
30. Vinnem, J.E., *Risk indicators for major hazards on offshore installations.* Safety Science, 2010. **48**: p. 770-787.
31. Herrera, I.A., *Proactive safety performance indicators – Resilience engineering perspective on safety management*, 2012, NTNU: Trondheim.
32. Braut, G.S., et al., *Risk images as basis for two categories of decisions.* Risk management: An International Journal, 2012. **14**: p. 60-76.
33. Watson, S.R., *The meaning of probability in probabilistic safety analysis.* Reliability Engineering and System Safety, 1994. **45**: p. 261-269.
34. Kjellen, U., *Prevention of Accidents Through Experience Feedback*2000, London: Taylor & Francis.
35. Aven, T., et al., *Samfunnssikkerhet [Societal safety]*2004, Oslo: Universitetsforlaget. 296 s.
36. Njå, O., *Issues and judgements in accident investigation*, in *The International Emergency Management Society (TIEMS), Waterloo, Canada*2002.
37. Øien, K., S. Massaiu, and R. Kviseth, *Guideline for implementing the REWI method*, 2012, SINTEF: Trondheim.
38. Tinmannsvik, R.K. and P. Hokstad, *Metodikk for måling av sikkerhetsmessig utvikling i og omkring storulykkesvirksomheter*, 2013, SINTEF: Trondheim.
39. HSE, *Developing process safety indicators: A step by step guide for chemical and major hazard industries*, 2006, UK Health and Safety Executive
40. OGP, *Process Safety - Recommended Practice on Key Performance Indicators*, 2011, International Association of Oil and Gas Producers.
41. Johnsen, S.O., et al., *Proactive Indicators to Control Risk in Operations of Oil and Gas Fields*, in *SPE International Conference on Health, Safety and Environment in Oil and Gas Exploration and Production*2012: Rio de Janeiro, Brazil.
42. Reiman, T. and E. Pietkäinen, *Indicators of safety culture - selection and utilization of leading safety performance indicators*, 2010, Swedish Radiation Safety Authority: Sweden.
43. Tinmannsvik, R.K., et al., *Deepwater Horizon - ulykken: Årsaker, lærepunkter og forbedringstiltak for norsk sokkel*, 2011, SINTEF: Trondheim.
44. Heide, B., *Proaktive risikoindikatorer: Aktivitetsindikatorer for storulykkesrisiko på offshoreinstallasjoner*, 2003, University of Stavanger: Stavanger.
45. Mearns, K. and R. Flin, *Risk perception and attitudes to safety by personnel in the offshore oil and gas industry: a review.* Journal of Loss Prevention in the Process Industries, 1995. **8**(5): p. 299-305.
46. Junge, M., *"Mangelfull risikoforståelse" – årsaksforklaringen som betyr alt og ingenting.*, 2010, University of Stavanger: Stavanger.
47. Rundmo, T., *Associations between risk perception and safety.* Safety Science, 1996. **24**: p. 197-209.
48. Dreyfus, H.L. and S.E. Dreyfus, *Five Steps from Novice to Expert*, in *Mind over machine: the power of human intuition and expertise in the era of the computer*1986, Free Press: New York.
49. Hopkins, A., *Thinking about process safety indicators*, in *Oil and Gas Industry Conference*2007: Manchester.
50. Hopkins, A., *Reply to comments.* Safety Science, 2009. **47**: p. 508-510.
51. Dyreborg, J., *The causal relation between lead and lag indicators.* Safety Science, 2009. **47**(474-475).
52. Cullen, W.D.L., *The public inquiry into the Piper Alpha disaster*, 1990, HMSO: London.
53. Hale, A., *Is Safety Training Worthwhile?* Journal of Occupational Accidents, 1984. **6**: p. 17-33.
54. Kongsvik, T., S.K. Johnsen, and S. Sklet, *Safety climate and hydrocarbon leaks: An empirical contribution to the leading-lagging indicator discussion.* Journal of Loss Prevention in the Process Industries, 2011. **24**(6).

Appendix A: Evaluation of identified indicators

Nr.	Identified indicators	Indicator criteria 1	2	3	4	5	Total score
1*	Number of hours backlog in maintenance for safety critical equipment (both PM and CM)	B	B	B	C	B	25
2*	Number of failures on safety critical equipment during testing	B	C	B	C	C	22
3	Total number of WPs in a specific area (process area)	B	C	D	C	C	16
4	Total number of WPs for hot work class A and B	B	C	D	C	C	16
5	Total number of WPs on HC-systems	B	C	D	C	C	16
6	Total number of WPs	B	C	D	C	C	16
7	Average number of years of experience with the specific system for personnel	C	D	C	C	D	15
8	Average number of years of experience on the specific installation for personnel	B	C	D	D	B	15
9	Maximum number of simultaneous activities in one specific area	C	C	D	C	C	15
10	Fraction of operating personnel that have received system training last 3 months	C	C	D	C	C	15
11*	Status/condition of technical barriers (E.g. Number of red traffic lights in the system for barrier control)	B	D	D	C	D	13
12*	Number of open findings from barrier verifications	B	C	D	D	D	13
13	Number of workers in each personnel category whose training/courses are overdue	C	C	D	D	C	13
14	Total number of dispensations on HC-systems	C	D	D	C	C	13
15	Number of plans sent onshore for reassessment and improvement.	C	D	D	C	D	12
16	Number of near-misses with major accident potential	C	D	D	D	D	12
17	Number of WPs approved outside of WP - meetings	C	D	D	C	D	12
18	Number of SJA operating personnel have attended during last 3 months	C	D	D	D	C	11
19	Turnover of personnel last 6 months	B	B	E	D	B	11
20	Number of overdue actions in Synergi with respect to HC-leaks	B	C	E	C	C	10
21	Number of dispensations exceeding design	C	D	D	D	D	10
22	Fraction of relevant personnel with formal training in use of SJA	B	C	E	C	C	10
23	Number of dispensations that are overdue	C	C	E	D	C	8
24†	Number of HC-leaks (<0,1kg/s)	B	C	E	D	C	8
25	Number of implemented operational arrangements to maintain approved dispensations.	D	D	E	C	D	5
26	Portion of operating personnel taking risk courses last 12 months	D	D	E	D	C	4
27†	Falling Object Frequency (FOF)	B	C	E	E	C	4
28†	Serious Incident Frequency (SIF)	B	C	E	E	C	4
29	Portion of operating personnel informed about risk analysis last 3 months	D	D	E	D	D	3
30	Number of reviews of major accidents/near misses on other installations/facilities (experience transfer)	E	E	E	C	D	-1
31	Number of cases of inadequate decision support from onshore last 3 months	E	E	E	E	E	-9

† Indicators highlighted by the OIMs and PIPs

The Role of NASA Safety Thresholds and Goals in Achieving Adequate Safety[*]

Homayoon Dezfuli[a†], Chris Everett[b], Allan Benjamin[c], Bob Youngblood[d], and Martin Feather[e]

[a] NASA, Washington, DC, USA
[b] Information Systems Laboratories, Rockville, MD, USA
[c] Independent Consultant, Albuquerque, NM, USA
[d] Idaho National Laboratory, Idaho Falls, ID, USA
[e] Jet Propulsion Laboratory, California Institute of Technology, Pasadena, CA, USA

Abstract: NASA has recently instituted requirements for establishing Agency-level safety thresholds and goals that define long-term targeted and maximum tolerable levels of risk to the crew as guidance to developers in evaluating "how safe is safe enough" for a given type of mission. This paper discusses some key concepts regarding the role of the Agency's safety thresholds and goals in achieving adequate safety, where *adequate safety* entails not only meeting a minimum tolerable level of safety (e.g., as determined from safety thresholds and goals), but being as safe as reasonably practicable (ASARP), regardless of how safe the system is in absolute terms.

Safety thresholds and goals are discussed in the context of the Risk-Informed Safety Case (RISC): A structured argument, supported by a body of evidence, that provides a compelling, comprehensible and valid case that a system is or will be adequately safe for a given application in a given environment. In this context, meeting of safety thresholds and goals is one of a number of distinct safety objectives, and the system safety analysis provides evidence to substantiate claims about the system with respect to satisfaction of the thresholds and goals.

Keywords: Safety Thresholds, Safety Goals, Safety Performance Margin, As Safe As Reasonably Practicable (ASARP), Risk-Informed Safety Case (RISC).

1. INTRODUCTION

NASA has recently instituted requirements for establishing Agency-level safety thresholds and goals that define "long-term targeted and maximum tolerable levels of risk to the crew as guidance to developers in evaluating "how safe is safe enough" for a given type of mission" [1]. Safety thresholds specify the minimum tolerable/allowable level of crew safety (maximum tolerable level of risk) for the design in the context of its design reference mission, and are to be used by the Agency as criteria for program acquisition decisions. Safety goals (and the accompanying requirements to implement safety upgrade and improvement programs) are motivated by the fact that the level of risk associated with initially flown designs is typically unacceptable in the long term and the fact that human spaceflight programs, informed by flight experience and analysis, can achieve significant reductions of risk over the life of a program.

This paper discusses some key concepts regarding the role of the Agency's safety thresholds and goals in achieving adequate safety. As discussed in the recently-released NASA/SP-2010-580, NASA System Safety Handbook [2], adequate safety involves not only meeting the minimum tolerable level of safety (e.g., as determined from safety thresholds and goals), but it also involves being as safe as reasonably practicable (ASARP), regardless of how safe the system is in absolute terms.

---

[*] This paper is based on the work performed by the NASA Office of Safety and Mission Assurance (OSMA) in support of the development activities for Volume 2 of NASA System Safety Handbook (NASA/SP-2014-612),

[†] hdezfuli@nasa.gov

This paper also provides a framework for implementing the safety thresholds and goals in a way that reflects expectations that the safety of a new system will improve over time, and is consistent with the technical challenges inherent in assessing the safety of such systems. In particular, synthetic[‡] methods of risk analysis are vulnerable to risk model incompleteness (i.e., only a subset of the system failure causes are identified and analyzed), especially in the early phases of the system life cycle. This paper outlines a method of accounting for these un-modeled sources of risk, providing a rational basis for determining whether or not a system meets the minimum tolerable level of safety.

Finally, this paper discusses the safety thresholds and goals in the context of the Risk-Informed Safety Case (RISC), defined in the NASA System Safety Handbook as "a structured argument, supported by a body of evidence, that provides a compelling, comprehensible and valid case that a system is or will be adequately safe for a given application in a given environment." In this context, meeting of safety thresholds and goals is one of a number of distinct safety objectives, and the system risk analysis provides evidence to substantiate claims about the system with respect to satisfaction of the thresholds and goals. The nature of a specific RISC depends upon the decision context in which it is developed. Different decision contexts produce different safety objectives, and use different sets of engineering observables as evidence to support the claim that the objectives have been met.

2. FUNDAMENTAL PRINCIPLES OF SAFETY

The NASA System Safety Handbook articulates two fundamental principles of safety that together constitute "adequate safety":

- An adequately safe system is assessed as meeting a minimum threshold level of safety, as determined by analysis, operating experience, or a combination of both. Below this level the system is considered unsafe. This minimum level of safety is not necessarily fixed over the life of a system. As a system is operated and information is gained as to its strengths and weaknesses, design (hardware and software), and operational modifications are typically made which, over the long run, improve its safety performance. In particular, an initial level of safety performance may be accepted for a developmental system, with the expectation that it will be improved as failure modes are "wrung out" over time.

- An adequately safe system is ASARP. The ASARP concept is closely related to the "as low as reasonably achievable" (ALARA) and "as low as reasonably practicable" (ALARP) concepts that are common in U.S. nuclear applications and U.K. Health and Safety law, respectively [3, 4]. A determination that a system is ASARP entails weighing its safety performance against the sacrifice needed to further improve it. The system is ASARP if an incremental improvement in safety would require a disproportionate deterioration of system performance in other areas including technical, cost, and schedule.

Figure 1, reproduced from the System Safety Handbook, illustrates application of these two principles of safety throughout the entire system lifecycle.

2.1. Meeting the Minimum Tolerable Level of Safety

In the context of the fundamental principles of adequate safety, Agency-level safety thresholds and goals express stakeholders' expectations about the minimum tolerable level of crew safety, where a safety threshold expresses an initial minimum tolerable level, and the goal expresses expectations about the safety growth of the system in the long term. As such, the safety thresholds and goals work together to establish a minimum tolerable level of safety that increases from the threshold to the goal

---

[‡] By "synthetic methods," we mean methods that produce risk estimates by explicitly constructing a scenario set and summing risk contributions to obtain an estimate of aggregate risk, as is typically done in probabilistic risk assessment (PRA).

over the life of the program. As discussed in NPR 8705.2B, safety thresholds are used by the Agency as criteria for program acquisition decisions, whereas safety goals specify the level of safety that is considered acceptable for repeated missions and serve as the long-term target for proactive safety upgrade and improvement programs that must be maintained for the duration of the program or until the safety goals have been met. The NPR does not specify the rate at which the minimum tolerable level of safety moves from threshold to goal, only that safety growth must be proactively pursued so long as the goal is unmet.

Figure 1: Fundamental Principles of Adequate Safety

```
                    Achieve an
                  adequately safe
                      system
                   /           \
    Achieve a system that      Achieve a system that is
    meets or exceeds the       as safe as reasonably
    minimum tolerable level    practicable (ASARP)
    of safety
    /      |      \             /      |      \
Design  Build  Operate      Design  Build  Operate
the     the    the system   the     the    the system
system  system to           system  system to
to meet to meet continuously to be  to be   continuously
or      or     meet or      as safe as safe be as safe
exceed  exceed exceed the   as       as     as reasonably
the     the    minimum      reasonably reasonably practicable
minimum minimum tolerable   practicable practicable
tolerable tolerable level of
level of level of safety
safety  safety
```

Figure 2, reproduced from the NASA System Safety Handbook, illustrates the NASA safety thresholds and goals.

Figure 2: NASA Safety Thresholds and Goals

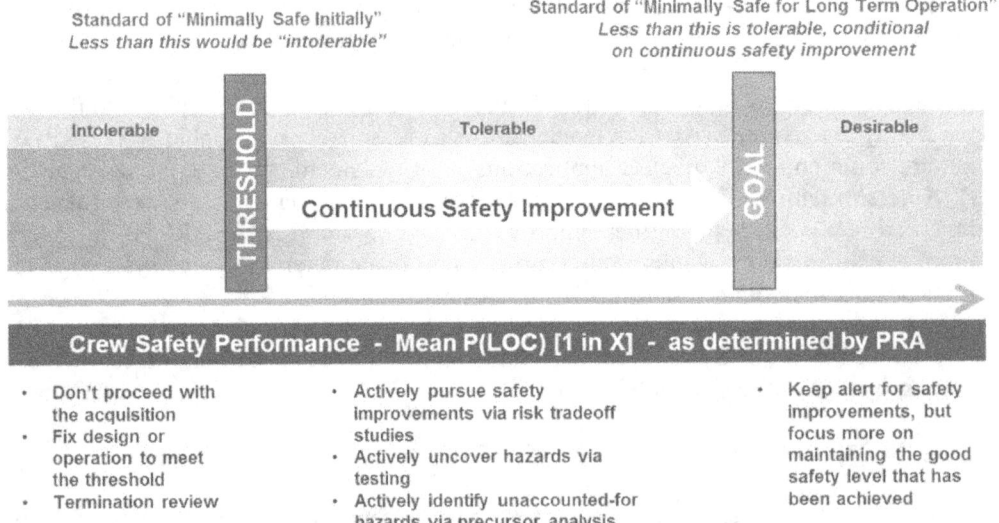

2.2. Being As Safe As Reasonably Practicable

It is ethically problematic to accept a given level of safety if improvements to safety can be easily made. An adequately safe system is therefore one that is ASARP. Being ASARP is a separate and distinct consideration from meeting a minimum tolerable level of safety. It reflects a mindset of continuous safety improvement regardless of the current level of safety. It is an integral aspect of good systems engineering process that guides risk-informed decision making throughout the system lifecycle, beginning in formulation.

The ASARP concept is illustrated graphically in Figure 3, which is adapted from the System Safety Handbook. The curve represents the efficient frontier of the trade space of identified alternatives, and shows the tradeoff between safety performance and performance in other mission execution domains (cost, schedule, technical).§ The ASARP region contains those alternatives whose safety performance is as high as can be achieved without resulting in intolerable performance in one or more of the other domains.

Figure 3: As Safe As Reasonably Practicable (ASARP)

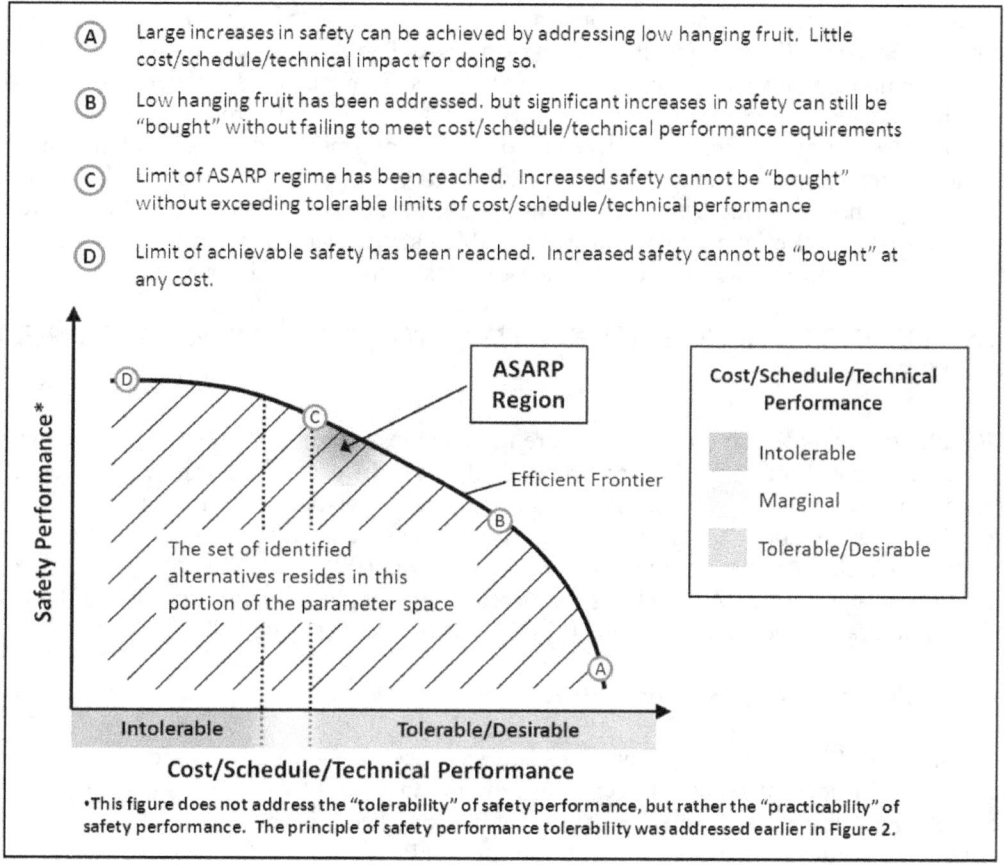

Figure 3 illustrates that:

- Improvements to cost, schedule, or technical performance beyond minimum tolerable levels are not ASARP if they come at significant expense in safety performance.

- The ASARP concept makes no explicit reference to the absolute value of a system's safety performance or the tolerability of that performance. It is strictly concerned with the system's safety performance relative to that of other identified alternatives.

- ASARP is a region of the trade space and can contain more than one specific alternative. Moreover, the boundaries of that region are not sharply defined. Determining that a system is ASARP entails the prudent application of engineering and management judgment.

---

§ Figure 3 is a two-dimensional representation of a space that involves four dimensions: safety, cost, schedule, and technical performance. For conceptual purposes, the cost, schedule, and technical performance dimensions are combined onto a single axis, since in this context they are all regarded as impacts incurred in order to increase safety. Each point on the efficient frontier may be interpreted as the maximum level of safety performance that may be achieved for a given level of cost, schedule, and technical performance.

The means of achieving a system that is ASARP include those for meeting the safety goal (i.e., implementation and maintenance of proactive safety upgrade and improvement programs), as well as meeting other applicable safety criteria; applying best practices in system design (e.g., appropriate margins and failure tolerance); high quality manufacture; robust operational procedures (including contingencies); and maintaining a safety analysis process to identify, evaluate, and, as appropriate, mitigate risks or deficiencies for the life of the program regardless of the satisfaction of the safety goal. In general, ASARP results from systems engineering activities that prioritize safety throughout the system lifecycle.

The concept of ASARP is necessarily in play not only during the development of a selected design alternative, which is usually the context within which ASARP is discussed, but also during the time in which the selection among competing design alternatives is made. In the latter case, the ASARP concept is imbued within the Risk-Informed Decision Making (RIDM) process [5], which has as its objective the selection of a design alternative based on risk-informed trade-offs involving safety, cost, schedule, and technical performance. The principle of RIDM calls for a decision maker who is risk-informed to apply his/her value system to the weighting of the trade-offs. In so doing, however, the decision maker is guided by the ASARP principle, which states that safety is to have a higher weight than any of the other three dimensions of the trade-off (cost, schedule, and performance).

3. ASSESSING THE SYSTEM RELATIVE TO THE MINIMUM TOLERABLE LEVEL OF SAFETY

NPR 8705.2B states that thresholds and goals are to be expressed in terms of an aggregate measure of risk such as the probability of a loss of crew (P(LOC))[**]. In order to compare a system's safety performance to the thresholds and goals, a risk analysis is performed that quantifies the relevant measure. The Administrator's letter on the Agency's safety thresholds and goals for crew transportation missions to the International Space Station (ISS) specifies using NASA-accepted probabilistic risk analysis (PRA) methods similar to those applied by the Space Shuttle, ISS, and Constellation programs, and using the mean P(LOC)[††] as the risk measure [6].

3.1. The Issue of Risk Analysis Incompleteness

However, although PRA methods [7] have a history of use at NASA for providing insight into the relative risk significance of potential accident scenarios that might occur in a system, and into the relative safety performance of different systems,[‡‡] it has long been recognized that there are challenges inherent in using synthetic methods such as those used in PRA to quantify a system's actual risk, due to the inherent incompleteness of the scenario sets identified by these methods [8-13]. The unaccounted-for (or insufficiently-accounted-for) scenarios typically involve organizational issues and/or complex intra-system interactions that may have little to do with the intentionally engineered functional relationships of the system. Such underappreciated interactions (along with other factors) were operative in both the Challenger and Columbia accidents. In the Challenger disaster, O-ring blow-by impinged on the external tank, leading to tank rupture and subsequent loss of crew. In the

---

[**] Other risk measures, such as the probability of loss of mission (P(LOM)), might be relevant in addition to or instead of P(LOC), for example in the case of robotic missions. However, for convenience, P(LOC) is used in this paper to refer to aggregate risk measures generally.

[††] PRA methods treat uncertainty quantitatively, e.g., by quantifying it at the basic event level and propagating it through the logic model to produce a probability distribution for the risk measure (e.g., P(LOC)). This enables statistics such as the mean value to be estimated. Additionally, the risk-driving uncertainties can be identified for prioritization of uncertainty reduction efforts.

[‡‡] PRA methods are generally effective at identifying system failures that result from combinations of failures events that propagate through the system due to the functional dependencies of the system that are represented in the risk model. Their use, especially during design, can contribute significantly to safety, since they enable risk reduction efforts to be focused on those known issues of greatest safety significance.

Columbia accident, insulating foam from the external tank impacted the wing leading edge reinforced carbon-carbon (RCC), puncturing it and allowing an entryway for hot plasma upon reentry into the Earth's atmosphere.[§§]

The situation is illustrated in Figure 4. On the left side of the figure is the risk due to known scenarios that are included in the risk analysis. Depending on the nature and magnitude of a given scenario, quantification may be based on synthetic, actuarial, or bounding methods. Synthetic methods are used when system-level risk data are scarce and must instead be constructed via probabilistic modeling of accident scenario initiation and propagation. Actuarial methods can be used when the volume of data supports quantification of demonstrated risk with reasonable degree of certainty. Bounding methods are typically used for residual risk contributors, provided that their aggregate contribution to explicitly quantified risk is small. In contrast, the right side of Figure 4 shows the risk due to the unknown and underappreciated scenarios (referred to throughout the remainder of this paper as "UU scenarios").[***] Their presence in the system is inferred by the historical observation that aerospace systems have consistently experienced scenarios that were either not identified during design, or whose probability and/or magnitude was underrepresented in analysis.

Figure 4: Characterizing the Contributors to Actual Risk

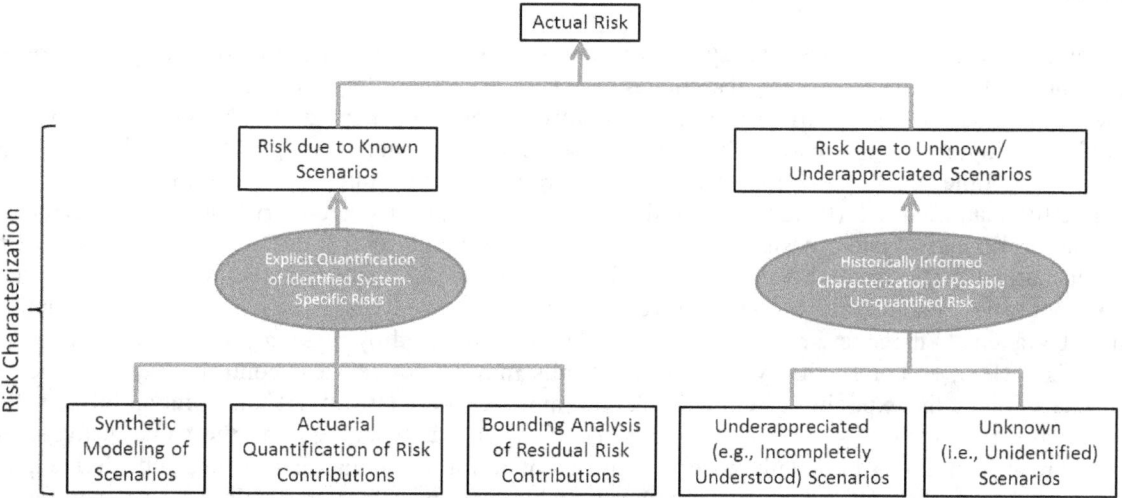

The Aerospace Safety Advisory Panel (ASAP) [8] and others have raised the need to consider the gap between actual risk and explicitly quantified risk when applying safety thresholds and goals. This concern reflects the expectation that during the early stages of operation there is likely to be significant risk due to UU scenarios. NASA's agency-level safety thresholds and goals do not explicitly address the question of how to account for UU scenarios. Nevertheless, the expectation is that the safety thresholds and goals refer to the system's actual risk, which includes both known, adequately modeled scenarios as well as UU scenarios.[†††]

[§§] Because of the often holistic and environment-dependent nature of such interactions, they tend not to be revealed by subsystem testing. Full-up testing has the potential to reveal them, but the cost of full-up testing in as-flown environments is generally too high to allow a sufficient volume of testing. Consequently, they tend to remain unknown (or known but underappreciated) until they manifest as an accident.

[***] Note that the term "underappreciated" refers here to scenarios whose likelihoods of occurrence or severity of impact are underestimated because of a deficiency in knowledge about the etymology of the risk. This differs from risks that are underestimated because they have not been analyzed in detail. The Challenger and Columbia accidents would fall in the category of underappreciated risks as we define the term here.

[†††] The position that UU scenarios should be acknowledged and accounted for when evaluating whether safety thresholds and goals are satisfied is also reflected in the ASAP Annual Report for 2011 [8].

3.2. Addressing Risk Analysis Incompleteness via Safety Performance Margin

One possible approach for characterizing the contribution of UU scenarios to the actual risk is imbedded in the concept of *safety performance margin* (referred to in the NASA System Safety Handbook as *safety risk reserve*). In the safety performance margin approach, the actual risk of a system is understood to be the sum of the risk from known scenarios, as explicitly quantified using traditional risk analysis methods, plus the risk from UU scenarios, as characterized by the safety performance margin. Just as the minimum tolerable level of safety starts at an initial value (the safety threshold) and diminishes to a lower value as flight experience is gained (the safety goal), the safety performance margin starts an initial value that is consistent with historical information about the magnitude of risks from UU scenarios and subsequently diminishes with time as safety performance information is gained through system operation (including tests). The concept is illustrated in Figure 5.

A limit on the allowable explicitly quantified risk can be derived by subtracting (in risk terms) the safety performance margin from the minimum tolerable level of safety. If the explicitly quantified risk is within this limit, then by implication there is reasonable assurance that the actual risk is within the minimum tolerable level of safety. Methods for establishing an initial safety performance margin and a margin draw-down profile based on historical data for similar systems are currently being investigated by the NASA Headquarters Office of Safety and Mission Assurance (OSMA).

As an example, Figure 5 illustrates system development and operation within a defined profile for the minimum tolerable level of safety that moves over time from the safety threshold (for initial flights) to the safety goal (for long-term operation). Initially, given a concept of operations, an analysis is performed to determine a reasonable value for the safety performance margin (the length of the light blue bar at time A). Subtraction of this margin from the safety threshold leads to the limit on the explicitly quantified risk (the point marked A1). In addition, a "first-order" risk analysis is performed on the preliminary system design to scope out the mean value of risk due to known scenarios (the point marked A2). Since the value of the safety measure at A2 exceeds the value at A1, the system does not satisfy the minimum tolerable level of safety. To remedy the situation, the following actions may be taken: 1) the safety performance margin may be reduced by making provisions to reduce the UU risks (B1); and/or 2) the system design details may be refined and controls may be added to further mitigate the explicitly quantified risk and improve safety (B2).‡‡‡ At this point an initial safety requirement for explicitly quantified risks is specified (B3) as being equal to the value of the safety measure at point B1. Additionally, a decreasing safety performance margin profile can be derived, and subtracted from the profile for the minimum tolerable level of safety, to obtain a safety requirement profile against which the explicitly quantified risk will be assessed over the operational life of the system. At time "C" the known risk of the as-built system satisfies the safety requirement. At time "D," newly discovered scenarios are added to the risk analysis, increasing the explicitly quantified risk beyond the safety requirement. Mitigations are introduced and the system is re-analyzed (E) to demonstrate that the known risk has been brought back within the requirement.§§§ During system operation, proactive safety upgrade and improvement programs reduce the risk in increments (F) until the safety goal is met (G).

---

‡‡‡ The two are not independent. Organizational, programmatic, and design philosophy changes can impact the quantitative assessment of risk, and design refinements can impact the factors that affect safety performance margin.

§§§ If the risk could not be brought within the requirement, the issue would be elevated according to the program/project risk management process, and could potentially be resolved by re-baselining the safety requirement, if organizational and/or programmatic changes provide an adequate basis for reducing the safety performance margin.

Figure 5: Conceptual View of Managing Safety Performance consistent with the Minimum Tolerable Level of Safety

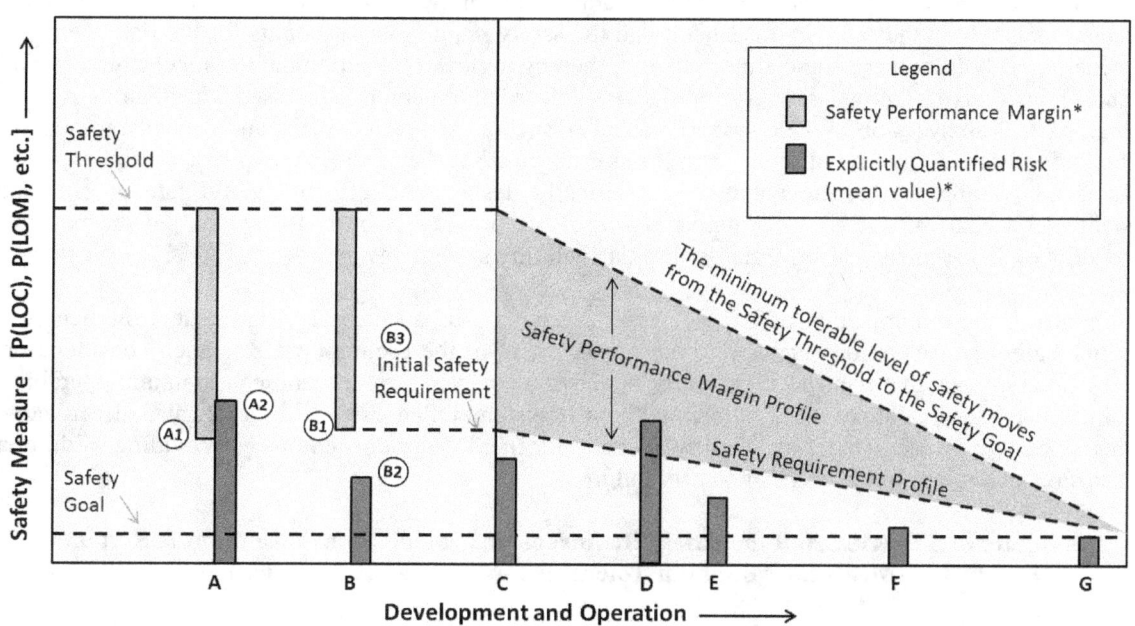

*The bars represent the value of an appropriate statistic that reflects an acceptable degree of certainty that the safety requirement is met.

4. MAKING THE CASE THAT THE SYSTEM MEETS THE MINIMUM TOLERABLE LEVEL OF SAFETY

The NASA System Safety Handbook introduces the construct of the risk-informed safety case (RISC), defined as "a structured argument, supported by a body of evidence, that provides a compelling, comprehensible and valid case that a system is or will be adequately safe for a given application in a given environment." The safety case addresses all aspects of safety, including the standing of the system relative to the minimum tolerable level of safety.**** The elements of the RISC are [14]:

- An explicit set of safety claims about the system(s), for example, the probability of an accident or a group of accidents is low.

- Supporting evidence for the claim, for example, representative operating history, redundancy in design, or results of analysis.

- Structured safety arguments that link claims to evidence and which use logically valid rules of inference.

The claims made (and defended) by the RISC dovetail with the safety objectives negotiated at the outset of system formulation. In other words, the satisfaction of each distinct safety objective is stated as a corresponding distinct claim in the RISC. By substantiating each claim with appropriate arguments and supporting evidence, the RISC demonstrates that the corresponding objective has been met and, thus, that the system is adequately safe.

**** For example, the claim that the system is ASARP is also within the scope of the RISC, as is the claim that proactive safety upgrade and improvement programs will enable the safety goal to be met at some (reasonable) point in the future.

Figure 6 builds on the decomposition of actual risk into known and UU components as shown in Figure 4, notionally illustrating the structure of the safety claim that the system meets the minimum tolerable level of safety. This claim is the conjunction of two sub-claims: the risk due to known scenarios is within the safety requirement; and the safety requirement accounts for the risk due to UU scenarios. If both of these sub-claims are valid, then by logical implication the top-level claim is valid. The claim that the known risk is within the safety requirement is addressed via quantitative risk analysis that analyzes observable system attributes, such as test results/plans and controls, to quantify the known risk. As indicated by the term "risk-informed" in the figure, known risk can be explicitly managed by applying controls that are specifically designed to effectively mitigate or eliminate identified scenarios. Indeed, quantitative risk analysis provides a basis for evaluating the potential benefits of different control strategies, particularly during system design.

In contrast, the minimization of UU risk chiefly relies on broad measures that reflect the degree to which safety considerations factor into the management of the program/project. Such considerations provide assurance that the system is robust against unexpected stresses, and can maintain operability (or at least safety) in the event of failures within the system. Of course, these factors also affect known risk, e.g., by assuring the functionality of risk-informed controls and/or by providing additional measures of defense in the event of control failure.

Figure 6: The RISC and the Safety Requirement Provide Assurance that the System Meets the Minimum Tolerable Level of Safety (notional)

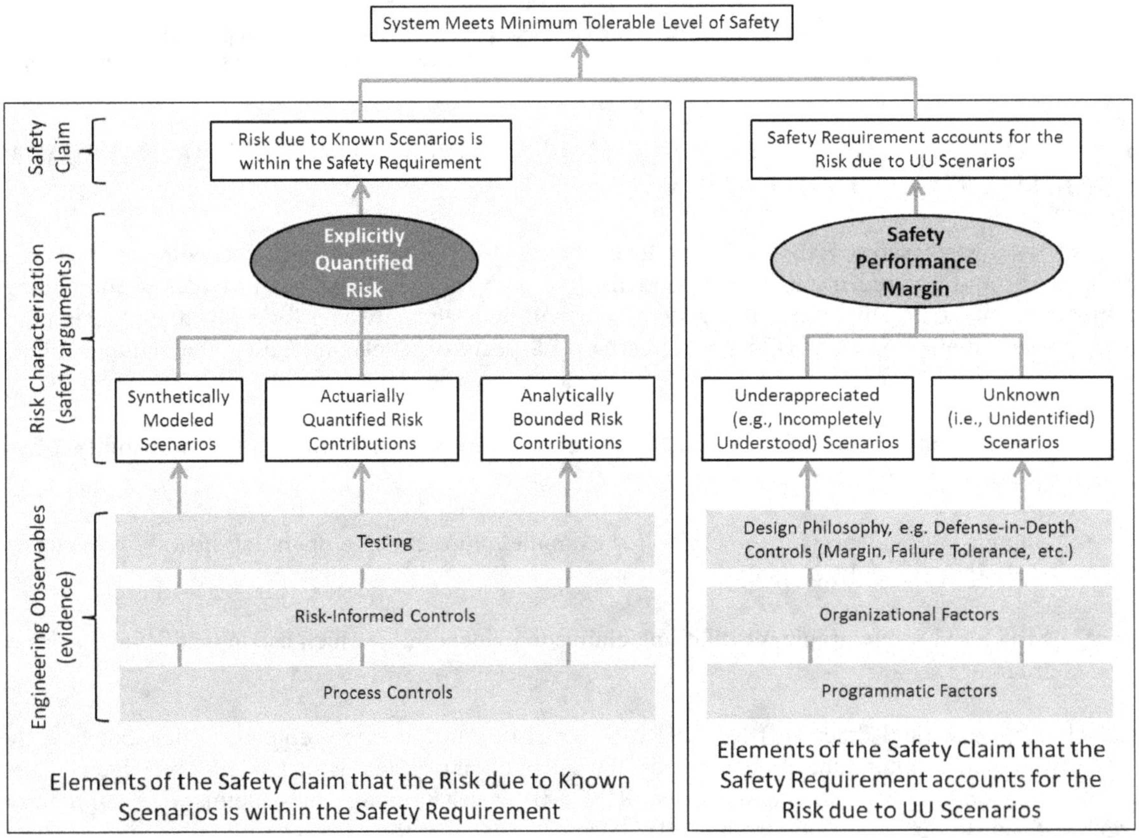

It is the responsibility of the organizational unit upon whom the safety requirement is levied to make the case that the system meets it. However, it is not the responsibility of that organization to make the case that the requirement itself is appropriately derived.†††† That responsibility rests with the organizational unit at the next higher level of the NASA hierarchy, who owns the requirement and is responsible for risk management oversight of the lower-level unit. Correspondingly, the case that the safety requirement accounts for the risk from UU scenarios is the responsibility of this higher-level unit.

The situation is illustrated in the notional example of Figure 7. In this figure, an upper-level organization establishes the minimum tolerable level of safety for a service of the type to be provided by a service provider. This minimum tolerable level of safety is levied on the program that is managing the service provider and focuses on actual safety risk. The managing program operationalizes this requirement by deriving a safety requirement that is levied on the service provider and focuses on explicitly quantified risks. This safety requirement, including quantitative risk analysis protocols for demonstrating that the requirement is met, is informed by a safety performance margin that is based on relevant historical data, supplemented with applicable system and organizational information from the service provider.

Figure 7: Activity Flowchart for Meeting the Minimum Tolerable Level of Safety

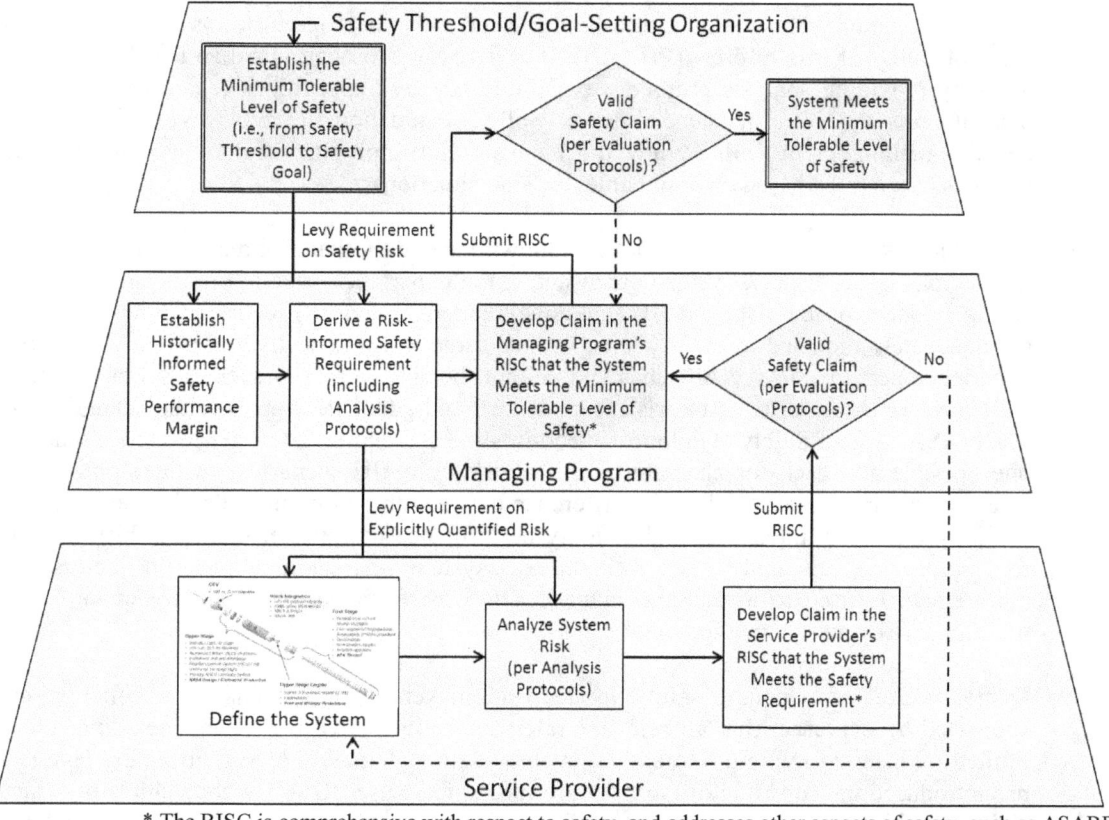

* The RISC is comprehensive with respect to safety, and addresses other aspects of safety, such as ASARP.

The service provider demonstrates adherence to the safety requirement by conducting a risk analysis per the specified analysis protocols, and, in the service-provider-generated-RISC, makes the claim that the safety requirement is met. This RISC is submitted to the managing program, who evaluates the claim according to established protocols. If it is sound, the managing program uses it, in conjunction with the claim that the safety requirement has been appropriately derived, to make the claim in its own RISC that the system meets the minimum tolerable level of safety established by the safety

---

†††† Although it is the responsibility of that organization to establish the feasibility of successfully meeting it, per NPR 8000.4A.

threshold/goal-setting organization. This RISC is submitted to the upper-level organization, who evaluates the managing program's claim according to appropriate protocols. If it is sound, then there is a sound basis for concluding that the system meets the minimum tolerable level of safety.

5. CONCLUSION

This paper discusses the role of the Agency's safety thresholds and goals in the context of "adequate safety" as discussed in the NASA System Safety Handbook. The following bullets summarize the discussion and reiterate some of the challenges and potential approaches to its implementation:

- "Adequate safety" for a given technology means both (1) meeting a minimum tolerable level of safety that is acceptable to the decision maker, and (2) being as safe as reasonably practicable (ASARP) such that safety performance is given priority relative to technical, cost, and schedule performance, given the subject technology and program constraints.

- NASA has instituted requirements for establishing safety thresholds to be used by the Agency as criteria for program acquisition decisions, and safety goals that reflect expectations about the long-term safety performance that is achievable from the system as its design and technology mature. The goal and threshold are currently specified in terms of a risk measure: the probability of loss of crew (P(LOC)). It is necessary to operationalize the thresholds and goals by providing analysis protocols explaining agency expectations regarding the analysis that supports the performance claims, as well as evaluation protocols explaining how the decision-making entity will review the analysis and apply its results. In principle, these protocols need to be based on applicable decision situations.

- Experience with various technologies has shown repeatedly that risk analysis incompleteness is a serious issue. Especially for new systems, the experience base is too meager to provide strong evidence of low risk, and UU (unknown/ underappreciated) scenarios may exist that are not adequately modeled in the analysis. In fact, a recent retrospective of Space Shuttle flight experience performed by NASA has shown that the risk from UU scenarios was initially at least three times as large as the risk from known and adequately appreciated scenarios [15]. The NASA System Safety Handbook introduces the concept of safety performance margin as one possible approach for characterizing the risk from UU scenarios. In this approach, the safety requirement is derived as the difference between the minimum tolerable level of safety (initially the safety threshold) and a safety performance margin derived from historical data for similar systems and adjusted based on system/organizational-specific factors. This requirement is used to assess and manage safety performance as explicitly quantified via traditional risk analysis methods.

- Demonstration of adequate safety to decision makers requires making a coherent case, supported by evidence, that all relevant safety objectives have been met, including, but not limited to, meeting minimum tolerable levels of safety. The NASA System Safety Handbook has introduced the construct of the risk-informed safety case (RISC) as the vehicle by which a claim of adequate safety is conveyed. The RISC serves as a comprehensive proxy for the safety of the system, stressing appropriate processes, clearly stating the assumptions that must be actualized if the safety claims are to remain valid, and committing to ongoing analysis of operating experience, so that safety performance improves continuously. The RISC is not a radically new idea. Rather, it is proposed as a formalization and integration of processes and ideas that are already in place or being incorporated to support decision contexts such as certification. The RISC is meant to subsume those processes, and furnish a coherent argument for how safe the system is or will be.

Acknowledgements

The research was carried out at NASA, and at Information Systems Laboratories, Idaho National Laboratory, and the Jet Propulsion Laboratory, California Institute of Technology, under contract with NASA.

References

[1] NASA. NPR 8705.2B, *Human-Rating Requirements for Space Systems*, Washington, DC. 2008.
[2] NASA. NASA/SP-2010-580, *NASA System Safety Handbook*, Washington, DC. 2010.
[3] U.S. Code of Federal Regulations, 10 CFR 20, *Standards for Protection Against Radiation*, Washington, DC. 1991.
[4] Parliament of the United Kingdom, *Health and Safety at Work etc. Act*, London, UK. 1974.
[5] NASA. NASA/SP-2010-576, *NASA Risk-informed Decision Making Handbook*, Washington, DC. 2010.
[6] NASA. Decision Memorandum for the Administrator, "Agency's Safety Goals and Thresholds for Crew Transportation Missions to the International Space Station (ISS)," Washington, DC. 2011.
[7] NASA. NASA/SP-2011-3421, *Probabilistic Risk Assessment Procedures Guide for NASA Managers and Practitioners*, Second Edition, Washington, DC. 2011.
[8] ASAP. *Aerospace Safety Advisory Panel Annual Report for 2011*, Washington, DC. 2012.
[9] Review of U.S. Human Spaceflight Plans Committee, "Seeking a Human Spaceflight Program Worthy of a Great Nation," Washington, DC, October 2009.
[10] Kaplan, S., et al., "On the Quantitative Definition of Risk," Risk Analysis, Vol. 1, No. 1. 1980.
[11] Leveson, N., "The Use of Safety Cases in Certification and Regulation," Journal of System Safety, Vol. 47, No. 6. 2011
[12] Vesely, W., et al., "Demonstrating the Safety and Reliability of a New System or Spacecraft: Incorporating Analyses and Reviews of the Design and Processing in Determining the Number of Tests to be Conducted," NASA Office of Safety and Mission Assurance, Washington, DC. 2010.
[13] Morse, E., et al., "Modeling Launch Vehicle Reliability Growth as Defect Elimination," Valador Inc., Herndon, VA. 2011.
[14] Bishop, P., and Bloomfield, R., "A Methodology for Safety Case Development, Safety-Critical Systems Symposium," Birmingham, UK. 1998.
[15] Hamlin, T., et al., "Shuttle Risk Progression: Use of the Shuttle PRA to Show Reliability Growth," AIAA SPACE Conference & Exposition. 2011.

Improving Consistency Checks between Safety Concepts and View Based Architecture Design

Pablo Oliveira Antonino[a,*], Mario Trapp[a]
[a] Fraunhofer IESE, Kaiserslautern, Germany

Abstract: Despite the early adoption of ISO 26262 by the automotive industry, managing functional safety in the early phases of system development remains a challenge. One key problem is how to efficiently keep safety assurance artifacts up-to-date considering the recurrent requirements changes during the system's lifecycle. Here, there is a real demand for means to support the creation, modification, and reuse of safety assurance documents, like the Safety Concepts described in ISO 26262. One major aspect of this challenge is inconsistency between safety concepts and system architecture. Usually created by different teams at different times and in different contexts of the development environment, these artifacts are often completely disassociated. This becomes even more evident when system maintenance is necessary; in this case, the inconsistencies result in intensive efforts to update the safety concepts impacted by the changes, and, consequently, significantly decrease the efficiency and efficacy of safety assurance. To overcome this challenge, we propose a model-based formalization approach for specifying safety concepts that allows creating precise traces to architectural elements while specifying safety concepts using natural language. We observed that our approach minimize the inconsistencies between safety models and architecture models, and offers basis to perform automated completeness and consistency checks.

Keywords: Safety Concepts, Safety Requirements, Architecture Design, Traceability.

1. INTRODUCTION

In 2011, ISO 26262 [1] was published as a safety standard in the automotive industry, emphasizing functional safety management in early phases of system development. Despite its early adoption, there are still open issues that limit the efficiency of assuring the safety of complex systems. Our experience has shown that two core contributors to these issues are (i) recurrent requirements changes during development time and over the system's lifetime, and (ii) the multitude of different artifacts to be considered. This intermittent dynamics leads to challenges regarding how to efficiently keep safety assurance artifacts up to date. One central safety assurance artifact defined in ISO 26262 is the Safety Concept. Safety concepts are requirements with a strong emphasis on the architectural elements that compose the measures to be used to prevent safety-critical failures [1].

In practice, safety concepts have been defined by means of natural text in documents, spreadsheets, or requirements databases. Sometimes, graphical notations like the Goal Structuring Notation (GSN) [2] or UML [3] are used to provide a more structured overview. Nevertheless, the lack of an underlying formalism of these approaches is a key factor contributing to the incompleteness and inconsistency of safety concept specifications.

One major aspect of this challenge is the relationship between safety concepts and the system architecture. As these two artifacts are usually created by different teams at different moments and in different contexts of the system development environment, they are, often, completely disassociated. However, by definition, the requirements defined in safety concepts, often result from a safety analysis of the preliminary architecture. Therefore, this lack of traces between safety concept and architecture

---
* pablo.antonino@iese.fraunhofer.de

is a key factor contributing to inconsistencies between safety concepts and the actual architecture design, and, consequently, to the incompleteness of safety concept specifications.

To overcome this challenge, we propose a model-based formalization technique for specifying safety concepts that supports safety engineers in creating precise traces to architectural elements while specifying safety requirements using natural language. Our approach consists of two items: (i) the Safety Concept Decomposition Pattern, which is a structural decomposition of elements that we understand to be mandatory in any safety concept specification, and (ii) the Parameterized Safety Concept Specification templates, which are generic parameterized textual templates that should be instantiated for the different elements of the Safety Concepts Decomposition pattern, and has the purpose of guide engineers during the specification of the safety concepts using natural language. With that, we ensure seamless integration of safety concepts and architectural design without the need to use formal specification languages like Lambda Calculus or Z, rather preserving intuitiveness during the specification of safety concepts.

The remainder of this paper is organized as follows: In Section 2, we provide a general overview of ISO 26262, particularly of safety concepts; in Section 3, we discuss the main challenges in specifying safety concepts in practice; in Section 4 we discuss the related works; in Section 5, we present our approach; in Section 6, we show how part of the safety concept of a Power Sliding Door Module is specified with our approach; and in Section 7, we conclude and present perspectives for future works.

2. ISO 26262 AND THE NOTION OF SAFETY CONCEPTS

ISO 26262 is an adaptation of IEC 61508 [4], addressing functional safety of electrical and/electronic (E/E) systems in the automotive industry. It defines a safety lifecycle that addresses safety-related aspects during the concept, development, and production phases (cf. Figure 1). It is important to mention that ISO 26262 deals only with possible hazards caused by malfunctions of E/E safety-related systems; it does not address hazards like electric shocks, radiation, or corrosion (a complete list can be found in [1]). Safety Concepts are particularly addressed in the Concept and Product Development phases (cf. Figure 1). Therefore, in this section we will focus only in these two phases of the safety lifecycle.

The Concept phase is where the following items are considered: (i) item definition, (ii) initiation of the safety lifecycle, (iii) hazard analysis and risk assessment, and (iv) the functional safety concept. The main item of interest for us in this phase is the Functional Safety Concept (FSC), which is the *"specification of the functional safety requirements, with associated information, their allocation to architectural elements, and their interaction necessary to achieve safety goals."* [1].

The Product Development at the system level phase is where technical safety requirements (refinement of the functional safety concept) are specified, taking into account not only the functional concept, but also technical aspects of the preliminary architecture. The specification of the technical safety requirements and their allocation to system elements (software and hardware) is called Technical Safety Concept (TSC) [1].

3. SAFETY CONCEPT SPECIFICATION IN PRACTICE

To better understand these challenges, let us consider the example of a Power Sliding Door Module (PSDM) system adapted from [5]. We adapted the function network (cf. Figure 2) and the deployment view (cf. Figure 3) from [5], and created a data view (cf. Figure 4) on our own, assuming, then, these three views as the PSDM preliminary architecture. To get a general understand on the PSDM, let us start with the central component *Open Door Computation* (cf. Figure 2), whose mission is to trigger the door opening based on two events:

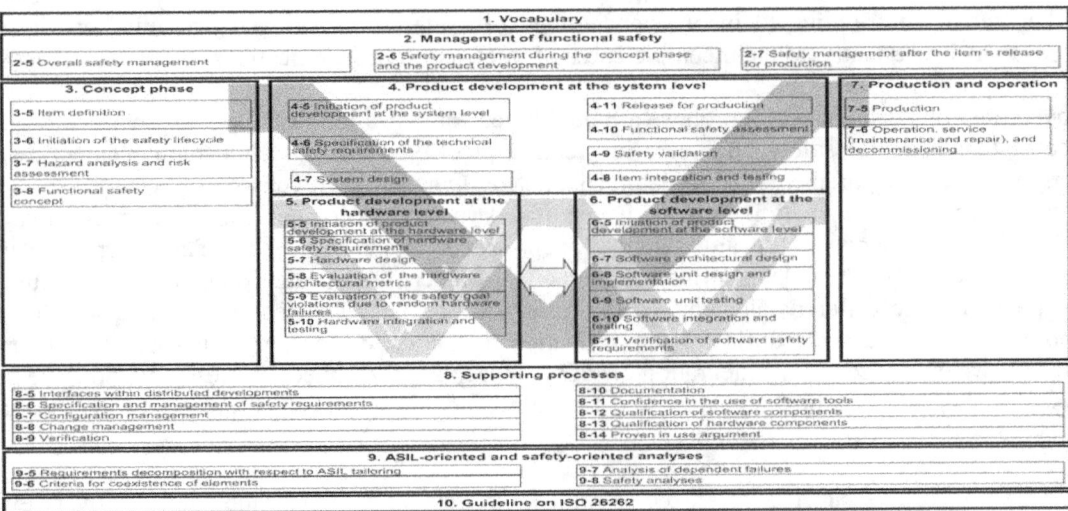

Figure 1 - ISO 26262 overview (extracted from [1]).

i Passenger/driver request: The request to open the sliding door is triggered when the driver or passenger presses the *Door Opener button* (cf. Figure 2). This sends a signal to the *Door Opener Request Processor*, which is responsible for converting the door opener request signal to a format that can be used by the *Open Door Computation* component.

ii Vehicle Speed: To determine the *vehicle speed* (cf. **Figure 4**), first the *wheel rotation speed* is read by the *Wheel Rotation Speed Sensor* (cf. Figure 2), which then sends this information to the *Wheel Rotation Speed Processor,* which is responsible for converting the sensed wheel rotation speed into a format that can be used by the other components. Next, the processed *wheel rotation speed* is sent to the *Computation Vehicle Speed* component, which computes the vehicle speed based on this information. From this point, the computed vehicle speed is sent to the *Open Door Computation* component and also to the *Vehicle Speed Integrity Checker*, which checks if that the vehicle speed is not corrupted or outside a predefined acceptable value range. If the speed value is not as expected, it will notify the *Open Door Computation* component that the value sent by the *Computation Vehicle Speed* component should not be accepted.

Once the *Open Door Computation* component receives the open door request and the vehicle speed, it evaluates if the vehicle is at a speed that allows the door to be opened, which, according to the specification available in [5], is 15km/h. If the vehicle is at 15km/h or less, the *Open Door computation* component notifies the *Open Door Signal Generator* component, which, then, sends a signal that triggers the *Sliding Door Actuator*, which is responsible for opening the sliding door. There is also a component called *Sliding Door Actuator Monitor*, which is responsible for verifying whether the door was indeed opened. If it detects that the actuator did not work properly, it notifies the *Open Door Computation* component to re-send the open door command.

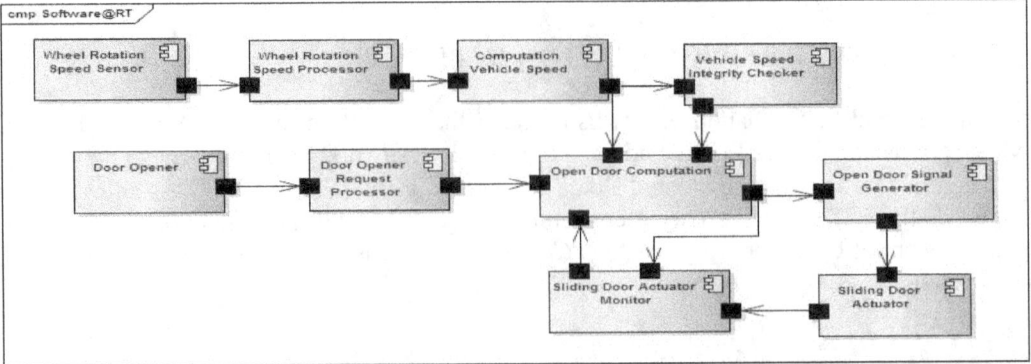

Figure 2 – Power Sliding Door Module functional network (adapted from [5]).

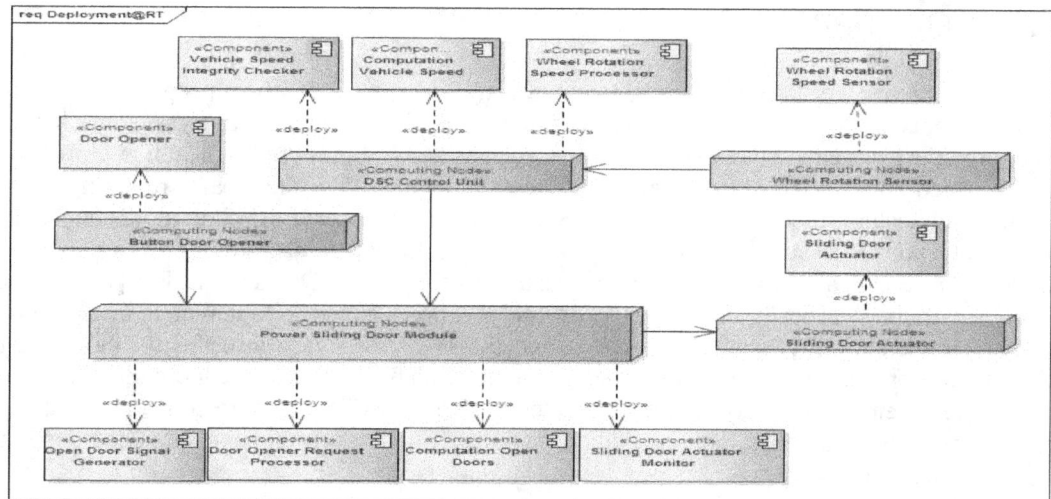

Figure 3 - Power Sliding Door Module deployment view (adapted from [5]).

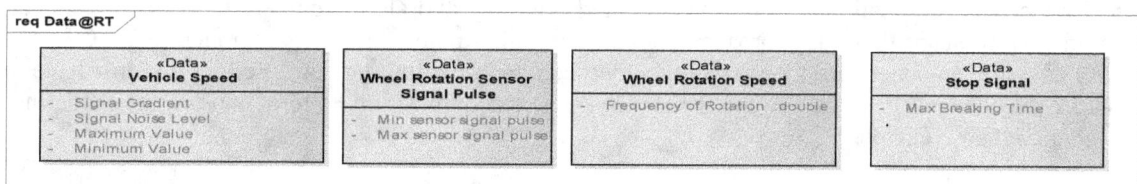

Figure 4 - Power Sliding Door Module data view.

Let us assume that the safety engineers identified that if the vehicle speed is not updated to the PSDM system within 100ms, the safety of the passengers might be compromised, because they might be able to open the door while the vehicle is at a high speed. To better structure this potential situation, let us assume that they have explicitly specified the hazard, safety goal, and a strategy (cf. Figure 5) that should be addressed by the system architecture to avoid the occurrence of this situation. The specification presented in Figure 5 is already decomposed to a much greater extent than in actual practice. In general, one will only find a statement like: *"The information about the actual vehicle speed should be updated within a cycle time of 100ms"* and a simple link from this textual statement to a component in the architecture model that addresses this safety requirement, and this link will be considered the safety concept (cf. [5]). This kind of traceability might be useful when the traces are simple, for instance, when the architectural elements that address a safety concept are located in only one architecture diagram. Now let us consider that to address this safety concept, more than one component is necessary; let's assume that it is also necessary to modify deployment items, data formats, and types, change the structure of a component in terms of adding and/or removing ports, change data flows, etc. Even appropriate automated support, such as the one provided by PREEvision[†] and MEDINI Analyze[‡], do not provide adequate support for specifying safety concepts and, at the same time, supporting the creation of proper trace links to architectural elements from different architecture views along the textual specification.

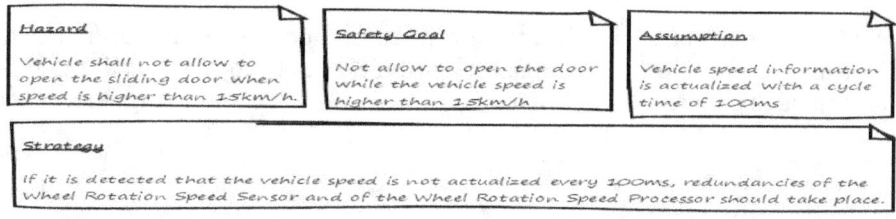

Figure 5 - PSDM Safety Concept specification (adapted from [5]).

[†] http://vector.com/vi_preevision_en.html
[‡] http://www.ikv.de/index.php/en/products/functional-safety

4. RELATED WORK

As already mentioned, in practice, the most common way to specify safety concepts is using natural text in documents, spreadsheets, or requirements databases. There are, nevertheless, some initiatives aimed at providing more structured specifications, such as the one proposed by Habli et al. [6], where it is investigated how traceability between safety cases modeled with the Goal Structuring Notation (GSN) and architectural models represented in SysML can be improved. Birch et al. [2] analyzed the implicit safety argument structure of ISO 26262 and conducted a case study showing how these arguments can be logically decomposed and structured using GSN. Denney and Pai [7] proposed a breakdown pattern specifically for safety case specifications aimed at automated instantiation of arguments. Domis et al. [8] introduced Safety Concepts Trees (SCTs) as a means for modeling how safety goals are broken down into safety requirements, which, in turn, are continuously refined by other safety requirements using typical logical gates. However, SCTs are more concerned with taking safety concepts to the modelling domain, and not with the content of the elements that compose safety concepts specifications as we do.

In a nutshell, we indicate how to structure and relate the elements that we consider important to be in safety concepts specification (cf. Section 5.1), and how to elaborate the content of each element (cf. Section 5.2). It means that our approach is not strictly tight to any modelling approach (e.g. SCTs and GSN); actually, engineers are free to use the approach and notation they are used to, and just have to consider the elements described in the Safety Concepts Decomposition Pattern, and use the guidelines to elaborate their contents.

With respect to the support to structure and relate the elements of safety concepts specification, the closest to our approach are those proposed by Birch et al. [2] and Denney and Pai [7]. Nevertheless, Denney and Pai focus on a structural decomposition of safety cases, and do not offer appropriate means for supporting safety concept decomposition. Birch et al. propose rational arguments strategies that are useful to justify *"why"* a safety requirement is, indeed, a Safety Requirement, using GSN to structure the justification argument. Our approach does care with the *"why"* as well, but in a more structured way (cf. Section 5), as we explicitly indicate elements that, when precisely described, provides enough basis to justify the existence of a safety concept, such as the potential causes of failures, their nature, and the associated failure mode. Additionally, we provide means to come up with precise descriptions of (i) these elements that justify the existence of safety concepts, and (ii) safety requirements specifications that indicate, already along their definition, *"how"* the architecture should be modified to address failure causes, and, consequently, satisfy the safety goal.

With respect to support in elaborating the content of the elements that compose safety concepts specifications, Firesmith [9] presents parameterized requirements for different types of safety requirements. He argues that, because safety requirements usually have the form of system specific quality criteria associated with different levels of quality measures, they can be written as instances of parameterized safety requirements templates. However, the parameterized templates proposed by Firesmith are far from appropriate to specify safety concepts elements. Therefore, we have created parameterized templates that are appropriate to the elements of our Safety Concepts Decomposition Pattern, strongly focusing on parameterization of architecture design elements.

Summarizing, to the best of our knowledge, there is no other model-based approach that allows semi-formal hierarchical decomposition of functional and technical safety concepts, and that also guides the creation of traces to architectural elements from multiple views while specifying safety concepts using natural language.

5. MODEL-BASED FORMALIZATION OF SAFETY CONCEPTS SUPPORTING THE USE OF NATURAL LANGUAGE

In this section we present our approach to specify safety concepts, which consists of two core items: (i) the Safety Concept Decomposition Pattern, which is a structural decomposition of elements that we consider important to be in any safety concept specification, and (ii) the Parameterized Safety Concept Specification templates, which are generic parameterized textual templates that should be instantiated for the different elements of the Safety Concepts Decomposition pattern

To properly specify safety concepts with our approach, we assume that the following three sets of artifacts should be already in place: (i) the result of the Hazard and Risk analysis, (ii) the preliminary architecture, and (iii) the results of Failure Mode and Effect Analysis – FMEA, where problems in the preliminary architecture are described.

5.1. Safety Concept Decomposition Pattern

ISO 26262 explicitly indicates the content of safety concept specifications, but it doesn't specify a defined structure for them. What happens in practice is that each safety concept specification has a different structure, most likely based on the understanding and experience of the engineers involved in the specification process. Furthermore, because of this lack of a structured definition, following the argumentation of safety concept specifications is not a trivial task, especially if the reader is not the author. The consequence is that it becomes difficult to ensure that all the aspects demanded by ISO 26262 are present in the specification, and it is also hard to ensure that the architecture elements referenced by the safety concepts are consistent with those specified in the architecture design. To overcome this challenge, we have specified the Safety Concept Decomposition Pattern, which comprises elements that we understand as being fundamental for ensuring the completeness of functional and technical safety concepts. The metamodel of the Safety Concept Decomposition Pattern is shown in Figure 6, which is followed by the descriptions of its elements.

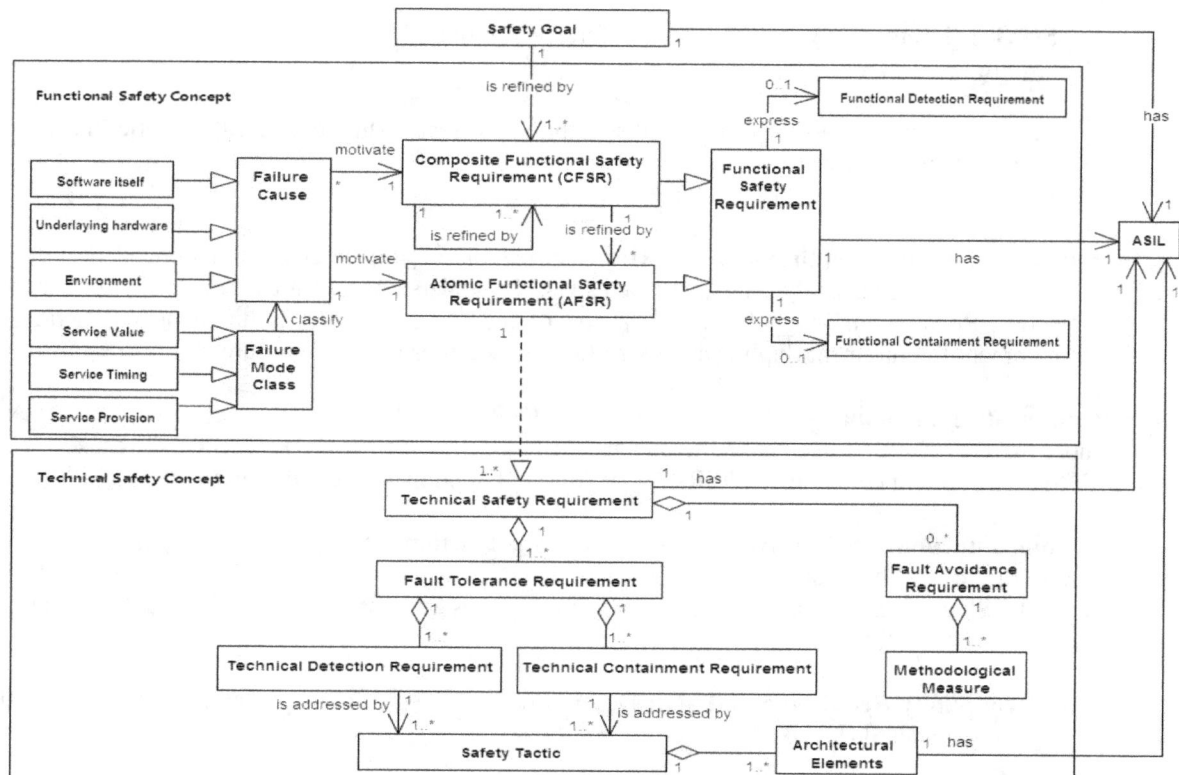

Figure 6 - Safety Concepts Decomposition pattern metamodel.

- **Safety Goal:** Top-level requirement resulted from the hazard and risk analysis assessment. For each hazardous event with an ASIL evaluated in the hazard analysis, a safety goal should be determined [1].

- **ASIL:** Levels to specify the necessary requirements and safety measures to avoid unreasonable residual risks, with D representing the most stringent and A the least stringent level [1].

- **Failure Cause:** Condition of a system (or parts of it) that motivates the existence of a safety requirement. It is usually identified when conducting a safety analysis to identify paths that can lead to a critical system failure. We adopted the convention proposed by Wu and Kelly [10], thus assuming that a cause can be categorized according to one of the three following abstract types:

    - **The software itself:** Incomplete or inaccurate specification, or incorrect design and implementation can cause unexpected behavior of the software.

    - **Underlying hardware:** Correct software can still misbehave because of unexpected behavior of the underlying hardware.

    - **Environment:** Also known as environment disturbances, causes can be originated outside the software and can be fed into the system in the form of inputs.

- **Failure mode class:** Manner in which an element or an item fails [1]. We understand that a failure mode class is a higher level of abstraction that classifies failure causes, in order to enrich them semantically. Following the example of Wu and Kelly [10], we also adopted the failure mode classification proposed by Fenelon et al. [11]:

    - **Service provision:** Omission (expected event does not occur), Commission (spurious occurrence of event).

    - **Service timing:** Early (event occurs before time required), Late (event occurs after the time required).

    - **Service value:** Coarse incorrect (detectable incorrect value delivered), Subtle incorrect (undetectable incorrect value delivered).

- **Functional Safety Requirement (FSR):** Specification of implementation independent safety measure, including its safety-related attributes [1]. We understand that it can express a Functional Detection Requirement (high level description of measures to detect failures) or a Functional Containment Requirement (high level description of measures to handle failures).

- **Composite Functional Safety Requirement (CFSR):** Functional Safety Requirement that has more than one failure cause that motivates its existence. i.e., $cfsr \in CFSR: \{FSR \| FSR.cause \geq 1\}$. CFSRs can be refined by other CFSRs, or by Atomic Functional Safety Requirement (AFSR).

- **Atomic Functional Safety Requirement (AFSR):** functional safety requirement that has only one failure cause that motivates its existence. i.e., $afsr \in AFSR: \{FSR \| FSR.cause = 1\}$. AFSRs refine CFSRs that cannot be decomposed into finer grains anymore. AFSR are realized by a set of Technical Safety Requirements.

- **Technical Safety Requirement (TSR):** Description of strategies to realize an Atomic Functional Safety Requirement [1].

- **Fault Avoidance Requirement:** We adopted the definition of Laprie, and consider Fault Avoidance Requirements as a group of means that aim for systems free of faults, and comprises fault prevention and fault removal mechanisms [12].

- **Fault Tolerance Requirement:** Description of means that allows *"living"* with systems that are susceptible to faults. [12].

- **Technical Detection Requirement:** Description of how Functional Detection Requirements will be realized by elements of the architecture design.

- **Technical Containment Requirement:** Description of how Functional Containment Requirements will be realized by elements of the technical architecture design. It describes means to take a system from a state containing errors and faults to a state without detected errors and without faults [12].

- **Safety tactics:** Architectural design decisions made to avoid or handle failures to which safety-critical systems are subject [10]. They become more concrete when they are realized by safety patterns indicating architectural elements (mainly components, connector, deployment units, and communication channels) and a set of constraints on how instances of these types should be combined into a system to detect or contain a failure [13].

5.2. Parameterized Safety Concept Specification Templates

Even though we do understand that the user should be free to write textual the safety concepts specifications, we believe that some guidelines can be useful to indicate items that must not be absent from the specification. Therefore, we have created Parameterized Safety Concepts Specification Templates that should be used to guide the engineers during the specification of some elements of the Safety Concepts Decomposition Pattern. The template elements delimited by square brackets are textual descriptions that should contain elements to be linked to the architecture design. We understand that these elements should be selected during the safety concept specification in order to ensure early traceability in the specification process. Moreover, these elements are mandatory in the safety concept specification because they are about the very core notion of safety concepts defined in ISO 26262, which is to assign architectural elements to the safety requirements. Therefore, if architectural elements are not included in the safety concept specification, its completeness is compromised. The textual descriptions between parentheses have no such strict constraints; however, they still address some constraints in the sense that they indicate important constraints that should be considered, as, for instance, an indication that a signal should not be sent later than within 2ms. The parameterized templates are as follows:

- **Safety Goal:** [System || Component Group || Component || Computing Node] shall (avoid || not cause || not allow || not be || not || no) (harm).

- **Functional Detection Safety Requirement:** The System shall detect (accidental harm | Safety incident | Hazard | Safety Risk). This template was reused from [9], and there it describes safety requirements of type Detection of Violation of Prevention.

- **Functional Containment Safety Requirement:** When the System shall detects (accidental harm | Safety incident | Hazard | Safety Risk), then the system shall (List of Actions). This template was also reused from [9], and there it describes safety requirements of type Reaction to Violation of Prevention.

- **Technical Safety Requirement:** The template for this element depends on the failure mode classification of the failure cause that motivates the existence of the Atomic Safety Requirement associated to the Technical Safety Requirement. The two possible templates are:

- **Service Value Failure Mode:** [System || Component Group || Component || Computing Node] shall (perform action) [artifact affected by action] (Values threshold of measurement: within || exactly with || not exceed || not less than) [Data constraint].

- **Service Timing Failure Mode:** [System || Component Group || Component || Computing Node] shall [perform action] [artifact affected by action] (timing threshold of measurement: within || before || after || exactly || no later than) [timing constraint].

> **Fault Tolerance Requirement:** Detect and Handle (type of violation) violation of [artifact affected by action].

> **Technical Detection Requirement:** It should be detected if [artifact affected by action] is not (action performed - past tense) (threshold of measurement) [Value Constraint||Timing Constraint].

> **Technical Containment Requirement:** This element should be described as free text so the engineer can describe his strategy in detail. However, it is important to highlight that this description must reference architectural elements.

> **Safety Tactics:** We have created specification templates for almost all the safety patterns described by Douglass [13], and examples are shown in Section 6.

It is important to highlight that the elements of the Safety Concepts Decomposition Pattern doesn't have associated parameterized templates are those that not necessarily have to reference architectural elements. However, the references can be created, whenever the engineers understand that such references will contribute to a clearer understand of the safety requirement. Another aspect to be considered is that we do not impose that the textual constructions have to be strictly formulated as indicated in the templates. However, it is strongly recommended that the specifications contain references to the elements indicated between the square brackets and the parentheses, since they are the key to ensuring the completeness and consistency of the safety concept specifications.

To illustrate the instantiation of these parameterized templates, consider the example below, where a Technical Safety Requirement and a Fault Tolerance Requirement are specified for an Airbag System. The color coding is intended to make it easier to understand the mapping between the template and the example:

> **Technical Safety Requirement:**

- **Template:** [System || Component Group || Component || Computing Node] shall (perform action) [artifact affected by action] (Values threshold of measurement: within || exactly with || not exceed || not less than) [Data constraint]

- **Example:** [Front Acceleration sensor] shall (send) [Front sensed acceleration signal amplitude] (with at least) [0,56dB]

> **Fault Tolerance Requirement:**

- **Template:** Detect and Handle (type of violation) violation of [artifact affected by action]
- **Example:** Detect and Handle (value range) violation of [front sensed acceleration signal amplitude]

6. SPECIFYING SAFETY CONCEPTS FOR A POWER SLIDING DOOR MODULE WITH OUR APPROACH

In this section we show how the Power Sliding Door Module example described in [5] look like when specified with our approach. However, due to space constraints we will focus in only two safety requirements: (i) *"The information about the actual vehicle speed should be actualized with a cycle time of 100ms."*, and (ii) *"The wheel rotation speed should be measured with an accuracy of at least 30 rad/min"*. Also due to space constraints, along this section we describe in detail only the first requirement; the model with the two requirements is show in Figure 7. It also important to mention that the the preliminary architecture considered was the one shown in Figure 2, Figure 3, and Figure 4. Please, also note that the items from the specification retain the square brackets and the parentheses to facilitate the understanding of the example. However, in practice, when the trace links are created, these signs should be removed to guarantee the natural flow of reading.

- **Safety Goal:** [Vehicle] shall (not allow) (door to be opened while the vehicle speed is in motion).

- **CFSR:** Control Unit shall send accurate vehicle speed information to power sliding door module.

- **Failure Cause 1:** Vehicle speed is not updated in time.

- **Failure Cause 2:** Wheel vehicle speed is not measured with the proper accuracy.

Note: Due to the lack of space, we will show the decomposition used to address only Failure Cause 1. The description of the items related to Failure Cause 2 can be seen in Figure 7.

- **AFSR (referring to Cause 1):** The information about the actual Vehicle Speed should be updated with an updating cycle of vehicle speed. *Violation: Timing Violation – Time to execute operation; **Failure Mode**: Service Timing.*

- **Technical Safety Concept (Service Timing Failure Mode):** [Computation Vehicle Speed Component] shall (update) [vehicle speed] (not later than) [a cycle time of 100 ms].

- **Fault Tolerance Requirement:** Detect and handle (timing accuracy) violation of [vehicle speed] updating.

- **Detection Requirement:** It should be detected if [Vehicle Speed] is not (updated) (not later than) [a cycle time of 100 ms] at the [**Computation Vehicle Speed component**].

- **Containment Requirement:** Redundancy - there should be a [redundant Wheel Rotation Speed Sensor] and a [redundant Rotation Speed Processor] that should substitute the [Wheel Rotation Speed Sensor] and [Rotation Speed Processor] if it is detected that the [vehicle speed] is not updated every [100ms].

- **Detection Safety Tactic:** Let's assume that the engineers decided to monitor the vehicle speed using a Watchdog. As previously mentioned, we have specified a grammar for most of the safety patterns described by Douglass [13].

    - **Template:** [Watchdog component] monitors [monitored architectural element] to check if [monitored aspect] is (action) (threshold of measurement) [Timing Constraint].

    - **Example:** [Watchdog component] monitors [computation Vehicle Speed component] to check if [vehicle speed] is (updated) (not later than) [a cycle time of 100ms].

- **Containment Safety Tactic:** Let's assume that the engineers decided to use homogeneous redundancies of Wheel Rotation Speed Sensor and Rotation Speed Processor.

    - **Template:** [Component], which is deployed to [Computing Node||Thread], have (n) homogeneous redundancy(ies), which is(are) deployed to: [Computing nodes||Threads] [n .. n-1].

    - **Example:** [Rotation Speed Processor], which is deployed to [DSC Control Unit], have (1) homogeneous redundancy, which is deployed to: [DSC Control Unit].

We observed that safety concepts bases on the Safety Concepts Decomposition Pattern offer great basis for safety engineers in identifying if all failure causes were properly safe-guarded. Another positive aspect is about the compliance created between functional and safety concepts, which is a valuable step towards safety concepts correctness. Another observed benefit is with respect to the consistency improvement between safety concepts and architecture design, because the preliminary architecture can be considered while the safety engineer are writing down the safety concept using the parameterized templates as basis, and not, as usual, first defining the safety concepts, and only later indicate which architecture element addresses them. In a sense, we observed improvements the overall system specification process, because engineers can make early detection if any rearrangement in the design is necessary, or if new architecture elements are required. Another positive aspect is that safety engineers are comfortable to use it because they can keep writing the specifications textually, and also can keep using the mechanisms they are used to for specifying safety concepts, because, in a nutshell, we only indicate the elements to be considered in the specification, how they should be structured, and suggests how the textual content should look like. Therefore, they can do it using well know modeling mechanisms such as GSN and SCTs, and tools like MEDINI Analyze and PREEvision.

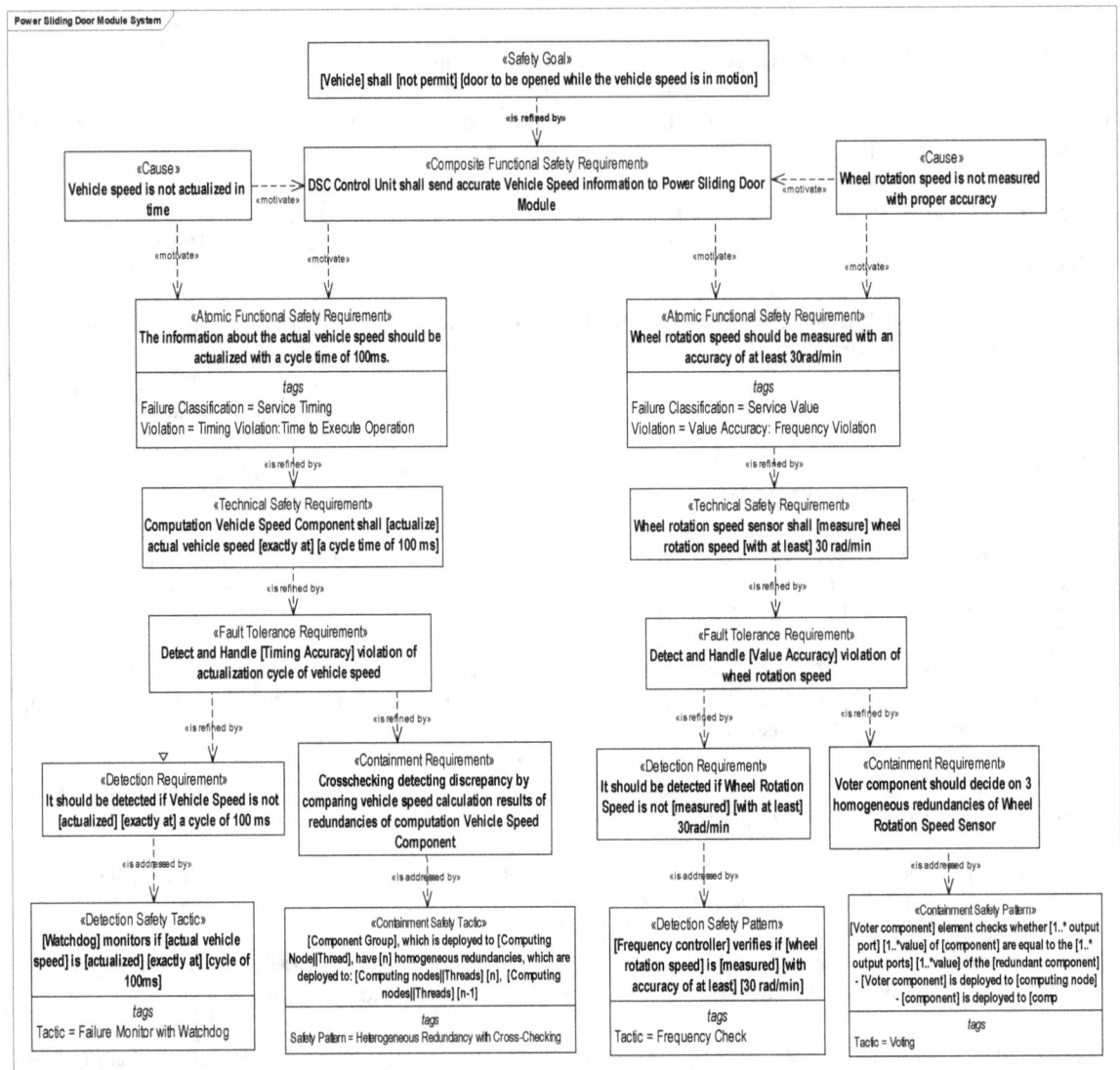

Figure 7 - PSDM Safety Concepts specified with our approach.

7. FUTURE WORK AND CONCLUSIONS

This work was motivated by the lack of approaches for semi-formal hierarchical decomposition of safety concepts and for the creation of traces to architectural elements while specifying safety requirements using natural language. To fill this gap, we proposed a model-based formalization technique for specifying safety concepts that supports safety engineers in creating precise traces to architectural elements while specifying safety requirements using natural language. The approach consists of a Safety Concept Decomposition Pattern and a Parameterized Safety Concepts Specification Templates, with the former specifying the elements to be considered in a safety concept specification and the latter specifying a grammar containing elements whose presence in textual specifications of safety concepts is strongly recommended to ensure completeness and consistency. This is our first step towards the realization of automated consistency and completeness checks of safety concepts.

Acknowledgments

This work is supported by the Fraunhofer-Innovation Cluster Digitale Nutzfahrzeugtechnologie (Digital Commercial Vehicle Technology), and by the Software Platform Embedded Systems "XT" - SPES XT project. We would also like to thank Sonnhild Namingha for proofreading.

References

[1] International Organization for Standardization. *"ISO/DIS 26262 - Road Vehicles – Functional Safety"*, Technical Committee 22 (ISO/TC 22), Geneva, Switzerland, 2011.

[2] J. Birch, R. Rivett, I. Habli, B. Bradshaw, J. Botham, D. Higham, P. Jesty, H. Monkhouse, and R. Palin. *"Safety Cases and Their Role in ISO 26262 Functional Safety Assessment"*. Springer, In Proceedings of 32nd SAFECOMP, 2013.

[3] Object Management Group. *"UML profile for modeling QoS and FT characteristics and mechanisms.* Technical report, April 2008.

[4] International Organization for Standardization. *"IEC 61508 - Functional safety of electrical/electronic/programmable electronic safety-related systems"*, The International Electrotechnical Commission, Geneva, Switzerland, 1998.

[5] M. Hillenbrand, M. Heinz, K. D. Müller-Glaser, N. Adler, J. Matheis, and C. Reichmann. *"An approach for rapidly adapting the demands of ISO/DIS 26262 to electric/electronic architecture modelling"*. International Symposium on Rapid System Prototyping, 2010.

[6] I. Habli, I. Ibarra, R. Rivett, and T. Kelly. "Model-Based Assurance for Justifying Automotive Functional Safety" SAE Technical Paper, 2010.

[7] E. Denney, G. Pai. *"A Formal Basis for Safety Case Patterns"*. Springer, In Proceedings of 32nd SAFECOMP, 2013.

[8] D. Domis, M. Forster, S. Kemmann, and M. Trapp. *"Safety concept trees"*. In Reliability and Maintainability Symposium, 2009. RAMS 2009. Annual, pages 212 - 217, jan. 2009.

[9] D. Firesmith. *"A Taxonomy of Safety-Related Requirements"*, Software Engineering Institute White Paper, 2004.

[10] W. Wu and T. Kelly, *"Safety Tactics for Software Architecture Design"*, in COMPSAC, 2004.

[11] P. Fenelon, J. A. McDermid, M. Nicolson, and D. J. Pumfrey. 1994. *"Towards integrated safety analysis and design."* SIGAPP Appl. Comput. Rev. 2, 1 (March 1994), 21-32. DOI=10.1145/381766.381770 http://doi.acm.org/10.1145/381766.381770

[12] A. Avizienis, J.-C. Laprie, B. Rendell and C. Landwehr, *"Basic Concepts and Taxonomy of Dependable and Secure Computing,"* IEEE Trans. Dependable Secur. Comput., vol. 1, pp. 11--33, Jan 2004.

[13] B. P. Douglass. *"Real-Time Design Patterns: Robust Scalable Architecture for Real-Time Systems"*. Addison-Wesley Longman Publishing Co., Inc., 2005, Boston, MA, USA.

Uncertainty Evaluation in Multi-State Physics Based Aging Assessment of Passive Components

Askin Guler[a,*], Tunc Aldemir[a], and Richard Denning[a]
[a] Nuclear Engineering Program
The Ohio State University, Columbus, OH, USA

Abstract:

A methodology is presented to evaluate aging degradation of passive components under uncertainty. Stress corrosion cracking (SCC) degradation is selected as the example aging phenomenon and the methodology is implemented on the pressurizer surge line pipe weld of a pressurized water reactor. The degradation is described as a multi-state model consisting of six differential equations with system history dependent transition rates. The input data to the model include operating temperature, weld residual stress, stress intensity factor, thermal activation energy for crack initiation and crack growth. The associated uncertainties are represented by probability distributions derived from historical data, experimental data, expert elicitation, physics, or a combination of these. Latin Hypercube Sampling is used to generate observations from the distributions governing these parameters with a two-step approach that distinguishes between aleatory and epistemic uncertainties. The degradation model is solved by a semi-Markov approach using the concept of sojourn time to account for system history dependence of transition rates. The results are compared to a single step sampling process. The results show highest sensitivity of damage to weld residual stress.

Keywords: Aging, Passive components, LHS, semi-Markov, Uncertainty analysis

1. INTRODUCTION

Long term reliability of systems, structures and components (SSCs) in the existing fleet of operating reactors should be ensured to prevent a reduction in component and system safety margins due to aging. In order to address multiple aging mechanisms involving large numbers of components (with possibly statistically dependent failures) in a computationally feasible manner, a methodology is being developed where the sequencing of events leading to damage is conditioned on the physical conditions predicted in a simulation environment [1].

A state transition model [2] was selected as a case-study to implement the methodology. This model is applied to the pressurizer surge line pipe weld of a pressurized water reactor (PWR) to model primary water stress corrosion cracking (PWSCC) degradation during extended operation life (80 years) of the plant. The degradation model described in this paper was originally developed in [2] and later improved by using the sojourn time approach [3] with operational history-dependent transition rates [4]. The model has many input parameters including temperature, weld residual stress, stress intensity factor, and thermal activation energy for crack initiation and crack growth. The associated uncertainties are represented by probability distributions derived from historical data, experimental data, expert elicitation, physics, or a combination of these. The model output is pipe rupture probability as a function of operating time.

Although the separation of uncertainties as aleatory versus epistemic can be subjective, it can be helpful in supporting decision-making. In this paper, the uncertainty in the PWSCC model (Section 2) input is propagated using Latin hypercube sampling (LHS) while distinguishing between the aleatory

---

[*] guler.11@osu.edu

and epistemic uncertainties (Section 3). The results are compared to a single step approach that does not make such a distinction (Section 3.1). A sensitivity analysis is performed using the response surface methodology (RSM) to identify the most important model parameter that affects the rupture probability (Section 3.2). Conclusions of the study are given in Section 4.

2. THE MULTI-STATE SEMI-MARKOV MODEL

The state transition model to predict piping system reliability that is used in the paper was originally proposed by Fleming [5] and later adapted for system history dependent transition rates by Unwin et al. [2]. This 6-state state transition model is implemented for a PWR pressurizer surge line pipe weld (Alloy 182) for the case of a PWSCC scenario. Section 2.1 describes the model and Section 2.2 describes the procedure that is used to convert the model to a semi-Markov process using the concept of sojourn time [3].

2.1. Multi-State Model and Transition Rates

The state transition model [2] that was selected as a case-study for passive component degradation involves six ordinary differential equations which are solved by a semi-Markov approach to account for system history and local thermal operating condition-dependent transition rates [3,4]. Fig. 1 shows the Markov model states for the crack initiation and growth. State evolutions are described through

$$dS/dt = -\phi_1 S + \omega_1 M + \omega_2 D + \omega_3 C + \omega_4 L \tag{1}$$

$$dM/dt = \phi_1 S - \omega_1 M - \phi_2 M - \phi_3 M \tag{2}$$

$$dC/dt = \phi_3 M - \omega_3 C - \phi_6 C \tag{3}$$

$$dD/dt = \phi_2 M - \omega_2 D - \phi_4 D \tag{4}$$

$$dL/dt = \phi_4 D - \omega_4 L - \phi_5 L \tag{5}$$

$$dR/dt = \phi_6 C + \phi_5 L \tag{6}$$

where $S(t)$, $M(t)$, $C(t)$, $D(t)$, $L(t)$, $R(t)$ denote the probability of being in the states shown in Fig.1 at time t. The transition rates ϕ_5, ϕ_6 and repair rates ω_i (i=1, 2, 3) are constant. Other transition and repair rates in Eqs.(1)-(6) are as defined below

ω_1 : Repair transition rate from micro-crack
ω_2 : Repair transition rate from radial macro-crack
ω_3 : Repair transition rate from circumferential macro-crack
ω_4 : Repair transition rate from leak
ϕ_5 : Leak to rupture transition rate
ϕ_6 : Macro-crack to rupture transition rate

A simplified model of the transition rates $\phi_i(t)$ (i=1,...,4) as presented in [2] are as follow:

$$\phi_1(t) = (b/\beta)(t/\beta)^{b-1} \tag{7}$$

$$\phi_2(t) = \begin{cases} 0 & \text{if } u \leq \frac{a_D}{\dot{a}_M} \\ a_D P_D / (u\, a_D P_D + u^2\, \dot{a}_M P_c) & \text{if } u > a_D/\dot{a}_M \text{ and } u \leq a_C/\dot{a}_M \\ \frac{a_D P_D}{u\, a_D P_D + u\, a_C P_c} & \text{if } u > \frac{a_D}{\dot{a}_M} \text{ and } u > a_C/\dot{a}_M \end{cases} \tag{8}$$

$$\phi_3(t) = \begin{cases} 0 & \text{if } u \leq \frac{a_C}{\dot{a}_M} \\ a_C P_C/(u\, a_D P_D + u^2\, \dot{a}_M P_c) & \text{if } u > a_C/\dot{a}_M \text{ and } u \leq a_D/\dot{a}_M \\ \frac{a_C P_C}{u\, a_D P_D + u\, a_C P_c} & \text{if } u > \frac{a_C}{\dot{a}_M} \text{ and } u > a_D/\dot{a}_M \end{cases} \quad (9)$$

$$\phi_4(t) = \begin{cases} 1/w & \text{if } w > (a_L - a_D)/\dot{a}_M \\ 0 & \text{otherwise} \end{cases} \quad (10)$$

In Eqs.(8) - (10), u is a time after crack initiation and w is time after macro-crack formation (see Fig.1). The other parameters in Eqs.(7) - (10) are the following

a_D : Crack length threshold for radial macro-crack
P_D : Probability that micro-crack evolves as radial crack
a_C : Crack length threshold for circumferential macro-crack
P_C : Probability that micro-crack evolves as circumferential crack
a_L : Crack length threshold for leak

Figure 1: Multi-state transition model for PWSCC [2].

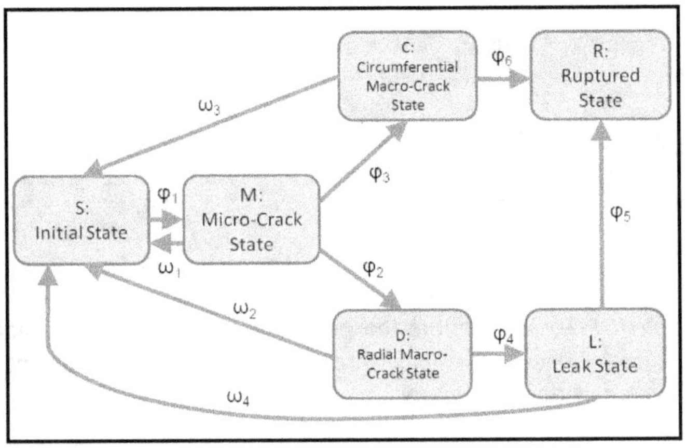

2.1.1. Crack Initiation and Growth Rate Equations

Of the alternative models that have been used to characterize initiation, the Weibull model is the most widely adopted [6] in which the cumulative probability $P(t)$ of crack initiation by time t is quantified through,

$$P(t) = 1 - e^{(t/\tau)^b} \quad (11)$$

$$\tau = A\sigma^n e^{(Q/RT)} \quad (12)$$

where

b : Weibull shape parameter for crack initiation model
τ : Weibull scale parameter for crack initiation model
A : Fitting parameter
σ : Explicit stress factor
n : Stress exponent factor

Q : Crack initiation activation energy
T : Absolute temperature at crack location
R : The universal gas constant.

Table 1 below defines the units for the variables in Eqs.(11) and (12), as well as their values/distributions as used in different studies. Base case values refer to the data which were used in [3].

Table 1: Definition of the Inputs in Crack Initiation Model

Symbol	Unit	Crack Initiation (Weibull Model) Inputs			
		xLPR [7]		Unwin[8]		Base Case [3]
T	K	Distribution Type	Normal	617	610
		Mean	617.9		
		Std. Deviation	0.0882		
		Deterministic	618		
b	ND	3		Distribution Type	Triangular	2
				Minimum	3.915	
				Mode	4.35	
				Maximum	4.785	
σ	MPa	Distribution Type	Normal	Distribution Type	Normal	106
		Mean	300.3	Mean	300.3	
		Std. Deviation	110	Std. Deviation	110	
		Deterministic	150	Deterministic	150	
n	ND	-4		Distribution Type	Triangular	-7
				Minimum	-7.7	
				Mode	-7	
				Maximum	-6.3	
A	ND	0.04		2.524×10^5		2.524×10^5
Q_I	kJ/mole	182.908		Distribution Type	Triangular	130
				Minimum	116.73	
				Mode	129.7	
				Maximum	142.67	

The maximum crack growth rate, $\dot{a}_M$, (see Eqs. 8-10) is calculated by using MRP-115 Model [9]

$$\dot{a}_M = \alpha f_{alloy} f_{orient} K^\beta e^{-\left(\left(\frac{Q_G}{R}\right)(T^{-1}-(T_{ref})^{-1})\right)} \tag{13}$$

where

α : Fitting constant – crack growth amplitude
T : Absolute operating temperature at crack location
T_{ref} : Absolute reference temperature used to normalize data (598.15 K)
Q_G : Thermal activation energy for crack growth
R : The universal gas constant
K : Crack tip stress intensity factor
f_{alloy} : 1.0 for Alloy 182 and 1/2.6 for Alloy 82
f_{orient} : 1.0, except 0.5 for crack propagation that is perpendicular to dendrite solidification direction.
β : Stress intensity exponent

Table 2 below defines the units for the variables in Eq.(13) as well as their values/distributions as used in different studies.

Table 2: Definition of the Inputs in Crack Growth Model

Symbol	Unit	Crack Growth (MRP-115) Model Inputs			
		Value			
		xLPR [7]	Unwin [8]		Base Case[3]
β	ND	1.6	Distribution Type	Triangular	1.6
			Minimum	1.44	
			Mode	1.6	
			Maximum	1.76	
α	(m/s) (MPa-m$^{0.5}$)$^{1.6}$	9.82 x 10^{-13}	Distribution Type	Normal	1.5 x 10^{-12}
			Threshold	-	
			Mean	8 x 10^{-13}	
			Std. Deviation	-	
Q_G	kJ/mole	Distribution Type	Normal	Distribution Type	Normal	130
		Mean	130	Mean	130	
		Std. Deviation	5	Std. Deviation	5	
		Deterministic	130	Deterministic	130	
f_{alloy}	ND	Distribution Type	Lognormal	1.0		1.0
		Mean	0.99894		
		Std. Deviation	1.83475		
		Deterministic	1.074897		
f_{orient}	ND	1.0	1.0		1.0

2.2. Sojourn Time Approach

The variables u and w in Eqs.(8)- (10) represent the residence time of the system in States M and C or D, respectively (see Fig.1). In that respect, Eqs.(1)-(10) do not have the Markov property of being independent of state history. These equations are converted into a semi-Markov process using the sojourn (or expected residence) time approach determined through

$$\frac{d\tau_{n,k}(t)}{dt} = (t - T_{k-1}) \left[-x_n(t) \sum_{\substack{m=1 \\ m \neq n}}^{N} mn(\tau_n) + \sum_{\substack{m=1 \\ m \neq n}}^{N} nm(\tau_m) x_m(t) \right] \quad (14)$$

where $\tau_{n,k}(t)$ is sojourn time of state n and for kth time interval, $x_n(t)$ is the probability of being in State n at time t, and $nm(\tau_m)$ is the transition rate from State n to State m as a function of sojourn time.

A sensitivity analysis of the aging model on local thermal operating conditions has been performed [4] since the transition rates $\phi_1(t)$ through $\phi_4(t)$ in Eqs.(7)-(10), respectively, are affected by the thermal-hydraulic conditions as they affect the time constant. Thermal-hydraulic data for this analysis have been obtained using the transient code RELAP/SCDAMSIM [10] for a simplified model of a four loop PWR.

3. QUANTIFYING UNCERTAINTY FOR THE STATE TRANSITION MODEL

As indicated in Tables 1 and 2, the multi-state transition model described in Section 2 is subject to considerable uncertainty and/or variability in both initial conditions and parameters. Most numerical approaches address such challenges by: a) computing local sensitivity indices (partial derivatives of the solution with respect to the input variables) [11], b) solving the model for a statistically large ensemble of random or quasi-random input values [12-14], or, c) by approximating the functional relationship of the input and output [15-17]. In this paper, Option (b) is used for uncertainty analysis, and Option (c) is used for sensitivity analysis, as described in Sections 3.1 and 3.2, respectively.

3.1. Uncertainty Analysis

Several random variable sampling techniques are employed in the literature [7], such as random sampling, Latin hypercube sampling (LHS), and discrete probability distribution (DPD) sampling schemes. The output of interest is the probability of rupture, which is expected to be extremely low for primary piping systems. Therefore random sampling may generate many runs without any rupture. In such a situation, a common strategy is to use Latin hypercube sampling to propagate the effects of aleatory uncertainty [12].

In this study, all parameters in Eqs.(1)-(13) except T and σ are treated as invariant. The T and σ were assigned uncertainty distributions based on the results of a preliminary sensitivity screening process among the parameters T, σ, n, b, Q_G and α in Eqs.(11)-(13) to identify those with the most significant effect on rupture probability.

In the initial analysis, temperature T and residual weld stress σ were importance sampled without distinguishing between epistemic and aleatory uncertainties. Confidence levels on the output variables were estimated as a function of time. The estimation was accomplished by computing the 95th and 5th percentiles of the distribution on the probability of leak $L(t)$ and rupture $R(t)$ at each time point. Figure 2 shows the results of this one-step uncertainty analysis with the 95th (red line) and 5th (blue line) percentiles.

As indicated in the xLPR study [7], when it is possible to differentiate between aleatory and epistemic uncertainties in a model, additional information can be obtained that could affect the interpretation of results by performing a two-loop simulation in which the inner loop and outer loop address the aleatory uncertainties and the epistemic uncertainties, respectively. In the outer loop, parameter values are sampled from epistemic uncertainty distributions and passed on to the inner loop. For each sample in the outer loop, LHS draws from aleatory uncertainty distributions are performed in the inner loop over the time-frame of interest accounting for the aleatory uncertainty. From these results an average rupture probability can be calculated over the variability associated with the input parameters. From the outer loop analysis, it is therefore possible to obtain an uncertainty distribution of the variability-averaged rupture probabilities. This distribution provides measures of the uncertainty in rupture probability that could be reduced by further experimentation or model development. As LHS is also used to generate epistemic uncertainty, the simple arithmetic mean of the rupture probability over epistemic uncertainty can be used to estimate an over-all expected value $<R>$ of the rupture probability from

$$\langle R \rangle = \frac{1}{N}\sum_{n=1}^{N}\left(\frac{1}{M}\sum_{m=1}^{M} R_{mn}\right) \qquad (15)$$

where M is the number of aleatory draws, N is the number of epistemic draws and R_{mn} are the rupture probabilities calculated from the state transition model for each draw combination. In general, it is expected that the overall average value will be the same regardless of whether the sampling is performed in a single-step or a two-step process. However, to test whether the overall average is affected by the sampling approach, a comparison was made of the two-step LHS versus single step LHS. Uncertainty in T and σ were characterized using normal PDFs: $T \sim$ Normal (617.9, 0.0882), $\sigma \sim$

Normal (300.3, 30)) for both cases. In the two-step LHS, the uncertainty for T and σ were characterized as epistemic and aleatory, respectively, instead of both as epistemic. Equation (15) was implemented with $N=100$ and $M=100$ resulting in a total of 10,000 realizations. The single step LHS realization was also performed for 10,000 draws to obtain $\langle R \rangle$. Fig. 3 shows the comparison of single step and two-step LHS sampling processes and Fig. 4 shows $\langle R \rangle$ as a function of time for the single-step and two-step LHS. The temperature draws T_N in Fig. 3 are for every 2 years. As can be seen in Fig. 4, difference between two methods is extremely small (on the order of 10^{-5}).

Figure 2: Leak (L) and Rupture (R) probabilities of PWR pressurizer surge line pipe (Alloy 182) for a PWSCC scenario over 80 years. The red line indicates 95th and blue line indicates 5th percentile.

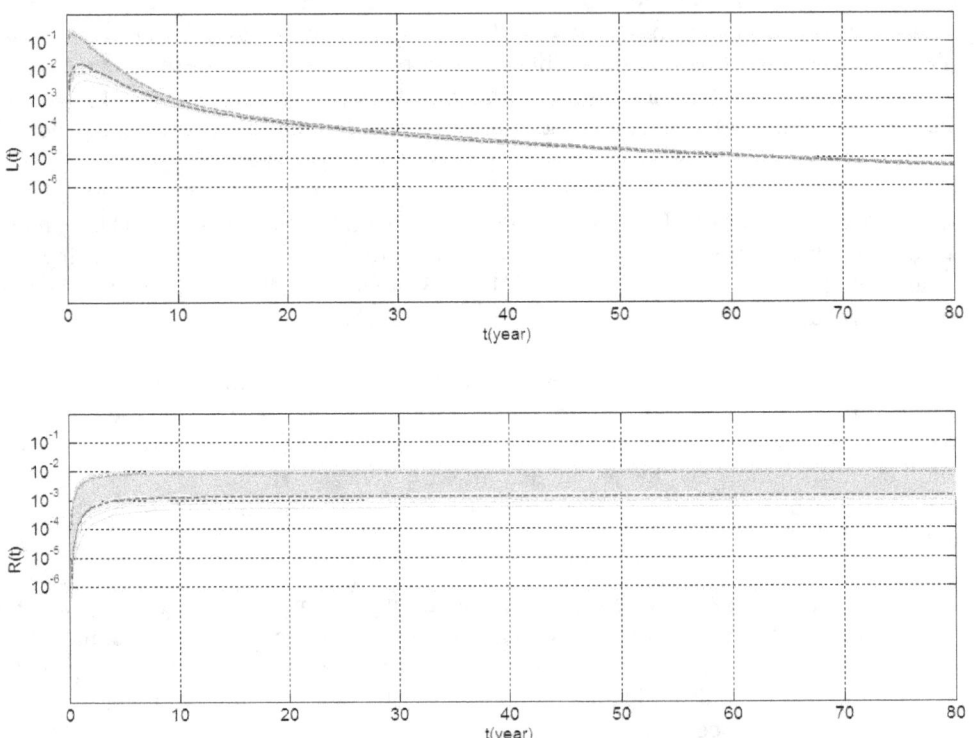

3.2. Sensitivity Analyses

In order to determine whether T or σ had more impact on the rupture probability, a response surface approach was used. The fitting was done by using the method of least squares and a second order fitting using

$$f(T,\sigma) = p_{00} + p_{10}T + p_{01}\sigma + p_{20}T^2 + p_{11}T\sigma + p_{02}\sigma^2 \tag{16}$$

Coefficients of Eq. (16) are listed in Table 3 and statistics goodness-of-fit data are summarized in Table 4.

Figure 3: Illustration of single step and two-step LHS comparisons

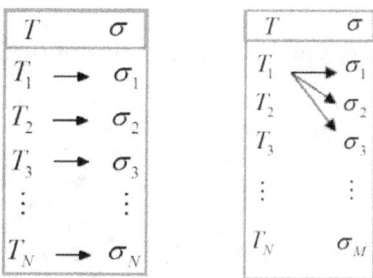

a. Single Step LHS   b. Two-Step LHS

As can be seen in Fig. 4, difference between two methods is extremely small (on the order of 10^{-5}).

Figure 4: Illustration of single step and two-step LHS comparisons

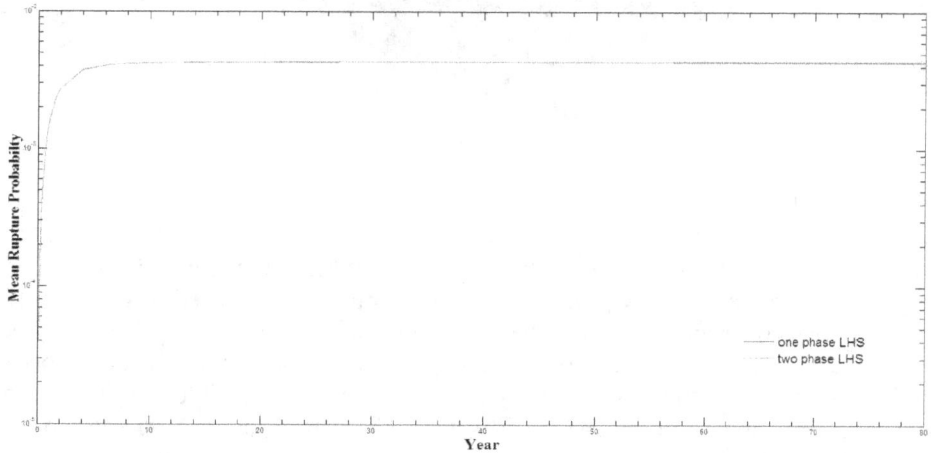

Table 3: Fitting equation coefficients and 95% confidence bounds

Coefficients	95% confidence bounds
$p_{00} = -24.39$	-363.2, 314.5
$p_{10} = -7.953 \times 10^{-5}$	-0.003004, 0.002845
$p_{01} = 0.07899$	-1.017, 1.175
$p_{20} = 3.302 \times 10^{-7}$	3.228×10^{-7}, 3.375×10^{-7}
$p_{11} = -7.747 \times 10^{-8}$	-4.81×10^{-6}, 4.655×10^{-6}
$p_{02} = -6.394 \times 10^{-5}$	-9.51×10^{-5}, 8.231×10^{-4}

Table 4: Statistic goodness-of-fit data

Goodness of fit
SSE: 2.457×10^{-7}
R-square: 0.9995
Adjusted R-square: 0.9995
RMSE: 5.113×10^{-5}

Fig. 5 shows the impact of weld residual stress and temperature variations on the rupture probability at $t=40$ years in case of 100 realizations and clearly indicates greater sensitivity of rupture probability to the uncertainty in stress than to the uncertainty in temperature.

Figure 5: Response Surface of Rupture Probability

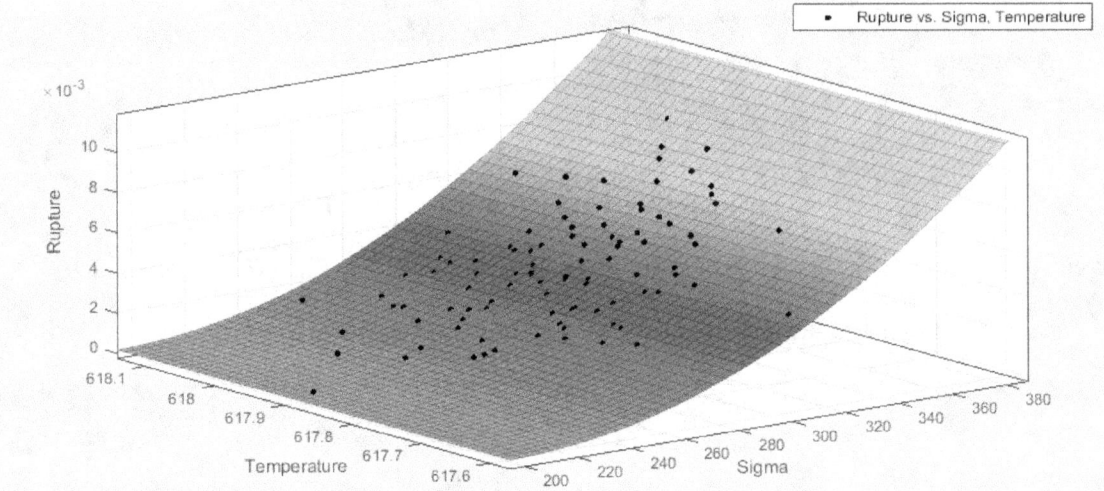

4. CONCLUSION

Using a state-transition model to describe (PWSCC), uncertainties in the input data for crack initiation and crack growth are represented by probability distributions. LHS is used to generate observations from the distributions governing T and σ with a two-step approach that distinguishes between aleatory and epistemic uncertainties. Comparison of the results to a single-step quantification process indicates that the differences between one-step and two-step approach are negligible with regard to the mean rupture probability (on the order of 10^{-5}). However, the separation into sources of aleatory and epistemic uncertainty could enable the decision-maker to determine the potential value of activities to reduce the epistemic uncertainty, such as by the performance of research. Indeed, for this example, nearly all of the uncertainty shown in Fig. 2 arises from the "epistemic" uncertainty associated with the weld residual stress σ.

Acknowledgements

This research has been performed using funding received from the DOE Office of Nuclear Energy's Nuclear Energy University Programs.

References

[1] R. Lewandowski, R. Denning, T. Aldemir, J. Zhang, *"Development of a Living, Plant-Condition Dependent Probabilistic Safety Assessment,"* these proceedings.

[2] S. D. Unwin, P.P. Lowry, R.F. Layton Jr., P. G. Heasler, and M. B. Toloczko, *"Multi-State Physics Models of Aging Passive Components in Probabilistic Risk Assessment,"* ANS PSA 2011 International Topical Meeting on Probabilistic Safety Assessment and Analysis, Wilmington, NC, (2011).

[3] A. Guler, T. Aldemir, R. Denning, *"The Sojourn Time Approach for Modeling Aging in Passive Components,"* Trans. Am. Nucl. Soc., 108, pp. 552-554, (2013).

[4] A. Guler, T. Aldemir, R. Denning, *"Multi-State Physics Based Aging Assessment of Passive Components,"* Proc. ANS PSA 2013 International Topical Meeting on Probabilistic Safety Assessment and Analysis, Columbia, SC, (2013).

[5] K. Fleming, *"Markov models for evaluating risk-informed in-service inspection strategies for nuclear power plant piping systems,"* Reliability Engineering and System Safety, 83, pp. 27–45, (2004).

[6] W.J. Shack and O.K. Chopra, *"Statistical Initiation and Crack Growth Models for Stress Corrosion Cracking,"* Proceedings of the ASME Pressure Vessels and Piping Conference (PVP2007), San Antonio, Texas, July 22–26, 2007, pp. 337-344, (2007).

[7] P.D. Mattie, D.A. Kalinich, C.J. Sallaberry, U.S. Nuclear Regulatory Commission, *"Extremely Low Probability of Rupture Pilot Study: xLPR Framework Model User's Guide"*, SAND2010-7131, Sandia National Laboratories, Albuquerque, New Mexico, (2010).

[8] S.D. Unwin, K.I. Johnson and P.W. Eslinger, *"Robustness of RISMC Insights under Alternative Aleatory/Epistemic Uncertainty Classifications,"* Draft Report PNNL-21810, Pacific Northwest National Laboratory, Richland, WA, (2012).

[9] Electric Power Research Institute, *"Materials Reliability Program Crack Growth Rates for Evaluating Primary Water Stress Corrosion Cracking of Alloy 82,182, and 132 Welds,"* EPRI Report MRP-115, 1006696, Palo Alto, CA (2003).

[10] SCDAP/RELAP5 Development Team, *"SCDAP/RELAP5/MOD3.2 Code Manual, Vol. 1–5,"* NUREG/CR-6150, INEL-96/0422, (1998).

[11] T. Turányi, *"Sensitivity Analysis of Complex Kinetic Systems. Tools and Applications,"* J Math Chem, 5, pp. 203-248, (1990).

[12] J.C. Helton, F.J. Davis, *"Latin hypercube sampling and the propagation of uncertainty in analyses of complex systems,"* Reliability Engineering and System Safety, 81, pp. 23–69, (2003).

[13] S. Marino et al., *"A methodology for performing global uncertainty and sensitivity analysis in systems biology."* J Theor Biol., 254, pp. 178-196, (2008).

[14] C.P. Robert, G. Casella, *"Monte Carlo statistical methods,"* New York: Springer, (2004).

[15] S. Fang et al., *"Improved generalized Fourier amplitude sensitivity test (FAST) for model assessment,"* Statist Comput, 13, pp. 221-226, (2003).

[16] W.J. Hill, W.G. Hunter, *"A Review of Response Surface Methodology: A Literature Survey,"* Technometrics, 8, pp. 571-590, (1966).

[17] A.I. Khuri., *"Response Surface Methodology and Related Topics,"* Singapore, World Scientific Publishing Co, (2006).

Passive system Evaluation by using integral thermal-hydraulic test facility in passive NPP(nuclear power plant) PSA(probabilistic safety assessment) process

Ruichang Zhao[a], Huajian Chang[a], Yang Xiang[a]
[a] State Nuclear Power Technology Research & Development Center, Beijing, China

Abstract: Passive safety engineered systems are designed to take effect by physical phenomena or passive procedure during the scenario of imaginary accident of AP600/AP1000/CAP1400 passive safety type nuclear power plant.
Generally, associative thermal-hydraulic experiments have been studied to support specific physical phenomenon research or evaluation model development in these scenarios.
Data from T-H experiment is credibility and direct. However, for the size of safety engineered systems of NPP are very huge, it's almost impractical to simulate a whole physical process by original scale test facilities. Some reduction scaled test facilities have been applied in design verification or safety research.
So, it's worthy to explore how to apply the research of scaled integral T-H experiment target at the specific physical process or phenomena in PSA procedure.
Scaling analysis method is usually applied in the integral test facility design, construction, and data evaluation especially. Through the scaling analysis and evaluation of experiment data, the uncertainty of every test result can be achieved. The result, trend or uncertainty of specific parameters of physical phenomena or process can be explicated. If test facilities and experiments are implemented by scaling analysis approprite, the most important result of test can present the prototype one in some degree (some uncertainty level). By these, the test can present the target physical phenomena simulated. And the prototype passive system can be explored by experiment result in some uncertainty degree level.
Containment Experiment via integral safety validation test facility (CERT) has been set up for design validation of passive containment cooling system(PCCS) of CAP1400 NPP. CERT can simulate LOCA or MSLB accident scenario by equal ratio power-volume. The figure of merit of CERT is the pressure inner containment. Different trends of figure of merit can be obtained by adjust the boundary or initial condition of experiment, such as total enthalpy of steam injection, flow rate/coverage of cooling water outer of containment, wind speed in the annulus(the structure between shield and steel containment), concentration of non-condensation gas(the helium which is used to simulate hydrogen during accident), and so on. Besides, it's also can be achieved which quantitatively defines the possibility of the figure of merit beyond the design criterion. According to the scaling analysis and the experiment results(by integral test facility) of corresponding important physical phenomena, the quantitative performance assessment of PCCS can be obtained. Afterwards, these evaluation can support II level PSA of passive safety NPP.

Keywords: Passive, PSA, Scaling Analysis, AP1000, CAP1400, CERT.

1. Research of Engineered Safety System in Large Passive Plant

Recently, passive systems applied in nuclear power plant became more and more widespread. In AP1000/CAP1400 type NPP, most of the engineered safety systems are designed by physical process functional characters without any power of electric or active mechanics. It's a necessary issue to evaluate these systems in a way of PSA method. The safety analysis of related physical processes or phenomena are usually the start of begin.
Although there are lots of works in PSA quantification, it's difficult to apply traditional PSA methods in evaluation of passive system. How to get the correct value which expresses the specific performance of the physical procedure must be the one of most difficulties.
In the guide of NUREG-5809 and relative materials, some separate and integral thermal-hydraulic test facilities have been set up the aiming to the technically researching of physical phenomena expecting in the passive system action during accident scenarios. It's valuable to develop an approach to achieve

the performance of prototype system by converting test results, in order to support PSA applied in passive system safety evaluations.

By deducing the data from thermal-hydraulic experiments, the physical performance quantification will be more reliable, and can be confirmed and validated in PSA processes.

2. T-H test method

The first passive safety characteristic nuclear power plant AP1000 has been constructed in China. Similar to AP1000, State Nuclear Power Technology Corporation (SNPTC) of China has designed CAP1400 to enhance the plant power and the ability of relative engineered safety system. Because of the lack of experiences and uncertainty in increasing plant power, a serial safety issues have been researched, especially the passive reactor core cooling system (PXS), passive containment cooling system (PCCS), and in-vessel retention (IVR). Focus to these issues, relative separate or integral experiments have been taking effect.

These experiments draw on the experiences of previous researches especially during the development of AP600/1000 design. Applying the spirits of scaling analysis and uncertainty methods, some innovations have been carried out in the experiments researches according to CAP1400 features.

2.1. Severe accident research and analysis

Paper [1] and [2] have set up a frame solution procedure of NPP severe accident safety issues. Zuber et. al., had developed the Integrated Structure for Technical Issue Resolution (ISTIR) in 1990's. Severe Accident Scaling Methodology (SASM) is one of the elements.

In ISTIR, there are 5 main elements combined the whole structure, specify the safety issue and important phenomena, scaling methodology and experiments, technical resolution by experiment and uncertainty quantification, analysis (and uncertainty quantified) by frozen code, and code development. In these 5 steps, specifying the critical physical phenomena will be the first and most important one. By achieved the phenomena identification and ranking table (PIRT), the relative experiment (separate or integral), analysis, code development can be set up and directed in process of SASM.

SASM can be regarded as the guide for thermal-hydraulic experiment research. There are some details. Three elements compose SASM by [1]:
1. Experimental requirements.
   According to the PIRT, the experimental objectives can be specified in this step.
2. Evaluation and specification for experiments and testing.
   In this step, scaling analyses should be applied. Similarity criteria should be set up. Separate or Integral experiment facility (experiments) should be specified. Relative models, correlations, facilities distortions, test design optimizing should be developed.
3. Data acquisition and documentation.
   By experiment operation, data can be achieved in different conditions. Data base and uncertainties by quantification can be established.
   With the whole procedure, facilities design requirements & specifications, test results, and experiment data etc., can be documented in this step.

The main frame of flow diagram of SASM is showed in [1] as Fig.1:

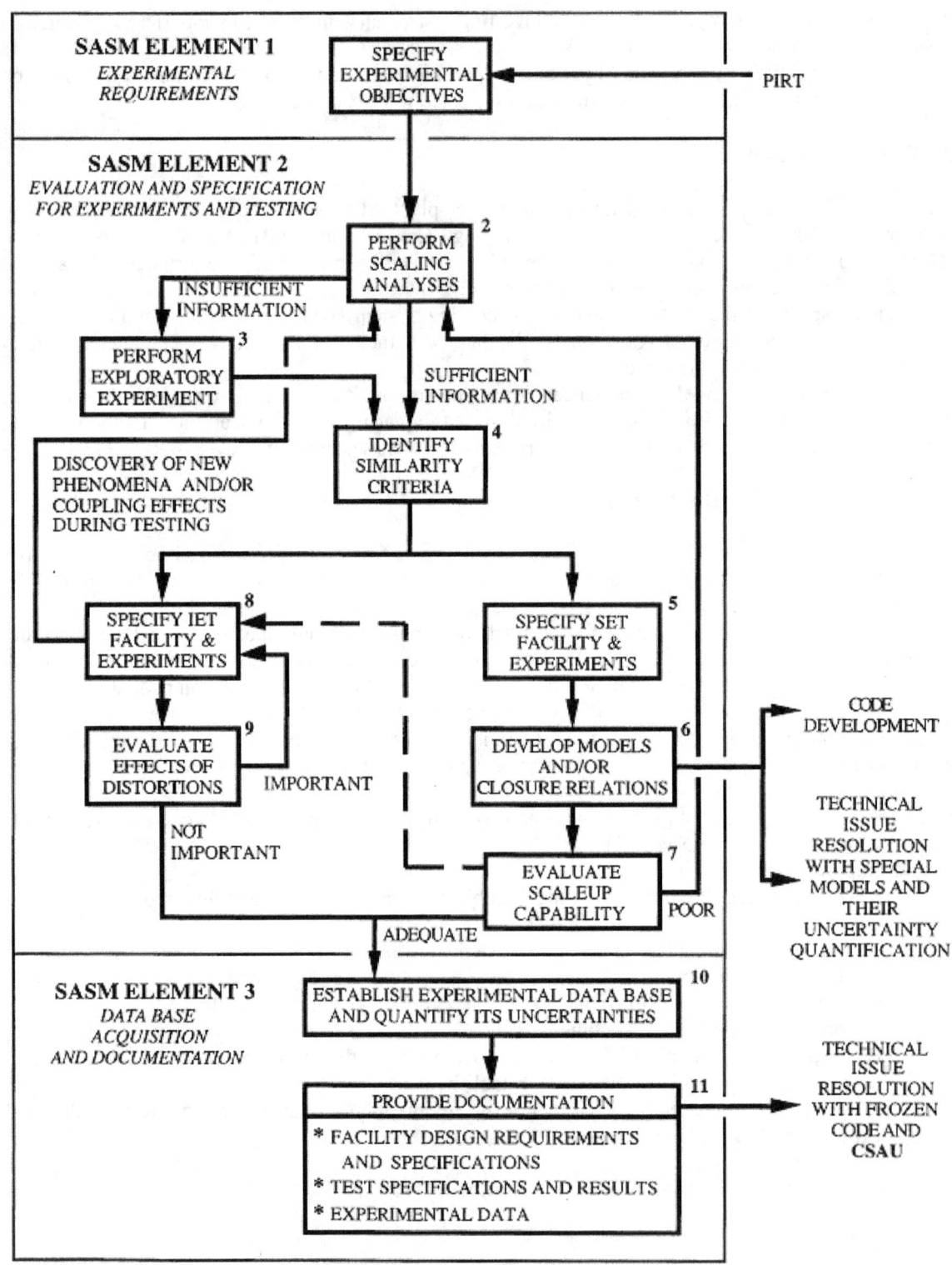

Fig.1 Flow diagram for SASM [1].
The details of the flow diagram will not be discussed here, see [1].

2.2. Scaling analysis methodology application

According to SASM execution process, there are several critical steps in the process of scaling analysis during thermal-hydraulic experiments design.

1. To specify PIRT according to the scenarios of the specific NPP by expert judgment.
2. To identify the critical physical processes or phenomena, and gather the conservation equations and relative correlations.
3. To deconstruct the system by subsystems and physical fields etc. by H2TS (hierarchical, two-tiered scaling) procedure.
4. To define the sub-scaled facilities and boundary conditions especially the characteristic dimensions.
5. To explicit the requirement of abilities and precisions of measurement and control system by feedback of application of scaling analyses or H2TS methods.

The flow diagram of scaling analysis method namely 'H2TS' (hierarchical two-tiered scaling analysis) of SASM is showed in [1] as Fig.2:

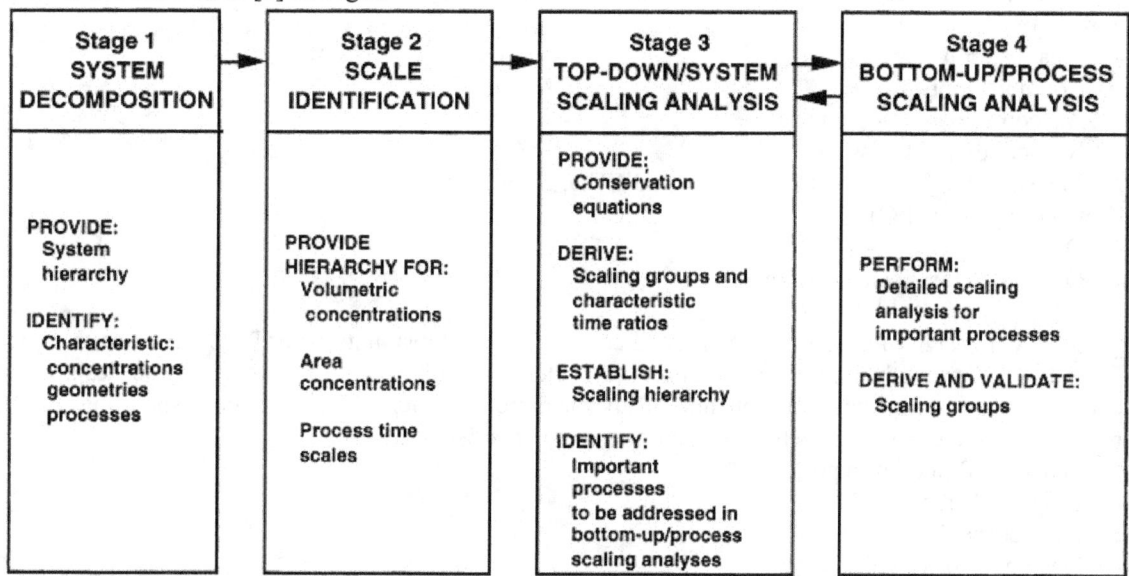

Fig.2 Flow diagram for H2TS [1].

2.3. Objectives of sub-scaled tests facilities

Reactor core cooling and containment integrity are the most two important safety issues in nuclear power plant design or safety research. For CAP1400, two integral experiment facilities have been constructed to the PXS (passive core cooling system) and PCCS (passive containment cooling system, also PCS) separately.

ACME (Advanced Core-Cooling Mechanism Experiment) simulates the situations of core covered by coolant in the act of PXS, during whole range of pressure in LOCA (Loss Of Coolant Accident) or SBLOCA (Small Break Loss Of Coolant Accident) accident scenarios.

The missions of ACME are:
- To simulate the operation of passive core cooling system of CAP1400 for SB-LOCA.
- To validate the engineering design of the passive core cooling system.
- To collect thermal-hydraulic data for safety code assessment.

CERT (Containment safety vErification via integRal Test) simulates the trend of containment inner pressure and energy transferred in act of PCCS. During steam mass & energy injection, the power to the test vessel volume is equal to the ratio of prototype during the accident scenarios.

The missions of CERT are:
- To validate the applicability of WGOTHIC(safety code for containment assessment)
- To verify the engineering design of the passive containment cooling system
- To scaled-simulate the physical process in accident scenario, and the performance of passive containment cooling system of CAP1400

These two integral test facilities are all under operation this time in SNPTRD (State Nuclear Power Technology Research & Development center). Some instruments are under adjustment in early 2014. The experiment which is simulating LOCA scenario of CAP1400 will be executed in the mid of the year.

3. CERT introduction

The test facility named 'CERT' is the abbreviation of "Containment safety verification via integral test". As mentioned above, the main purpose of CERT is to verify the PCCS ability in specific accident scenarios by sub-scaled model.

CERT test section is a bearing pressure steel vessel which length (include diameter) ratio (test to prototype) is 1:8. Apart from the test vessel, several dynamic or accessory system consist CERT, as below [3]:
- Steam supply sys.
- Coolant supply sys.
- Air compressor/vacuum sys.
- Non-condensable gas supply sys.
- Measure sys. (Common measure, Digital CCD, LDV[i], etc.)
- Control sys. (PLC)
- Database.
- Other auxiliary instruments.

By the PCCS design of CAP1400, the related components and installations of PCCS are all sub-scaled according to the length ratio. The operation parameters or ability of the systems are also designed to keep consistent with the requirements of the parameters after sub-scaled corresponding to the prototype characteristic parameters. Some important items list below:
- Scale: 1:8 (model to prototype)
- Height: 9.125 m
- Diameter: 5.375 m
- Design pressure: 0.7 MPa(a)
- Design temperature: <200℃
- Shell thickness: 18.0~20.0 mm
- Total height: ~15 m
- Steam mass rate: 1.5 t/h~108.0 t/h (four branches supply)
- Coolant mass flow rate (outer containment): 0~8.0 t/h

The key character of CERT is the performance of mass & energy release, in other words, the ability of steam supply system (SSS) operation. So the most important accessory system is the steam supply system. By considering the ratio of power to volume / area, SSS can simulate the LOCA and MSLB scenarios in a desirable state. By the synthetic action of other systems, the conditions of boundary can be achieved, such as inner containment temperature, pressure, gas flow rate, humidity, non-condensable component, outer containment water film flow rate, wind velocity in annulus, etc.

4. Process of test result apply to PSA evaluation

The figure of merit in CERT experiment is *the pressure inner containment*. The main target of experiment is to achieve the pressure trend in the specified boundary and initial conditions. Futhermore, the inner pressure of prototype NPP can be inversely deduced by scaling analysis methodology with a sort of uncertainty of course.

Here we assign the inner pressure is the criterion of containment failure. Namely, if the inner pressure is greater than a specific constant value P_{crit} the containment is assumed to failure.

In fact, the peak pressure will be achieved in every test, however it's a little different even given the nearly same conditions (since attaining the exactly same conditions in an integral T-H experiment is very hard, little bias that do not affect results can be accepted. However this conclusion should be confirmed by sensitive analysis,). Because every parameter has a specific distribution, the inner pressure which is the function of these parameters also has a distribution. And this distribution or uncertainty range can be seemed as the uncertainty propagation of relative parameters'.

The data of LOCA or MSLB scenario experiments is used to evaluate the failure of containment.

As the peak pressure values from all previous tests can be treated as the space of containment performance sampling. The sample of failure occurred because of inner pressure over the criterion.

During this process, the pressure value is fluctuated in a range, because of the other relative parameter's variation as mentioned above. By majority times of repeat tests, the probability distribution of the peak pressure value can be concluded by a number of results. And the probability of containment failure can be deemed as the probability integrated by counting the value greater than the criterion.

The expression of inner containment is deduced from energy conservation equation. The pressure is the integration

$$\pi_{p,\tau} X V \frac{dP}{dt} = \dot{m}_{brk}(h_{brk} - h_{stm}) - \sum_{i=1}^{N}(\Lambda(\pi_{P,cond,i}\dot{m}_{stm,i}) + (\pi_{P,q,i}h_{q,i}A_i \Delta T_{if,i})) \qquad (1)$$

here P : containment (or test model) inner pressure value,

t : time,

$\pi_{p,\tau}$: dimensionless value represent the pressure development of time because of mass/energy injection. $\pi_{p,\tau} = X\frac{P}{\rho_{g,brk}}$, while $\rho_{g,brk}$ is the density of steam blowing from the nozzle (to simulate breach). While X is the coefficient of atmosphere existence in the containment..

Λ : the pressure develop because of the mass increasing, $\Lambda = \frac{C_p P_{stm}}{ZR\rho_{stm}}$, while C_p is heat capacity at constant pressure, Z is compressible factor, R is gas constant, ρ_{stm} is density of steam in containment, P_{stm} is partial pressure of steam in containment.

$\pi_{P,cond,i}$: dimensionless value represent the pressure development affected with the condensation on the i th component inner containment.

$\dot{m}_{stm,i}$: mass flow rate of the condensation on the surface of the i th component inner containment.

$\pi_{P,q,i}$: dimensionless value represent the pressure development affected with the radiation and convection on the surface of i th component inner containment.

$h_{q,i}$: the equivalent heat transfer coefficient of radiation or convection heat transfer mechanism on i th component surface.

A_i : the equivalent area of heat transfer by radiation or convection.

$\Delta T_{if,i}$: the equivalent different of temperature of the atmosphere to the film on the i th component surface.

V : the freevolume of containment or containment model (test model/section).

Some parameters above are the function of other more particular variables, such as $\pi_{p,\tau}$ or Λ mentioned above.

So the pressure equation can be written as:

$$p(t, x_i) = \int_0^t f(x_i) \qquad (2)$$

$f(x_i)$ is the function of x_i, while x_i represents the i th parameter of relative measurement variables.
The peak pressure appeared in one test can be expressed as:

$$p_{max} = \max[p(t, x_i)] \qquad (3)$$

By the equations and definitions above, the failure probability of the PCCS can be indicated as the probability of the event indicating that PCCS cannot reduce inner pressure lower than the criterion value during the DBA accidents.

$$P[\text{fail of PCCS}] = \text{Prob}[p_{max} > p_{crit}] \qquad (4)$$

While $P[X]$ is the probability of random event X. p_{crit} is the criterion of containment (const value).

The uncertainty evaluation of "p(t)" will be deduced from scaling analysis by dimensionless equations and propagation of parameter bias, and by dealing with data from test results.

The main flow of evaluation process lists below. Because of CERT tests haven't finished yet, the real values haven't been substituted.

1. To establish the quantitative relationships between the test model and prototype PCCS of NPP, by using conservation equations and dimensionless correlations.

    The most important one is the energy equation in the form of pressure (equation 1). The dimensionless process can be applied to this equation. After dimensionless process, the equation can be simplified as:

$$\left(\frac{dP}{dt}\right)_P = \frac{\left(\pi_{p,\tau} XV \frac{dP}{dt}\right)_T}{(\pi_{p,\tau} XV)_P} - \frac{\left(\left(\Lambda \sum_{i=1}^{N}(\pi_{P,cond,i} \dot{m}_{stm,i})\right)_P - \left(\Lambda \sum_{i=1}^{N}(\pi_{P,cond,i} \dot{m}_{stm,i})\right)_T\right)}{(\pi_{p,\tau} XV)_P} - \frac{\left(\left(\sum_{i=1}^{N}(\pi_{P,q,i} h_{q,i} A_i \Delta T_{if,i})\right)_P - \left(\sum_{i=1}^{N}(\pi_{P,q,i} h_{q,i} A_i \Delta T_{if,i})\right)_T\right)}{(\pi_{p,\tau} XV)_P}$$

(5)

Here, $(...)_P$: the relative parameter values of prototype system.
$(...)_T$: the relative parameter valures of test model.

The relative correlations can be used to close the equation above. such as:

$$h_{q,i} = 0.13 \frac{k}{\left(v^2/g\right)^{1/3}} \left(\frac{\Delta \rho}{\rho}\right)^{1/3} Pr^{1/3}$$

and etc (didn't describe them all).

And much more parameters as $\dot{m}_{stm,i}$ or $\pi_{p,\tau}$ ..., are all can be deduced from lots of basis measured variables, such as the specific local temperature\pressure and so on.

2. To quantify the value of inner pressure and its uncertainty/ distribution etc.

    The most parameter values of $(...)_T$ in the equation can be obtained from test, while others can be predicted by T-H correlations, or general database. Then the value of pressure can be deduced by given the parameters' value above.

    Here, π_T or π_P means a specific physical process/mechanism corresponding to the test (T) or prototype (P). And $\frac{\pi_T}{\pi_P}$ indicates the ratio of relative parameter belongs to test and to prototype.

    According to the spirit of scaling analysis, the nearly $\frac{\pi_T}{\pi_P}$ approaches to 1, the similarly of the physical phenomena in the test facility to the one in the prototype system.

    In fact, it's impossible to keep every π in the dimensionless equations to 1. If a deviation appeared, it means the corresponding physical phenomena or process is not exactly the same. Then PIRT can be used to determine whether the phenomenon (π represents) is important to the figure of merit or not. However, it's difficult to quantify the influence of Π deviation to the result, because PIRT has just been identified in high/mid/low ranks by expert judgments.

3. To obtain the figure of merit (the pressure value) and the failure probability with a specific uncertainty of prototype PCCS. The pressure value can be expressed as the integration of equation (5).

    The uncertainty of the peak pressure value can be expressed as:

$$\Delta p_{max} = \Delta p(t)|_{t=t_{pmax}} = \sqrt{\sum_{i=1}^{N} \left(\frac{\partial p(t, x_i)}{\partial x_i} * \Delta x_i\right)^2}\bigg|_{t=t_{pmax}}$$

(6)

The partial differential expression in the right of the equation seems very difficult to calculate (because the expression of $p(t, x_i)$ is very complex). However the numerical method can be applied in the calculation. In fact the important of variables can be achieved by test (most of x_i or Δx_i), then others can be digested from general database or expert judgments. And there exists uncertainties of general database & experts judgments too, which is not considered here.

5. Conclusion

To sum up, in CAP1400 NPP safety verification experiment, the process of applying the thermal-hydraulic integral test result to evaluate the system failure probability. And the method of sub-scaling analysis experiment design has been introduced.

CAP1400 PCCS test facility CERT has taken nearly 4 years from design to bring into service. During the design period scaling analysis and relative key methodologies were draw from ISTIR and SASM. Some tricks were deeply revised or modified in order to consist with physical process anticipated in the containment. The measurement and control devices all have calibrated and adjusted to achieve a high degree of accuracy. High quality & reliability test data have been obtained unfailing nowadays.

The main steps of the process to evaluate the PCCS fail probability which can be applied to the PSA assessment further, by scaled T-H experiment is summarized below:

1. To digest the key physical phenomena or transport processes from PIRT of target NPP (CAP1400).
2. To use H2TS method to set up the frame of the analysis of these important physical processes considered, top to down combined with bottom up (conservation equations and specific correlations).
3. To establish the quantitative connection between the test model (1:8 scaled) and the prototype PCCS of CAP1400 NPP, by scaling analysis -- dimensionless equations.
4. To evaluate the value and the uncertainty or distortion of test result for each parameter. From which the relative parameter value and uncertainty of prototype system of NPP can be deduced (also dimensionless correlations should be used).
5. To obtain the distribution of specific key parameters by repeating experiments for times. Then the distribution or characteristic value can be achieved through the quantitative connections with some kind of uncertainties.

From above basis, a further uncertainty analysis to test results is necessary to proceed. In order to make an integrated similarity or distortions about the test and prototype. And the test result of the T-H tests can be evaluated quantitatively, completely, reliably to predict the performance of prototype engineered safety system in some confidence level.

For PCCS of CAP1400, the result of the test data can be applied in the assessment of II level PSA.

References

[1] Novak Zuber and etc., "*An integrated structure and scaling methodology for sever accident technical issue resolution: development and methodology*", Nuclear Engineering and Design, 186, pp. 1-21, (1998).
[2] Idaho NEL., "*An integrated structure and scaling methodology for sever accident technical issue resolution*", Nuclear Regulatory Commission Report NUREG/CR-5809, EGG- 2659 , 1991, Idaho Falls.
[3] Ruichang Zhao, Xiang Yang and etc., "*Specification of CERT test facility (P)*", SNPTRD Reporrd, 2011, Beijing.

---

[i] LDV: Laser Doppler Velocimetry (LDV) measurement.

Probabilistic Assessment of Composite Plate Failure Behavior under Specific Mechanical Stresses

Somayeh Oftadeh[a], Mohammad Pourgol-Mohammad[a], and Mojtaba Yazdani[a]

[a] Sahand University of Technology, Tabriz, Iran

Abstract: This research focuses on determination of composite materials reliability and probabilistic assessment of their failure models. The principal task is to determine the probability distribution function for the composite behaviour in order to explain scatter and size effect and to describe composite reliability. A model for the statistical failure of composite materials is presented. As the first step of reliability evaluation, it is essential to understand the candidate failures modes of composite materials and their influence on structural performance. Failure mode and effect analysis (FMEA) is conducted. Based on the FMEA results, failure of a lamina is the main cause of a composite laminate failure. By considering only the failure of lamina, reliability analysis is done by utilizing the Monte Carlo simulation. Also a process is proposed to evaluate the reliability of composite structures. A composite structure of $[0_2/\pm45/90]_4$ graphite-fibre/epoxy-matrix is selected as the case study for the methodology presentation. These result analysis concludes that the Weibull distribution is fitted with enough confidence to represent composite behaviour. In addition to sample size which affects directly accuracy of evaluated reliability, the input variance magnitude is another factor that plays an important role in uncertainty of analysis and converging the results.

Keywords: Composite Materials, Reliability, Monte Carlo Simulation, Wiebull Distribution, Uncertainty

1. INTRODUCTION

Composites are an important engineering material in construction of automobile, mechanical, space and marine structures in recent years. It resulted in a significant increase in payload, weight reduction, speed, manoeuvrability and durability of products using these materials. In pursuing these achievements, the reliability analysis has thus become an important topic of research. There is considerable statistical variation in mechanical and material properties of composites. Despite years of extensive research around the world, a complete and validated methodology has not yet been fully achieved for predicting the behaviour of composite structures including the effects of damage. This is largely due to the complex nature, so that for any composite structure the performance and the development of damage leading to failure are dependent on a range of parameters including the geometry, material, lay-up, loading conditions, load history and failure modes.

Traditional design methods use global safety factors to take into account the uncertainties in manufacturing, loads, materials properties. Their values have been established after many years of experiments and calibration by judgment, but they are not suited to new materials with particular features. Over the years, a range of stochastic analysis methods have been developed to account for the uncertainties at different scales. Researchers have modelled uncertainty at the micro-scale, as well as macro-scale [1].

Reliability techniques have been in developing since the 1920's. Cassenti [2] furthered deterministic methods by developing the probabilistic static failure analysis procedure of unidirectional laminated composite structures. Kam [3] predicted the reliability of simply supported angle-ply and cantilever symmetric laminated plates subject to large deflections within the context of first-ply-failure and also developed an analysis procedure for clamped symmetric laminated plates subjected to central point loads based on the first-ply-failure analysis. Chen et al. [4] investigated the reliability of composite tophat stiffened plates for ship hulls providing a rapid analysis for hull girders. Whiteside et al. [5]

showed the effects of using a stochastic failure envelope on uni-directionally stiffened carbon/epoxy composites. Eamon and Rais-Rohani [6] performed a reliability analysis on a full composite boat hull as part of a sizing optimization. An excellent summary and comparison of different reliability studies is given by Sutherland and Guedes Sorares [7]. Also Sorares provides a review of different formulations that have been used to assess the reliability of laminates under plane stress conditions, assuming that they do not fail by delamination.

In this research, an algorithm was presented for reliability evaluation of composite materials. The demonstrated algorithm includes the past researches features and it is arrayed logically and user friendly. For this aim, a special composite laminate was selected as a case study. Since physics of failure method was used for evaluating the reliability because of lack of failure data, FMEA is utilized for identification of effective failure modes in the composite laminate. A composite structure of $[0_2/\pm45/90]_4$ graphite-fibre/epoxy-matrix is selected as the case study. The structure is a rectangular plate with 1×1 dimension in simply support exposure. Since the problem is faced to variety and uncertainty in material properties as input data, Monte Carlo method was implemented to accomplish the non-deterministic calculations. As results of this study, the reliability variations were presented versus different loads, probability distribution function for composite plate failure and the factors affecting it.

2. COMPOSITE FAILURE CRITERIA

Strength of a laminate depends upon the strength of each individual lamina. Therefore the strength of all lamina and arrangement style of them on each other provides a laminate specifications. Based on failure mechanisms governing in composite materials mechanics, it is more appropriate to consider the composite as a structure rather than as a material. It is vital to have required knowledge about failure mechanisms of composite materials before any analysis. In the following sections, it is explained briefly the main causes/ mechanisms of composite materials failure. Since the main failure mode of a laminate is failure of lamina, only this mode is analysed in FMEA and the simulation. Further details about failure modes of a lamina is collected in FMEA table.

2.1. Ply Failure

Composite materials consist of at least two constituents: a series of purposefully oriented fibers, surrounded by a solid matrix. Typically, the fibers act as load-carrying members while the matrix transfers the load between them while fixing the fibers in the desired orientation and location within the composite. The resulting material is both strong and stiff. Composite materials display a wide variety of failure mechanisms as a result of their complex structure and manufacturing processes, which include fiber failure, matrix cracking, buckling and delamination [8].

2.1.1 Fiber Failure

Fiber failure is one of the simplest failure mechanisms to identify and quantify. It occurs when the loads applied to a composite structure cause fracture in the fibers. Fiber failure in tension occurs due to the accumulation of individual fiber failures within plies, which becomes critical when there are not enough intact fibers remaining to carry the required loads.

Fiber failure in compression occurs due to micro buckling and the formation of kink bands, and though there is still debate over whether these phenomena are separate failure modes, micro buckling is a more global failure mode whilst kinking seems to be initiated by local microstructural defects and is the most common failure feature observed after testing.

2.1.2 Matrix Failure

Matrix cracks are an intralaminar form of damage, and involve cracks or voids between fibers within a single composite layer, or lamina. Matrix failure is a complex phenomenon in laminated composites, in which matrix cracks initiate typically at defects or fiber–matrix interfaces, accumulate throughout the laminate, and coalesce leading to failure across a critical fracture plane. Failure modes of a composite lamina under mechanical loads, has been collected in table 1.The FMEA is limited to identification of the potential failure mode, their effect on the composite structure and the causes/mechanisms for such failure development.

Buckling is a structural phenomenon that occurs in compression or shear, and though not necessarily resulting in failure, the large deformations, bending and loss of structural capacity involved typically promotes other types of damage and leads to structural collapse. Delamination are separations between internal layers of a composite laminate caused by high through-thickness stresses, and cause significant structural damage, particularly in compression [9]. As it is mentioned in Failure effect column of table 1, matrix cracking can be considered as the most crucial failure mode of a composite lamina.

Table 1: FMEA of a composite lamina [8-9]

Component	Potential Failure Mode	Failure Effect(s)	Failure Cause(s)/ Mechanisms
Fiber	fiber fracture	- Stiffness reduction - Performance and payload reduction	-longitudinal tensile - transverse tensile - Longitudinal Compressive - in plane shear
	fiber fracture with pullout		-longitudinal tensile - in plane shear
	Micro buckling of fibers		- longitudinal compressive
Matrix	Matrix cracking	- the applied load, results in a crack and the crack grows till the matrix fracture	- longitudinal compressive - transverse tensile
	Delamination	-Structural life reduction - laminate rupture	- interlaminate stress -longitudinal tensile
Fiber-Matrix	Fiber pullout with fiber–matrix debonding	- stiffness reduction	- Longitudinal Tensile - shear in plane
	Fracture of fiber-matrix interface		- transverse tensile

3. COMPOSITES RELIABILITY ANALYSIS AND MODELLING

The need to incorporate uncertainties in an engineering design has long been recognized. The traditional approach, the so-called "deterministic design", makes use of safety coefficients in order to prevent unpredicted failures due to the variability of the data. As a consequence, it is not possible to quantify the reliability of the structure, defined as the probability that the structure does not experience a failure. On the other side, a relatively new trend, named "probabilistic design", allowing the estimation of the reliability of the design, considers the stochastic variability of the data. The performance is generally evaluated by means of a variable such as the displacement of a point, the maximum stress, etc... Due to many reasons (e.g., unpredictability of future loading conditions, inability to express the material properties accurately, simplifications in the modelling of the behavior of the structure, limitations in the numerical methods, human errors or omissions, etc...),the 100% reliability cannot be guaranteed. However, the design can be conducted in order to raise the reliability up to a chosen level. Totally based on the researches have been done in this field, the reliability evaluation process can be divided into five major following steps which includes the past researches background and it is arrayed logically and user friendly. These steps was explained in each part and assumptions have been qualified.

3.1. Random Variables

The first and the most important step for analyzing the reliability of composite materials, is selection of random variables and their statistical distributions. Design parameters are defined by n-dimensional vector $X=(X_1, X_2 \ldots X_n)^T$ which its elements are uncertain [10].

Random variables considered in this study are shown in Table 2. According to reference [11], the properties and geometry of materials followed normal distribution and a standard deviation of 5% to 20% is allowed range to be assumed. In this article, it is considered 5% as coefficient of variation. The longitudinal tensile strength according to [12], is a Weibull distribution function. The magnitudes of all parameters was brought in table 2.

Table 2: Strength Data of Composite Material

Random variable	Parameters of normal distribution		Parameters of Weibull distribution	
	μ	σ	α	B
E_1	142.25	7.11	-	-
E_2	8.69	0.43	-	-
G_{12}	4.38	0.219	-	-
υ_{12}	0.24	0.012	-	-
Y_t	51.70	2.585	-	-
h	4.7	0.235	-	-
X_t	-	-	1507.86	75.39

3.2. Failure Criteria

The development of failure criteria for composite materials has been actively pursued for over 30 years by researchers around the world, and there are enormous number of theories available in the literature. In this study, Tsai–Hill criterion of composite laminate plate's failure was selected. This criterion is used for determination of an orthotropic material failure. An orthotropic material has different mechanical properties in three mutually perpendicular directions denoted as 1, 2, and 3, respectively. Composite materials are considered orthotropic in the principal material coordinate system. The Tsai–Hill criterion is expressed as Eq. (1) for a composite material plane layer element subject to stresses in its principal directions [9]:

$$\left[\frac{\sigma_1}{X_1}\right]^2 - \left[\left(\frac{\sigma_1}{X_2}\right)\left(\frac{\sigma_2}{X_2}\right)\right] + \left[\frac{\sigma_2}{Y}\right]^2 + \left[\frac{\tau_{12}}{S}\right]^2 < 1 \qquad (1)$$

In the Eq. (1), X_1, X_2 and Y obtain one of X_C, X_T, Y_C and Y_T depending upon compressive or tensile magnitudes of σ. The next step is to utilize this information in a reliability model. Unfortunately, principal stresses calculation methods and the Tsai–Hill criterion lead to a very complex expression to compute the probability of failure analytically. Fig.1 expresses the algorithm of calculations utilized in composite materials mechanics. All the steps illustrated in Fig.1 is essential to be performed and as the results of that algorithm, it can be accessible the input parameters of Eq. (1). Since the problem is faced to large amount of data, it is better to utilize a computer program to analyze them. For this reason, MATLAB is selected and the algorithm is applied in it.

3.3. Limit State Function

Limit State Function (LSF) is used to model the reliability. This function is defined by failure scenario of the material and according to the definitions [10], negative magnitudes of this function represents failure. In another word, it is represented as the differences of challenge and strength of material and is defined as Eq. (2). Fig.2 shows the separation of safe region from unsafe region:

$$G(R, S) = G(X_1, X_2 \ldots X_n) = R-S \qquad (2)$$

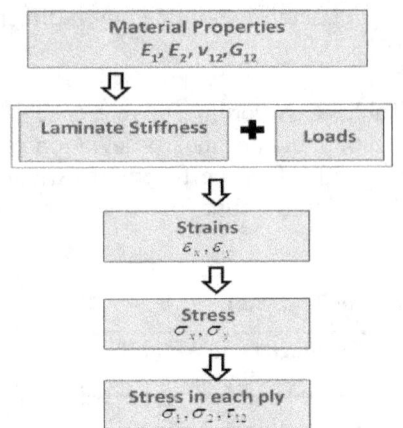
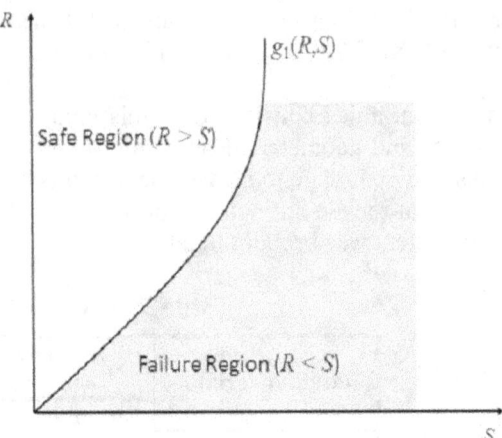

Figure.1: Determine the stress and strain

Figure.2: The separation of safe region from failure region by LSF [10]

In this study the limit state function is represented as differences of stress and stength of structure and using Tsai-Hill failure critera. The LSF was defined like Eq. (3):

$$G = 1 - \left[\frac{\sigma_1}{X_1}\right]^2 - \left[\left(\frac{\sigma_1}{X_2}\right)\left(\frac{\sigma_2}{X_2}\right)\right] + \left[\frac{\sigma_2}{Y}\right]^2 + \left[\frac{\tau_{12}}{S}\right]^2 \qquad (3)$$

As mentioned before, when G<0, the structure is considered failure.

3.4. Determine the Extent of Structural Strength

From the strength limit point of view, composite laminate failure is classified in two major categories:
1- Last Ply Failure: which is caused by crack in matrix.
2- First Ply failure which is caused by delamination, crack in matrix and fiber failure.
In this study the First ply failure approach was used because of simplicity in Safe Mode determination and more prudence in reliability.

3.5. Reliability Evaluation

The aim of structural reliability analysis is to get the probability of structural failure, while the failure state is denoted by the limit state function. In structural reliability analysis, reliability is defined as a multidimensional nonlinear integral. In the structural reliability analysis, the reliability is defined as:

$$P_f = prob[G(X) \le 0] = \int_{G(X) \le 0} f(X) dX \qquad (4)$$

Where G is a performance function or the limit state function, X a vector consisted of the random variables, and f(x) the joint probability density function (PDF) of the random variables X.

Direct evaluation of such an integral is unfeasible or even impossible in most cases. Therefore, some approximation or simulation methods for probabilistic uncertainty analysis have been developed. A direct way to compute this probability of failure is by Monte Carlo simulation. For this particular study, Monte Carlo simulation is preferable to first and second order reliability methods since non-linear complex behavior does not complicate the basic procedure.

Monte Carlo simulation method is sampling procedures for estimating the probability of failure of a component or system. The basic random variables are randomly generated and then inserted a fraction of the overall conditions that lead to failure, as they are considered likely to fail, production and substitution of random variables is in the algorithm illustrated in Fig.3.

Using Monte Carlo simulation, the approximate value of reliability is achieved by the following equation:

$$\hat{R} = \frac{N_S}{N_T} \qquad (5)$$

Where N_s is Total number of successful iterations, and N_T is number of sample size. The algorithm which is illustrated in Fig.3 is related to random variable generation. The random variables are provided for Monte Carlo Method as input data which is brought in table 2. It is essential to use an appropriate sample size in random variable generation in the illustrated algorithm.

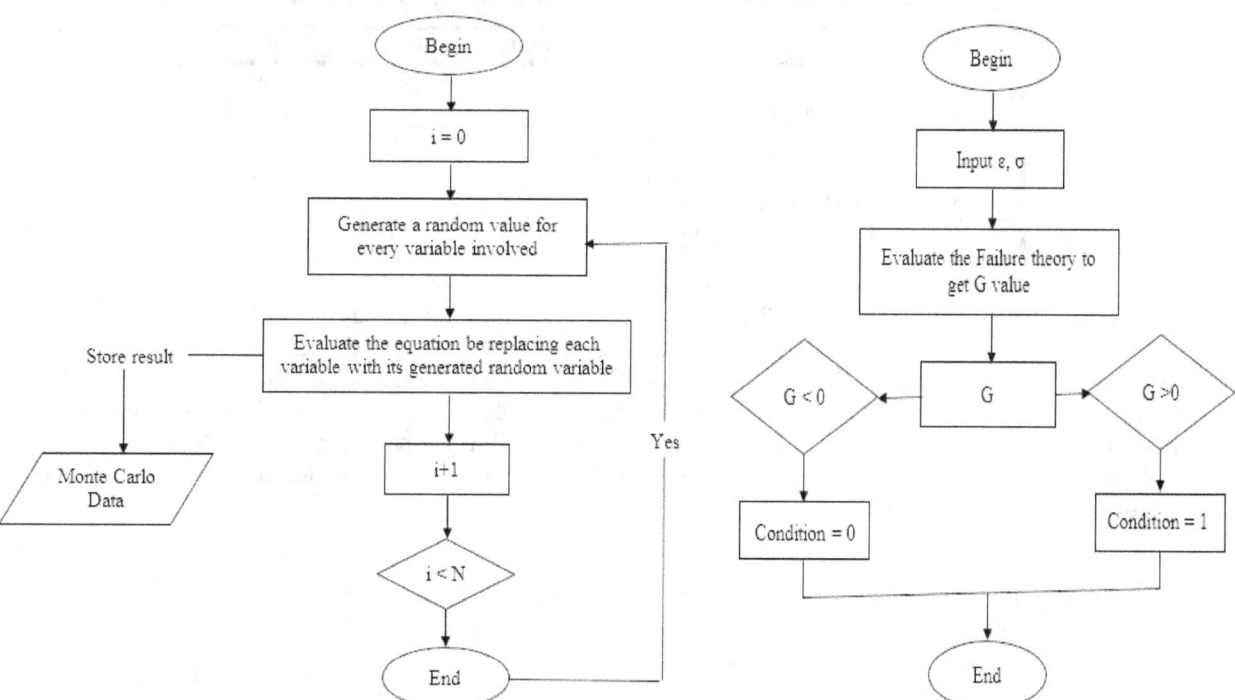

Figure 3: Random variable generation algorithm   **Figure 4: Determination of LSF measure**

4. RESULTS

The probability of failure of the plate assumed as a weakest-link system, was calculated by Monte Carlo simulation. The results indicate that p is almost equal to the probability of failure of first layer. By implementation of the presented algorithms illustrated in Fig.3,4, the results is derived including reliability evaluations and fitted failure probability distributions. The sample of 10^6 iteration is determined adequate as the source of input data for MCM. The simulation is performed in MATLAB. Boundary conditions are assumed as simply support and it is considered tensile loads in X and Y direction that are increasingly changed by the fixed step of 200 KN in each loading condition. It was obtained that the model is converged after almost 10^6 iterations. The convergence of the simulation model is illustrated in Fig.5, under [F_x=1500, F_y=1000] KN loading condition.

Utilizing a computer program in MATLAB lead to the conclusion that the 2P-Weibull distribution is the best alternative to describe the failure behavior of the laminate. Easyfit program is used for benchmarking and the results evident the postulate. In this direction, shape and scale parameters of Weibull distribution has been estimated. They are in good agreement as illustrated in Fig.8. The estimated Weibull parameters including the scale and shape parameters for failure probability distribution of composite laminate was collected in table 3. The parameter α represents the reliability. If α is considered as time, when system reaches that time, the probability of failure would be 63.5 percent. In this case study, it is factor of stress as shown in table 3, while declining the load, stress dropped down and consequently the scale parameter plummeted. Also there is a soft increase in β values. Based on bathtub curve can be represented that the increment in load would result in deterioration of the structure.

Table 3: Evaluated reliabilities and Weibull parameters

Condition	Load	Reliability	α	β
1	Fx=800 , Fy=400	≈1	10.2	0.39
2	Fx=1000 , Fy=500	0.99	9.9	0.54
3	Fx=1300 , Fy=800	0.9612	9.2	0.97
4	Fx=1400 , Fy=900	0.8211	9.01	1.21
5	Fx=1500 , Fy=1000	0.5314	8.06	1.28
6	Fx=1800 , Fy=1300	0.011	8.08	1.81

As a result of the simulation, reliability measures was obtained for different loading conditions and collected in Table 3. This is illustrated in Fig.6 where the variation is demonstrated for reliability under each loading condition. The results reveals that the trend is downward gradually following with a sudden decline after the 3rd loading condition. It is justifiable with β measures. There is a jump in β from condition 3 to 4 as illustrated in table 3. In the 4th condition, shape parameter is greater than 1 declaring that it is in deterioration. Therefore the laminate is faced to an impressive reduction in reliability after the 3rd condition.

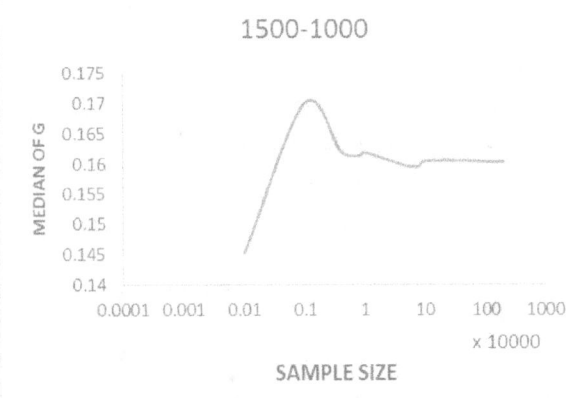

Figure 5: convergence of the simulation under a loading condition

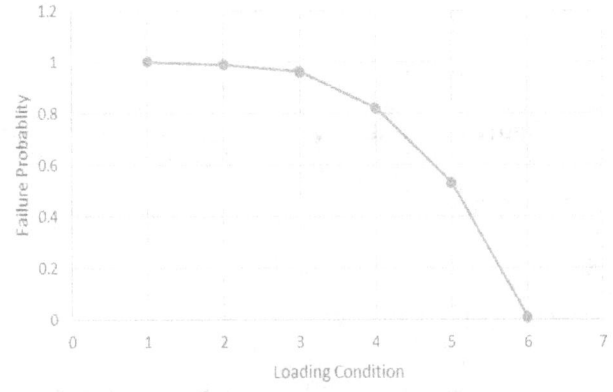

Figure 6: Reliability gradients versus loading conditions

Fig.7 illustrates another concept of failure. Incensement in load, results in incensement in α and the probability of failure which is the area under each probability distribution function diagram is getting large and larger. Furthermore the diagrams by load increment, are driven to left side and according to Weibull distribution concepts, the occurrence of failure has more probability.

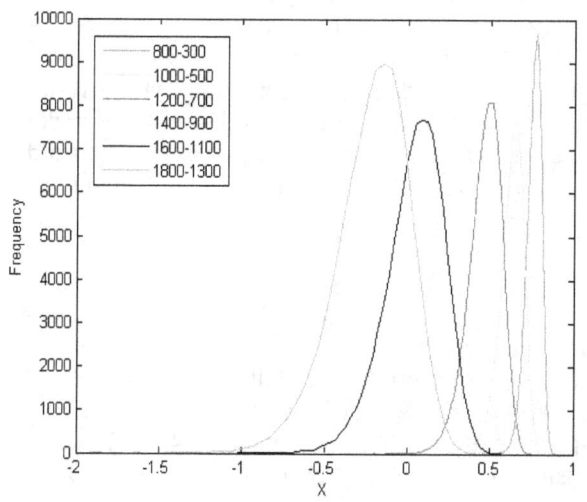

Figure 7: MATLAB output for different loading condition

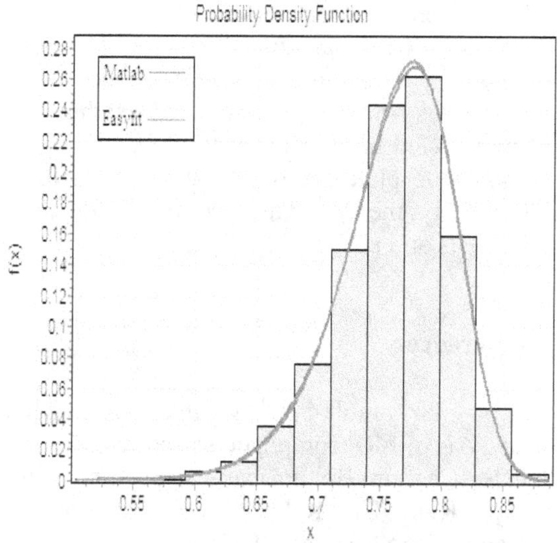

Figure 8: Benchmarking of MATLAB result with Easyfit

Uncertainties always play an important role in reliability assessments. As mentioned before, the standard deviation of input data varies based on various factors such as production process, material deficiencies and so on. In this part, it is assumed that the standard deviation can change almost from 5 to 7.5 percent of median magnitudes. By analyzing the data again, the graph shown in the Fig.9 is resulted. This figure gives good information about the evaluated reliability magnitudes and the probability distribution of composite plate failure. By raising the variance magnitudes of strength factors, the uncertainty trend in data is obviously upwards. This jump is sharper in upper loads. As presented in Fig.9, the last load condition [F_x=1800, F_y=1600] diagram shot up dramatically by contrast of other diagrams and it is while the lower loads diagrams rose gradually at slower pace. This finding, indicates that more uncertainty in input strength data caused by coefficient of variation magnitudes, results in more uncertainty in reliability evaluation.

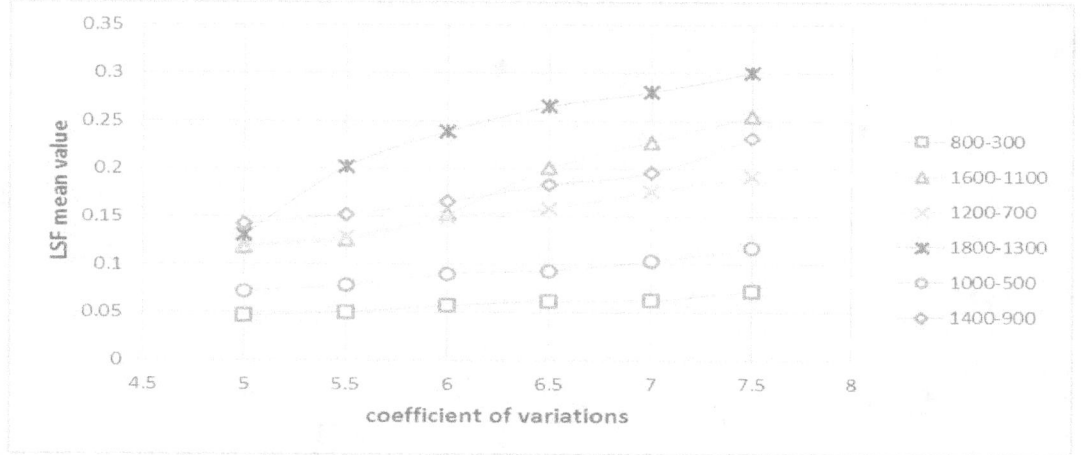

Figure 9: Strength input data Coefficient of variations effect on LSF values

5. CONCLUSION

Since there is uncertainties in composite material strength data, it is essential to utilize the probabilistic assessment in analyzing their specifications. In this paper, an algorithm is presented to evaluate the reliability of composite materials under uncertainty. A composite structure of $[0_2/\pm45/90]_4$ graphite-fiber/epoxy-matrix was selected as the case study for the methodology presentation. These result analysis led to conclusion that the Weibull distribution is fitted with enough confidence to represent composite plate behaviour. In addition to sample size which affects directly accuracy of evaluated reliability, the input variance magnitude is another factor that plays an important role in uncertainty of analysis and converging the results.

References

[1]   Sriramula S, Chryssanthopoulos MK. Quantification of uncertainty modelling in stochastic analysis of FRP composite structures. Composites Part A: Appl Sci Manuf 2009; 40:1673–84.
[2]   Cassenti BN. Probabilistic static failure of composite materials. AIAA Journal 1984; 22:103–10.
[3]   Kam TY. Reliability formulation for composite laminates subjected to first- ply failure. Composite Structures 1997; 38:447–52.
[4]   Chen N-Z, Sun H-H, Guedes Soares C. Reliability analysis of a ship hill in composite materials. Composites Structures 2003; 62:59–66.
[5]   Whiteside MB, Pinho ST, Robinson P. Stochastic failure modelling of unidir- ectional composite ply failure. Reliability Engineering and System Safety 2012; 108:1–9.
[6]   Eamon CD, Rais-Rohani M. Integrated reliability and sizing optimization of a large composite structure. Marine Structures 2009; 22:315–34.
[7]   Guesdes Soares C. Reliability of components in composite materials. Relia- bility Engineering and System Safety 1997; 55:171–7.
[8]   R. M. Jones, "Mechanics of composite material," *Taylor & Francis, New York*, 1999.
[9]   A. K. Kaw, *Mechanics of composite materials*: CRC press, 2010.
[10]   M. Modarres, M. Kaminskiz, and V. Krivstov, *Reliability Engineering and Risk Analysis: A Practical Guide* vol. 55: CRC press, 1999.
[11]   N. E. Dowling, K. S. Prasad, and R. Narayanasamy, *Mechanical behavior of materials: engineering methods for deformation, fracture, and fatigue*: prentice Hall Upper Saddle River^ eNJ NJ, 1999.
[12]   S. W. Tsai, *Composite Materials, Testing and Design*: ASTM International, 1979.

Development of Feedwater Line & Main Steam Line Break Initiating Event Frequencies for Ringhals Pressurized Water Reactors

Anders Olsson[*,a], Erik Persson Sunde[a], and Cilla Andersson[b]
[a] Lloyd's Register Consulting, Stockholm, Sweden
[b] Ringhals NPP, Väröbacka, Sweden

Abstract: During the last years the LOCA initiating event frequencies in the PSA's for the three Ringhals PWR units has been updated using the piping reliability data provided in the R-Book. Since the data currently presented in the R-Book only covers ASME Code Class 1 and 2 it cannot be used for initiating event frequency update for ASME Code Class 3 and 4 (the intention is though that the R-Book shall also cover Code Class 3 and 4 in the future). In order to proceed with initiating pipe break frequency update for the Ringhals PWR units a project has been started with the purpose to develop updated initiating event frequencies for certain Feed Water Line Break and Main Steam Line Break scenarios. The updated initiating event frequencies shall account for the known piping damage and degradation mechanisms, applicable industry-wide and plant-specific service experience data, the plant specific piping layout and material specifications, as well as the plant-specific risk-informed in-service inspection (RI-ISI) program currently implemented for the Main Steam Line and Main Feed Water systems. The updated frequencies shall reflect state-of-the-art piping reliability models that explicitly address aleatory and epistemic uncertainties. Also, the data analysis that underlies this frequency calculation shall be consistent with the requirements of the ASME/ANS PRA Standard Capability Category II. The causes of pipe failure (e.g., loss of structural integrity) are attributed to damage or degradation mechanisms. Oftentimes a failure occurs to synergistic effects involving operating environment and loading conditions. In piping reliability analysis, two classes of failure are considered. The first class is so called "Event-Driven Failures". These failures are pipe stress driven and attributed to conditions involving combinations of equipment failures (other than piping itself; e.g., loose/failed pipe support, leaking valve) and unanticipated loading (e.g., hydraulic transient or operator error). Examples of event-based failures include various fatigue failures (high-cycle vibration fatigue, thermal fatigue). The second class is defined as "Failures Attributed to Environmental Degradation". Environmental degradation is defined by unique sets of conjoint requirements that include operating environment, material and loading conditions. These conjoint requirements differ extensively across different piping designs (material, diameter, wall thickness, method of construction/fabrications). Similarly, pipe flaw incubation time growth rates differ extensively across the different combinations of degradation susceptibility and operating environments. For the piping systems included in the scope (i.e. Main Steam and Main Feed Water systems), flow accelerated corrosion constitutes a potentially key degradation mechanism. The initiating event frequency calculation will be based on a methodology similar to the one used in previous applications of R-book. This means that service experience data together with a Bayesian analysis framework will be utilized to derive piping reliability parameters for input to PSA models and PSA model applications. The piping service experience data input to the pipe failure rate and rupture frequency calculations will be taken from the Lloyd's Register Consulting proprietary PIPExp database which includes detailed information on piping damage and degradation mechanisms in Code Class 1, 2 and 3 and non-Code piping systems. The paper will present the work that has been performed together with conclusions and insights achieved during the project.

Keywords: LOCA, Feedwater, Steam, Degradation, Bayesian

---

[*] anders.olsson@lr.org

1. INTRODUCTION

In the current version of the Ringhals-3 PSA model the FWLB and MSLB initiating event frequencies are adapted from SKI Report 94:12 [1]; Initiating Events at the Nordic Nuclear Power Plants (the "I-Book"). Appendix III of Reactor Safety Study (WASH-1400) [2] provided the technical basis for "pipe break initiating event frequencies" in the I-Book. Therefore, the current Ringhals-3 FWLB and MSLB initiating event frequencies reflect the state-of-knowledge of the WASH-1400 era (early 1970s).

Major progress has been made in the advancement of piping reliability analysis methodology. This progress is, in part, due to R&D in probabilistic fracture mechanics as well as in statistical models of piping reliability. Significant work has addressed pipe failure data collection and evaluation. Finally, the application of risk-informed in-service inspection (RI-ISI) methodologies has enabled highly realistic characterization of the consequences of pipe failures as a function of degradation mechanism (DM), break location and inspection strategy. During the past ten years, RI-ISI has been implemented at Ringhals Units 2, 3 and 4. The RI-ISI implementation project at Ringhals is referred to as RIVAL.

The technical challenge that is to be addressed here is to enhance and update the characterization of the five initiating event frequencies. Specifically, the project will address the following technical aspects:

- Apply state-of-the-art piping reliability models. This includes application of DM-centric conditional rupture probability (CRP) models. Also, DM-centric and location-specific failure rate distributions will be developed through a DM analysis coupled with exposure term definitions that account for plant-to-plant variability in component populations.
- Derivation of piping reliability parameters based on the applicable, current body of industry-wide and plant-specific service experience data.
- Completeness and modeling uncertainty will be addressed according to existing industry guidelines.
- Consideration of the Ringhals piping integrity management practices and procedures.

1.1. Task breakdown and work flow

The scope of work in this proposed project includes six (6) tasks. Included are technical tasks to calculate the initiating event frequencies, as well as additional tasks to facilitate the management of the project and to address specific items unique to the Ringhals PWR Units 2, 3 and 4:

- Task 1. Project Mobilization; this task includes the kick-off meeting, work scope definition and information collection.
- Task 2. Definition of Calculation Cases
- Task 3. Parameter Estimation
- Task 4. Documentation
- Task 5. Review & Comment Resolution
- Task 6. Technology Transfer

The scope of the proposed work includes calculating new IE frequencies associated with a feedwater line break inside and outside containment (T_{Fx}), and steam line break (T_{Sx}).

2. TECHNICAL APPROACH

The technical approach to estimate FWLB and MSLB initiating event (IE) frequencies is based on the model expressed by Equations (1) and (2) for estimating the frequency of an IE of a given magnitude. Oftentimes, the magnitude is expressed by an equivalent break size (EBS) and corresponding through-wall flow rate. The parameter x is treated as a discrete variable representing different equivalent break-size ranges.

$$F(IE_x) = \sum_i m_i \rho_{ix} \qquad (1)$$

$$\rho_{ix} = \sum_k \lambda_{ik} P(R_x|F_{ik}) I_{ik} \qquad (2)$$

Where:

- $F(IE_x) =$ Frequency of pipe break of size x, per reactor calendar-year, subject to epistemic uncertainty calculated via Monte Carlo simulation
- $m_i =$ Number of pipe welds (or fittings or inspection locations of type i; each type determined by pipe size, weld type, applicable damage or degradation mechanisms, and inspection status (leak test and NDE); no significant uncertainty
- $\rho_{ix} =$ Frequency of rupture of component type i with break size x, subject to epistemic uncertainty calculated via Monte Carlo simulation or lognormal formulas
- $\lambda_{ik} =$ Failure rate per "location-year" for pipe component type i due to failure mechanism k, subject to epistemic uncertainty determined by RI-ISI Bayes method and Eq. (3) below
- $P(R_x|F_{ik}) =$ Conditional probability of rupture of size x given failure of pipe component type i due to damage or degradation mechanism k, subject to epistemic uncertainty. This parameter may be determined on the basis of expert elicitation or service experience insights.
- $I_{ik} =$ Integrity management factor for weld type i and failure mechanism k, subject to epistemic uncertainty determined by Monte Carlo simulation and Markov model

Point estimates of the failure rate for type i and failure mechanism k is calculated according to formula (3):

$$\lambda_{ik} = \frac{n_{ik}}{\tau_{ik}} = \frac{n_{ik}}{f_{ik} N_i T_i} \qquad (3)$$

Where:

- $n_{ik} =$ Number of failures in pipe component (i.e., weld) type i due to failure mechanism k; very little epistemic uncertainty. The component boundary used in defining exposure terms is a function of DM.
- $\tau_{ik} =$ Component exposure population for welds of type i susceptible to failure mechanism k, subject to epistemic uncertainty determined by expert opinion
- $f_{ik} =$ Estimate of the fraction of the component exposure population for weld type i that is susceptible to failure mechanism k, subject to epistemic uncertainty, estimated from results of RI-ISI for population of plants and expert opinion
- $N_i =$ Estimate of the average number of pipe welds of type i per reactor in the reactor years exposure for the data query used to determine n_{ik}, subject to epistemic uncertainty, estimated from results of RI-ISI for population of plants and expert knowledge of damage mechanisms
- $T_i =$ Total exposure in reactor-years for the data collection for component type i; little or no uncertainty

For a Bayes' estimate, a prior distribution for the failure rate is updated using n_{ik} and τ_{ik} with a Poisson likelihood function. The formulation of Equation (3) enables the quantification of conditional failure rates, given the known susceptibility to the given damage or degradation mechanism. When the parameter f_{ik} is applied, the units of the failure rate are failures per welds susceptible to the damage or degradation mechanism. This formulation of the failure rate estimate is done because the susceptible damage or degradation mechanisms typically are known from the results of a previously performed degradation mechanism analyses that are part of RI-ISI evaluations. If the parameter f_{ik} is set to 1.0, the failure rates become unconditional failure rates, i.e., independent of any knowledge about the susceptibility of damage or degradation mechanism, or alternatively that 100% of the components in the population exposure estimate are known to be susceptible to a certain degradation mechanism (e.g., flow-accelerated corrosion, FAC).

2.1. Analysis Convention & Nomenclature

The causes of pipe failure (e.g., loss of structural integrity) are attributed to damage or degradation mechanisms. Oftentimes a failure occurs to synergistic effects involving operating environment and loading conditions. In piping reliability analysis, two classes of failure are considered:

- Event-Driven Failures. These failures are pipe stress driven and attributed to conditions involving combinations of equipment failures (other than piping itself; e.g., loose/failed pipe support, leaking valve) and unanticipated loading (e.g., hydraulic transient or operator error). Examples of event-based failures include various fatigue failures (high-cycle vibration fatigue, thermal fatigue).
- Failures Attributed to Environmental Degradation. Environmental degradation is defined by unique sets of conjoint requirements that include operating environment, material and loading conditions. These conjoint requirements differ extensively across different piping designs (material, diameter, wall thickness, method of construction/fabrications). Similarly, pipe flaw incubation times flaw growth rates differ extensively across the different combinations of degradation susceptibility and operating environments. For the piping systems that are in the scope of work (i.e., System 411, Main Steam, and System 415, Main Feedwater), flow accelerated corrosion constitutes a potentially key degradation mechanism.

Synthesized in Figure 1 is the existing worldwide service experience with metallic piping in commercial nuclear power plants; in excess of 8,820 failure records. Included in this figure are the unique failure manifestations of concern. In piping reliability, a "failure" is any degraded condition that necessitates repair or replacement. The "magnitude" of a failure manifestation can be measured through non-destructive or destructive examination. Through-wall defects are characterized by the size of a flaw and resulting leak or flow rate (from perceptible leakage to gross leakage). As depicted in Figure 1, certain combinations of material, operating environments have produced "major structural failures." The high-level database summary in Figure 1 is used to formulate specifications for a quantitative analysis of pipe failure parameters.

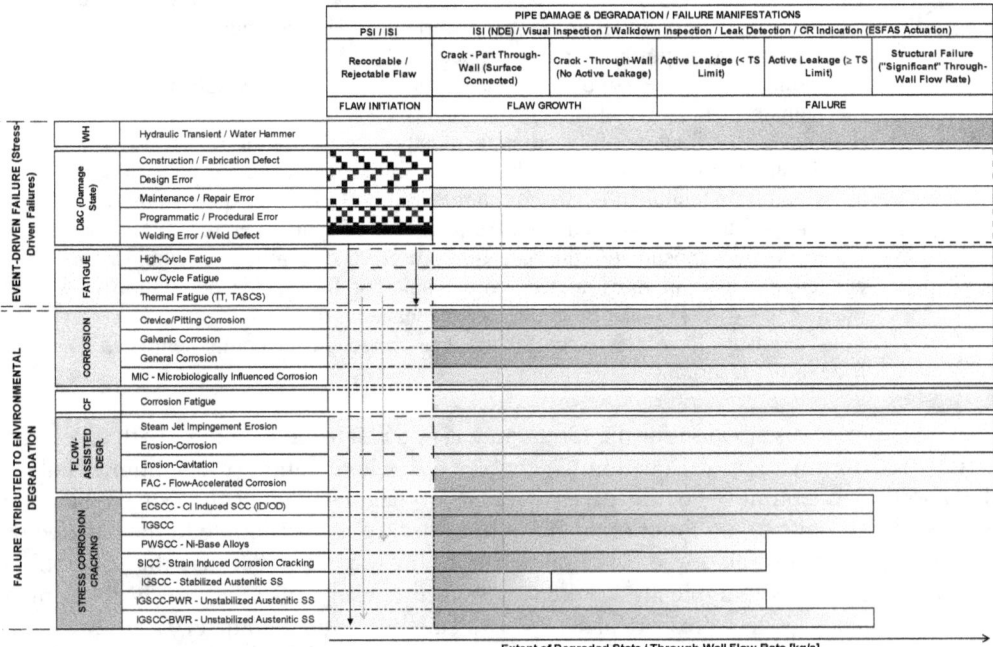

Figure 1 – Pipe Failure Manifestations.

2.2. Definition of Calculation Cases

Task 2 is concerned with the desired degree of IE frequency model refinement. For example, an underlying assumption is that a single set of "representative IE frequencies" for the three PWR units is to be developed. It should also be defined if the IE frequency models should address the positive effect on IE frequency by different ISI strategies or degradation mitigation strategies. For illustrative purposes, the chart in Figure 2 shows results from a sensitivity study performed within the scope of the Kewaunee HELB IE Frequency Study. This sensitivity study addressed three cases:

1) no or inadequate implementation of a FAC program,
2) effect on Extraction Steam piping reliability through implementation of an effective FAC program consistent with NSAC-202L [3] guidelines, and
3) effect on Extraction Steam piping reliability by replacement of original carbon steel piping with a FAC-resistant material (e.g., stainless steel).

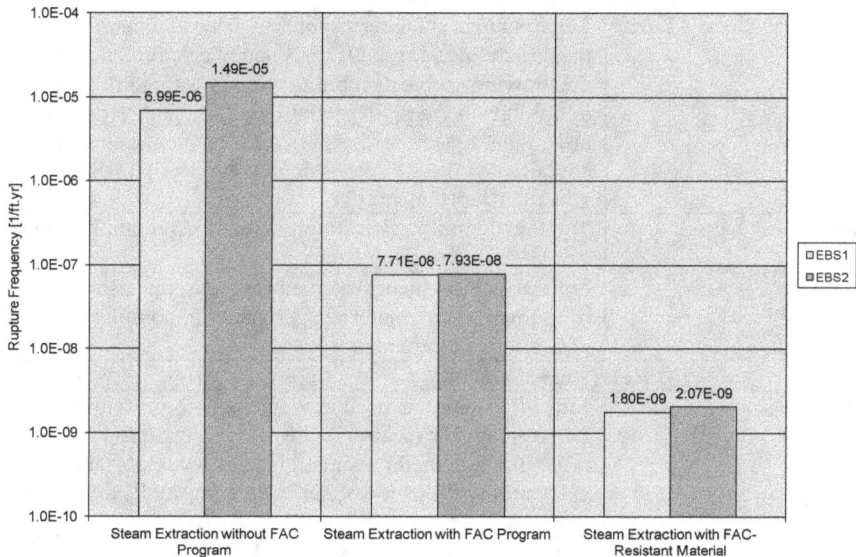

Figure 2 – Example of HELB IE Frequency Sensitivity Analysis.

Another technical consideration is the correlation of IE frequencies with equivalent break sizes as required by a PSA model. Potentially, the IE frequency models require the consideration of a spectrum of pipe break sizes; e.g., ESFAS (Engineered Safety Features Actuation System) set points. The piping reliability analysis methodology includes the application of a conditional rupture probability (CRP) model for a range of pipe break sizes. The break sizes to be considered range from the minimum break sizes triggering ESFAS actuation at the upper end and a double-ended guillotine break (DEGB) at the lower end of the "actuation distribution".

2.3. Parameter estimation

The parameter estimation task encompasses the estimation of the pipe failure parameters that are necessary to quantify the IE frequency models. The IE frequency is calculated by multiplying the number of components in a given system by a rupture frequency per unique component category. The parameter estimation task consists of the following subtasks:

- Finalization of Calculation Workbooks. All calculations will be performed in MS-Excel with Oracle Crystal Ball^{TM} (for uncertainty propagation) and R-DAT Plus^{TM} (for data specialization). The workbook format originally developed for the Kewaunee HELB IE Frequency Analysis will be modified to include a new CRP model and to fit the requirements specific to this study.

- Preparation of Calculation Input Parameters. This subtask establishes event populations and corresponding exposure terms.
- Parameter Estimation. This subtask consists of parameter estimation using the aforementioned MS-Excel workbook and quantification of the IE frequency models.

2.4. Information requirements

The detailed information requirements are itemized in Table 1. Three types of technical information are needed: 1) drawings (i.e., isometric drawings and P&IDs) that clearly define the system boundaries for the IE frequency calculation, 2) in-service inspection (ISI) information, and 3) plant-specific service experience data for the data specialization subtask.

Table 1: List of Information Requirements

Information Item	Comment / Question
System Design Information / Drawings	
Full set of isometric drawings for Systems 411 & 415	The drawings shall correspond with system boundaries that are defined for the IE Frequency calculation. Also, the drawings serve as input to the exposure term definitions in the piping reliability parameter estimation task. What System 415 piping routing changes inside containment were implemented in connection with steam generator replacements?
Full set of P&IDs for Systems 411 & 415	The P&ID information supports the interpretation of the isometric drawing information
Piping material & pipe size data	For the pipe segments selected for this scope of work, need line ID with material specification, nominal pipe size, and wall thickness
ISI Reference Material	
Itemized list of inspection locations	This information is used in identifying the potentially DM-susceptible locations. It is used as input to the reliability parameter estimation tasks to support the calculation of location-specific data for the IE frequency model
RIVAL DM Evaluation Report	This document identifies the DM-susceptible locations in Systems 411 & 415
ISI Reference Material	
List of flaw indications (R2, R3, R4)	This list itemizes the observed flaws (i.e., wall thinning in excess of minimum allowable wall thickness, and through-wall flaws), location, date. Included in the list are events that have required power reduction (to affect repairs) and forced outages.
List of piping replacements	This list itemizes the replacements made as a result of degraded conditions. For System 415 and sections inside containment, what is the service experience pre- and post-S/G replacement?
Applicable Reportable Occurrence (RO) reports submitted to SKI/SSM	

3. DAMAGE MECHANISM EVALUATION

In developing the FWLB and MSLB initiating event frequency models, piping reliability parameters are estimated that are conditional on the presence of certain active degradation mechanisms. The process of estimating these reliability parameters begins by performing a systematic degradation mechanism (DM) evaluation of all pipe segments within the evaluation boundary. The DM evaluation addresses the conjoint requirements for degradation; the material, operating environment and pipe stress conditions necessary for material degradation (cracking, wall thinning, leakage or rupture) to occur.

3.1. Scope of DM evaluation

The following five initiating events involve plant transients caused by a pipe break inside or outside the containment:

TFI: A break of the feed water line between one of the check valves 415-1243, 415-1244 and 415-1245 and the respective steam generator.

TFY: Transient TFY models a break of the feed water line inside/outside the containment up to the check valves 415-1243, 415-1244, and 415-1245.

TSI: Steam line break inside the containment is defined as a break of a main steam line between the steam generator steam line outlet and the main steam isolation valves 411-1131, 411-1132, and 411-1133.

TSY: A steam line break outside the containment is a break downstream of the steam isolation valves 411-1131, 411-1132, 411-1133 up to the turbine stop valves and the steam dump valves to the condenser 411V113, -114, -115, -116.

TSH: Transient TSH models a break, which occurs on the steam piping to the steam-driven auxiliary feed water pump.

The subject DM evaluation addresses piping material damage and degradation mechanisms potentially affecting the Feedwater (System 415) and Main Steam (System 411) pressure boundary integrity. The evaluation utilizes industry wide and plant specific service experience insights applicable to the five initiating events.

3.2. High-Level Service Experience Overview

The service experience related to the main feedwater piping includes cracking and wall thinning of feedwater nozzles and piping. The cause of the cracking was determined to be thermal fatigue, possibly corrosion assisted, and the root cause was generally attributed to thermal stratification of the coolant in the nozzle. Although fatigue cracking from thermal stratification (TASCS) has been found only in the feedwater nozzle area, the presence of thermal stratification has been verified from temperature measurements of the piping wall far upstream of the nozzle but inside the containment. The high loads from thermal stratification have bowed pipes and damaged feedwater piping. Insights from DM evaluations performed in support of RI-ISI program development confirm that the piping may experience thermal gradients between the top and bottom of the main feedwater piping at hot standby conditions and startup due to flow low enough to cause the Richardson number[†] (Ri) to be greater than 4. This is considered cyclic in the horizontal piping near the SG. Fatigue is not the only degradation mechanism that has caused damage to the PWR feedwater systems. Flow-accelerated corrosion has caused wall thinning in feedwater lines. Also, the feedwater nozzle thermal sleeves have experienced thinning at the leading edge, possibly because of flow-accelerated corrosion.

Water hammer can fracture piping at areas degraded by fatigue or flow-accelerated corrosion, and have caused through-wall cracks and ruptures of nozzles and piping. Water hammer events have also damaged piping supports.

[†] A dimensionless parameter used to determine whether stratification will occur. If $Ri < 4$ stratification will not occur. This occurs if the temperature difference between two fluids is low or flow velocity is high, causing mixing. Additional details can be found in EPRI TR-103581, "Thermal Stratification, Cycling and Striping" (1994).

Illustrated in Figure 3 is the MS-specific, worldwide piping service experience involving all damage and degradation mechanisms. The steam exiting the steam generators is dry, superheated steam. Assuming long-term base load operation of the reactor, the MS piping is less likely to be exposed to FAC than the FW-specific piping.

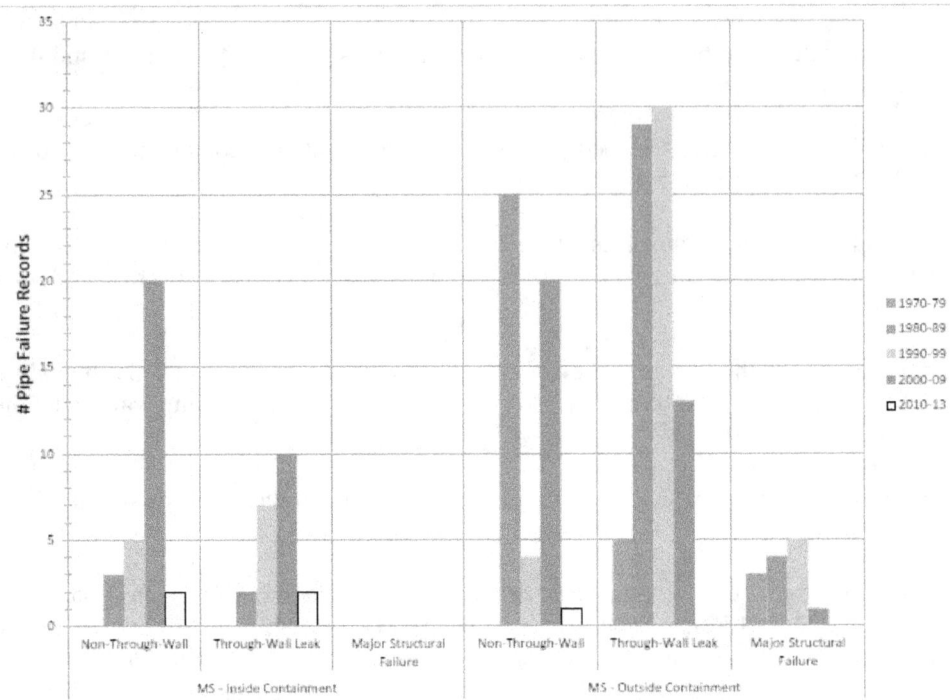

Figure 3 – Service experience with MS piping.

3.3. DM evaluation guidance

According to the EPRI RI-ISI methodology, all inspection locations in the assessed systems must be classified by failure potential. This classification is accomplished by determining those damage and degradation mechanisms that might apply to each assessed piping system location. The damage and degradation mechanisms to be assessed are given below:

- TASCS Thermal Stratification, Cycling, Striping
- TT Thermal Transient
- IGSCC Intergranular Stress Corrosion Cracking
- TGSCC Transgranular Stress Corrosion Cracking
- ECSCC External Chloride Stress Corrosion Cracking
- PWSCC Primary Water Stress Corrosion Cracking
- MIC Microbiologically-Influenced Corrosion
- PIT Pitting
- CC Crevice Corrosion
- E/C Erosion-Corrosion
- E-C Erosion-Cavitation
- FAC Flow-Accelerated Corrosion
- SH Steam Hammer
- WH Water Hammer

3.4. Evaluation of FW and MS systems DM susceptibility

The following conclusions were made for the two systems on their susceptibility for different DMs:

- The piping between the steam generator and first elbow is therefore considered susceptible to TASCS.
- Thermal transient (TT) can occur in the piping between the steam generator and first elbow during hot standby conditions due to the large temperature differences between mixing fluids.
- The FW piping inside and outside containment is susceptible to flow-sensitive (FS) wall thinning. Erosion cavitation (E-C) is a concern for areas immediately downstream flow control valves. All FW piping falls under the scope of the managed FAC program.
- With reference to Ringhals-3, it is noted that certain sections of the FW piping consists of low-alloy steel (LAS) of type 15Mo3 with a chromium content of 0.3% (weight %), which is resistant to FAC. Therefore these pipe sections are assigned low-to-very-low failure potential.
- According to the available service experience data, water hammer events have been found to occur in six areas in the FW system: 1) Upstream of the steam generator, 2) downstream of the FW flow control valves, 3) adjacent to the FW recirculation valves, 4) at the FW pump turbine steam supply line, 5) piping adjacent to the FW pump suction and discharge piping, and 6) preheater bypass line.
- According to the available service experience data, water/steam hammer events have been found to occur in three areas: 1) Adjacent to the main steam isolation valves, 2) piping downstream of the turbine bypass valve, and 3) in the main steam relief valve piping.

The overall conclusion from the DM evaluation was that credible degradation mechanisms potentially affecting the structural integrity of the R3 FW and MS piping include FAC of carbon steel piping, and LDIE of low-alloy steel piping. FAC "entrance effects" may impact localized areas at or near the interface between carbon steel and low-alloy steel pipe sections.

4. INPUT PARAMETERS

This section summarizes the Service Experience Data and Exposure Term Data, the results of these two tasks provide the input to the subsequent piping reliability parameter estimation.

4.1. Pipe failure data source

The source of all piping service experience data supporting this study is the proprietary PIPExp Database. Established in 1993, it is a continuously maintained and updated database on the service experience with piping in commercial nuclear power plants worldwide. It covers the period 1970 to date. An early (1998) version of this database serves as the 'parent database' of the OECD Nuclear Energy Agency (NEA) international database projects OPDE (2002-2011) and CODAP (2011-2014)[‡].

The PIPExp Database is a relational database developed to support a broad range of piping reliability analysis tasks. The database structure is documented in a Coding Guideline and the basic database functionalities are documented in an Applications Handbook. An integral element of all database applications involves defining data screening rules and data query functions. There are two types of queries. A 'topical query' is performed to identify those data records of direct relevance to an application. Application-specific queries are performed to invoke data screening rules to address unique combinations of piping reliability attributes and influence factors.

[‡] Additional information is available on the Internet at www.oecd-nea.org.

4.1. Calculation case summary

The definition of calculation cases that are needed for the five initiating event frequency models address unique combinations of piping reliability attributes and influence factors. Pipe failure rates are to be generated for each calculation case. All FW and MS piping system components in the respective IE model are assigned a Calculation Case. A derived failure rate is common to all piping component IDs within the calculation case. Key inputs that are needed to define calculation cases include:

- Results of the degradation mechanism (DM) evaluation;
- Assignment of a DM to each of the piping components in respective IE model;
- Pipe materials, pipe size and component type.

Tables 2 and 3 identify the proposed calculation cases.

Table 2: FW System Calculation Cases

Calculation Case ID	DM	CRP Model Basis	Comment
1.a	TASC	NUREG-1829 - PWR Thermal Fatigue - CRP technical basis is presented in STP GSI-191 Technical Report	This calculation case applies to the FW pipe-to-SG-nozzle welds; one location per FW loop
1.b	TASC	Same as 1.a	This case applies to first straight section between SG nozzle and check valve - one pipe section per loop
2	LC-FAT	Specific CRP model is developed	Case 2 applies to welds in pipe sections for which no active DM is identified.
3.a	FAC	Specific CRP model is developed. The model applies to single-phase flow conditions.	Case 3.a applies to bends and elbows in carbon steel pipe sections
3.b	FAC	See Case 3.a	Case 3.b applies to locations immediately downstream of welds between low alloy steel and carbon steel (interface between FAC-resistant and FAC-susceptible piping. This calculation case addresses potential FAC-entrance effects.
4	LDIE	Specific CRP model is developed	Case 4 applies to bends and elbows in low alloy steel pipe sections. This calculation case addresses potential localized erosion attributed to cavitation[§].
5	VF	Service experience data	Case 5 applies to small-bore branch connection weld locations.

Table 3: MS System Calculation Cases

Calculation Case ID	DM	CRP Model Basis	Comment
1	SH/WH	Specific CRP model is developed	This calculation case applies to welds at MS relief valves and MSIVs
	LC-FAT	Specific CRP model is developed	Case 2 applies to welds in pipe sections for which no active DM is identified.
3	LDIE	Specific CRP model is developed	Case 3 applies to bends and elbows – the MS piping is considerably less susceptible to FAC than the FW piping

---

[§] Low alloy steels, such as 15Mo3, have less FAC wear rate than carbon steels. 15Mo3 has a carbon content of 0.3%. For chromium contents greater than 0.04% the wear rate decreases linearly on a log-log-scale. According to experimental data, with a 0.1% of chromium, the wear rate is approximately 2.5 times lower than standard carbon steel.

4.1. Data reduction principles

The FW service experience data is applicable to the FWLB evaluation boundary as well as Westinghouse type PWR plants. Because differences in piping design/layout and material selections, service experience from Babcock & Wilcox (B&W), Combustion Engineering (CE) and KWU/Siemens plants was not considered. The MS service experience data only considers pipe sections subjected to single-flow conditions (super-heated steam). Excluded from the service experience evaluation are Main Steam Cross-under, Extraction Steam, and Moisture Separator Reheater piping.

5. CONDITIONAL RUPTURE PROBABILITY MODELS

This section documents the development of the conditional rupture probabilities (CRPs; $P(R_x | F_x)$) given presence of a degraded condition for different pipe break sizes for each of the component types defined. A degraded condition can be an embedded flaw (not connected to the inside diameter of a piping component), a non-through-wall crack, a thinned pipe wall, or a through-wall flaw.

The likelihood of a pipe flaw propagating to a significant structural failure is expressed by the conditional failure probability $P(R_x / F_{ik})$). With no service data available to support a direct statistical estimation of the conditional probability the assessment can be based on probabilistic fracture mechanics (PFM), expert judgment, or a combination of service data insights, expert judgment and PFM. Different PFM algorithms have been developed, but with a focus on fatigue growth through vibration fatigue, thermal fatigue) and stress corrosion cracking in Reactor Coolant Pressure Boundary piping. There remain issues of dispute with respect to reconciliation of results obtained through statistical estimation and extrapolation versus the physical models of PFM, however.

The approach taken in this study is to utilize service experience insights and results from the expert elicitation documented in NUREG-1829 [4]. For certain combinations of material, loading conditions and degradation susceptibility, sufficient service experience exists to support direct CRP estimation. As an example, extensive data exists on FAC-induced pipe rupture. Correlating, the statistical evidence on equivalent break size (EBS), material and flow conditions (e.g., single-phase vs. two-phase flow) provides an empirical CRP-EBS correlation.

For other types of degradation mechanisms (DMs), only "precursor data" is available. That is, the service experience data is limited to observations of rejectable non-through-wall flaws and minor through-wall flaws. The technical approach to CRP development for DMs other than FAC is structured to capture the current state of knowledge of LOCA frequencies as documented in NUREG-1829. The expert elicitation as documented in NUREG-1829 synthesizes inputs from experts representing two schools of thought on how to best quantify pipe break frequencies: one based on statistical analysis of service data and simple models, and another based on probabilistic fracture mechanics approaches. The 12 experts that participated in this expert elicitation provided a balanced perspective on these two approaches and produced estimates of the LOCA frequencies vs. break size for use in risk-informed evaluations. NUREG-1829 included some "base case" analyses that were performed on selected components to inform the expert elicitation. The technical approach to CRP model development used herein is designed to make use of both sets of information developed in NUREG-1829, namely, the base case analyses and the inputs provided by the nine experts and documented in NUREG-1829.

Details on the technical approach are documented in ASME-PVP-2007-26281 [5] and References [6][**] and [7].

[**] Available on the Internet at http://nrc.gov/wba/ (accession No. ML112770237).

5.1. CRP models for FWLB & MSLB initiating event Frequency analysis

Based on the results of preceding tasks, the FWLB and MSLB initiating event frequency analysis require the following CRP models:

- FAC-centric CRP model that reflects single-phase flow conditions in carbon steel piping. The proposed CRP model is based on empirical data;
- LC-FAT-centric CRP model that reflects the through-wall growth of a pre-existing weld flaw under the influence of low-cycle fatigue loading such as normal cooldown and heatup cycles. This CRP model is derived from NUREG-1829 and the South Texas Project and Vogtle Electric Power Station GSI-191 LOCA frequency evaluations.
- LDIE-centric CRP model that reflects localized erosion in low alloy steel (LAS) piping. The proposed CRP model is derived from NUREG-1829 using the technical approach documented in Section 5 above.
- TASCS-centric CRP that reflects thermal fatigue acting on the Code Class 2 FW nozzles. This CRP model is derived from NUREG-1829 using the technical approach documented in Section 5 above.
- SH/WH-centric CRP, which is based on empirical data.

6. CONCLUSIONS

Given the analysis framework, the knowledge about possible damage mechanisms with respective conditional rupture probabilities, the mapping of piping components (welds, bends etc), and the service experience gained from more than 4000 reactor years of operation updated initiating event frequencies have been derived during the project. At this stage the calculations are ongoing but we hope that the results will be possible to present during the PSAM12 conference.

Acknowledgements

We would like to acknowledge the Ringhals NPP for their support during the project and also Mr. Bengt Lydell for his outstanding guidance to the project team throughout the entire project.

References

[1] Pörn, K., Shen, K. and Nyman, R., *Initiating Events at the Nordic Nuclear Power Plants*, SKI Report 94-12, Swedish Nuclear Power Inspectorate1
[2] U.S. Nuclear Regulatory Commission, Failure Data. Appendix III to the Reactor Safety Study, WASH-1400 (NUREG-75/014), Washington, DC, pp III-74—78, October 1975. , Stockholm, Sweden, October 1994.
[3] Electric Power Research Institute, *Recommendations for an Effective Flow-Accelerated Corrosion Program* (NSAC-202L-R3), 1011838, Palo Alto (CA), May 2006.
[4] Tregoning, R., Abramson, L. and Scott, P., *Estimating Loss-of-Coolant Accident (LOCA) Frequencies Through the Elicitation Process*, NUREG-1829, U.S. Nuclear Regulatory Commission, Washington, DC, April 2008.
[5] Lydell, B., "*The Probability of Pipe Failure on the Basis of Service Experience,*" Proc. 2007 ASME Pressure Vessel and Piping Division Conference PVP-2007, Paper ASME-PVP-2007-26281, American Society of Mechanical Engineers, New York, NY, July 2007
[6] Fleming, K. and Lydell, B., *Development of LOCA Initiating Event Frequencies for South Texas Project GSI-191*, Final Report for 2011 Work Scope, Revision 1, Scandpower Inc., Houston, TX, October 2011.
[7] Fleming, K. and Lydell, B., *Development of LOCA Initiating Event Frequencies for Vogtle Electric Generating Plant GSI-191*, Draft Report, Lloyd's Register Consulting, Houston, TX, September 2013.

Improvement of the Reliability and Robustness of Variance-Based Sensitivity Analysis of Final Repository Models by Application of Output Transformation

Dirk-Alexander Becker,
Gesellschaft fuer Anlagen- und Reaktorsicherheit (GRS) mbH, Braunschweig, Germany
dirk-alexander.becker@grs.de

Abstract: Long-time performance assessment models for final repositories for radioactive waste typically produce heavily tailed output distributions that extend over several orders of magnitude and under specific circumstances can even include a significant number of exact zeros. A variance-based sensitivity analysis gives a strong overweight to the typically very few values that are far away from the expected value of the distribution, which can lead to a low robustness of the evaluation. Moreover, while a variation of the model output, even over orders of magnitude, is of little interest if it happens on a radiologically irrelevant level, a mere factor of 2 near the permissible dose limits can be very important. Both types of problems can be mitigated by applying appropriate output transformations before performing the sensitivity analysis. The effects of different transformations on the sensitivity analysis results for typical final repository model systems are demonstrated.

Keywords: Variance-based sensitivity analysis, transformation, radioactive waste disposal

1. INTRODUCTION

The long-term performance of final repositories for radioactive waste has to be investigated using computational models describing the release of radionuclides from waste containers and their transport through the near field, the geosphere and the biosphere. Probabilistic sensitivity analysis is an important tool for improving the understanding of the model behavior as well as for identifying research needs. With realistic parameter ranges, however, such models typically produce heavily tailed output distributions that extend over several orders of magnitude and, under specific circumstances, can even include a significant number of exact zeros. This causes two types of problems:

a) A variance-based evaluation gives a strong overweight to values that are far away from the expected value of the distribution. Therefore, the total output variance is typically dominated by very few individual values, which can lead to low robustness of the evaluation.

b) In view of radiological safety, the highest output values are the most relevant ones. While a variation of the model output, even over orders of magnitude, is of little interest if it happens on a radiologically irrelevant level, a mere factor of 2 near the permissible dose limits can be very important. Usual sensitivity analysis methods do not by themselves take account of such non-mathematical asymmetries.

Both types of problems can be mitigated by applying appropriate output transformations before performing the sensitivity analysis. The goal pursued with such transformations is to map the model output, typically the effective annual dose to a human individual, to some magnitude that better represents the actual harmful effects to the environment on a scale that is well consistent with the mathematical method of evaluation.

In cases of model output values distributed over several orders of magnitude, it is sometimes recommended to analyze the logarithms of the output instead of the calculated values themselves. This, however, does not always solve the problem and can even make it worse. Zero values are mathematically excluded from a log-transformation, and very low values are highly overvalued. A change of, say, three orders of magnitude would be given the same weight, regardless of whether it occurs at a near-zero level, maybe due to some more or less arbitrary numerical specifics of the model, or at a numerically significant and radiologically relevant level. The results of a sensitivity analysis on a log-transformed model output can therefore become more or less useless.

To avoid this problem, we have to look for transformations that handle low values differently from high values. Of course, the essential precondition for such an approach is a sensible criterion that discriminates "low" from "high" values. One possibility is that we take a strictly mathematical point of view and define anything below the mean or the median of all values as low and anything above as high. Another, more physically motivated approach is to use a threshold value that is based on radiological considerations and does not take account of the actual distribution of the model output.

It seems sensible to select a continuous transformation that maps
- zero to zero,
- very low values to values near zero,
- the threshold value to one
- and high values to moderately increasing values above one, without giving them an undue overweight.

In this paper three such transformations are proposed and their effects on the results of a variance-based sensitivity analysis are demonstrated using two different final repository models.

2. THE MODELS

This chapter gives a short overview of the basic characteristics of the two models investigated here. Both models represent hypothetic final repositories in rock salt formations in Germany.

2.1. HLW Repository

The first model is based on a former planning for a possible final repository for high-level radioactive waste in Germany, which was foreseen to be installed in excavations specifically mined for this purpose in a salt formation with a high creeping capability. The concept envisaged boreholes for emplacement of canisters with vitrified waste from reprocessing as well as drifts for emplacement of containers with spent fuel elements. Each borehole or drift is sealed by a plug, and the same applies to the loading drifts. Since under the high temperatures in the vicinity of heat-producing waste the salt creeps rather fast, the wastes will, in most cases, be tightly included in the salt within several decades. Then each contact of brine with parts of the wastes is excluded for the future and no contaminant release is possible. This situation leads to a zero-output of the model. It is, however, possible that a brine inclusion in the salt formation opens during the early phase. Then the fluid can reach the waste containers and dissolve radionuclides. The creep-induced convergence of the remaining voids in the mine has then a disadvantageous effect, since it presses the contaminated brine through the seals and into the geosphere.

The calculation results analyzed here were produced in 2003 using the code package EMOS [1]. The model consists of parts for the near field, the geosphere and the biosphere, and it finally yields the hypothetical time-dependent dose rates to a human individual. 31 parameters were varied statistically according to their specific distributions. Due to the high probability of tight inclusion of all wastes, only some 15 % of the model runs yield a non-zero output.

The HLW repository model was calculated 3000 times with a random sample. Only 491 of the runs yielded a non-zero output, but some of them reach maxima above 10^{-4} Sv per year. The time curves of the six runs with the highest maxima are shown in Figure 1 (left).

2.2 LILW Repository

The second model investigated here represents a hypothetic final repository for low- and intermediate-level radioactive waste that is assumed to be installed in an abandoned salt production mine. Its main features are loosely based on a real site of this type in Germany, the Morsleben repository. The salt formation is inhomogeneous and has a low creeping capability. It is assumed that the mine openings are filled with brine from the overburden after some time. At that point in time, some short-lived wastes, which are disposed of in one of the openings, start to release contaminants. These are dissolved in the brine and pressed out to the geosphere by the convergence process. In order to protect the longer-lived and more radiotoxic wastes from the brine, the main waste emplacement area is isolated from the rest of the mine by a specific seal, which, however, can be chemically corroded by magnesium. Depending on the magnesium content of the brine and the initial permeability of the seal material, the seal can nearly suddenly fail at some point in time. This leads to a short-lasting decrease, followed by a fast, significant increase of the contaminant release. The decrease is due to the fact that after seal failure it takes a little time to fill up the emplacement area, during which the brine outflow from the mine is reduced.

Like the HLW model, the LILW model consists of three parts, describing the near field, the geosphere and the biosphere. The calculations were done with the software package RepoTREND [2], which contains modules for each of the three parts and is specifically designed for calculating the transport of radionuclides through and release from a repository system. The model output is the annual dose to a human individual. In the investigations presented here, 11 uncertain parameters, all pertaining to the near field, were varied according to appropriate distributions. In contrast to the HLW case the LILW model does not produce zero output runs.

Three different random samples, each containing 3000 parameter sets, were drawn. The right side of Figure 1 displays some typical time curves calculated by the LILW model, which show the seal failure and the subsequent increase of the dose rate. The highest maxima reach about 10^{-5} Sv/yr.

Figure 1: Typical Time Development Curves of Both Models (Left: HLW, Right: LILW)

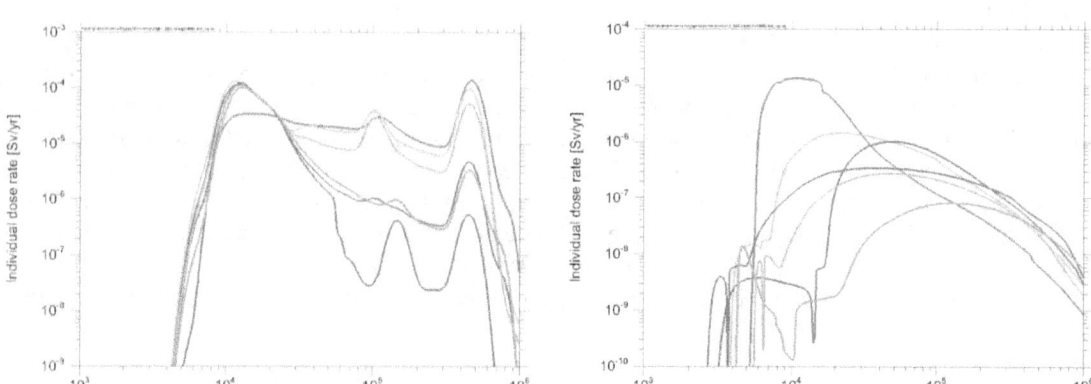

3. THE CONSIDERED OUTPUT TRANSFORMATIONS AND THEIR EFFECTS

In chapter 1, four requirements were formulated that should be fulfilled by an output transformation. According to these requirements, we chose the following three transformations for investigation in this paper (the model output value is denoted by y):

$$\text{transformation 1:} \quad y \mapsto \log_2(1 + y/a) ,$$
$$\text{transformation 2:} \quad y \mapsto (y/a)^{0.2} ,$$
$$\text{transformation 3:} \quad y \mapsto (y/a)^{0.3} .$$

Figure 2 shows how these transformations act on the model output on a linear and on a logarithmic scale. All three transformations map low values to values near 0, the threshold value a to 1 and high values to values that seem "not too far above" 1. The threshold parameter a should be selected to discriminate "low" from "high" values in a sensible way. For the investigations presented here it seems reasonable to fix it about three orders of magnitude below the highest occurring dose rates. Therefore, we chose $a = 10^{-7}$ Sv/yr for the HLW model and $a = 10^{-8}$ Sv/yr for the LILW model. Additionally, the effects of transformation 1 are demonstrated for a threshold value a that is adaptively calculated as the median of the analyzed data for each point in time. Since the transformation is monotonic, it maps the median of the original data to the median of the transformed data, which is automatically equal to 1.

Figure 2: Effects of the Transformations ($a = 10^{-7}$)

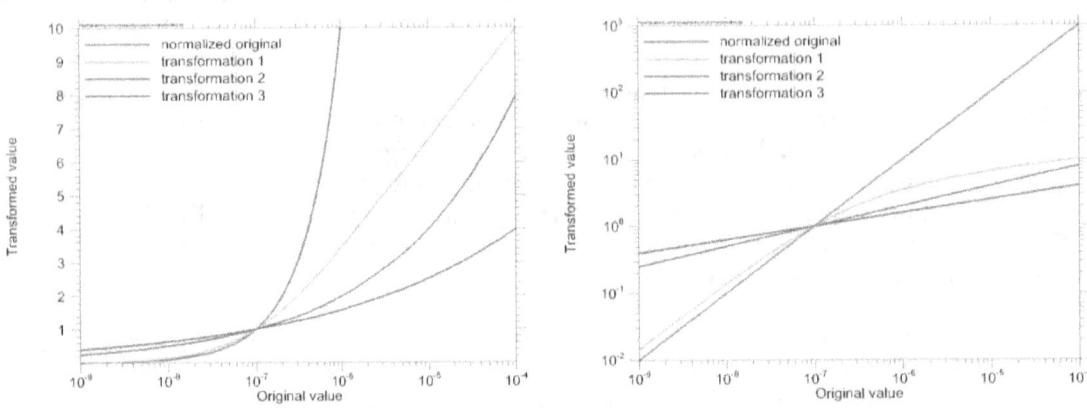

Figure 3 demonstrates how the three transformations affect the output of the HLW model in view of the variance. The histograms show the distribution of the peak values of all runs as red columns and the respective contributions to the total variance in blue. Due to the high number of zero runs the lowest bin is by far most populated. It can be seen that for the original data the figure looks rather unbalanced. While there are 2689 (89%) zeros or very low values with a common contribution to the total variance of about 5.5%, the two (0.067%) highest peak output values contribute 7.4% to the variance. The highest 25 values, that is 0.83%, are responsible for 50% of the variance. This disproportion is considerably mitigated by any of the three transformations. The figures show that in all cases the relation between the frequencies and the contributions to the variance is much more balanced than for the original data. The percentage of the highest values commonly responsible for 50% of the variance is about 4 to 5%. The highest contributions to the variance do no longer result from very few outliers at the upper end, but from the zeros and from the most populated bins in the region of higher values. It is therefore expected that a variance-based sensitivity analysis gives more robust and reliable results if performed on the transformed data.

Figure 3: Histograms for Frequency and Contribution to the Total Variance for the Original and the Transformed Data (Peak Dose Values)

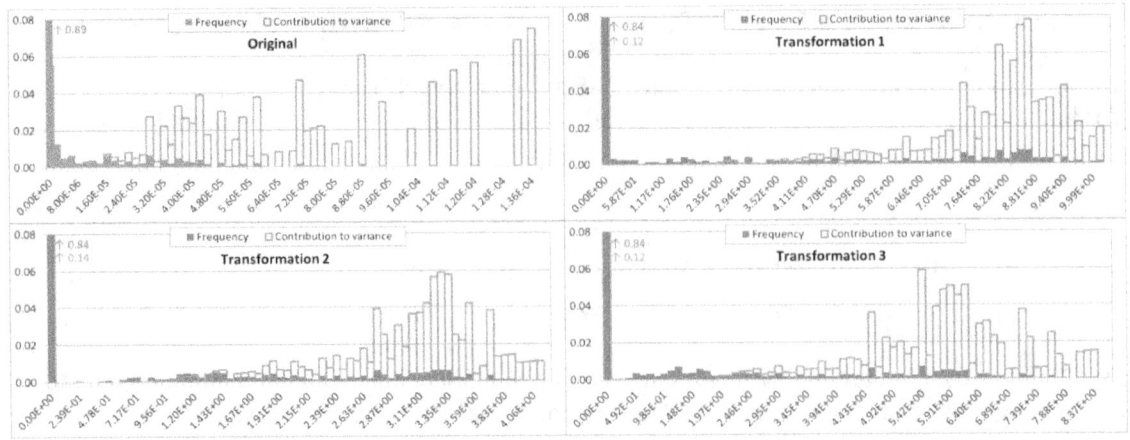

4. RESULTS

In the following some results are presented. With the HLW model 3000 runs were performed; with the LILW model three times 3000 runs were performed with different random samples. The transformations were applied to the time-dependent output of the calculations, and afterwards, a simple variance-based sensitivity analysis was performed. This means that the first-order sensitivity indices

$$SI1_j = \frac{\operatorname{Var}(\operatorname{E}(Y \mid X_j))}{\operatorname{Var}(Y)}$$

were calculated for all input parameters X_j and 300 points in time. Y means the entirety of the model output values for a specific point in time, $\operatorname{E}(Y|X_j)$ is the expectation of Y under the condition that X_j is hold constant. For the details of the underlying theory see, e.g. [3]. For calculation of the sensitivity indices the EASI method [4] was applied using a MATLAB script by E. Plischke (available under http://ipsc.jrc.ec.europa.eu/?id=756). EASI is a simple effective algorithm for calculating global sensitivity indices of first order using Fast Fourier Transformation. It is very quick, can be applied with any kind of sample and seems to yield results of similar robustness and reliability as other, much more numerically expensive methods [5].

4.1. Results for the HLW Model

In Figure 4 the time-development of the SI1 values for the HLW model, calculated with EASI from the original model output as well as from the transformed data, is presented. The curves of all 31 parameters are plotted, but for clarity reasons, only the six most important ones are given in the legend.

Figure 4: Time Development of SI1 for the HLW Model, Calculated with EASI from the Original and the Transformed Data

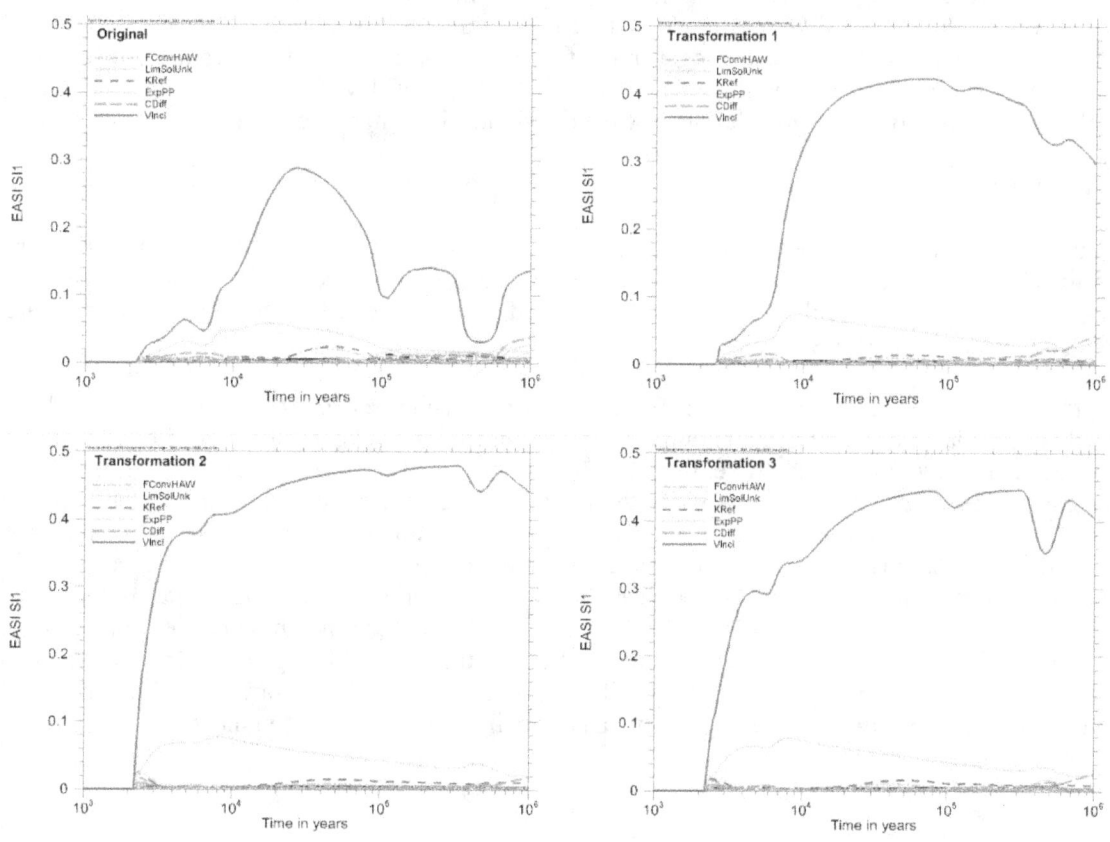

It can be seen from all figures that the model is clearly dominated at all times by the parameter *VIncl*, which is the volume of the brine inclusion in the salt formation. Additionally the parameter *ExpPP*, the exponent in the permeability-porosity relation, plays a certain role. Moreover, the curves of *CDiff* (Diffusion coefficient), *KRef* (reference convergence rate), *FConvHAW* (reduction of convergence in HLW fields) and *LimSolUnk* (solubility limits under unknown chemical conditions) reach, at least at some points in time, values that can be optically distinguished from the zero line.

While all four figures agree on these general facts, there are essential differences in the details. The dominance of *VIncl* is still more pronounced if transformed data are evaluated, no matter which of the transformations is applied. The calculated SI1 values for this parameter are at all times considerably higher for the transformed data than for the original output. The SI1 values of all other parameters but *ExpPP* seem to decrease, except at the very end of the simulation period. Looking at the time phase between about $4 \cdot 10^5$ and $6 \cdot 10^5$ years we see that, by reasons we do not discuss here, the SI1 of *VIncl*, if calculated for the original data, decreases below 0.04, which is only slightly above the values of the other parameters. From this figure alone we would conclude that during this period the parameter is nearly as insignificant for the output as all the other parameters. The figures for the transformed data, however, give a completely different impression. In all of them there is only a slight decrease of the SI1 of *VIncl* during the mentioned period and it remains between 0.32 and 0.44, depending on the applied transformation.

In the evaluation of the original data, the SI1 curve of *ExpPP* nearly vanishes in the mess of insignificant curves at about 10^5 years until the end of the simulation period. For the transformed output, however, it remains clearly silhouetted against the other curves and *ExpPP* is identified as the second most important parameter for nearly the entire simulation period. The small differences seem to be amplified by performing a transformation before the evaluation.

It is noticeable that in the early phase, up to about 8000 years, the results obtained with transformation 1 resemble more those of the original data than those of transformations 2 and 3. This is obviously due to the fact that during early times the model output is generally below the threshold of 10^{-7} Sv/yr, as can be seen in Figure 1 (left), and the logarithmic transformation leaves low values more or less unchanged, except from a factor (see Figure 2). In contrast, the power law transformations 2 and 3 specifically pronounce differences in values below the threshold, since their exponents are smaller than 1.

4.2. Results for the LILW Model

In Figure 5 the time-development of the SI1 values for the LILW model, calculated with EASI from the original model output as well as from the transformed data, is presented. All parameters are shown, distinguished by different colors. The results obtained using three different random samples are marked by different line styles.

It is predominantly conspicuous that the parameter *TBrine*, which does not produce considerably high SI1 values calculated from the original output, becomes much more important if transformed data are evaluated, especially in the early phase. For transformation 2 its SI1 reaches a maximum of 0.74. This parameter represents the point in time when the mine openings are filled with brine, which happens around 10 000 years with a log-normal distribution. Before this point in time there is no contaminant release at all, so that there are lots of zeros at early times. It is clear that *TBrine* is dominant in this phase, because it decides about zero or non-zero results. This dominance is much better reflected by the SI1 curves for the transformed output. Since in many cases, in the time phase between brine intrusion and seal failure the calculated dose rate remains below the threshold value of 10^{-8} Sv/yr, we have the same effect as described for the HLW model: the power law transformations 2 and 3 amplify variations in the range of low values and therefore additionally emphasize the dominance of the parameter *TBrine* in this phase.

Figure 5: Time Development of SI1 for the LILW Model, Calculated with EASI from the Original and the Transformed Data

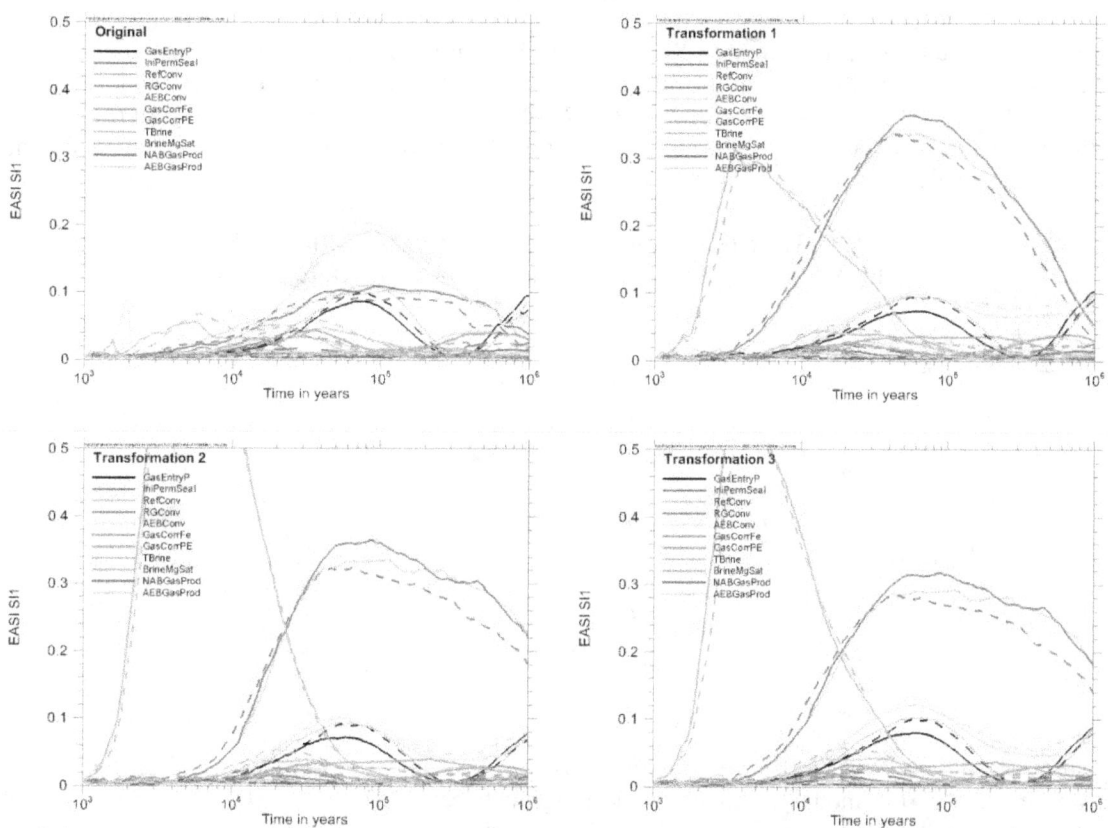

Apart from the eye-catching increase of the importance of *TBrine* due to the transformation, there is another noteworthy effect. While in the original data the parameter *AEBConv*, which represents the convergence rate of the sealed emplacement area, dominates for most of the simulation period, reaching a maximum SI1 of 0.2, this parameter is only of secondary importance in the transformed data with SI1 values no higher than 0.12. Instead, the parameter *IniPermSeal*, representing the initial permeability of the dissolving seal, assumes the dominating role with a maximum SI1 between 0.3 and 0.38, depending on the sample and the transformation. *IniPermSeal* determines the initial flow rate of corroding brine through the intact seal and is predominantly responsible for the time of seal failure. So, for each point in time, this parameter decides about whether the seal has already failed or not, which typically means a difference in dose rate of two or three orders of magnitude. We call this model behavior *quasi-discrete*, because it leads to separate congregations of output values. The parameter *BrineMgSat*, which represents the magnesium saturation of the brine and with it its corrosive potential, also has an influence on the time of seal failure and contributes to triggering this behavior. While, however, its SI1 curve is nearly invisible in the evaluation of the original model output, it becomes more relevant if transformed data are evaluated. Obviously, transformations of the considered kind emphasize the sensitivity of the model against parameters that cause a quasi-discrete model behavior.

The SI1 curves of all other parameters progress close to the zero line and show only little differences between the four evaluations.

With the output of the LILW model an additional evaluation was performed using transformation 1 with a threshold value a that was determined adaptively. This means that, instead of using a more or less subjectively fixed threshold, for each point in time the value a was calculated independently as the median of all 3000 output values. On the one hand, this makes sure that the determination of a follows an objective procedure, always keeping one half of the data below and the other half above the threshold. As the transformation is monotonic and maps a to 1, the median of the transformed data is constantly equal to 1. On the other hand, a non-constant threshold between values considered as "low" or "high" is a bit hard to understand and must not lead to misinterpretation of results.

Figure 6 shows the time development of SI1, calculated from the adaptively transformed data, on the left side. As long as more than half of the data are zeros, the median is also exactly zero and cannot be used for a; therefore the curves start only at about 7000 years. On the right side of Figure 6, the time development of the median, which was used as the transformation parameter a, is presented for the three random samples.

Figure 6: Time Development of SI1, Calculated with EASI from the Adaptively Transformed Data, and of the Data Median

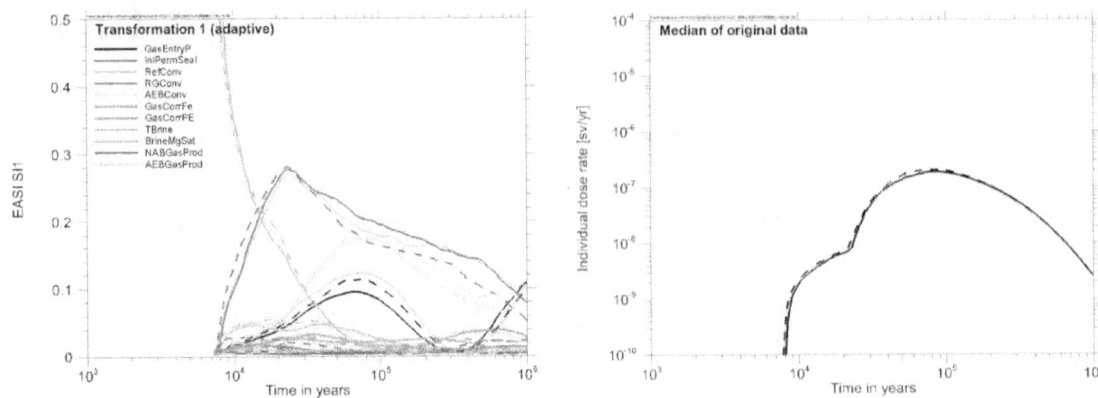

There are very little differences between the medians calculated for the three samples. For most of the time, the median of the model output is in the range between $2 \cdot 10^{-9}$ and $2 \cdot 10^{-7}$ Sv/yr. After about 22000 years it reaches the value of 10^{-8} Sv/yr and afterwards increases faster. At this time, in most of the cases the seal has failed, so that the higher results become dominating. This is also visible in the SI1 curve for *IniPermSeal*, which is the most relevant parameter for the time of seal failure. The curve reaches its maximum exactly at that time where the median curves show the bend. The then increasing threshold value obviously leads to a decrease of the SI1 of *IniPermSeal*, but does not seem to have a comparable effect to the other parameters.

With the adaptive transformation it becomes clearly visible that in the very early phase below 10^4 years *TBrine* is by far the dominating parameter. The curve starts at a value of 0.95. The importance of *AEBConv* appears more pronounced with the adaptive transformation than it was the case with any of the fixed transformations. Its SI1 curve reaches a maximum of about 0.2 and resembles more that one calculated from the original data (see Figure 5). For all other parameters there are only little differences to the curves obtained from the original data or the other transformations.

5. SUMMARY AND CONCLUSIONS

We have applied three different transformations to the time-dependent output of two different models for long-term performance assessment of final repositories and calculated the variance-based first-order sensitivity indices using the EASI algorithm. All transformations make use of a threshold value a, which is used for normalization of the model output data and discriminates "low" from "high" values. Additionally, a transformation with adaptive determination of a was applied to the LILW model (this could not be done for the HLW model, because it produces 85% zero output so that the median is always zero and cannot be used as the threshold value). The transformations are monotonic and map zero to zero, a to one and "high" values to values in the range of 1 to about 10. The motivation for this was to project the widely distributed model output to a range that is better adequate for a variance-based evaluation and to reduce the overvaluation of a few high outliers. We investigated a logarithm-based transformation and two power transformations with different exponents smaller than 1.

Transformations of the considered type seem to amplify the differences and accentuate the relevant results of a variance-based sensitivity analysis. Generally, compared to an evaluation of the original model output, the SI1 values of the important parameters seem to increase if calculated from the transformed output, while those of less relevant parameters remain more or less unchanged or even decrease. Thus, by performing an adequate transformation prior to a variance-based sensitivity analysis, one can obtain more unique results.

While the log-based transformation does not significantly change low values and consequently, nearly the same SI1 values as for the non-transformed output are calculated if most of the data are below the threshold, the power-based transformations amplify differences in low values. Parameters that are specifically relevant for low values may therefore be better identified using a power-based transformation.

The threshold parameter a has to be selected adequately for the intended investigation. It can either be chosen from a radiological point of view and set to a value that is considered to be a threshold for radiological relevance, or according to more mathematical aspects. A possible choice of the latter type would be the overall median of all analyzed data. An adaptive transformation that uses the time-dependent median of the model output at each point in time as threshold value can also provide interesting results. In time phases with many exceptionally high or low values this approach inhibits over- or undervaluation of data. This kind of transformation can be of specific interest for the investigation of models that produce very different output distributions and medians at different times. When interpreting the results one should keep in mind, which threshold value has been used and why.

In a different investigation [6] we found that the SI1 calculated from the original model output yield results that qualitatively differ from those obtained with regression-based sensitivity analysis. This is no longer the case if the transformed output is analyzed, no matter which of the transformations is applied. The results of the variance-based sensitivity analysis of transformed data seem to be well in line with those of the regression-based evaluation. From this we conclude that such transformations might improve the reliability of a variance-based evaluation.

With the LILW model three different random samples were investigated. The differences between the SI1 time curves obtained with these three samples do not significantly decrease if a transformation is applied, except from the very early time phase, during which only a few parameter sets yield a non-zero model output. From this observation we conclude that transformations can improve the robustness of the variance-based evaluation if it is based on only a few non-zero data.

Acknowledgements

The work presented in this paper is financed by the German Federal Ministry for Economic Affairs and Energy (BMWi) under sign 02E10941.

References

[1] D. Buhmann, *"Das Programmpaket EMOS – Ein Instrumentarium zur Analyse der Langzeitsicherheit von Endlagern"*, GRS-159, Gesellschaft fuer Anlagen- und Reaktorsicherheit (GRS) mbH, 1999

[2] T. Reiche, D.-A. Becker, D. Buhmann and T. Lauke, *"Anpassung des Programmpakets EMOS an moderne Softwareanforderungen"*, GRS-A-3623, Gesellschaft fuer Anlagen- und Reaktorsicherheit (GRS) mbH, 2011

[3] A. Saltelli, K. Chan and E. M. Scott (ed.), *"Sensitivity Analysis"*, Wiley, 2000

[4] E. Plischke. "*An effective algorithm for computing global sensitivity indices (EASI)*", Reliability Engineering and System Safety 95, pp. 354–360, (2010)

[5] S. Spiessl and D.-A. Becker, "*Sensitivity Analysis of a Final Repository Model with Quasi-Discrete Behaviour Using Quasi-Random Sampling and a Metamodel Approach in Comparison to Other Variance-Based Techniques*", Reliability Engineering and System Safety (submitted for publication)

[6] S. Spiessl and D.-A. Becker, "*Investigation of Different Sampling and Sensitivity Analysis Methods Applied to a Complex Model for a Final Repository for Radioactive Waste*", PSAM12, Honolulu, (2014)

Bayesian Approach Implementation on Quick Access Recorder Data for Estimating Parameters and Model Validation

Javensius Sembiring[a*], Lukas Höhndorf[a], and Florian Holzapfel[a]
Institute of Flight System Dynamics TUM, München, Germany

Abstract: This paper presents the implementation of Bayesian inference on Quick Access Recorder data for parameter estimation purpose. Posterior density is sampled by employing Markov Chain Monte Carlo method. The reason for employing the Bayesian inference, instead of classical method such as Maximum Likelihood is because the data used in this paper has more uncertainties than the data obtained from a flight testing. These uncertainties come from the facts that Quick Access Recorder data obtained from untailored flight maneuvers, variables are measured/recorded at low and different sampling rates, control inputs such as elevator, rudder, aileron are not optimized, and flight is performed based on daily operational activities (wind and turbulence might disturb the measured variables). Results show that this approach is capable of capturing the uncertainties in the data since the estimated parameters are presented in the distribution forms. The flight data used as a case study are obtained from Airbus 320 Quick Access Recorder device. Some parameters to be estimated in this study consist of thrust and the effect of spoiler and flap deflection on lift and drag coefficient during approach phase.

Keywords: QAR Data, Bayesian Inference, Model Validation, MCMC

1. INTRODUCTION

Flight safety is one main issue in aviation area. Many efforts are conducted to improve safety level both from aircraft design and operational side. From operational point of view, airlines are interested in improving safety by utilizing flight data which are recorded during daily operational flight. The flight data called Quick Access Recorder (QAR) data is monitored or analyzed by flight safety crew and the result will be given to the related parties as a feedback. To deal with huge daily flight data, the flight safety department is provided with Flight Data Monitoring (FDM) program which is available commercially. These FDM programs work based on recorded parameter, analytical, and simple computation only. For instance, if there is a runway overrun incident occur, some possible parameters to investigate are spoiler deployment, thrust reverser, brakes – *are they working properly or not?*, and other related contributing factors to the incident (Figure 1).

However, sometimes these recorded parameters do not provide the flight safety crew with enough information to determine the cause of the incident. Parameter estimation technique comes into the picture by providing more parameters to be investigated in which these parameters are not recorded/measured directly in QAR data. As example of the incident mentioned above, the additional parameter which might be estimated is runway friction coefficient. This parameter is not recorded in QAR device but can be estimated by employing the parameter estimation method. Not only parameters during ground phase but also parameters during air-phase can be estimated such as lift and drag coefficient increment/decrement due to flap or spoiler deflection during approach phase. Implementation of the parameter estimation technique along with current FDM program would give a great benefit to FDM crew since more parameters are obtained and the cause of incident can be revealed

---

* Corresponding author: javensius.sembiring@tum.de

with more solid foundation. In this paper, the Bayesian inference is employed as estimation technique which presents the estimated parameter in a distribution form. The distribution of posterior density is sampled by using Markov Chain Monte Carlo (MCMC) method.

Figure 1. Runway Overrun Incident

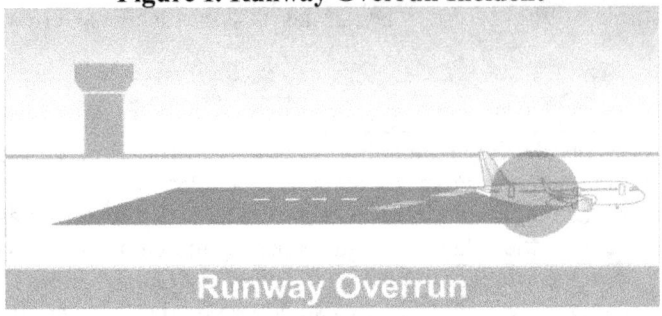

The following sections of this paper are organized as follows: Section 2 introduces the problem formulation. In particular, the flight phases and related parameters to be estimated are thoroughly introduced. Section 3 briefly reviews the Bayesian method as well as technique for sampling the posterior density. This is followed by Section 4 which presents the implementation and result obtained by implementing the Bayesian inference on QAR data. Finally, Section 5 draws the conclusion for the paper.

2. PROBLEM FORMULATION

The number of parameters to be estimated is related to the postulated model and flight phase of the aircraft. In this paper, the approach flight (air phase) is selected as flight phase to be investigated (see Figure 1 and Table 1). During the approach phase, parameters such as lift and drag coefficient increment/decrement due to flap or spoiler deflection are estimated. Along with these aerodynamic parameters, thrust produced by aircraft engine is also estimated during this phase. The mathematical formulation during the flight phase is postulated as a linear relation as shown in the following:

$$a_x = \frac{1}{m}(\bar{q}S \sin \alpha \, C_L - \bar{q}S \cos \alpha \, C_D + \delta_T T) \quad (1)$$

where $a_x, m, \alpha, \bar{q}, \delta_T, S$ consecutively denote the acceleration along longitudinal axes, mass of the aircraft, angle of attack, dynamic pressure $\left(\frac{1}{2}\rho v^2\right)$, throttle input, and wing area. All these parameters are obtained from QAR data except S parameter which is obtained from A320 technical data [3].

Aerodynamic coefficients denoted by C_L (lift coefficient), and C_D (drag coefficient) as well as thrust (T) are parameters to be estimated. The aerodynamic coefficients during the selected flight phase are affected by flap and spoiler deflection and are modeled as incremental changes in the lift and drag coefficients, i.e., as ΔC_{LF}, ΔC_{LS}, ΔC_{DF}, and ΔC_{DS} (the subscript F and S denote flap and spoiler deflection) [1]. The effects of flap and spoiler deflection on lift and drag coefficient are investigated for three different flap settings, namely $\delta_F = 0$, 15, and 35 degrees and four different spoiler settings, i.e., $\delta_S = 0$, 18, 22, and 27 degrees.

Figure 1: Selected Approach Flight Phase
(Flap, Spoiler Deflection and Longitudinal Acceleration Plots)

Table 1: Selected Flight Phase Description

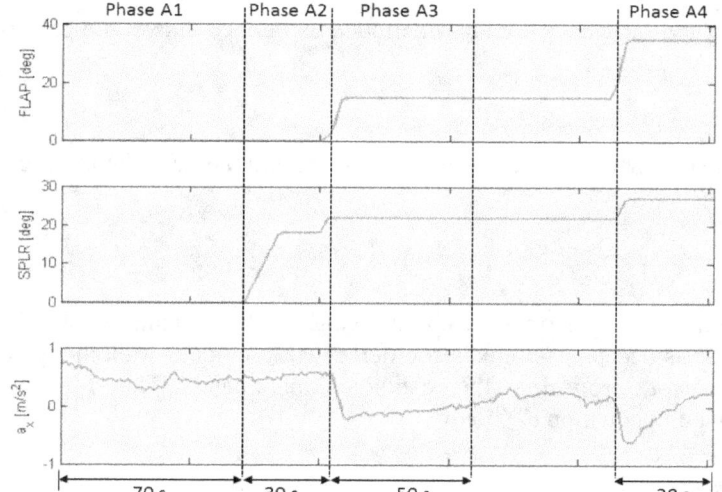

Phase	Description
A1	FLAP = 0 SPLR = 0 Duration = 70 s
A2	FLAP = 0 SPLR = 18 Duration = 30 s
A3	FLAP = 15 SPLR = 22 Duration = 50 s
A4	FLAP = 35 SPLR = 27 Duration = 38 s

3. BAYESIAN INFERENCE AND POSTERIOR SAMPLING METHOD

3.1. The Bayes Formula

In the Bayesian context, the probability is represented as distribution of possible values. This approach is based on prior and likelihood distributions of parameters. The prior distribution describes our belief about the problem beforehand (subjective judgment). The likelihood represents the probabilities of observing a certain set of parameter values. Both of these distributions are updated to a posterior distribution, which represents the parameter distribution given on the observed data, formulated as follows:

$$p(\theta|Z) = \frac{p(Z|\theta) \cdot p(\theta)}{\int p(Z|\theta) \cdot p(\theta) d\theta} \quad (2)$$

where, $p(\theta|Z)$, $p(data|\theta)$, $p(\theta)$, and $\int p(data|\theta) \cdot p(\theta)d\theta$ consecutively denote posterior, likelihood, prior and normalizing constant, while Z and θ denote data and unknown parameters consecutively. The Bayesian solution for parameter estimation is the posterior distribution of parameters (conditional probability of unknown parameters given the data). This posterior distribution is the distribution we are interested in knowing since it represents the distribution directly of the unknown parameter. In this paper, the prior of unknown parameter is assumed to be *uninformative prior*, i.e. $p(\theta) = 1$, whereas the likehood is formulated as:

$$p(Z_i|\theta) = \frac{1}{(2\pi)^{n/2}|C|^{1/2}} e^{-\frac{1}{2}(Z_i - y(t_i;\theta))^T C^{-1}(Z_i - y(t_i;\theta))/\sigma^2} \quad (3)$$

Variable $y(t_i;\theta)$ denotes the postulated model which depends on the unknown parameters. The likelihood formulation in equation (3) is based on assumptions that the measurement error is distributed as Gaussian with mean zero and covariance C, that is $\epsilon \sim N(0,C)$. Furthermore, if the measurement error ($\epsilon_i = Z_i - y(t_i;\theta)$) is assumed to be independent and normally distributed, that is $\epsilon_i \sim N(0,\sigma^2)$ and $\epsilon \sim N(0,\sigma^2 I)$, equation (3) is simplified to (4):

$$p(Z_i|\theta) = \frac{1}{\sigma\sqrt{2\pi}} e^{-\frac{1}{2}(Z_i - y(t_i;\theta))^2/\sigma^2} \quad (4)$$

or in the combined likelihood of all the measurements can be written as a product:

$$p(Z|\theta) = \prod_{i=1}^{n} p(Z_i|\theta) \frac{1}{\sigma\sqrt{2\pi}} e^{-\frac{1}{2}(SS_\theta)^2/\sigma^2} \tag{5}$$

where $SS_\theta = \sum_{i}^{n}(Z_i - y(t_i;\theta))^2$. By combining the prior and likelihood as defined above, the posterior up to the normalizing constant can be written as:

$$p(\theta|Z) \propto p(Z|\theta) \tag{6}$$

Equation (6) is the distribution of interest which will be sampled using technique described in section 3.2 below.

3.2. Posterior Sampling Method

Sampling from posterior density function is performed by employing Markov Chain Monte Carlo method. Metropolis algorithm is used as acceptance/rejection criteria, while random walk along with component wise method is used for proposing the candidate parameter. The Metropolis acceptance/rejection criterion is defined in equation (7) below.

$$\alpha(\theta^{i-1}, \theta^*) = min\left\{1, \frac{p(\theta^*)}{p(\theta^{i-1})}\right\} \tag{7}$$

The candidate parameter(θ^*) will be accepted if $(\theta^{i-1}, \theta^*) > u$, where $u \sim U[0,1]$. The pseudo-code of this algorithm is presented in Section 4.

3.3. Initial Parameter and Covariance Estimation

Using Random Walk Metropolis algorithm with Gaussian proposal distribution in posterior sampling process requires a guess for the covariance matrix C as well as starting values of parameters. The starting value of parameters can be estimated by using a least square sense as formulated in equation (8).

$$\theta_0 = \min_\theta \sum_{i=1}^{n} (Z_i - y(t_i;\theta))^2 \tag{8}$$

Taking parameters obtained from equation (8) as starting values in MCMC process avoid a long burn-in time and speed-up the convergence rate of sampling process. The parameter covariance matrix C is obtained by employing equation (9) below:

$$C \approx \sigma^2 [X^T X]^{-1} \tag{9}$$

where σ^2 and X consecutively represent variance of the residual and Jacobian matrix calculated at θ_0. In details, the step obtaining the initial parameter and covariance matrix can be found in [4].

3.4. Model Validation

Model validation is performed by comparing model output and measurement. In the context of Bayesian inference, the model output is described as predictive distributions of model output. The predictions of model output are naturally based on the posterior distribution of estimated parameters, as defined as [2]:

$$p(Z^*|Z) = \int p(Z^*, \theta|Z) d\theta = \int p(Z^*|\theta) p(\theta|Z) d\theta \tag{10}$$

Here $p(Z^*|Z)$ denotes the prediction of future observations Z^* given Z as the current one. Since the model output based on distribution of estimated parameter, the model output will be in distribution form as well as forming a confidence region. This approach is different with the classical estimation approach in which the model output is based on a single value of estimated parameter hence forms the model output without distribution [5, 6].

4. IMPLEMENTATION AND RESULTS

4.1. Implementation

The pseudo code of Random Walk Metropolis algorithm is presented below. It shows the initial value of parameter and covariance are first estimated in the least square sense. After the initialization, the posterior sampling process is done iteratively until reaching the maximum number of samples. The candidate parameters will be accepted/rejected based on criterion defined in equation (7). The accepted parameters are then stored and form the posterior distribution.

Random Walk Metropolis pseudo code with estimated initial parameter and covariance

1. Initialize
- Choose $\theta^0 = \arg\min_{\theta \in Q} \sum_{j=1}^{n}(z_j - y(t_j, \theta))^2 = \arg\min_{\theta \in Q} SS_\theta$ using optimization routine.
- Compute $SS_{\theta^0} = \sum_{j=1}^{n}(z_j - y(t_j, \theta^0))^2$
- Estimate error variance $\sigma^2 = \frac{1}{n-p}\sum_{j=1}^{n}(z_j - y(t_j, \theta^0))^2$
- Construct covariance matrix $C = \sigma^2 [X(\theta^0)^T \cdot X(\theta^0)]^{-1}$
- Compute $R = \text{cholesky decomposition}(C)$

2. Do sampling
- for i = 1, 2, ...n (n = number of samples)
    - \# sample $z \sim N(0,1)$
    - \# construct candidate $\theta^* = \theta^{i-1} + Rz$
    - \# sample $u_\alpha \sim U(0,1)$
    - \# compute $SS_{\theta^*} = \sum_{j=1}^{n}(z_j - y(t_j, \theta^*))^2$
    - \# compute $\alpha(\theta^*, \theta^{i-1}) = \min\left(1, e^{-[SS_{\theta^*} - SS_{\theta^{i-1}}]/2\sigma^2}\right)$
    - if $u_\alpha < \alpha$
        - \# set $\theta^i = \theta^*$, $SS_{\theta^i} = SS_{\theta^*}$
    - else
        - \# set $\theta^i = \theta^{i-1}$
    - endif
- endfor

The pseudo code of Random Walk Metropolis algorithm shown above is then implemented based on the postulated model and the selected flight phase as defined in Section 2. Number of samples is set to 80,000 with acceptance ratio for each of parameters varies between 3% - 18%. The related results are shown in the Figure 3 to 6 below.

During Phase A1, the thrust varies in range between 0.8 – 1 kN, lift coefficient deflection varies in range between 0.4 – 0.5, and drag coefficient has value distributed in range 0.03 – 0.05. Both of these aerodynamic coefficients are obtained with no flap and spoiler deflection. The predictive model output distribution is also presented along with measurement (left side). From Figure 3, it shows that the predictive model output match with a good agreement with that of the measurement.

In Phase A2, the flap deflection is still in the same state as Phase A1 but spoiler now deflected to 18 degrees. This spoiler deflection decreases the lift coefficient and increases the drag coefficient. These effects can be seen in Figure 4, i.e. the lift coefficient decreases to value between 0.24 – 0.28, whereas drag coefficient increases in range 0.08 – 0.1. Predictive output plot is also presented along with measurement (Figure 4, left side), but the distribution of predictive plot is wider than that of Phase A1. This indicates the estimated parameters in Phase A2 have more uncertainties than the parameters estimated in Phase A1. The change in thrust parameter is not caused by flap/spoiler deflection but by the flight condition and throttle command from pilot.

Figure 2: Phase A1 Predictive Model Output and Related Estimated Parameters Distribution

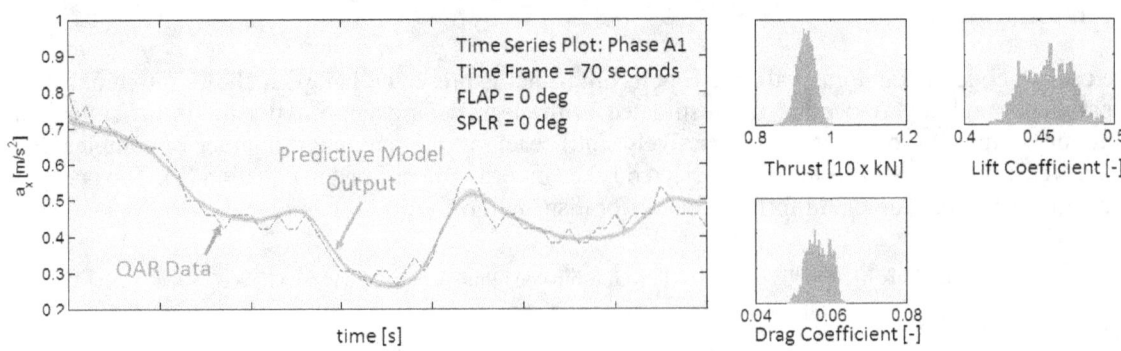

Figure 4: Phase A2 Predictive Model Output and Related Estimated Parameters Distribution

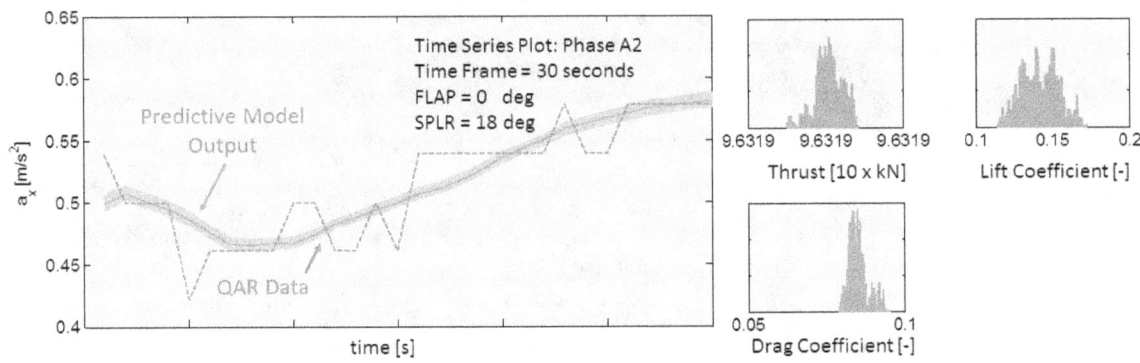

During Phase A3, both flap and spoiler are in deflection state of 15 and 18 degrees consecutively. This deflection affects both lift and drag coefficient. The lift coefficient increases in range 0.24 – 0.28, whereas the drag coefficient varies in range 0.125 – 0.135 as shown in Figure 5 (right side). The predictive plot distribution of model output is also plotted together with the measurement. From Figure 5, it can be seen that the trend of the measurement can be captured by the model output.

Figure 5: Phase A3 Predictive Model Output and Related Estimated Parameters Distribution

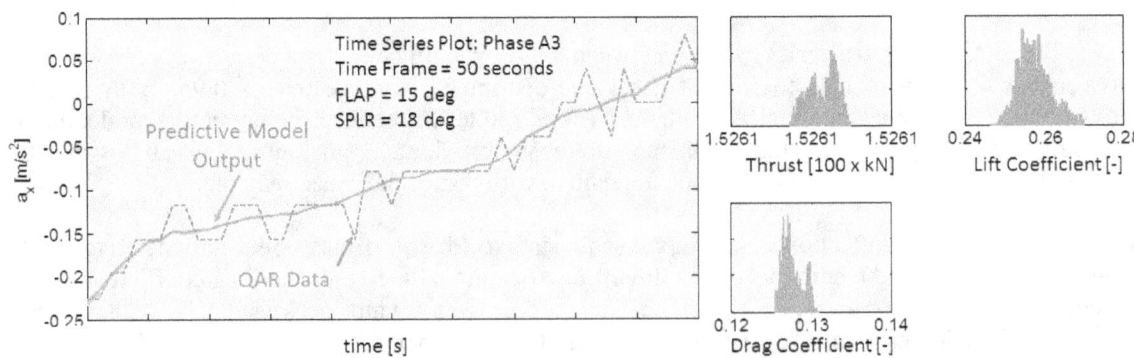

The results during Phase A4, presented in Figure 6 below, shows that the lift coefficient is now increasing with values around 0.8. This increment might come mostly due to the flap deflection (35 degrees). On the other hand, the drag coefficient slightly change and remain in value around 0.09 – 0.1. The predictive model output distribution is also plotted together with measurement as shown in Figure

6 (left side). It shows that they are in a good agreement with small spread of the predictive distribution of model output. This indicates that the estimated parameters have a high accuracy in the postulated model being employed.

Figure 6: Phase A4 Predictive Model Output and Related Estimated Parameters Distribution

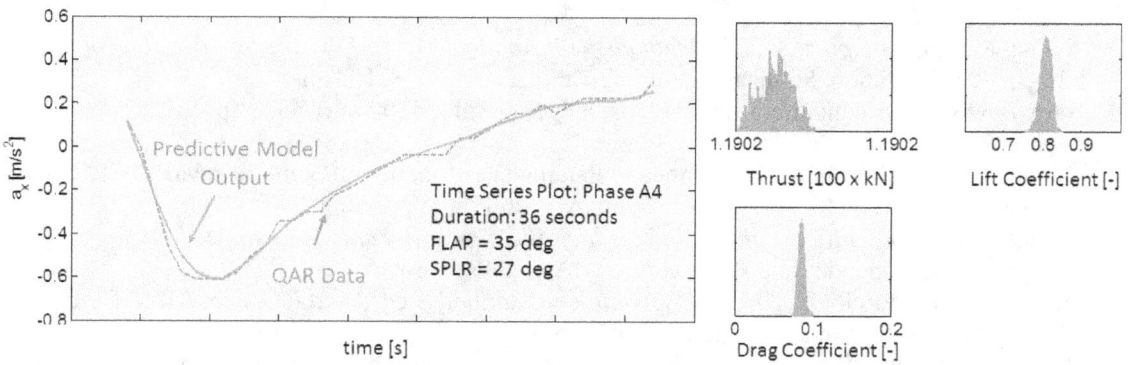

5. CONCLUSIONS

The Bayesian inference along with MCMC technique are successfully implemented on Quick Access Recorder data. The estimated parameters are represented as distribution form which gives information about the uncertainties in parameters. The algorithm implemented in this study opens the possibility to be integrated into current FDM program hence can extend the capability and functionality of the program. The benefit of this implementation provides parameters which are not recorded/measured in QAR device. These estimated parameters can be used as additional information by FDM crew to investigate a specific event or incident so that the cause of incident can be determined with more solid foundation. The output of this work is also used as an input in one of active research related to incident prediction in Institute of Flight System Dynamics, TUM.

References

[1] Jategaonkar, R.V., Mönnich, W., Fischenberg, D., and Krag, B., "*Identification of Speed Brake, Air-Drop, and Landing Gear Effects from Flight Data*", Journal of Aircraft, Volume 32, No. 2, (1997).
[2] Laine, Marko, "*Adaptive MCMC Methods with Applications in Environmental and Geophysical Models*", Dissertation, Finnish Meteorological Institute, 2008, Helsinki.
[3] http://www.airbus.com/aircraftfamilies/passengeraircraft/a320family/a320/specifications/ Feb. 2 2014.
[4] Solonen, Antti, "Monte Carlo Methods in Parameter Estimation of Nonlinear Models", Master's Thesis, Lappeenranta University of Technology, 2006, Lappeenranta.
[5] Jategaonkar, R.V., *Flight Vehicle System Identification – A Time Domain Methodology*, AIAA Progress in Astronautics and Aeronautics, AIAA, 2006, New York.
[6] Klein, V., and Morelli, E., *Aircraft System Identification: Theory and Practice*, AIAA Progress in Astronautics and Aeronautics, AIAA, 2006, New York.
[7] Bolstad, William M., *Understanding Computational Bayesian Statistics*, John Wiley & Sons, 2010, New Jersey.

Comparative Assessment of Severe Accidents Risk in the Energy Sector: Uncertainty Estimation Using a Combination of Weighting Tree and Bayesian Hierarchical Models

M. Spada[a]*, P. Burgherr[a] and S. Hirschberg[a]

[a] Laboratory for Energy Systems Analysis, Paul Scherrer Institute (PSI), 5232, Villigen PSI, Switzerland

Abstract: This study analyzes the risk of severe fatal accidents within the full fossil energy chains causing five or more fatalities. The risk is quantified separately for OECD and non-OECD countries. In addition for the Coal chain, Chinese data are analyzed separately because it has been shown that data prior to 1994 were subject to strong underreporting. In order to assess the risk and its uncertainty, a Bayesian hierarchical model was applied. This allows yielding analytical functions for frequency and severity distributions. Furthermore, Bayesian data analysis inherently delivers a measure of a combination of epistemic and aleatory uncertainties, through the *a priori* distribution and likelihood function that compose the Bayes theorem. In this study, in order to reduce the epistemic uncertainty related to the subjective choice of the likelihood function, Bayesian Model Averaging (BMA) is applied. In BMA the final posterior distribution is a weighted combination of the posterior distributions assessed for different likelihood functions (models). The proposed approach provides a unified framework that comprehensively covers accident risks in energy chains, and allows calculating specific risk indicators, including their uncertainties, to be used in a holistic evaluation of energy technologies.

Keywords: Comparative Risk Assessment, Accident Risk, Bayesian Hierarchical Model, Bayesian Model Averaging, Fossil Energy Chains

1. INTRODUCTION

Risk assessment of severe accidents in the energy sector is an important aspect that contributes to improve safety performance of technologies, but is also essential in the broader context of sustainability, energy security and policy formulation by decision makers. Accidents in the energy sector are not only occurring in the production phase, but along the entire energy chain (e.g., [1]). Therefore, a comprehensive analytical framework is needed (e.g., [2]).

The classical approach to assess the risk of severe accidents in fossil energy chains is based on the use of metrics such as aggregated risk indicators focusing on human health impacts, i.e., fatality rates, or frequency-consequence curves (e.g., [1], [2]). However, when dealing with risk, uncertainty estimation is of great importance in order to take into account possible random fluctuations, for example due to the lack of data. Furthermore, uncertainty levels cannot be fully addressed using the aforementioned standard approach (e.g., [3]).

In this study, the risk and its uncertainty levels are assessed through a Bayesian Hierarchical model. In this way separate analytical functions for frequency and severity distributions can be calculated. In Bayesian data analysis, the posterior distribution is given by the product of an *a priori* distribution, describing how the data are distributed before introducing them into the analysis, and a likelihood function (e.g., [4]). The former defines the lack of knowledge and thus is intrinsically related to the epistemic uncertainties. The likelihood function is one of the fundamental parts of the Bayes theorem (e.g., [4]). It describes the randomness of the data and thus defines the aleatory uncertainties. The likelihood function is commonly defined following expert judgment and/or is selected following scientific community agreement. Therefore, besides describing the aleatory uncertainty, the likelihood function is as well a source of epistemic uncertainty due to subjectivity involved in its choice.

The concept of Bayesian Model Averaging (BMA) has been proposed in order to assess posterior distributions and to increase their robustness by considering a set of possible models that could describe a dataset (e.g., [5]). Therefore, BMA can serve as a tool to reduce the subjectivity in the

*matteo.spada@psi.ch

choice of the model used as likelihood function for the Bayesian analysis. In BMA, the posterior distribution of the parameters of interest is given by the sum of the product of the models belonging to the model space, and the corresponding weights for these models. BMA has been used in different scientific fields, such as for example for dose-response risk assessment (e.g., [6]). Moreover, different methods have been proposed to estimate the weight of each model belonging to the model space to the final posterior distribution, for example using the Bayesian Information Criterion (BIC) (e.g., [7]), or Markov Chain Monte Carlo methods (e.g., [8]).

In this study, BMA is applied to the severity distributions in the risk assessment. The BIC method is used to a set of possible models common in hazard and risk assessment as well as survival analysis (e.g., [9], [10]). Therefore, the final posterior distribution for the parameter of interest, e.g., fatalities, is averaged over the entire model space. The final result is a severity distribution where both aleatory, from the likelihood functions, and epistemic uncertainties, from both *a priori* distribution and likelihood functions, are taken into account. Finally, the estimation of frequencies in the risk assessment is modeled following a common Bayesian analysis, since they can be described by a Poissonian distribution.

The above-described model is applied to the energy sector, and specifically to assess accident risks in fossil energy chains. The current analysis covers severe (≥ 5 fatalities) accidents in fossil (coal, oil, natural gas) energy chains for the years 1970-2008, which are contained in PSI's Energy-related Severe Accident Database (ENSAD). First, various risk indicators for different energy chains and country groups (e.g., OECD, non-OECD) are calculated. Second, results from the BMA and the standard approach are compared. Finally, a comparative evaluation for average and extreme risk is undertaken across energy chains and country groups.

2. DATA
2.1 ENSAD database
The ENergy-related Severe Accidents Database (ENSAD) (e.g., [11]) comprehensively covers energy related severe accidents worldwide. There exist numerous databases that look at accidents related to various industrial activities (e.g. FACTS online, OSH Update), but in contrast to ENSAD none of them is clearly focused on accidents attributable to the energy sector. Furthermore, ENSAD takes a full-chain approach because accidents can occur at all stages of an energy chain and not only at the actual power generation step. In ENSAD, data on all energy-related accidents is collected and classified into energy chains and activities within those chains. In addition, information on location, accident type, and different types of consequences (e.g. human health, environmental and economic impacts) is coded for to achieve a comprehensive global coverage of severe accidents. Finally, the accidents and severity reported in ENSAD for fossil chains are divided into three major groups, namely the Coal (incl. Lignite) chain, the Oil chain and the Natural Gas chain.

ENSAD has been developed using a wide variety of commercial and non-commercial information sources, ranging from specialized databases to technical reports, journal and newspaper articles, websites, etc. In the literature no commonly accepted definition can be found of what constitutes a so-called severe accident (e.g., [1]). The database ENSAD uses seven criteria to distinguish between severe and smaller accidents (e.g., [2]). Whenever one or more of the following consequences is met, an accident is considered to be severe:

- at least 5 fatalities or
- at least 10 injured or
- at least 200 evacuees or
- extensive ban on consumption of food or
- releases of hydrocarbons exceeding 10,000 (metric) tons or
- enforced cleanup of land and water over an area of at least 25 km^2 or
- economic loss of at least 5 million USD (2000)

We considered the energy-related accidents for the entire chain (from exploration and extraction through processing to end use) of each group.

2.2 Frequency and Severity

Risk can be decomposed into the product of the frequency and severity. The number of accidents per year gives the frequency, while severity measures the extent of the consequences of each accident. In this study, the number of fatalities describes severity. The reason for this choice is that fatalities generally comprise the most reliable indicator with regard to the completeness and accuracy of the data (e.g., [1], [12]). Furthermore, fatality information is superior to injured or evacuated persons information because often the severity of an injury or the duration of an evacuation is not reported (e.g., [1]).

Frequency and fatality distributions in ENSAD exhibit very different statistical behavior in each energy chain (Figure 1). The frequency distribution is influenced by temporal trends within each chain as well as country groups. These trends are commonly related to technological and regulatory differences and changes.

The fatality distribution, on the other hand, follows a very different pattern, and stretches over a broad range (Figure 1). The severity distribution is influenced by a number of parameters such as the material involved in the accident or the different products and the amount of material present, or the number of people in the vicinity of an accident. In addition, the data set is composed of accidents under a large range of different circumstances, meaning that we can thus assume also a large number of drivers for the risk exists (e.g., [3]).

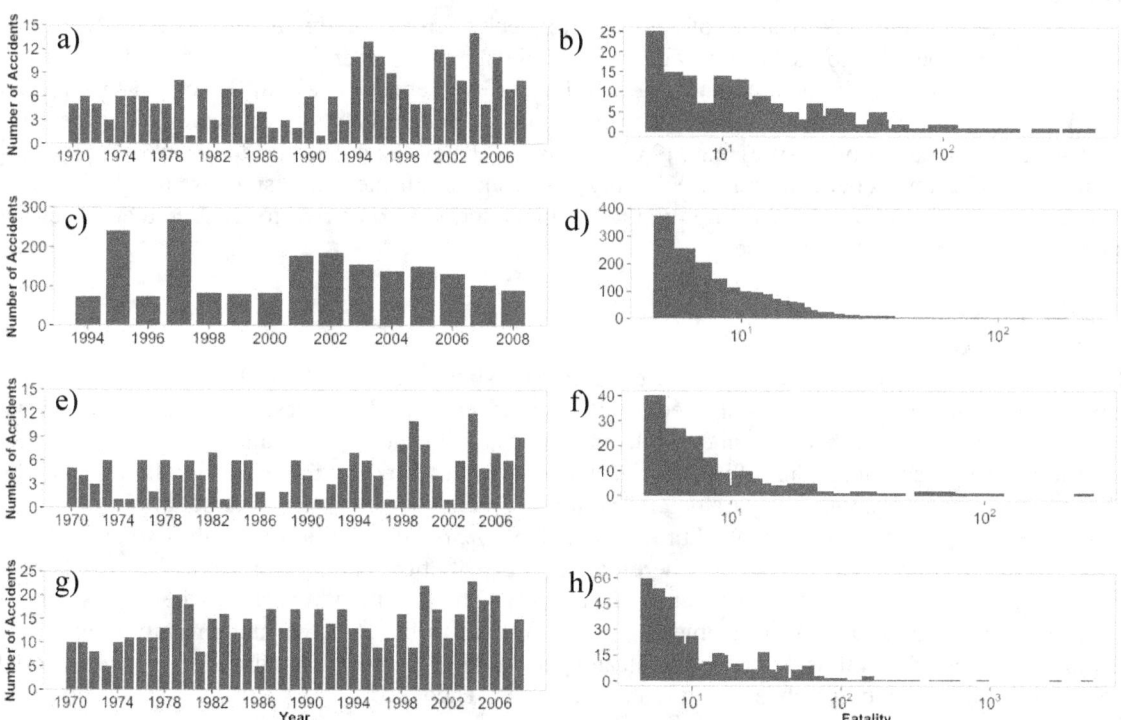

Figure 1 Number of accidents per year (Frequency) and number of accidents per fatalities (Consequence) for Coal (a and b), Coal China 94-08 (c and d), Natural Gas (e and f) and Oil (g and h).

2.3 Country Aggregates

Comparative results can be provided at the level of individual countries or for different country aggregates [12]. Based on the substantial difference in management, regulatory frameworks and general safety culture between industrial countries and developing ones, different energy chains are assigned to two major country groups, namely OECD and non-OECD. In addition for the Coal chain, Chinese data are analyzed separately because it has been shown that data prior to 1994 are subjected to strong underreporting (e.g., [16]). Furthermore, due to different data completeness for Chinese coal in 1994-1999 and 2000-2012, we subdivided the dataset into two subgroups related to the observation period. The summary of severe accidents for each substance energy chain and country group is shown in Table 1.

Table 1 Overview of the analyzed subsets per energy chain and country group for the years 1970–2008. Numbers of severe (≥ 5 fatalities) accidents, corresponding fatalities, and production used for normalization are given.

Energy Chain	Country Group	Number of Accident	Number of Fatalities	Production (GWeyr)
Coal	OECD	87	2259	18792
	non-OECD w/o China	162	5788	10071
	China 1994-1999	818	11302	1908
	China 2000-2008	1214	15750	7459
Oil	OECD	181	3430	36606
	non-OECD	350	19334	20524
Natural Gas	OECD	109	1258	17504
	non-OECD	77	1549	13459

2.4 Normalization

To derive a comparable risk measure of accident frequency across energy chains and country groups, the accident frequency should be normalized to a relevant unit of energy. Therefore, the accident frequency is normalized by the amount of production for each energy chain in GWeyr (Table 1). The Gigawatt-electric-year (GWeyr) is chosen because large individual plants have capacities of the order of 1 GW of electrical output (GWe), e.g., [2]. This makes GWeyr a natural unit to use when presenting normalized indicators generated within technology assessment. Furthermore, since we are dealing with data collected for fossil energy chains only, the thermal energy is converted to an equivalent electrical output using a generic efficiency factor of 0.35 (e.g., [2]).

3. METHOD
3.1 Bayesian Analysis

Bayesian inference is an alternative to the classical statistical inference (e.g., [4]). In the latter, also known as frequentist inference, only repeatable events have probabilities, while in the Bayesian inference, probability simply describes both epistemic and aleatory uncertainty (e.g., [4]). In fact, Bayesian analysis combines the information in the data represented by the entire likelihood function with prior knowledge about the parameters, which may come from other data sets or a modeler's experience and physical intuition [9]. Furthermore, the *a priori* distribution describes what is known before observing any data (e.g., [4]). Therefore, this distribution mainly contributes to the lack of knowledge and thus describes the epistemic uncertainty. The likelihood describes the process giving rise to data in terms of unknown parameter (e.g., [4]). It contributes to the random variability of the unknown parameter, and thus describes the aleatory uncertainty. Parameter estimation is made through the posterior distribution, which is computed using Bayes' Theorem:

$$p(\theta \mid y) = \frac{L(y;\theta)p(\theta)}{\int L(y;\theta)p(\theta)d\theta} \qquad (1)$$

where $p(\theta \mid y)$ is the posterior distribution for the parameter θ given the observed data y, $L(y; \theta)$ is the likelihood function, and $p(\theta)$ is the *a priori* distribution of the parameter θ. The denominator is a normalizing constant that scales the posterior so that the area under the posterior probability distribution function equals one, i.e. make it "proper" meaning that it must converge (e.g., [4]). The main issue in equation (1) is that computing the integral may not be easy in cases when the parameter vector θ is large (e.g., [9]). In order to overcome this issue, Markov Chain Monte Carlo (MCMC) methods are commonly used (e.g., [13]). In fact, MCMC algorithm samples values of the parameters from the posterior distribution without computing the normalizing constant (e.g., [14]). Therefore, equation (1) can be written as

$$p(\theta \mid y) \propto L(y; \theta) p(\theta) \qquad (2)$$

Among different type of MCMC algorithms (e.g., [13]), in this study the MCMC Gibbs algorithm is used in the sampling of the posterior distribution (e.g., [14]). This choice is made in order to avoid possible issues related to the incorrect choice of the jumping distribution, which is used for sampling the posterior in the other widely applied sampler, the Metropolis–Hastings algorithm (e.g., [13]).

3.2 Bayesian Model Averaging

The likelihood function in equation (2) is one of the fundamental parts of the Bayes theorem (e.g., [4]). It describes the probability of the evidence, i.e. the data, given the unknown parameter θ. The likelihood function is commonly defined following expert judgment and/or selected following scientific community agreement (e.g., [15], [3]). Therefore, besides describing the aleatory uncertainty, the likelihood function is source of epistemic uncertainty due to the level of subjectivity added in the choice of it. In this context, in order to reduce uncertainties related to the subjective choice of the likelihood, a possible solution, known as Bayesian Model Averaging (BMA), is given by [5] and modified by others, e.g., [8].

The basic idea of BMA is that the distribution of some interested quantity of a model, such as fatalities in our case, is derived over some space of possible models instead of only one, e.g., [5]. In other words, suppose that $\boldsymbol{M}$ is the set of all possible models of interest M, that is $M \in \boldsymbol{M}$. If θ is the parameter of interest, and the likelihood corresponding to the model $M_j \in \boldsymbol{M}$ is given by $f(y|\theta, M_j)$, then the formal Bayesian calculation, as given in equation (2), that summarizes the inference about θ is given by, e.g. [5]:

$$p(\theta \mid y) = \sum_{j=1}^{J} p(\theta \mid y, M_j) \, p(M_j \mid y) \qquad (3)$$

where $p(\theta \mid y, M_j)$ is the posterior density under M_j and $p(M_j \mid y)$ is the posterior probability of M_j. The former can be rewritten as:

$$p(\theta \mid y, M_j) \propto f(y|\theta, M_j) p(\theta \mid M_j) \qquad (4)$$

where $p(\theta \mid M_j)$ is the *a priori* density under M_j. Equation (4) is describing a Bayesian analysis. In fact, it has the same structure as equation (2) in case of the sampling of posterior distribution through MCMC algorithms (section 3.1). Therefore, equation (4) in this study is computed through the MCMC Gibbs sampler.

In Equation (3), $p(M_j \mid y)$ is the posterior probability of M_j, also known as posterior model weight. In fact, it describes the weight of the model M_j with respect to all the others belonging to the model space $\boldsymbol{M}$ in the posterior distribution of the parameter under interests:

$$p(M_j \mid y) = \frac{m(y \mid M_j) p(M_j)}{\sum_{t=1}^{J} m(y \mid M_t) p(M_t)} \qquad (5)$$

where $p(M_j)$ is the prior probability of the j-th model in the model space reflecting the expert beliefs in the relative correctness of this model. A common choice is $p(M_j) = 1/J$, with $j = 1, ..., J$, which means that each model considered is equally likely before the data are observed. Furthermore, $m(y \mid M_j)$ is the marginal density of the observations under M_j, e.g., [6], that is, the probability computed by integrating the likelihood multiplied by the prior distribution of the parameters over the parameter space:

$$m(\mathbf{y}|\mathbf{M_j}) = \int f(\mathbf{y} | \theta, \mathbf{M_j}) p(\theta | \mathbf{M_j}) \qquad (6)$$

Equation (6) is similar to the integral in equation (1), except that the model itself becomes a variable of the problem. Therefore, the integral in equation (6) can be difficult to compute, because a closed form might not be always available. In order to overcome this issue, researchers proposed different methods, from the use of Bayesian Information Criterion (BIC) (e.g., [6]) to MCMC algorithms (e.g., [8]). In this study, the former method is employed to estimate the posterior model weight $p(\mathbf{M_j} | \mathbf{y})$, e.g., [10].

3.2.1 BIC Method

The Bayesian Information Criterion has been proposed by researchers to provide an approximation of $p(\mathbf{M_j} | \mathbf{y})$, e.g., [8]. Such an approximation is adequate when a non-informative prior is assumed over the model space, e.g.,[10]. In fact, based on equation (1), if the prior is non-informative, the posterior distribution is strongly related to the likelihood function $(p(\theta | \mathbf{y}) \sim L(\theta; \mathbf{y}) / \int L(\theta; \mathbf{y}) d\theta)$. Thus, the introduction of the *a priori* distribution, $p(\theta | \mathbf{M_j})$, in equation (6) can be avoided. Under the aforementioned conditions, it has been shown, e.g., [8], that the posterior model weight can be described in terms of BIC as follow:

$$p(\mathbf{M_j} | \mathbf{y}) = \frac{exp(-0.5\ BIC_j)}{\sum_{t=1}^{J} exp(-0.5\ BIC_t)} \qquad (7)$$

where $BIC_j = -2L_j + p_j \log N$. N is the sample size of the training set, p is the total number of parameters and L is the log-likelihood. Moreover, the lower BIC score the better the model is fitting the dataset.

4 APPLICATION TO THE DATA

4.1 Frequency

Frequency denotes the number of accidents per year (Figure 1). Essentially in the ENSAD database, accidents can be considered rare, independent events so that the frequency can be modeled as a Poisson distribution. Therefore, the frequency is modeled applying the common Bayesian procedure described in section 3.1. In equation (2), the likelihood is described by the Poisson model, while the *a priori* distribution for the parameter of interest, the frequency rate λ, is set to a non-informative, very broad Γ distribution ($\lambda \sim \Gamma\ (\alpha = 0.001, \beta = 0.001)$, with α and β describing the shape and rate of the distribution, respectively). Thus, the posterior distribution would be mainly influenced by the data, since the *a priori* distributions are weak (e.g., [3]).

The MCMC algorithm is run for 30,000 iterations, following a burn-in of 1,000 updates. Furthermore, the latter is also used to train the model. According to the Gelman-Rubin diagnostic (e.g., [16]) the simulated chains converged adequately in the MCMC practice implemented in this study. Once the posterior distribution for the mean frequency is estimated, it is normalized by the corresponding energy production in GWeyr (Table 1).

4.2 Severity

Severity measures the extent of the consequences of each accident (Figure 1). The fatality distribution is right-skewed (skewness > 0) meaning that most of the accidents are located at the left side of the mean, with catastrophic (extreme) events located to the right of the distribution. A unique model possibly describing the fatality distribution is difficult to establish, since different probability distribution functions exhibit right skewness. Therefore, in order to model the fatality distribution, the BMA method is applied (section 3.2).

In this study, the model space is arranged by a group of possible right skewed models that are commonly used in hazard, risk assessment and survival analysis (e.g., [9], [10]). Furthermore, only models described by a maximum of three parameters (location, shape and scale) are considered. This choice is made in order to avoid overfitting due to a high number of parameters in the model. The models used are shown in Table 2.

According to the BIC method, the posterior model weight for the BMA is estimated for all energy chains disaggregated by country groups. Table 2 shows that in all considered datasets, the same two models described the data best, meaning they had the lowest BIC scores. Therefore, the Inverse Gaussian (IG) and the Lognormal (LOGNO) distributions are used to model all the datasets. In addition, in case of Coal China 1994-1999, the Weibull distribution (weight = 0.01) has also to be considered in the assessment of the posterior distribution.

Table 2 Summary of goodness of fit (BIC score) and relative posterior model weight (Weight) for the fatality distributions collected for different fossil energy chains disaggregated by country groups.

Distributions	Coal							
	China 00-08		China 94-99		non-OECD w/o China		OECD	
	BIC	Weight	BIC	Weight	BIC	Weight	BIC	Weight
Logistic	624	0.00	506	0.00	160	0.00	293	0.00
Reverse Gumbel	588	0.00	481	0.00	143	0.00	264	0.00
Generalized Pareto	416	0.00	349	0.00	152	0.00	245	0.00
Lognormal	406	0.09	336	0.97	124	0.30	216	0.05
Weibull	424	0.00	346	0.01	139	0.00	238	0.00
Inverse Gaussian	402	0.91	344	0.02	122	0.70	210	0.95

Distributions	Natural Gas				Oil			
	non-OECD		OECD		non-OECD		OECD	
	BIC	Weight	BIC	Weight	BIC	Weight	BIC	Weight
Logistic	159	0.00	135	0.00	264	0.00	507	0.00
Reverse Gumbel	147	0.00	124	0.00	242	0.00	466	0.00
Generalized Pareto	119	0.00	116	0.00	205	0.00	400	0.00
Lognormal	113	0.08	105	0.22	189	0.13	379	0.01
Weibull	119	0.00	113	0.00	205	0.00	408	0.00
Inverse Gaussian	107	0.92	102	0.78	185	0.87	369	0.99

Once the posterior model weight is estimated (Table 2), the MCMC algorithm is used to assess the posterior distribution, for each model, of the parameters of interest, namely the expected value and the expected extreme value. According to the BIC method applied to BMA, non-informative, very broad prior distributions have to be defined e.g., [8]. For the location parameter (μ) the prior is defined as a normal distribution with mean 0 and standard deviation 0.01. For the shape parameter (σ) the prior is defined as a Γ distribution with shape and rate both equal to 0.001. Finally, for distributions described by three parameters, such as the Weibull distribution, the scale is defined by a Γ distribution with shape and rate both equal to 0.001.

Finally, for each energy chain and country group, the posterior distribution is calculated according to equation (3). Then for each model with posterior model weight > 0 (see Table 2), the posterior distribution is assessed using an MCMC Gibbs sampler, e.g., [14]. The MCMC algorithm is run for 100,000 iterations, following a burn-in of 10.000 updates. Furthermore, the latter is also used to train the model. According to the Gelman-Rubin diagnostic (e.g., [16]) the simulated chains converged adequately in the MCMC practice implemented in this study. The final posterior distribution is then evaluated as the sum of the weighted posterior distributions associated to the different models (equation (3)).

5 RESULTS

The aforementioned models allow us to compare frequencies and severity distribution as well as the total risk, being the product of the two components, between energy chains and country groups (OECD and non-OECD). For each parameter, the mean and the 5 and 95% quantiles are extracted from the posterior distribution.

5.1 Frequency

Figure 2 shows the average accident frequency per GWeyr over the period 1970-2008 for all fossil energy chains, and OECD and non-OECD countries. In addition for the Coal chain, Chinese data are analysed separately because it has been shown that data prior to 1994 were subjected to strong underreporting (e.g., [16]).

The normalized accident frequency is clearly highest for the Chinese coal chain. However, a comparison of the periods 1994-1999 and 2000-2008 indicates that the frequency is decreasing, and thus slowly approaching other non-OECD countries. This result could be possibly explained by the fact that the Chinese government, in the last decade, undertook a large effort to close small private mines in order to move the entire production to large mines, which are under the safety and regulatory policy of the government. Consequently, Coal china should be treated separately at least with regards to analysis of accident frequency. Finally, accident frequencies are generally lower in OECD than non-OECD countries for the coal and oil chains, whereas for natural gas no significant difference is found. The latter could be possibly explained by the lack of data for both OECD and non-OECD country groups. In fact, as shown in Figure 1e, in the natural gas energy chain fewer accidents per year happened with respect to oil and coal energy chains (Figure 1a, c, g).

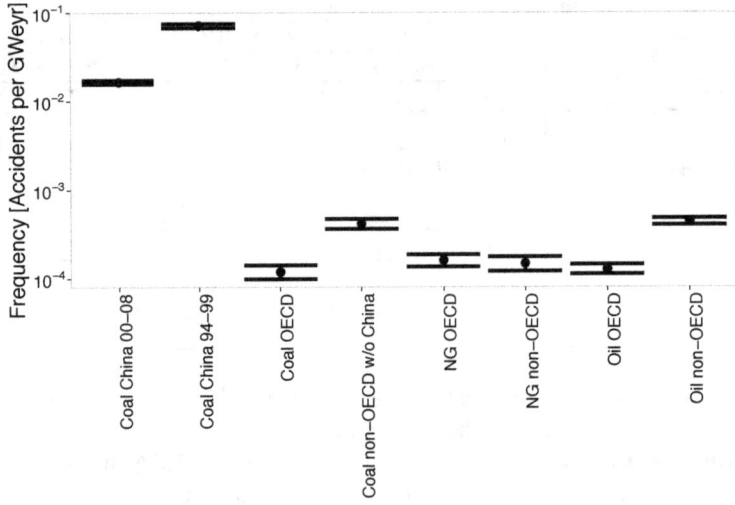

Figure 2 Mean frequencies (accidents per GWeyr), 5% and 95% of the posterior distribution, averaged over time.

5.2 Severity

Figure 3 shows the mean and 5-95% error bars for the expected fatalities per accidents as well as the number of fatalities exceeded in 1% of all accidents for various fossil energy chain and country group combinations. Additionally, the figure includes the contribution of each posterior distribution to the final BMA, described by the mean and 5% and 95% quantiles. Overall, the inverse Gaussian (IG) model was the dominant contributor to the final result in most of the cases (Table 2), except for Coal China 94-99 where it was only 2%. In the case of fatalities exceeded in 1% of accidents, the IG model shows a very broad uncertainty range compared to the lognormal and Weibull models. The same effect is to a lesser extent also visible for Coal China 00-08, although IG replaces LOGNO as the dominant contributor. This effect could be possibly related to the lack of events in the tail of the historical distribution of the accidents (Figure 1d). In fact, in case of Coal China 94-99, all events are clustered in the range 5-50 fatalities, with only a data point in the tail at 114 fatalities. Moreover, in case of Coal China 00-08, the maximum fatality is twice as big as the time period 94-99. However, the fatality distribution is similar to the ones for oil, natural gas and coal without China, where more than one observation is located in the tail of the distribution. This lack of data could cause large fluctuations in the inverse Gaussian tail, since IG needs relative high probability, meaning number of observations, in the tail in order to be able to model it (e.g., [17]). Therefore, the large fluctuations would increase the randomness in modeling the tail of the distribution and, thus, increase the aleatory uncertainty.

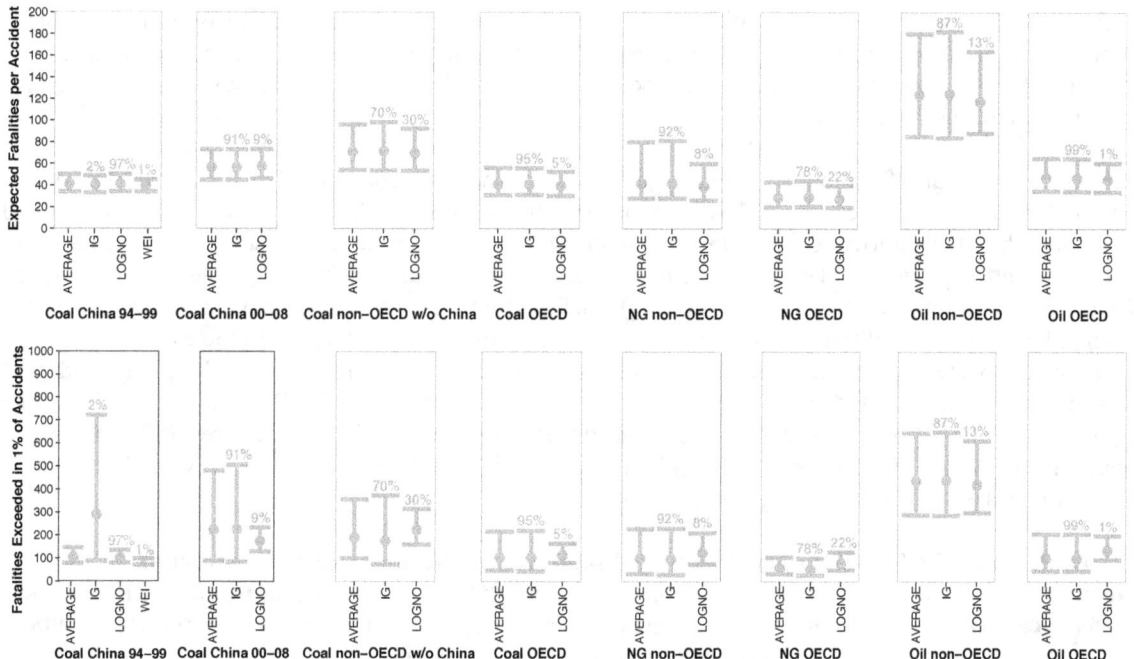

Figure 3 Mean, 5% and 95% quantiles for all the energy chains and country groups analysed. For each case, the BMA (AVERAGE) result is shown for the various distributions including their percent contribution or weight). Results are shown for the expected fatalities and fatalities exceeded in 1% of all accidents

For the expected fatalities per accident (upper panel in Figure 3), the final BMA shows no statistically significant differences between different energy chains and country groups, except in the cases of oil non-OECD and Coal energy chains. In these two cases the expected fatalities per accident are about twice as high for non-OECD compared to OECD countries. Concerning natural gas, similar to the frequency case, there is no significant difference between OECD and non-OECD country groups. However, in the latter, the shape of the major contributor's distribution (IG) clearly affects the final result. In fact, the posterior distribution of the expected value exhibits a long tail, resulting in large values at the 95% quantile. It is interestingly in the case of coal China that both time ranges taken into account are not significantly different with respect to other considered energy chains. However, the mean number of expected fatalities is larger for coal China 2000-2008 than 1994-1999. This could be explained by the fact that moving the production from small private mines to big mines, as was done by the Chinese government in the last decade, the number of accidents (Figure 2) could be reduced, but at the same time the potential consequences can be more severe due to the larger number of workers present in these mines.

Overall, the aforementioned behaviour for the number of expected fatalities well described the behaviour of the number of fatalities exceeded in 1% of the total accidents (lower panel in Figure 3).

However, the main differences are related to the large uncertainty in Coal China 00-08 and to the slightly similar behaviour of Coal non-OECD without China with respect to the other cases. The former can be possibly described by the increase of randomness in modelling the tail of the inverse Gaussian distribution due to the presence of a large number of events in the tail of the historical observations (Figure 1). Therefore, these would increase the aleatory uncertainty and, thus, affect the 95% quantile of the posterior distribution.

5.3 Risk Indicators: Mean and Exceedance

By definition, risk is the product of the frequency and severity. In order to compare the risk between different energy chains and country groups, two risk indicators are used in this study. The first one addresses the mean risk and is expressed as the expected number of fatalities per accident. The second one represents the extreme risk, defined by the threshold exceeded for a specific return frequency, and is given by the total number of fatalities exceeded at 1% frequency per accident (e.g., [5]). Table 3 summarizes the results for these risk indicators. In addition, results from the standard approach, non-normalized mean and 1% exceedance per accident based on frequency-consequence curves, are also shown in order to compare the results.

Generally, the results for the mean aggregate indicators show a good accordance between BMA and the standard approach. The main difference concerns the Oil non-OECD case. This could be explained by the fact that the historical observations are distributed with a very long tail (see Figure 1), due to an extreme event happened in 1987 in the Philippines, where the tanker *Victor* collided with the Ferry *Dona Paz* resulting in 4386 fatalities (e.g., [1]). Such extreme events can have a strong impact on the mean value in the case of the standard approach, resulting in a large difference to the expected value in BMA, where posterior distributions are more resistant to outliers (e.g., [6]). In addition, the 5 and 95% quantiles are significantly different between the standard approach and the BMA model. This can be explained by the fact that in the former case no uncertainty analysis is included, resulting in a broad range in the entire fatality space of the dataset, while in BMA both aleatory and epistemic uncertainties are modeled (e.g., [7]).

Table 3 Results for the full risk, expected fatalities per accident and 1% exceedances per accident. Each value is given by the mean with 5 and 95% intervals. In addition, the mean and 1% exceedance is calculated for frequency and production level in the time range 1970-2008 following the standard approach.

Country Group	BMA Model		Standard Approach	
	Mean per Accident	1% Exceedance per Accident	Mean per Accident	1% Exceedance per Accident
Coal China 00-08	57(45; 73)	222(89; 482)	55(8; 124)	60(5; 211)
Coal China 94-99	41(34; 50)	112(84; 153)	39(8; 88)	44(9; 112)
Coal non-OECD w/o China	71(54; 96)	191(100; 357)	67(8; 199)	183(120; 413)
Coal OECD	41(31; 56)	105(47; 217)	39(7; 93)	123(29; 231)
NG non-OECD	42(28; 80)	101(34; 228)	38(6; 97)	88(81; 215)
NG OECD	29(20; 43)	58(33; 106)	26(6; 87)	37(10; 104)
Oil non-OECD	124(85; 180)	439(291; 645)	173(9; 554)	419(28; 3812)
Oil OECD	47(35; 65)	102(46; 206)	44(7; 139)	48(37; 226)

For the risk indicator 1% exceedance per accident, the comparison between the BMA and standard approach results shows a different behavior for the 5 and 95% quantiles due to the fact that in the second case no uncertainty analysis is assessed. In case of the average, the results differ in all the cases. In most of them the average values estimated using the standard approach are lower than the modeled extreme. This is related to the fact that a significant number of the historical observations have small consequences. Furthermore, they compensate the presence of few data points, such as the extreme value, resulting in a shift of all the quantiles towards small number of fatalities. Therefore, the value for 1% exceedances is close to the mean of the distribution, resulting in a different value with respect to the BMA result, which is accounting for the uncertainty and the outliers that strongly affect the standard approach. In cases where the standard approach shows larger averages compared to BMA, the former's results is strongly affected by the presence of outliers. These extreme values strongly affect, in terms of number, the distribution, shifting the higher quantiles toward them. This results in a larger value of the fatalities exceeding 1% of accidents in the standard approach with respect to BMA. In fact, in the latter, the posterior distribution is resistant to outliers (e.g., [6]), while in case of the standard approach the result is strongly affected by them.

In Figure 4 the visualization of the risk is shown for the average risk (Figure 4a) and for the risk of extreme events (Figure 4b). The overall highest risk is found for coal China 94-99 with an expected number of 41 fatalities per accident at current consumption levels and a 1% probability that an accident with more than 112 fatalities takes place. However, as described above, this result is different from all other energy chains and country groups due to its much larger historical dataset than in any other case (e.g., [16]). Furthermore, in all other cases, Oil non-OECD clearly shows the highest risk. In fact for Oil non-OECD 124 fatalities are expected per year at current consumption levels, and at a 1% probability per year an accident with more than 439 fatalities is expected. It is important to note that the result for Coal non-OECD w/o China in terms of number of fatalities or fatalities exceeding 1% of the accidents is comparable with Coal China. However, based on the frequency it is not, since in case of China, many more accidents occurred. Overall, Natural gas is the least risky energy chain and, more specifically, the Natural gas OECD group performs best. Finally, OECD generally exhibits lower risk levels than non-OECD, and even more pronounced than coal China.

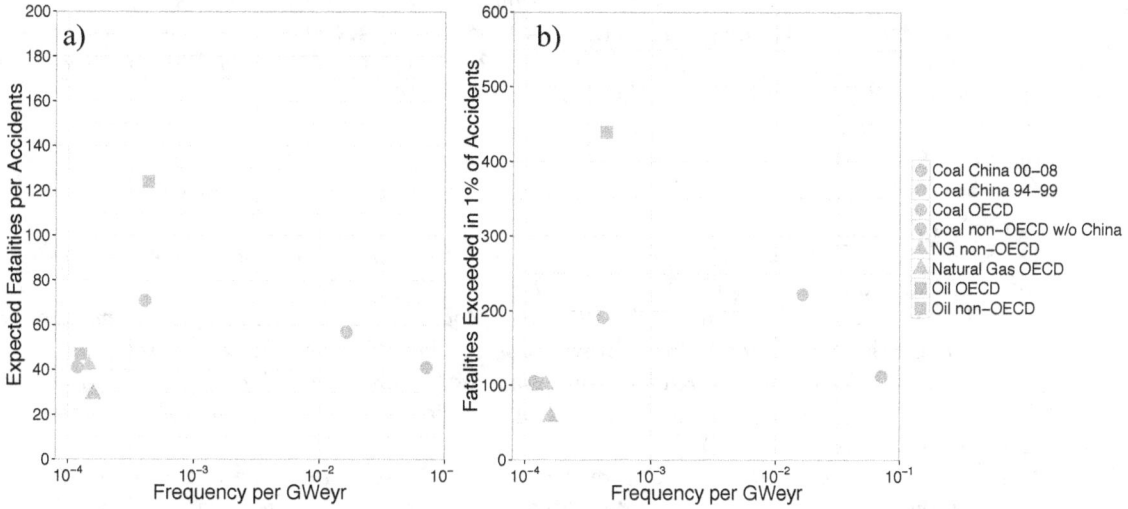

Figure 4 a) Mean number of fatalities per severe accident versus frequency of severe accidents per GWeyr. b) Fatalities exceeded in 1% of accidents versus total accident frequency per GWeyr.

6 CONCLUSIONS

This study presented a first-of-its-kind implementation of the Bayesian Model Averaging (BMA) method in a comparative risk assessment framework to comprehensively quantify the risk of severe accidents in fossil energy chains. This framework allows estimating uncertainty and dealing with lack of data and lack of knowledge by averaging the posterior distribution over a pre-defined model space. This "top down" approach can also be useful to complement conventional, detailed "bottom-up" models of risk quantification that are conducted for individual plants with specific physical processing and site conditions. Therefore, the proposed approach provides a unified framework that comprehensively covers accident risks in energy chains, and allows estimating specific risk indicators, including their uncertainties. This information provides an essential element in a holistic sustainability and energy security evaluation of energy technologies. The overall risk is found to be highest in Coal China for the time range 94-99. Among Coal China, non-OECD country groups for all energy chains show higher risk in terms of expected number of fatalities per accident as well as for extreme cases. Furthermore, results show that Natural gas is the least risky energy chain. In future work based on the database ENSAD, both the scope of this model will be expanded towards incorporating other types of consequences (e.g. injured) and other energy chains (e.g. hydropower), and the resolution of risk will be increased, that is, to differentiate the risk for more activities or regions.

Acknowledgements

This study was partially performed using data collected within the Collaborative Project SECURE (Security of Energy Considering its Uncertainty, Risk and Economic implications), Contract No. 213744 of the 7th Framework Programme of European Commission.

References

[1] P. Burgherr and S. Hirschberg. *A Comparative Analysis of Accident Risks in Fossil, Hydro, and Nuclear Energy Chains.* Human and Ecological Risk Assessment: an International Journal, 14, pp. 947–973, (2008).

[2] S. Hirschberg, G. Spiekerman and R. Dones. *Severe accidents in the energy sector.* PSI Report No. 98-16, Paul Scherrer Institut, (1998), Villigen PSI, Switzerland.

[3] P. Eckle and P. Burgherr. *Bayesian Data Analysis of Severe Fatal Accident Risk in the Oil Chain.* Risk Analysis, 33, pp. 146–160, (2013).

[4] A. O'Hagan. Bayesian statistics: principles and benefits. *Bayesian Statistics and Quality Modelling in Agro-Food Production Chain*, Dordrecht: Kluwer Academic Publishers, (2003).

[5] L. R. Pericchi and M. E. Pérez. *Posterior robustness with more than one sampling model.* Journal of Statistical Planning and Inference, 40, pp. 279–294, (1994).

[6] K. Shao and J. S. Gift. *Model Uncertainty and Bayesian Model Averaged Benchmark Dose Estimation for Continuous Data.* Risk Analysis, 34, pp. 101–120, (2013).

[7] A. E. Raftery. *Bayesian model selection in social research.* Sociological Methodology, 25, pp. 111–163, (1995).

[8] L. Wasserman. *Bayesian Model Selection and Model Averaging.* Journal of Mathematical Psychology, 44, pp. 92–107, (2000).

[9] D. S. Reis and J. R. Stedinger. *Bayesian MCMC flood frequency analysis with historical information.* Journal of Hydrology, 313, pp. 97–116, (2005).

[10] A. J. Bailer, R. B. Noble and M. W. Wheeler. *Model Uncertainty and Risk Estimation for Experimental Studies of Quantal Responses.* Risk Analysis, 25, pp. 291–299, (2005).

[11] P. Burgherr, P. Eckle and S. Hirschberg. *Final Report on Severe Accident Risks including Key Indicators.* Deliverable No 5.7.2a SECURE Project, 7th Framework Programme of European Commission, (2010).

[12] P. Burgherr, P. Eckle and S. Hirschberg. *Comparative assessment of severe accident risks in the coal, oil and natural gas chains.* Reliability Engineering and System Safety, 105, pp. 97–103, (2012).

[13] S. Brooks, A. Gelman, G. L. Jones and X.-L. Meng. *Handbook of Markov Chain Monte Carlo.* Taylor and Francis Group, LLC, (2011).

[14] C. Andrieu, N. de Freitas, A. Doucet and M. I. Jordan. *An introduction to MCMC for machine learning.* Machine Learning, 50, pp. 5–43, (2003).

[15] P. J. Birrell, G. Ketsetzis, N. J. Gay, B. S. Cooper, A. M. Presanis, R. J. Harris, A. Charlett, X.-U. Zhang, P. J. White, R. G. Pebody and D. De Angelis. *Bayesian modeling to unmask and predict influenza A/H1N1pdm dynamics in London.* Proceeding of the National Academy of Sciences, 108, pp. 18238–18243, (2011).

[16] A. Gelman and D. B. Rubin. *Inference from Iterative Simulation Using Multiple Sequences.* Statistical Science, 7, pp. 457–511, (1992).

[17] I. J. Good. *The Population Frequencies of Species and the Estimation of Population Parameters.* Biometrika, 40, pp. 237–264, (1953).

Investigation of Different Sampling and Sensitivity Analysis Methods Applied to a Complex Model for a Final Repository for Radioactive Waste

Sabine M. Spiessl[a], and Dirk-A. Becker[a]
[a] Gesellschaft fuer Anlagen- und Reaktorsicherheit (GRS) mbH, Braunschweig, Germany

Abstract: The performance of different types of sensitivity analysis methods in combination with different sampling methods on the basis of a Performance Assessment model for a repository for Low and Intermediate Level radioactive Waste (LILW) in rock salt has been investigated. This paper provides an insight into the results obtained with the following methods for sensitivity analysis: (i) a graphical method (CSM plot), (ii) a rank regression based method (SRRC) and (iii) a simple first-order SI calculations scheme (EASI). These methods were combined with random and LpTau sampling. The most robust results were obtained using LpTau sampling. The results obtained with CSM and SRRC analysis are fairly comparable. The EASI results, however, assign the dominating role to a parameter that seemed to be of secondary importance according to the results of the two other methods before. In addition, in the early phase below 10^4 years, the EASI results seem to be of low robustness.

Keywords: CSM plot, EASI, rank regression based method (SRRC), LpTau sampling, quasi-random sampling.

1. INTRODUCTION

For the assessment of the long-term safety of a geological repository for radioactive waste, adequate handling of the various uncertainties within the system and the available data is essential. Computational models for the Performance Assessment (PA) of final repositories, especially in a rock salt environment, typically include a number of interacting physical and chemical effects, which leads to a non-linear, non-monotonic and sometimes even virtually non-continuous behavior. A robust and reliable global sensitivity analysis of such models can be a demanding task, which requires a sufficiently high number of runs with a good coverage of the parameter space.

Simple random sampling is often not the best choice, since it tends to developing clusters and gaps in the parameter space. More sophisticated types of sampling techniques for Monte-Carlo methods such as Quasi-Monte-Carlo sampling have been developed to improve convergence and/or accuracy of probabilistic evaluations. Recent literature studies indicate that Quasi-Monte-Carlo sampling schemes can give more robust sensitivity measures with a lower number of simulations than random sequences, as the parameter space is covered more homogeneously.

The most adequate combination of a sampling scheme and a sensitivity analysis method may not only depend on the CPU cost and time of the analysis and the required accuracy but also on the system behavior (degree of linearity and monotonicity, discrete nature), the number of parameters and parameter interaction. This may be in particular important for comprehensive computational models that take account of a variety of different coupled processes. For instance, variance-based methods are recommended for computational models showing a non-linear and non-monotonic system behavior.

In combination with different sampling methods, the performance of different types of sensitivity analysis methods on the basis of a Performance Assessment model for a repository for Low and Intermediate Level radioactive Waste (LILW) in rock salt has been studied. This PA model shows a nearly non-continuous or quasi-discrete behavior as result of the dissolution and nearly sudden failure of the seal in the near field. When this barrier fails, a sudden high release of radionuclides, i.e., a jump in the model output occurs.

This paper provides an insight into the results of the following methods for sensitivity analysis:

- a graphical method (CSM plot),
- a rank-regression-based method (SRRC) and
- a simple first-order SI calculations scheme (EASI).

These methods were combined with random and LpTau sampling. The LpTau scheme belongs to the group of Quasi-Monte-Carlo sampling schemes. Numerous sample sets of different sizes generated by the different sampling schemes were examined.

2. METHODS

2.1. Sampling Techniques

With the random method, the sample values are randomly selected within the parameter space following given probabilistic density functions (pdfs). Random samples typically show some clustering and gaps in the parameter space. The quasi-random LP-Tau sequence (LpTau) belongs to the group of low-discrepancy sequences. It is designed to prevent clustering and gaps of sampling points as much as possible even for fairly small samples by placing the points as homogeneous as possible within the space. The LpTau sequence starts with the generation of Sobol sequences. The Sobol sequences establish successively finer partitions of the [0,1] interval on the base of two and then rearrange the coordinates in each dimension. The generated sample is then transformed to the desired intervals for each parameter. It was found to be a very fast computational algorithm.

2.2. Methods for Sensitivity Analysis

CSM (Contribution to Sample Mean) plots are obtained by sorting the model output values according to increasing values of the considered input parameter and then plotting the total contribution to the mean of the output versus the proportional size of increasing subsets. The more the curve deviates from the diagonal the higher is the sensitivity of the model against the respective parameter. A left-curved progression, normally below the diagonal, means a positive influence (parameter increase → output increase); a right-curved progression, normally above the diagonal, represents a negative influence (parameter increase → output decrease). More details about CSM plots can be obtained from [1,2].

SRRC (Standardized Rank Regression Coefficients) are calculated as the coefficients of a multilinear regression between the input and the output ranks [3]. By the rank transformation the nonlinear, but largely monotonic model is better adapted to linear regression. Positive SRRC values mean a positive influence, negative values a negative influence.

EASI stands for an effective algorithm for estimating sensitivity indices of first order [4]. In contrast to the FAST/EFAST methods, which require specific frequency data for the input parameters, EASI can introduce these into existing sample sets. This is accomplished by sorting and shuffling the values of the different input parameters. The output is arranged according to the input data. The arranged data are then analyzed using the power spectrum of the output as it is done in FAST/EFAST. A big advantage of EASI is that any sampling scheme can be applied, existing samples can be extended and model evaluations can be re-used.

As EASI belongs to the group of variance-based sensitivity analysis methods, changes in the model output are squared in the computation of SI1. For this reason, the direction of influence cannot be found by such methods. Moreover, they tend to overvaluing very high values, especially if the output varies over orders of magnitude. A characteristic of PA models is that a small fraction of output values may be rather high compared to the rest of the values.

3. MODEL AND SOFTWARE

The investigated model consists of three parts, describing the near field, the geosphere and the biosphere. In the near field part it is assumed that the mine openings are filled with brine from the overburden after some time. At that point in time, some short-lived wastes, which are disposed of in one of the openings, start to release contaminants. These are dissolved in the brine and pressed out to the geosphere by the convergence process. In order to protect the longer-lived and more radiotoxic wastes from the brine, the main waste emplacement area is isolated from the rest of the mine by a specific seal, which, however, can be chemically corroded by magnesium. Depending on the magnesium content of the brine and the initial permeability of the seal material, the seal can nearly suddenly fail at some point in time. This leads to a short-lasting decrease, followed by a fast, significant increase of the contaminant release. The decrease is due to the fact that after seal failure it takes some time to fill up the emplacement area, during which the brine outflow from the mine is reduced.

As model output, the annual effective dose to an adult human individual is calculated with the software package RepoTREND [5]. This package contains independent modules for the near field, the far field and the biosphere to calculate the transport of brine and radionuclides through the repository system. The three modules for the near field, the far field and the biosphere are called LOPOS, GeoTREND-POSA and BioTREND, respectively.

All samples were generated using the software package SIMLAB 3.2.6, developed by the Joint Research Centre (JRC) in Ispra (http://simlab.jrc.ec.europa.eu), within the MATLAB environment. The EASI analysis was done using the MATLAB script by [4]. This script can also be downloaded from the JRC website.

4. PERFORMANCE ASSESSMENT (PA) TEST CASE

The model structure of the near field is schematically shown in Figure 1. It consists of two emplacement chambers (EC) with emplaced radioactive waste, one of which is sealed and the other one is not (AEB and NAB, respectively), a mixing region (MB) and the partially backfilled residual mine without waste (RG). The mixing region is connected to both ECs and the residual mine and acts as the interface to the far field.

Figure 1: Illustration of the Near Field Model of the LILW Repository System in Salt

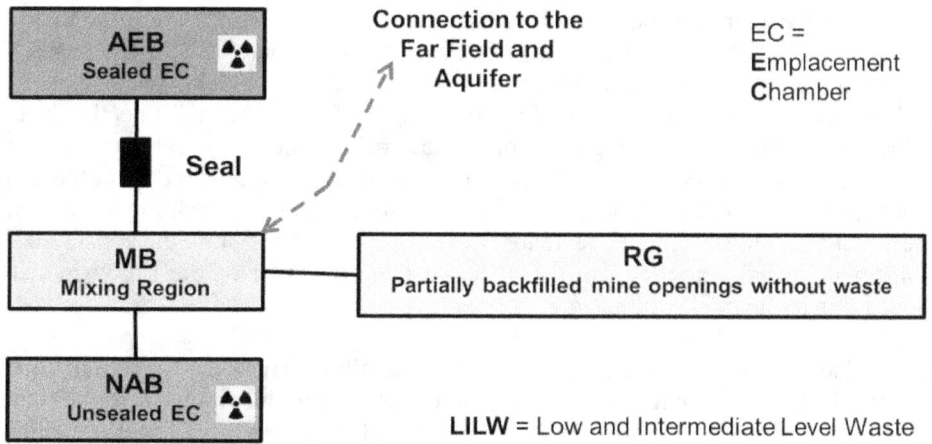

For the probabilistic investigations of the model, in total, 11 parameters, all pertaining to the near field, were varied with adequate probability density functions (pdfs). These parameters, along with their abbreviations, distribution types and ranges, are listed in Table 1.

Table 1. Distributions and Ranges of the Parameters of the PA Model for an LILW Repository in Rock Salt

Parameter	Unit	Description	Distribution Type	Minimum $\mu^{(1)}$ Peak$^{(3)}$	Maximum $\sigma^{(2)}$
IniPermSeal	[m^2]	Initial Permeability of Dissolving Seal	Normal	$3.23 \cdot 10^{-21}$ 41.0605$^{(1)}$	$6.7 \cdot 10^{-16}$ 1.9809$^{(2)}$
AEBConv	[-]	Factor of Local Convergence Variation in the Sealed Emplacement Chamber	Log Uniform	0.05	5
GasEntryP	[MPa]	Gas Entry Pressure	Uniform	0	2.5
GasCorrPE	[1/yr]	Corrosion Rate of Organics	Log Normal	10^{-7} -12.6642$^{(1)}$	10^{-4} 1.1177$^{(2)}$
RefConv	[1/yr]	Reference Convergence Rate	Log Uniform	10^{-5}	10^{-4}
TBrine	[yr]	Brine Intrusion Time	Log Normal	848.4 8.8857$^{(1)}$	61573 0.6933$^{(2)}$
MgBrineSat	[-]	Relative Magnesium Saturation of Brine	Triangular	0 0.1$^{(3)}$	1
RGConv	[-]	Factor of Local Convergence Variation in the Partially Backfilled Mine Openings without Waste	Log Uniform	0.25	2.5
GasCorrFe	[1/yr]	Corrosion Rate of Metal	Log Normal	$4 \cdot 10^{-5}$ -6.6728$^{(1)}$	$4 \cdot 10^{-2}$ 1.1177$^{(2)}$
AEBGasProd	[-]	Proportion of the Material Involved in Gas Production in the Sealed Emplacement Chamber	Triangular	0.1 0.8$^{(3)}$	1
NABGasProd	[-]	Proportion of the Material Involved in Gas Production in the Unsealed Emplacement Chamber	Triangular	0.1 0.8$^{(3)}$	1

$^{(1)}$ μ value$^{(*)}$
$^{(2)}$ σ value with quantiles of 0.001 and 0.999$^{(*)}$
$^{(*)}$ μ and σ values describe mean value and standard deviation of a normal or lognormal distribution
$^{(3)}$ Peak value of the triangular distribution

The seal of the emplacement chamber AEB can dissolve over time. Since the dissolution front progresses linearly through the seal and even a short piece of intact material still has a notable sealing capability, the seal fails nearly suddenly when the front reaches its end. This happens after a specific time, dependent on certain parameters. Therefore, two stages of model behavior need to be considered. As long as the seal is still functioning, the radionuclides are retained to a great extent in the emplacement chamber AEB. After failure of the seal, the radionuclides can be released from the emplacement chamber much quicker and the PA model may consequently bring forth much higher dose values. The transition between both kinds of model behavior, which happens very fast on the considered time scale, is the reason for the quasi non-continuous or quasi-discrete behavior of the LILW model, as it can cause a drastic jump in the model output.

Two parameters have a major influence on the time of seal failure. These are the initial permeability of the seal (IniPermSeal) and the magnesium saturation of the brine (MgBrineSat). Higher values of one or both of these parameters mean faster dissolution and earlier seal failure. Therefore, slight changes of these values can cause considerable jumps of the model output if investigated at a specific point in time, which makes the parameters act like switches. Additionally, the gas entry pressure (GasEntryP) behaves similarly as a switch, since the model behavior essentially changes depending on whether the gas entry pressure is below or above the value of 1.0 MPa. This is the threshold value for release of gas from the top of the sealed emplacement area AEB. This effect can lead to an increase of maximum release by more than one order of magnitude, caused by a very slight increase of GasEntryP.

5. USED SAMPLES

In total, five different samples were generated for the set with 11 parameters using random and LpTau sampling. The four random samples have sizes of 2048, 4096, 8192 and 16384 simulations. Only one LpTau sample with a size of 16384 was generated, which was broken into sets of 2048, 4096, 8192 and 16384. Table 2 lists the samples and sample sizes for each sampling technique.

Table 2. Investigated Numbers of Samples and Sample Sizes Used for the Different Sampling Schemes

Sampling Technique	Number of Samples	Sample Sizes (Number of Simulations)
Random	4	2048, 4096, 8192 and 16384
LpTau	1	

6. RESULTS

For each sampling and analysis method, time-dependent sensitivity measures (SM's) were determined. These are computed from 301 discrete annual dose values distributed over the simulation interval of 10^6 years. For the time-dependent CSM analysis, CSM curves were generated from the annual dose for each parameter at 201 time points and plotted in one figure in different colors. In this way, the time evolution of the curves is visualized. Time points below 10^4 years are not analyzed with the concept of the CSM plot as those do not provide representative CSM curves.

Results of the CSM and SRRC analysis identify the parameters IniPermSeal, AEBConv and TBrine as most important, though TBrine plays a role only in the early phase. Additionally, the parameters GasEntryP, RefConv and GasCorrPE seem to have some importance, at least in specific time phases.

The EASI analysis gives a similar impression, but assigns an outstanding importance to AEBConv. As EASI belongs to the group of variance-based methods, changes in the model output are squared in the computation of first-order sensitivity indices (SI1). This can lead to results significantly different from those obtained by linear methods, especially if the output varies over orders of magnitude. An approach to mitigating possible overvaluation of extreme values by output transformation is presented in [6].

Moreover, unlike regression- or correlation based methods, variance-based methods are unable to determine the direction of influence. This can be seen in particular at the parameters GasEntryP and GasCorrPE, which show a zero-crossing in their SRRC curves and have CSM plots extending over both sides of the diagonal. Obviously, there are opposing influences of the same parameter to the model output, annihilating each other at some point in time. The EASI SI1 curves reach zero at this point and then increase again or show a plateau.

The results obtained with the different sensitivity analysis methods using random sampling show that the deviations between the different sets with the same number of runs are bigger compared to the ones using LpTau sampling. This is revealed in particular by the results of the EASI method.

It seems that the methods investigated in this paper do not provide complete picture of the sensitivity. In another investigation [7] we showed for the same PA model that the parameter MgBrineSat, which appeared to be nearly non-important by itself, has a significant impact on the sensitivities of two other parameters (IniPermSeal, AEBConv). This parameter of indirect importance could only be identified by investigating different sets of parameters.

6.1. CSM plot

In Figure 2, color-coded time dependent CSM plots for all 11 parameters, calculated from a set of 16384 runs with LpTau sampling, are presented. For some of the parameters, the black curves show an extreme deviation, indicating that these parameters play a role mainly in the early phase. This applies in particular for TBrine, but also for IniPermSeal. The parameter GasFeConv has significant curves only for very early times. Disregarding the black curves, the parameter AEBConv shows the most conspicuous diagram with a widely expanded set of curves.

The parameter GasEntryP also produces a wide set of curves, all of which have a sharp bend at 0.4. This is due to the fact that there is a critical value of GasEntryP at which gas release from the emplacement area becomes impossible, which leads to a completely different model behavior. The parameter is distributed in such a way that 40% of the drawn values are below this critical value and 60% above. It is interesting that the CSM curves of GasEntryP expand over both sides of the diagonal, which means that the parameter changes its direction of influence to the model output at some point in time.

The time-dependent CSM plots attained from the parameters GasCorrPE and RefConv show clearly expanded but narrower sets of curves, which suggests a reduced significance of these parameters. The parameters BrineMgSat, RGConv and AEBGasProd have rather narrow sets of curves close to the diagonal, indicating a low relevance of these parameters. NABGasProd seems to have no importance as the CSM curves attained from this parameter show nearly no deviation from the diagonal. The same applies to GasCorrFe if the black curves for very early times are disregarded.

The sets of curves for all parameters except AEBConv, GasEntryP and CasCorrPE expand over only one side of the diagonal, which means that they have unique monotonic influence on the model. This influence is positive for all parameters except TBrine and GasCorrFe.

Figure 3 shows that Quasi-Random LpTau sampling produces more robust CSM plots than random sampling. The figure depicts the CSM plots of LpTau and random sets for all 11 parameters using 2048 and 4096 runs at 10^5 years. The CSM plots obtained using random sampling show that the deviations between the different sets with the same number of runs are bigger compared to the ones using LpTau sampling. For the set with LpTau sampling, a good convergence can be reached with about 2000 runs as the curves produced from the sets with 2000 and 4000 runs for most parameters closely agree. Figure 4 shows that a similar agreement of the CSM curves using random sampling is only achieved with two to four times higher sample sizes.

Figure 2: Time-Dependent CSM Analysis for Each Parameter

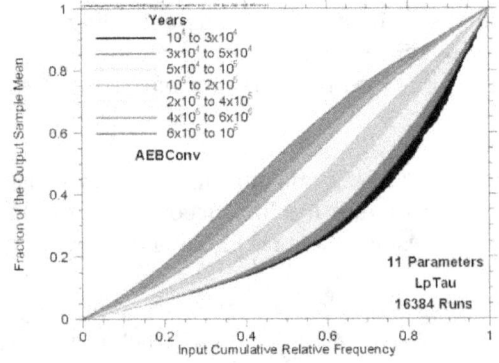
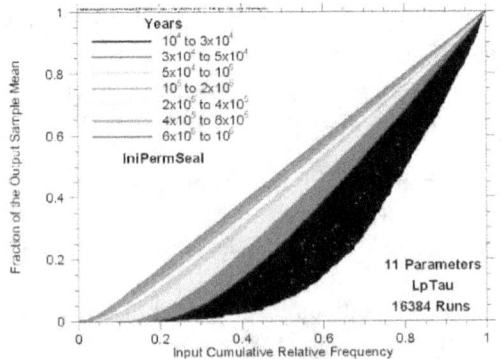

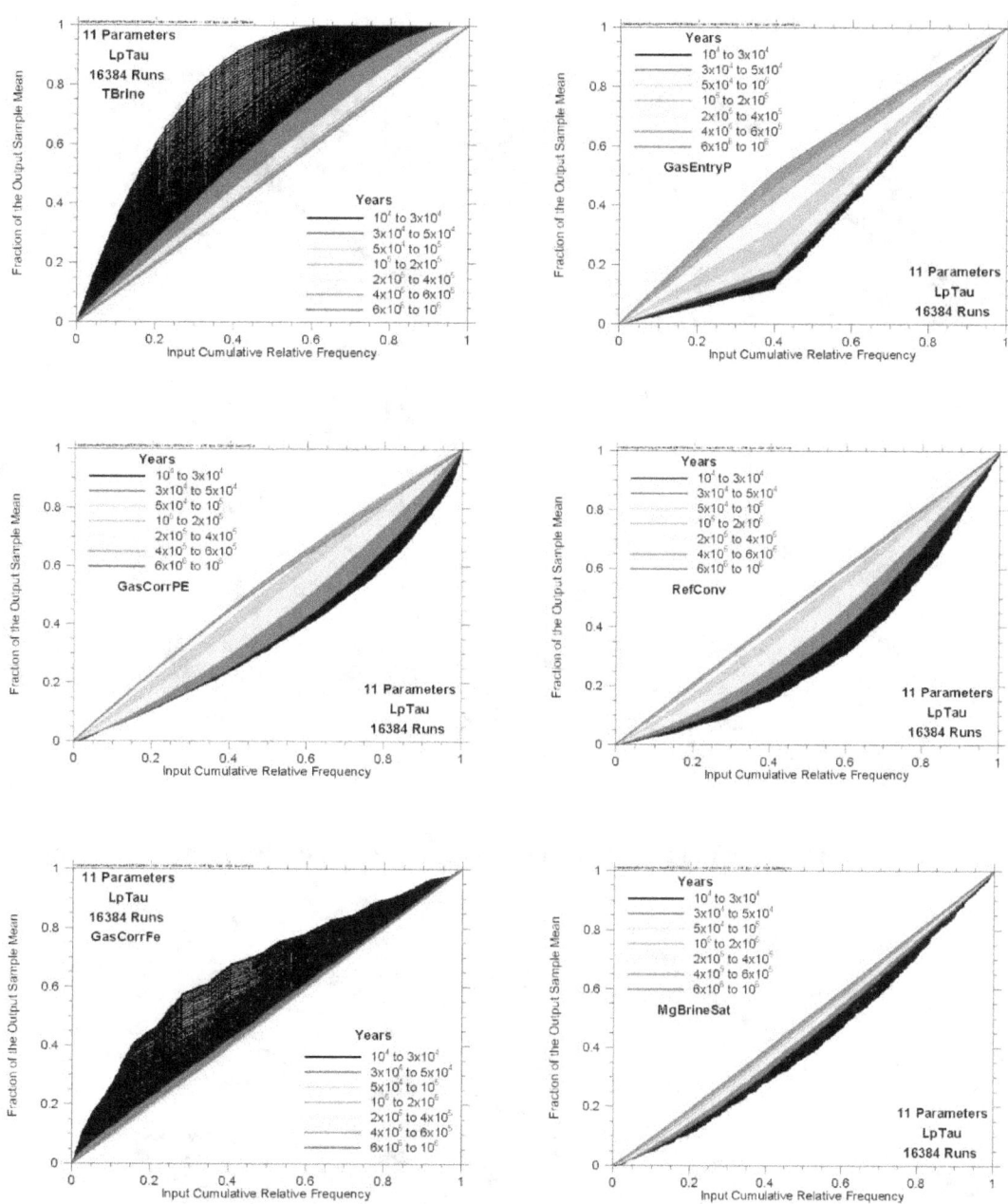

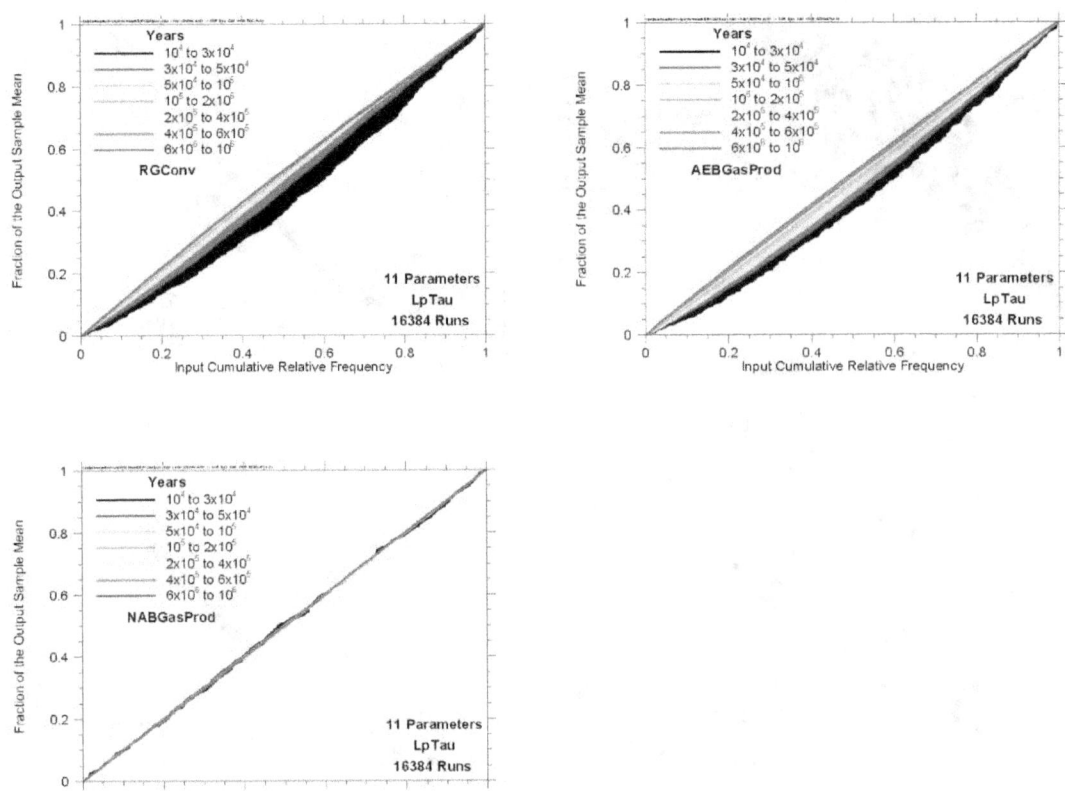

Figure 3: CSM Plots of the LpTau and Random Sets using 2048 and 4096 Runs at 10^5 Years

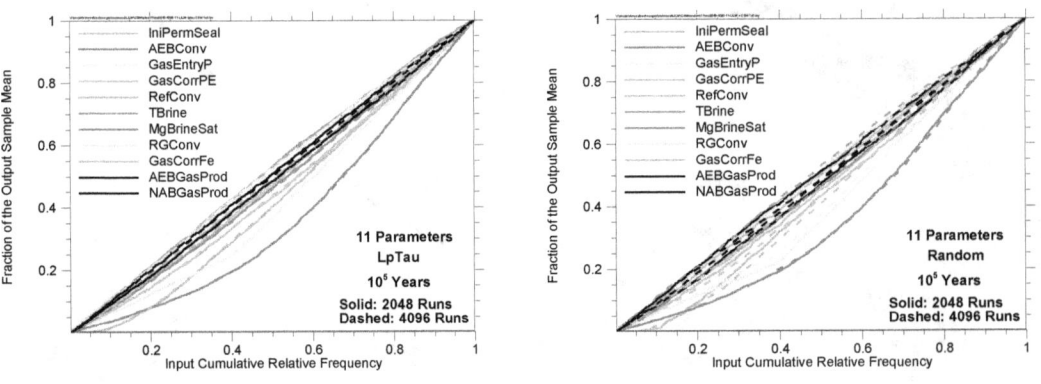

Figure 4: CSM Plots of the Random Set using 4096, 8192 and 16384 Runs at 10^5 Years

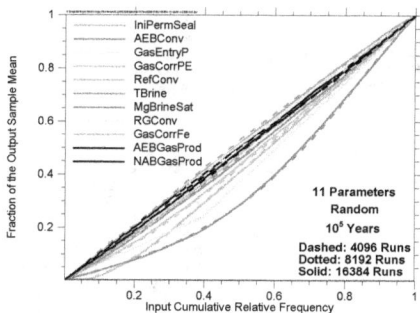

6.2. SRRC

Figure 5 shows the time-dependent SRRC coefficient values for all 11 parameters with LpTau and random sampling using 2048 and 4096 runs. The findings obtained with the SRRC method are similar to those derived from the CSM plots. The parameters IniPermSeal, AEBConv and TBrine and GasEntryP, which produce the most outstanding CSM plots, also have the most conspicuous SRRC curves and reach the highest absolute maxima. Like the CSM plot, the SRRC values obtained for TBrine indicate the highest importance in the early phase. The importance of IniPermSeal decreases with time, but the parameter still plays a role at the end of the simulation. The SRRC curves of GasCorrPE, RefConv, BrineMgSat and RGConv have less pronounced shapes. GasCorrFe, AEBGasProd and NABGasProd seem to have little importance as their SRRC coefficient values are close to zero.

The SRRC curves of AEBConv, GasEntryP, GasCorrPE, RefConv and TBrine cross the zero line, which means that, according to the SRRC evaluation, these parameters change their direction of influence at some point in time. Only AEBConv, GasEntryP and GasCorrPE, however, reach significant values on the opposite side. This is in line with the findings from the CSM plots.

As with the CSM analysis, the deviations of the SRRC coefficients between the different sets with the same number of runs are bigger using random sampling compared to the ones utilizing LpTau sampling (see Figure 5). For random sampling, there are still bigger deviations between the sets with about 8000 and 16000 runs compared to LpTau sampling using about 2000 and 4000 runs (compare Figure 5 and Figure 6). Figure 6 shows the SRRC curves using random sampling and about 4000, 8000 and 16000 simulations.

Figure 5: Time-dependent SRRC Ranking Coefficients of the LpTau and Random Sets using 2048 and 4096 Runs

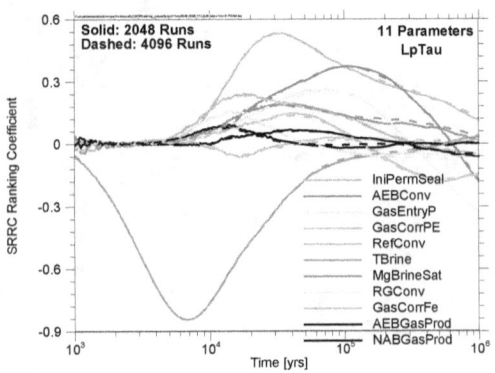
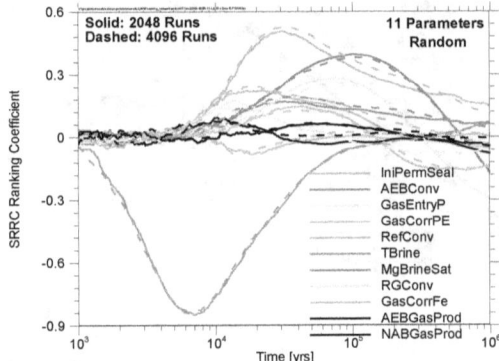

Figure 6: Time-dependent SRRC Ranking Coefficients of the Random Set using 4096, 8192 and 16384 Runs

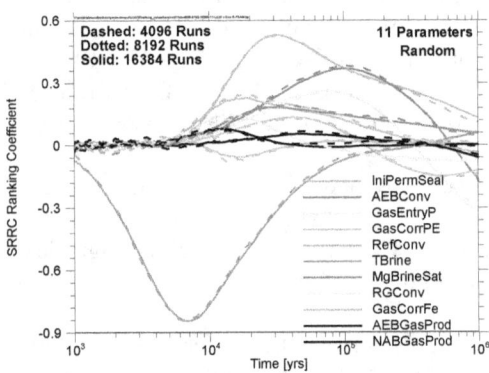

6.3. EASI

In Figure 7, the time curves of SI1 obtained with EASI for all 11 parameters are shown. The parameter AEBConv dominates the figure from about $2 \cdot 10^4$ years until the end of the scenario. The maximum SI1 values of IniPermSeal and GasEntryP are about half as high as that of AEBConv. The SI1 of TBrine is even lower in maximum, though dominating in the early phase. These results are different from those obtained with the CSM and SRRC analysis, which identified IniPermSeal and TBrine as more important than AEBConv, at least for the first 50 000 years of the simulation period.

The SI1's of RefConv and GasCorrPE go up to 0.05 and 0.04, respectively, which can be interpreted as a reduced importance of these parameters. The SI1 of GasCorrFe goes a little above 0.015 at the beginning of the simulation, which can be taken as some little significance. The SI1's of the rest of the parameters (BrineMgSat, AEBGasProd and NABGasProd) remain below 0.01, which seems to indicate that these parameters are insignificant.

As with the CSM and SRRC analysis, the deviations of the EASI SI1 indices between the different sets using random sampling are bigger compared to the ones obtained using LpTau sampling. Figure 7 also shows the results for the sets using random sampling and same number of runs as for LpTau sampling in this figure. It seems that with about 8000 runs, good convergence and smoothness of the SI1 indices can be achieved using LpTau sampling.

Figure 7: Time-dependent EASI SI1 of the LpTau and Random Sets using 4096, 8192 and 16384 Runs

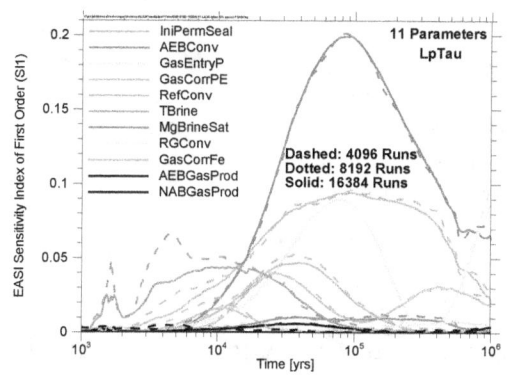
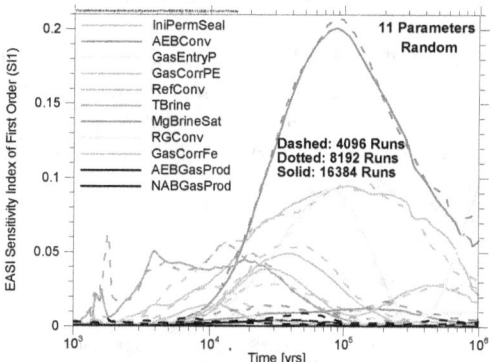

7. SUMMARY AND CONCLUSIONS

In this paper, we investigated the performance of three different types of methods for calculation of sensitivity measures in combination with different sampling schemes on the basis of a PA model for low- and intermediate-level radioactive waste in rock salt. This PA model behaves nearly non-continuously or quasi-discretely as a result of the dissolution and nearly sudden failure of the seal in the near field. When this barrier fails, a sudden high release of radionuclides, i.e., a jump in the model output occurs. The investigated sensitivity analysis methods include a graphical method (CSM plot), a regression-based (SRRC) method and a variance-based method (EASI). With these different methods, a time-dependent analysis was performed, in which the annual dose values for a number of discrete time points, distributed over a time interval of 10^6 years, were analyzed. In addition, results obtained using LpTau sampling were compared to the ones using random sampling. Samples of different sizes with 11 parameters were examined.

For random sampling, the differences between the sensitivity measures obtained with different numbers of runs are bigger compared to those calculated using LpTau sampling. This means that the LpTau-based investigations show satisfying convergence at smaller sample sizes than random-based investigations. For the model under consideration, this confirms the assumption that quasi-random sequences are advantageous for sensitivity analysis. For CSM and SRRC analysis and with LpTau sampling, a sample size of about 2000 for the considered PA model with 11 parameters seems to be sufficient for good convergence of the sensitivity measures. For the EASI method, it appears that the sample size should be about 8000 to obtain comparably smooth curves and good convergence.

The results obtained with CSM and SRRC analysis are fairly comparable. Both methods identify the same parameters as important. While CSM plots give an optical visualization of the sensitivity, the SRRC provide a quantitative measure, although it should be kept in mind that the rank transformation destroys some quantitative information.

The EASI results, however, differ from those obtained with CSM and SRRC. They assign the dominating role to a parameter that seemed to be of secondary importance in the other evaluations. Specifically in the early phase below 10^4 years, the EASI results seem to be of low robustness, and therefore, higher confidence might be given to the SRRC calculations. In this phase, the release of radionuclides is just starting and only a few simulations lead to a non-zero release at all.

The differences between the evaluations with EASI and with SRRC/CSM may be due to the fact that changes in the model output are squared in the computation of variance-based sensitivity measures and that only a few output values may dominate the computed sensitivity indices. In a different investigation we demonstrate that the EASI results get more similar to those obtained with SRRC and more robust in the early phase if an output transformation is applied [6]. However, despite of the transformation, one parameter that obviously has some influence on the model sensitivities was not identified as important by any of the methods under investigation. This parameter is one of those parameters that control the quasi-discrete behavior of the PA model.

From each type of evaluation valuable information could be obtained in terms of parameter ranking and understanding of the system behavior. Consequently, it can be inferred that for the investigated PA model with strong nonlinear system behavior, it is very important to explore the parameter space as homogeneously as possible to obtain robust sensitivity measures. In addition, it is very helpful to perform sensitivity analysis of the model with different types of methods.

Acknowledgements

The work presented in this paper is financed by the German Federal Ministry for Economic Affairs and Energy (BMWi) under sign 02E10941.

References

[1] R. Bolado-Lavin, W. Castaings and S. Tarantola. *"Contribution to the sample mean plot for graphical and numerical sensitivity analysis"*, Journal of Reliability Engineering and System Safety, 94, pp. 1041-1049, (2009).

[2] J. Sinclair. *"Response to the PSACOIN level S exercise. PSACOIN level S intercomparison"*, Nuclear Energy Agency, Organisation for Economic Cooperation and Development, (1993).

[3] A. Saltelli, K. Chan and E. M. Scott. *"Sensitivity analysis"*, Wiley Series in Probability and Statistics, (2000).

[4] E. Plischke *"An effective algorithm for computing global sensitivity indices (EASI)"*. Reliability Engineering and System Safety, 95, pp. 354–360, (2010).

[5] T. Reiche, D.-A. Becker, D. Buhmann and T. Lauke. „*Anpassung des Programmpaketes EMOS an moderne Softwareanforderungen*", Gesellschaft fuer Anlagen und Reaktorsicherheit (GRS) mbH, GRS-A-3623, (2011).

[6] D.-A. Becker. *"Improvement of the reliability and robustness of variance-based sensitivity analysis of final repository models by application of output transformation"*, PSAM12, Honolulu, (2014).

[7] S. M. Spiessl and D.-A. Becker. „*Sensitivity analysis of a final repository model with quasi-discrete behaviour using quasi-random sampling and a metamodel approach in comparison to other variance-based techniques*", Submitted to Journal of Reliability Engineering and System Safety, (2014).

Importance Analysis for Uncertain Thermal-Hydraulics Transient Computations

Mohammad Pourgol-Mohammad[*a], Seyed Mohsen Hoseyni[b]

[a] Department of Mechanical Engineering, Sahand University of Technology, Tabriz, Iran
[b] Department of Basic Sciences, East Tehran Branch, Islamic Azad University, Tehran, Iran

Abstract: Results of the codes simulating transients and abnormal conditions in nuclear power plants are inevitably uncertain. In application to thermal-hydraulic calculations by thermal-hydraulics codes, uncertainty importance analysis can be used to quantitatively confirm the results of qualitative phenomena identification and ranking table (PIRT). Several methodologies have been developed to address uncertainty importance assessment. Existing uncertainty importance measures which are mainly devised for the PRA applications are not suitable for tedious calculations of the complex codes like RELAP. On the other hand, for the quantification of the degree of the contribution of each phenomenon to the total uncertainty of the output, a new uncertainty importance measure that needs affordable computational cost is very promising. A new uncertainty importance measure is introduced in this article to cope with the aforementioned deficiencies of the TH uncertainty importance analysis. Important parameters are identified qualitatively by the modified PIRT approach while their uncertainty importance is quantified by the proposed index. Application of the proposed methodology is demonstrated on LOFT-LB1 test facility.

Keywords: Uncertainty Importance, Uncertainty Analysis, IMTHUA, Thermal-Hydraulics.

1. INTRODUCTION

TH codes are tools for the calculation of the response of nuclear power plant to abnormal and accident conditions. The approach is to compare the figure of merit (as the code output) to the regulator's criteria. However, these predictions are uncertain due to significant sources of uncertainty in fully understanding of physical phenomena occurring during the accident, uncertainties in models due to simplification (including model form and parameter uncertainties), and computational numerical methods approximations. The first step in conducting uncertainty analysis is to identify these sources.

The previous article of the authors (reference [1]) demonstrated a hybrid qualitative/quantitative framework was proposed for the uncertainty analysis plus importance in severe accident calculations. The qualitative phase identifies, ranks and screens the important phenomena in the course of severe accident progression. The quantitative phase covers the contribution of the parameters obtained through the first phase to the total uncertainty of the output variable of interest. To overcome high computational cost in this phase, the code is emulated by using a metamodel of the code model. The obtained metamodel of the complex model could then be easily utilized for calculation of uncertainty importance measures.

However the RSM approach has some limitations. Drawbacks to RSM include:
- Difficulty of developing an appropriate experimental design
- Use of a limited number of values for each input variable
- Possible need for a large number of design points; Ineffective as the number of uncertain parameters increases requiring larger number of code executions
- Difficulties in detecting thresholds, discontinuities, and nonlinearities
- Difficulty in including correlations and restrictions between input variables

---
[*] pourgol-mohamadm2@asme.org

- Difficulty in constructing an appropriate response surface approximation to the model under consideration

The authors concluded in reference [1] that the extension of this area could be devising a new effective uncertainty importance measure that is more suitable for thermal-hydraulics and severe accident uncertainty analysis considering the large computational cost of the calculations. Consequently, the existing methodologies for uncertainty importance are not practical for thermal-hydraulic calculations (e.g., RELAP5 code calculations) due to required computational time and resources. The aim of the present paper is to devise a new effective uncertainty importance measure that is more suitable for TH uncertainty analysis considering the high cost of the calculations.

2. DESCRIPTION OF THE PROPOSED METHODOLOGY

Several methodologies have been developed to address uncertainty importance assessment in general. Existing methodologies for uncertainty importance are not efficient for TH applications which require significant amount of computational time and resources. Due to large uncertainty resources and time consuming nature of TH code calculation, TH uncertainty and sensitivity calculations require enormous number of code calculation and significant computational resources to estimate their uncertainty importance. An efficient uncertainty importance ranking method is developed here for comprehensive TH code uncertainty assessment [1-5]. The proposed uncertainty importance methodology is a hybrid two-phase qualitative/quantitative method. The first phase is qualitative step to identify and rank phenomena and processes based on their TH and uncertainty importance. The qualitative step, itself two stages, (so called modified PIRT) identifies, ranks and monitors the sources of uncertainties based on their impact and uncertainty importance. The second phase is a quantitative step to measure the effect of uncertainty sources on code output uncertainty distribution. The steps of the methodology are discussed in following sections. A flow chart of the hybrid methodology is shown in Figure 1.

2.1 Qualitative Phase (Modified PIRT)

With many physical phenomena involved, TH analyses deal with various sources of uncertainties. While ideally all sources of uncertainties should be considered in the analysis explicitly [4], it is neither practical nor necessary to evaluate all processes and components in detail. The original PIRT process aims to identify and rank phenomena and processes based on their safety importance only. For the purpose of uncertainty analysis, this step is necessary but not adequate. The phenomena may be important from the TH as well as in respect to uncertainty importance. The degree of knowledge about phenomena and credibility of models must be characterized, and when possible quantified. This paper suggests a methodology for involving level of knowledge of each phenomenon into the problem for more effective uncertainty assessment.

The proposed two-step PIRT methodology here called "modified PIRT" provides a process for more precise uncertainty analysis. The process identifies and ranks phenomena based on TH importance as well as uncertainty importance. Experience with TH phenomena shows that phenomena with TH and uncertainty importance contribute more significantly to output uncertainty than those based on either TH importance or uncertainty importance alone. The analytical hierarchical process (AHP) has been used as a formal approach for TH identification and ranking. AHP [6] is a powerful tool for ranking of alternatives and attributes of a decision, especially when limited experts are available. A formal uncertainty importance technique is used to estimate the degree of credibility of the TH model(s) for the important phenomena. This part uses subjective assessment on the basis of evaluating available information and data from experiments on code predictions. The idea is shown in Figure 1a.

Figure 1b shows several phenomena with their TH and uncertainty importance. By uncertainty importance, we mean the level of contribution of the phenomena to the uncertainty in the code prediction (for a given figure of merit). For example, decay heat power is considered high in its TH importance due to its impact on PCT itself. The phenomenon is well known, and correlations to predict it are well developed. Therefore, low uncertainty is assigned to it, indicating a high confidence in the phenomena model used in TH codes. Loosely speaking, TH importance impacts the output's mean value, while uncertainty importance affects its variance. There are different qualitative and quantitative approaches to assigning ranks to phenomena. Rankings of high, medium, and low are used in some studies, while others use ranking on scale of 1 to 9, where 9 means the highest importance and 1 is the lowest. A detailed description of the modified PIRT is provided in [1-5].

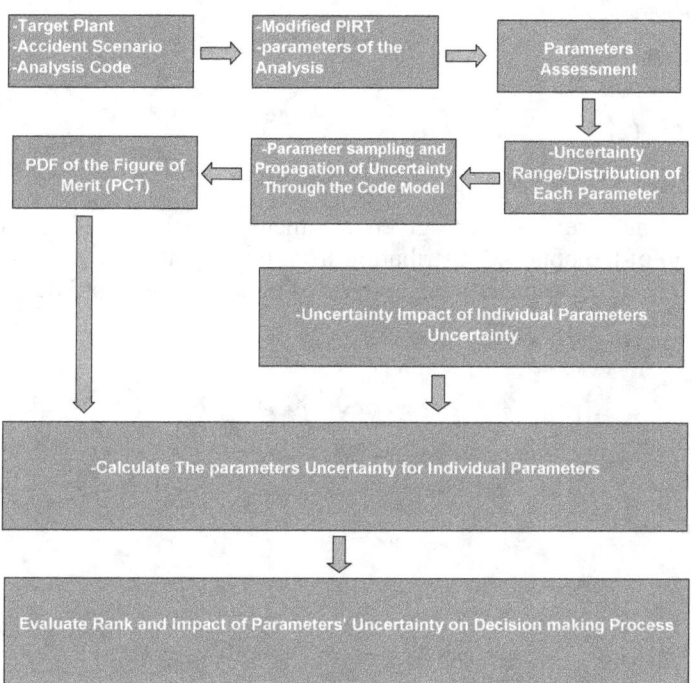

Figure 1: The Methodology Flow Chart

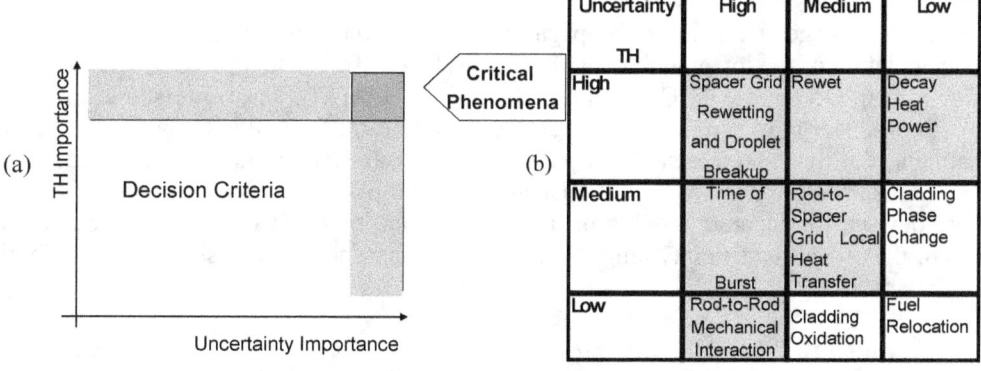

Figure 2: TH Importance vs. Uncertainty Importance in Chosen Criteria b) Some Phenomena with their TH and Uncertainty Ranks

2.2 Quantitative Phase (Uncertainty importance Calculation)

Every model of interest, Thermo-hydraulics code here, can be represented as a function of the form: $y = f(x)$, where $x=[x_1, x_2, ..., x_n]$ is a vector of uncertain analysis inputs and $y=[y_1, y_2, ..., y_n]$ is a vector of analysis results.

The proposed uncertainty importance measure is defined in multiples of standard deviation ($x\sigma$, $x=...,-2,-1,0,1,2,...$) changes in a given input parameter over the change in the output (Figure of Merit) as shown in Figure 3. For example, x is the number of σ's in the FOM (e.g., ΔPCT) resulting from running the code for +1σ and -1σ change in the input parameter nominal values. An average of 2, 4, or 6σ importance measure can also be used for the analysis but this should be applied uniformly for all uncertainty parameters. The measure can be defined as ratio of standard deviations.

$$IM = \frac{Uncertainty\,of\,Papametri}{Uncertainty\,of\,the\,Ouput} = \frac{\rho_{Pi}}{\rho_{out}} \qquad (1)$$

The term ρ is defined as:

$$\rho = \frac{\sigma}{\mu} \qquad (2)$$

where μ and σ are variable's mean and standard deviation.

IM is the importance measure, ρ_{Pi} is the given parameter coefficient of variation, and σ_{out} is the coefficient of variation of the obtained distribution from uncertainty propagation and assessment [3]. The uncertainty measure can also be defined the ratio of parameter standard deviation (σ_{P_i}) to overall FOM standard deviation (σ_{out}).

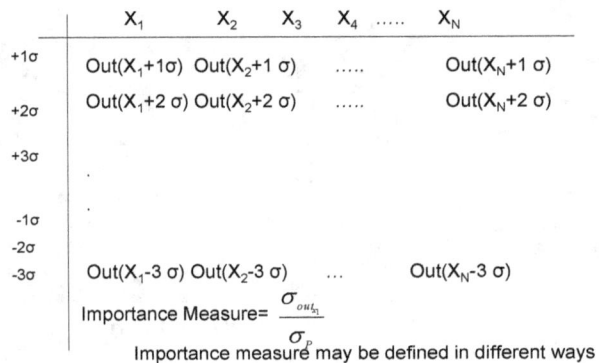

Figure 3: Perturbation of the input parameters for uncertainty importance assessment

The total uncertainty range resulted from propagation of uncertainties is obtained from the input-based uncertainty calculation of the integrated methodology IMTHUA [4-5], developed by the author or any other available methodologies e.g., CSAU [7], GRS [7], UMAE [7]. There are some difficulties in assessment of non-linearity of some input changes vs. variations in the output variables, which require special treatment. For more precise study of parameter uncertainty importance, the method proposed by Iman [8] furnishes more accurate results. Different levels of input change (multiples of standard variation) are devised for accurate ranking of uncertainty contributors. Comparing the output change as a fraction of the overall uncertainty range will result in a ranking index to show the contribution of each uncertainty source.

3. Application on LOFT LB-1 Experiment

A schematic view of LOFT test facility is shown in Figure 4. Components used in LOFT are similar in design to those of a PWR. Because of scaling and component design, the LOFT is expected to closely model a PWR LOCA. The facility is designed and scaled to represent a 1/60-scale model of a typical 1000-MWe commercial four-loop PWR. Three PWR primary-coolant loops are simulated by a single intact loop in LOFT scaled to have the same volume-to-power ratio. A broken loop in LOFT simulates the fourth PWR primary-coolant loop where a break may be postulated to occur. The facility includes most of components in a typical 4-loop nuclear power plant consisting of five major systems of: 1) Primary Coolant System, 2) The Reactor System with 1.68m nuclear core, 3) Blowdown Suppression System, 4) Emergency Core Cooling System, and 5) Secondary Coolant System.

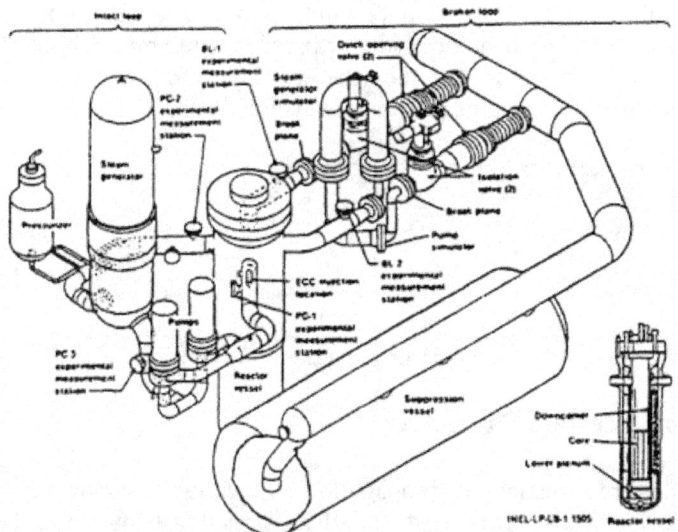

Figure 4: The LOFT Test Facility [14]

With recognition of the differences in commercial PWR designs and inherent distortions in reduced scale systems, the design objective for the LOFT facility was to produce the significant thermal-hydraulics phenomena that would occur in commercial PWR systems in the same sequence and with approximately the same frames and magnitudes [15].

3.1 Parametric Uncertainty Quantification

The input uncertainty quantification is focused on the identification of uncertainties in code structure (including model and parameters). These uncertainties are propagated through code calculations to arrive at a distribution of output uncertainty on specified figures of merit. Sources of uncertainty in "input" include values of model parameters, boundary/initial conditions, and uncertainties in structure of sub-models (sub-model uncertainty).

In the first step of uncertainty propagation, for each of the identified sources of uncertainty from the previous steps, a probability distribution is assigned. Figure 5 schematically illustrates the process of uncertainty propagation.

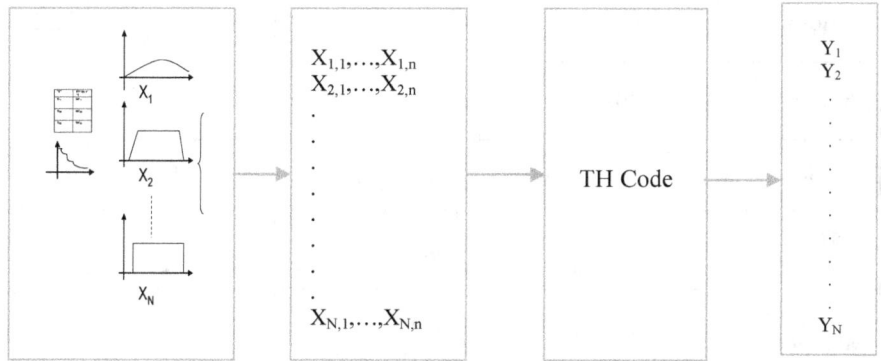

Figure 5 Sampling and Propagation of Uncertainties in Order Statistics Based Frameworks

3.2 Distribution Assignment for Uncertain Input Parameters

By input, models form and parameters are meant in this research. The main calculation for uncertainty assessment is performed in this stage of the work. In the first step of uncertainty propagation, for each of the identified sources of uncertainty from the previous steps [5-6], a probability distribution is assigned. For any parameter of interest, the range and the form of the distribution is determined. This range is used for the sampling in the next stages of the work and is one of the major steps in the quantification of uncertainties. Based on our information and the available data and knowledge on the phenomena or model or parameter the range of uncertainty is identified. If the available data and information is little then the uncertainty range will be large in opposite to the case of information abundance about the phenomena which results in the smaller uncertainty range.

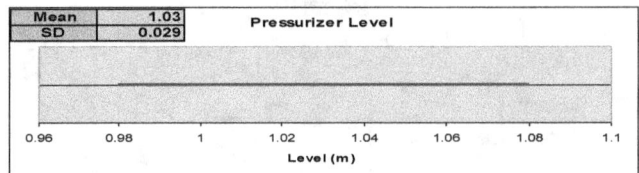

Figure 6: Uncertainty Range for Pressurizer Level in LOFT Test Facility with μ= 1.03 m and σ=0.029 m

3.3 Sampling from Uncertain Parameters

Total of 100 samples are generated for propagation of uncertainty parameters to the output. Figure 7 demonstrates how the samples are generated from the assigned distribution of the parameter. Pair-wise dependency between parameters is not considered in generating the samples in this stage of the research. If there was a significant dependency between parameters, it is included in obtaining the samples. Data-informed dependency calculation is the most common way to calculate dependency in domain of complex code calculation.

Table 1 lists all LOFT-LB1 uncertain parameters with their uncertainty characteristics.

Table 1: Uncertain Parameters for LOFT LB-1 Experiment

Parameter Name	Distribution Type	Nominal Value	Lower Bound	Higher Bound	Mean Value	Standard Deviation
Pressurizer Level (m)	Uniform	1.04	0.98	1.08	1.03	0.03
Pressurizer Pressure (MPa)	Uniform	14.92	14.81	15.03	14.92	0.06
Initial Core Power (MWt)	Uniform	49.3	48.1	50.5	49.3	0.61
Accumulator Level (m)	Uniform	2.362	2.337	2.387	2.362	0.01
Accumulator Pressure (MPa)	Uniform	4.22	4.05	4.39	4.22	0.09
Safety Injection Temperature (°K)	Uniform	302	296	308	302	3.06
Break discharge coefficient	Uniform	1.0 (default)	RC x 0.70	RC x 1.15	1.0	0.13
Peaking factor	Normal Multiplier	1.0	0.95	1.05		0.0255
Gap size	Normal Multiplier	1.0	0.8	1.2		0.102
UO2 conductivity	Normal Multiplier [0.9, 1.1] (Tfuel <2000 K) [0.8,1.2] (Tfuel >2000 K)	1.0	0.9	1.1		0.051 for (Tfuel <2000 K) 0.102 for (Tfuel >2000 K)

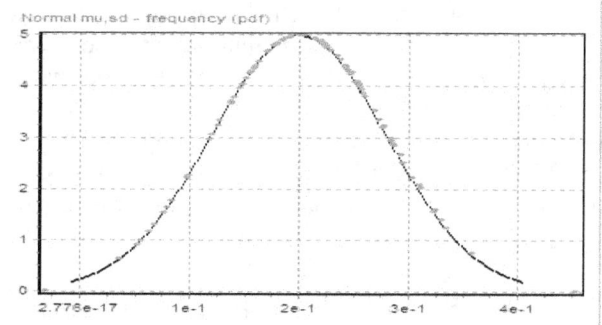

Figure 7 100 Samples from Normal distribution; an example

3.4 Uncertainty Propagation

In this application, we start with the results of the "input phase uncertainty propagation," developed in detail in reference [4]. In that exercise, the RELAP5 code structure and parameter uncertainties were explicitly propagated to obtain uncertainty scatters for the hottest fuel rod at 0.66m height in the active core. Figure 8 illustrates peak clad temperature (PCT) profile and compares the results of 100 code runs for different values of uncertain parameters and sub models.

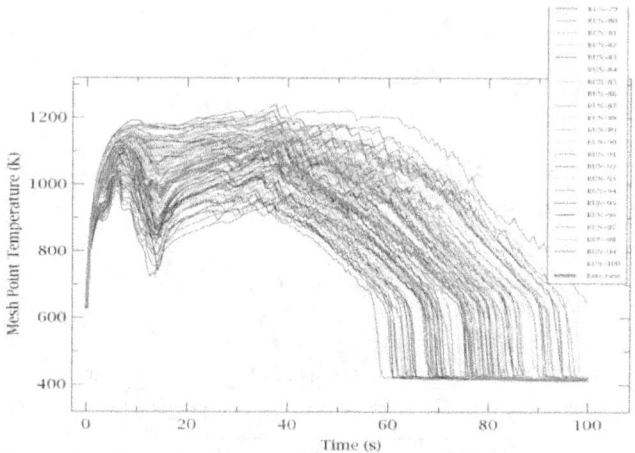

Figure 8: Uncertainty Propagation Results from Input Phase vs. Experimental Data

Figure 9 shows scatter plots for data obtained from RELAP5 uncertainty calculation. The scatter points are used to develop an "input phase" uncertainty distribution of PCT.

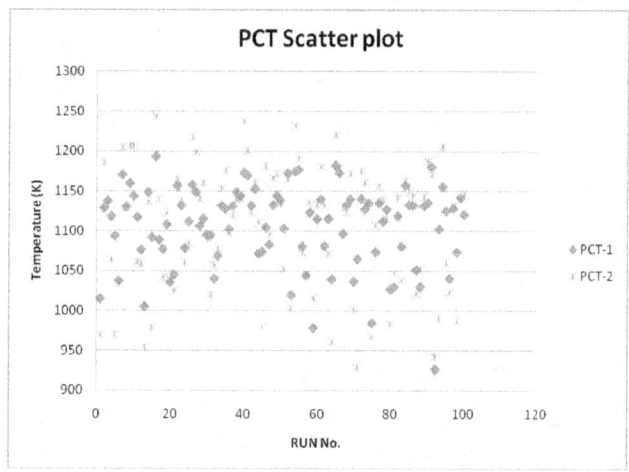

Figure 9: Scatter Plot for Peak Clad Temperature

Figure 10 shows the results of fitting truncated normal distributions separately to the RELAP5 code data. Two methods can be used to fit a parametric distribution to the code calculations:

I. A distribution shape that best fits the data is assumed (e.g., normal or lognormal distribution). With the distribution considered, we estimate parameters from the data. The range of the distribution from tolerance interval is assigned to distribution quantiles based on coverage (e.g. the smallest used as 2.5% and the largest as 97.5%, see Figure 4). The parameters of distributions are obtained from these quantiles. By doing so we are trying to preserve the information from order statistics based tolerance limit.

$$\frac{x_{0.95} - \mu}{\sigma} = 1.96$$
$$\frac{x_{0.05} - \mu}{\sigma} = -1.96 \quad (3)$$

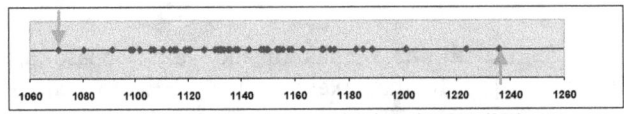

Figure 10: RELAP5 Calculated PCT (°K)

II. After a distribution shape is assumed for the data, Bayes' theorem is used to estimate the distributions of the parameters of the distribution based on the code-calculated date points. This is formally expressed as

$$\pi(\mu,\sigma \mid T_1, T_2, \dots, T_N) = \frac{L(T_1, T_2, \dots, T_N \mid \mu, \sigma)\pi_0(\mu,\sigma)}{\iint_{\mu,\sigma} L(T_1, T_2, \dots, T_N)\pi_0(\mu,\sigma)} \quad (4)$$

Where μ and σ are parameters of the assumed distribution, N is number of code runs, L is likelihood function of data and $\pi_0(\mu,\sigma)$ is the prior distribution of the parameters. The Bayes fit is then obtained through

$$\pi(T \mid T_1, T_2, \dots, T_N) = \iint_{\mu,\sigma} \pi(T \mid \mu, \sigma)\pi(\mu, \sigma \mid T_1, T_2, \dots, T_N) \quad (5)$$

Table 2: Statistical parameters in LOFT LB1 Uncertainty Analysis

Statistical Parameter	PCT1	PCT2
Sigma	50.62	76.44
Mean	1105.89	1106.75
Min	926.16	928.37
Max	1192.76	1242.69
Ref.	1116.27	1101.45

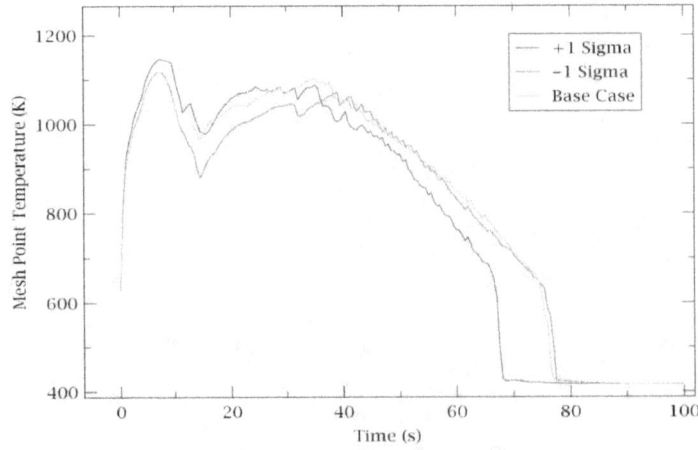

Figure 11: Axial PPF Effect on PCT

The approach is to use smallest and biggest value in the ordered sample as 5th and 95th percentiles as shown in equation (1), where μ and σ are two parameters of the normal distribution.

The results for axial power peaking factor parameter uncertainty importance analysis are given in Figure 11 which shows the uncertainty range introduced by this parameter. Table 3 lists the calculations of the effects of the changes in each input parameter on "PCT1", "PCT2" and "end of quench time" as the code calculated outputs. Here the proposed uncertainty importance measure is only applied on the PCT as the figure of merit of the problem.

Each parameter was perturbed 2 times in multiples of standard deviation values and the resulting changes in the PCT1 and PCT2 were recorded. With total uncertainty range for PCT from uncertainty quantification, uncertainty is calculated for the parameter.

Table 3: Effect of each parameter on different FOMs

Parameter	-1 Sigma			+1 Sigma		
	PCT1 (K)	PCT2 (K)	End of Quench (s)	PCT1 (K)	PCT2 (K)	End of Quench (s)
Pressurizer Level (m)	1116.2	1068.8	68.0	1115.5	1106.5	77.5
Pressurizer Pressure (MPa)	1118.7	1116.79	79.5	1116.03	1073.46	69.02
Initial Core Power (MWt)	1109.7	1095.5	72.5	1120.7	1136.89	77.0
Accumulator Level (m)	1116.27	1109.8	76.0	1116.27	1098.79	77.0
Accumulator Pressure (MPa)	1116.27	1107.8	77.0	1116.27	1111.58	78.0
Safety Injection Temperature (°K)	1116.27	1105.14	76.0	1116.27	1108.25	76.0
Break discharge coefficient	1117.5	1102.27	69.0	1112.43	1123.87	85.5
Peaking factor	1117.25	1070.36	77.5	1146.10	1087.13	68.0
Gap size	1098.37	1030.02	67.0	1130.71	1100.43	77.0
UO2 conductivity	1120.71	1088.52	69.5	1104.95	1040.58	67.5

4. DISCUSSION ON THE OBTAINED RESULTS

Importance of each parameter could be calculated in multiples of 2, 4, 6 sigma. The average value is calculated for uncertainty importance of the parameter. Each single value can also be used as the importance measure. The values for importance measure are relative (importance measure value for a parameter is compared with the importance of other components). Non-linearity of the TH code is another reason for calculation in different level of variation. In some cases non-proportionality of variation level in parameter with output variation was experienced.

Calculated uncertainty importance values are summarized in Table 4. We well know from LBLOCA phenomenology that PCT1 is during blow down phase while PCT2 occurs in refill phase of the accident. With this physical representation, accumulator parameters are not affecting PCT1 that is confirmed by the uncertainty importance measure. The highest value of uncertainty importance is for gap size meaning that if we need to invest in uncertainty reduction, gap size parameter should be a higher priority than others.

There are some difficulties in assessment of non-linearity of some input changes vs. variations in the output variables, which require special treatment. We believe that although the MC based approaches could furnish more accurate results but because of their large calculation cost are not applicable for TH applications unless the surrogate model of the complex model is used that it could be itself as an another uncertainty source in some cases. The proposed uncertainty importance measure calculates locally the effect of each parameter and gives promising results in TH application.

Table 4: Calculated uncertainty importance measure values

Parameter	PCT1 Importance %	PCT2 Importance %
Pressurizer Level	0.73	12.67
Pressurizer Pressure	2.76	14.467
Initial Core Power	11.46	13.557
Accumulator Level	0	3.64
Accumulator Pressure	0	1.25
Safety Injection Temperature	0	1.03
Break discharge coefficient	5.28	7.09
Peaking factor	29.61	5.68
Gap size	33.70	24.16
UO2 conductivity	16.45	16.46

5. CONCLUDING REMARKS

This paper summarizes a new framework for the quantification of the effect of uncertainty sources on the code output distribution. It was discussed that for uncertainty analysis plus importance in TH problems, the existing uncertainty importance measures are not computationally affordable. A new uncertainty importance measure is proposed here to overcome this limitation with minimal computational burden. Successful application of the proposed framework is demonstrated for LOFT LB1 large LOCA experiment. The extension of this work could be development of the mathematical proof of the proposed measure.

6. NOMENCLATURE

CSAU	Code Scaling, Applicability and
FOM	Figure of Merit
IMTHUA	Integrated Methodology on TH
LWR	Light Water Reactor
LOCA	Loss of Coolant Accident
LOFT	Loss of Flow test
MPIRT	Modified Phenomena Identification and
MC	Monte Carlo
NPP	Nuclear Power Plant
PCT	Peak Clad Temperature
PEC	Primary Evaluation Criteria
PIRT	Phenomena Identification and ranking
PSA	Probabilistic Safety Assessment
PWR	Pressurized Water Reactor
RSM	Response Surface Methodology
TH	Thermal-Hydraulics
USNRC	United States Nuclear Regulatory Commission

References

[1] Hoseyni S.M., Pourgol-Mohammad M., Abbaspour Tehranifard A. Yousefpour F.. "A systematic framework for effective uncertainty assessment of severe accident calculations; Hybrid qualitative and quantitative Methodology", Reliability Engineering and System Safety 125 (2014) 22-35, Special issue of selected articles from ESREL 2012.

[2] Hosseini S.M., Pourgol-Mohamad M., Abbaspour Tehranifard A., Yousefpour F. "Phenomena Identification and Ranking for Severe Accident Uncertainty Assessment; a Systematic, Two-Dimensional Approach." PSAM11, ESREL2012. 2012.

[3] Pourgol-Mohamad M., et. al. "Modified Phenomena Identification and Ranking Table (PIRT) for Uncertainty Analysis." *14th Int. Conf. Nuclear Engineering (ICONE 14).* 2006.

[4] Pourgol-Mohamad, M. *Integrated Methodology on Thermal Hydraulics Uncertainty Analysis (IMTHUA).* PhD Thesis, University of Maryland, 2007.

[5] Pourgol-Mohamad, M. "Integrated Methodology for Thermal-Hydraulic Code Uncertainty Analysis with Application." *Nuclear Technology* 165 (2007): 333-359.

[6] Saaty TL. Models, methods, concepts & applications of the analytic hierarchy process. Boston: Kluwer Academic Publishers; 2001.

[7] Pourgol-Mohammad M., 2009, "Thermal–hydraulics system codes uncertainty assessment: A review of the methodologies", Annals of Nuclear Energy 36 (2009) 1774–1786

[8] Iman, R.L., M.E. Johnson, and T.A. Schroeder. "Assessing hurricane effects. Part 2: Uncertainty analysis." *Reliability Engineering and System Safety* 78, no. 2 (2002): 145–153.

[9] OECD/NEA. "CSNI-R2006-2, BEMUSE Phase II Report", 2006.

[10] OECD/NEA. "CSNI-R2004-4, BEMUSE Phase III Report", 2004.

[11] OECD/NEA. "CSNI-R2008-6, BEMUSE Phase IV Report", 2008.

[12] OECD/NEA. "CSNI-R2009-13, BEMUSE Phase V Report", 2009.

[13] OECD/NEA. "CSNI-R2011-4, BEMUSE Phase VI Report", 2011.

[14] FZKA 6424, "Validation of the Reflood Model of RELAP5/MOD3.2 Gamma Using Experimental Data from the Integral Facility LOFT LP-LB-1"

[15] US-NRC, NUREG/IA-0089, "Post-Test-Analysis and Nodalization Studies of OECD LOFT Experiment LP-LB-1 With RELAP5/MOD2"

Insights from an Integrated Deterministic Probabilistic Safety Analysis (IDPSA) of a Fire Scenario

M. Kloos[a], J. Peschke[a], B. Forell[b]
[a] GRS mbH, Boltzmannstr. 14, 85748 Garching, Germany
[b] GRS mbH, Schwertnergasse 1, 50667 Cologne, Germany

Abstract: For assessing the performance of fire fighting means with emphasis on human actions, an integrated deterministic probabilistic safety analysis (IDPSA) was performed. This analysis allows for a quite realistic modelling and simulation of the interaction of the fire dynamics with relevant stochastic influences which refer, in the presented application, to the timing and outcome of human actions as well as to the operability of technical systems. For the analysis, the MCDET (*M*onte *C*arlo *D*ynamic *E*vent *T*ree) tool was combined with the FDS (*F*ire *D*ynamics *S*imulator) code. The combination provided a sample of dynamic event trees comprising many different time series of quantities of the fire evolution associated with corresponding conditional occurrence probabilities. These results were used to derive exemplary probabilistic fire safety assessments from criteria such as the temperatures of cable targets or the time periods with target temperatures exceeding critical values. The paper outlines the analysis steps and presents a selection of results which also includes a quantification of the influence of epistemic uncertainties. Insights and lessons learned from the analysis are discussed.

Keywords: IDPSA, Aleatory and Epistemic Uncertainty, Fire PSA, MCDET, FDS.

1. INTRODUCTION

Fire in a nuclear power plant (NPP) can deteriorate or even damage the required function of systems, structures and components (SSC) which are important to safety. Therefore, the investigation of the fire protection concept of a plant is an important issue of nuclear safety analyses. A recent research project at GRS aimed at demonstrating how a combination of the MCDET tool [1] and the deterministic fire simulation code FDS [2] can be used to perform an IDPSA in the frame of a probabilistic fire risk analysis.

An IDPSA is particularly suitable for those aspects of a probabilistic fire risk analysis which require a realistic modelling and assessment of the interaction between the fire dynamics and stochastic influencing factors in the course of time. Such stochastic factors refer to the timing and outcome of the human actions performed for fire fighting as well as to the operability of those systems and components which are designated to be applied for fire detection, alarm, confinement and suppression. An IDPSA permits to derive safety assessments from process criteria such as the temperatures of cable targets or the time periods with target temperatures exceeding critical values.

For the IDPSA presented in this paper, a combination of the MCDET tool and the FDS code was applied. With MCDET, a mixture of Monte Carlo (MC) simulation and the Dynamic Event Tree (DET) approach can be performed. What makes the MCDET tool particularly useful for fire risk analysis is its Crew Module which allows for considering the human actions performed for fire fighting as a dynamic process [3].

The fire scenario selected to be analyzed deals with the fire fighting means designated to be applied in a reference NPP once a fire occurs for any reason. One major assumption is the failure of the automatic actuation of the fixed fire extinguishing system in the compartment where the fire starts. This implies that fire fighting mainly depends on the operability of the fire detection and alarm system, the performance of human actions as well as on the operability of active fire barrier elements (e.g. fire dampers and fire doors) and of the fire extinguishing systems which can be manually actuated.

The first steps of the analysis which were already presented in [4] mainly focused on the performance of the human actions to be applied for fire fighting without running the FDS code. In those situations where human actions had to be considered as dependent on the fire evolution, case-by-case analyses were performed. For instance, human actions were analyzed for the cases that smoke is visible or not, when the shift personnel reach the fire compartment. The computational effort of those case-by-case analyses was negligible, since the MCDET tool making use just of the Crew Module without applying another dynamics code runs very fast. Results of the analyses were conditional distributions referring to the timing of human actions such as the distribution of the time period between fire alarm and the arrival of fire fighters at the fire compartment door or the distribution of the time period between the arrival at the fire compartment door and the beginning of fire extinguishing. These distributions express the stochastic variability (aleatory uncertainty) of corresponding time variables and were used as input to the analysis presented in this paper.

Subject of the exemplary IDPSA presented here was the interaction of the fire dynamics as simulated by FDS with the main stochastic factors affecting the fire over time. Results provided by the analysis include the distribution of the time elapsed from fire ignition to its successful suppression and, what is even more interesting, distributions referring to the temporal evolution of the temperatures of safety related targets (e.g. cable targets). The influence of epistemic uncertainties (due to lack of knowledge) on these distributions is considered.

Chapter 2 of this paper gives an overview on the method and the modules of the MCDET tool. Chapter 3 presents information on the fire scenario including the scenario-specific model specified as input to FDS and the fire fighting means investigated in detail. The steps of the IDPSA and a selection of analysis results are described in Chapter 4. Conclusions are presented in the last chapter.

2. MCDET TOOL

2.1. Method

Coupled with a deterministic dynamics code such as FDS, the MCDET tool performs a combination of the Dynamic Event Tree (DET) approach and Monte Carlo (MC) simulation [1]. This combination is able to treat any kind of uncertainty. Aleatory uncertainties due to stochastic influences referring to discrete quantities (e.g. success/failure of the fire detection and alarm system to operate; success/error of human actions) are handled by the DET approach, whereas aleatory uncertainties referring to continuous quantities (e.g. execution times of human actions) are handled by MC simulation. MC simulation is also applied to consider epistemic uncertainties (due to lack of knowledge). The sampling of values for continuous aleatory variables is not performed a priori, i.e. before the calculation of a DET is launched. It is performed when needed in the course of the calculation. In this way, it is possible to treat the influence of the dynamic process on aleatory uncertainties and to consider, for instance, a higher failure rate of a component, if a high temperature seriously aggravates the condition of the component.

Output of MCDET simulations is a large spectrum of different sequences, for instance, of the evolution of the fire. It can be considered as a sample of individual DETs, each constructed from a set of values sampled for the continuous aleatory variables. From the time series of safety relevant process quantities (e.g. temperatures of cable targets) and the corresponding likelihoods available for each sequence, conditional DET-specific and, by the application of statistical methods, unconditional scenario-specific likelihoods of damage states can be calculated.

2.2. Modules

The Probabilistic Module of the MCDET tool implements algorithms referring to the evaluation of the simulation state, the sampling of points in time for branching points with run-time failures, the generation of the initial states of new simulation paths and the calculation of corresponding occurrence probabilities. The module is linked to the Scheduler Module supervising all simulation processes of a deterministic code. The MCDET Scheduler can be coupled with any dynamics code which allows somehow for reading and modifying the simulation state and which is able to terminate a simulation early, because for instance a probabilistic cut-off criterion is pending. The Driver Module implements a generic interface for communicating with the simulation process and maps the abstract interface commands to whatever is needed, making the simulator process perform the requested action.

While the new version of the Scheduler Module allows for performing parallel and distributed simulations on multicore workstations and cluster environments [5], the previous version just permits to calculate the branches of a DET in one process. The sequential calculation of the branches of a DET is performed according to the last-in-first-out principle. Different DETs can be calculated in parallel – each on a separate computing node. Simulations with the old version of the Scheduler Module make extensive use of the restart capabilities of the applied dynamics code.

The extra Crew Module of the MCDET tool allows for simulating human actions as a dynamic process which evolves over time within a context given, for instance, by the intermediate results of a deterministic dynamics code [3], [4]. With the combination of the Crew Module and the deterministic code, an integral simulation of the mutual dependencies between human actions, the system behavior and the process dynamics can be performed. Parameters of the Crew Module which are subjected to epistemic or aleatory uncertainty can be easily handled by the Probabilistic module of MCDET.

The execution of the Crew Module requires a compilation of the action sequences potentially applied during a scenario. An action sequence is composed of one or more simple basic actions, each defined by an ID and additional information on the crew member performing the action, the time needed to execute the action, and on the system, component or operator somehow affected by the action. An action sequence is activated, if the condition attached to the sequence is fulfilled. A condition is defined, for instance, by the state of alarms and indicators in the control room, the system state or by the action sequence previously performed. Information on alarms and indicators has to be provided in an appropriate dataset of the Crew Module.

The post-processing modules of the MCDET tool are useful for evaluating the huge amount of data from MCDET simulations. They can provide graphical representations of the sequences of a DET in the time-event space (temporal order of the events of each sequence) or in the time-state space (temporal evolution of a sequence with respect to a process quantity). In addition, they can calculate and graphically represent the conditional DET-specific distribution and the unconditional application-specific distribution of a process quantity.

3. FIRE SCENARIO

Subject of investigation are the fire fighting means designated to be applied in a German reference NPP in case of fire. The fire was assumed to initially occur in a compartment with cooling and filtering equipment for pump lubrication oil. Besides this process equipment, the compartment is equipped with electrical cables routed below the ceiling. These cables partly carry out safety related functions.

It was assumed that malfunction of the oil-heating system designated to heat up the pump lubrication oil in the start-up phase of the NPP leads to an ignition of the oil. The evolution of the fire was considered to depend on the leakage rate of the oil. The heat release rate may be limited by the oxygen available in the compartment.

The dimensions of the compartment are about w x l x h = 8 m x 6.2 m x 6 m. Compartment walls are made from concrete. The compartment is divided into a lower and an upper level by a steel platform at a level of 2.4 m. The steel platform can be reached by steel stairs. The three compartment doors lead into the lower level of the compartment. It was assumed that one of those doors is left in open position with a reference probability of 0.005 (see Table 1). The mechanical air exchange by an inlet air duct and an outlet duct was supposed to be 800 m^3/h. The fire damper at the outlet duct was supposed to close after melting of a fusible link at 72 °C. This mechanism was assumed to fail with a reference probability of 0.01 (see Table 1). If the outlet damper is closed, the mechanical air supply into the room was considered to be reduced to 400 m^3/h. That value was chosen to account for increased pressure losses, if the inlet air leaves the room by other leaks.

The fire was assumed to start on the steel platform at the electrical oil heater. The evolution of the fire was considered to be linear with the time. The characteristic time to reach 1 MW heat release rate was varied from 250 s to 700 s (see Table 1). In the upper layer cable tray, the cable targets were exposed to hot smoke and radiation.

3.1. Scenario-specific Model

The fire simulation was performed with FDS (FDS 6.0) [2]. The compartment was discretized in one mesh with a grid solution of 0.2 m in all three directions (see Figure 1).

Figure 1: Snapshot of the Compartment Layout by FDS

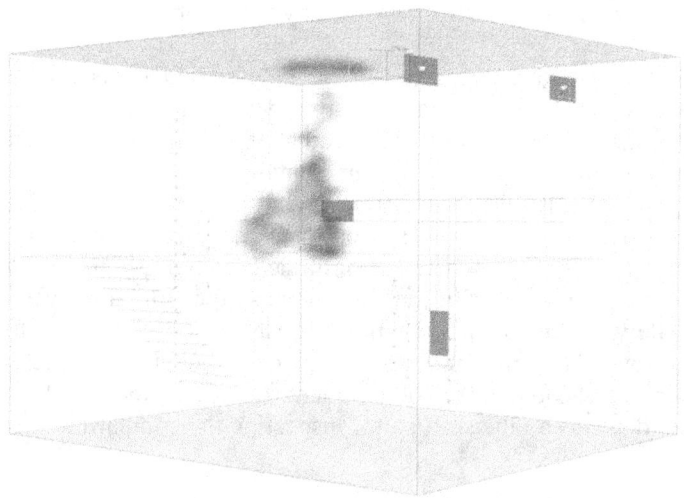

The air inlet duct (violet in Figure 1) has one diffusor above the fire and one into the lower level. The outlet duct (yellow) sucks from the upper layer by two diffusors which can be closed by the fire damper. The mechanical air supply was assumed to allow for a heat release rate of about 800 kW inside the compartment that is reduced to 400 kW after shutting the damper. The inlet air damper was supposed not to close, because the soldered strut is kept cold.

The thermal penetration of the cable material was described by the model for thermally induced electrical failure (THIEF) implemented in FDS. The THIEF model predicts the temperature of the inner cable jacket under the assumption that the cable is a homogeneous cylinder with one-dimensional heat transfer. The thermal properties – conductivity, specific heat, and density – of the assumed cable are independent of the temperature. In reality, both the thermal conductivity and the specific heat of polymers are temperature-dependent. In the analysis, conductivity, specific heat, density, and the depth of the cable insolation were considered as uncertain parameters with relevant influence (Table 1).

The local optical density D of the smoke at 3.20 m height of the fire compartment was assumed to be an important factor affecting the time needed to suppress the fire. For optical densities below $D = 0.1$ m^{-1}, it was assumed that the personnel in charge of fire fighting can detect the fire and start to suppress it by means of a portable fire extinguisher after a delay of 10 s. For $0.1\ m^{-1} < D < 0.4\ m^{-1}$, it is assumed that the delay time until the fire source is detected and the suppression can be started increases, when the optical density rises. The delay time was assumed to be D times 100 in seconds. For $D \geq 0.4\ m^{-1}$, fire suppression with a portable fire extinguisher and without any personal protective equipment was supposed to be impossible. More details are given in section 3.2. The threshold value of $0.4\ m^{-1}$ for the optical density was considered as a relevant parameter subjected to epistemic uncertainty (Table 1).

Once the fire suppression was started, the simulation of a fire sequence was stopped as soon as the temperature inside the cable jacket fell below 120 °C. Otherwise, a sequence was calculated up to 1800 s after fire ignition.

3.2. Fire Fighting Means

If equipment and procedures work as intended, fire fighting is a rather short process, because the compartment where the fire is assumed to occur is equipped with a fixed fire extinguishing system which suppresses the fire with a sufficiently large amount of water after actuation by the fire detection and alarm system. However, one major assumption of the analysis presented here was the failure of the automatic actuation of the fixed fire extinguishing system. This implies that the fire has to be extinguished by manual fire fighting means performed by the plant personnel in charge.

Three states of the fire detection and alarm system can be assumed as decisive for the human actions to be performed for fire fighting, namely at least two detectors, only one detector or none of the detectors indicating an alarm signal to the control room. If at least two fire detectors send an alarm signal, the control room operator (shift leader) immediately instructs the shift fire patrol and the on-site plant fire brigade to inspect the compartment and to perform the necessary steps for fire suppression. If there is a signal by only one detector, the signal might be a faulty or spurious one (e.g. due to dust, steam, etc.). This is why the fire patrol trained for fighting incipient fires is instructed to inspect the fire compartment and to verify the fire. Suppose the fire patrol verifies the fire, the shift leader who is immediately informed calls the fire brigade. In the meantime, the fire patrol tries to suppress the fire either by a portable fire extinguisher or by manually actuating the stationary fire extinguishing system from outside the fire compartment. If none of the fire detectors sends an alarm, the detection of the fire depends on the shift patrol inspecting the compartment at a random time once during a shift.

The fire patrol is always the first person who arrives at the fire compartment. His/her success of suppressing the fire with a portable fire extinguisher was assumed to be depending on the local optical density D of the smoke at 3.20 m height (0.80 m above the level of the platform). If the optical density is too high (see section 3.1), fire suppression by the fire patrol inside the compartment is assumed to be impossible due to reduced visibility and irritant smoke effects on eyes and breathing organs. The fire patrol does not wear personal protective equipment. In this case, his only chance of successfully fighting the fire is to manually actuate the fixed fire extinguishing system. If it does not operate as intended, the fire brigade has to extinguish the fire with their equipment.

Besides the reliability of the fire detection and alarm system and the performance of human actions, the success of fire fighting mainly depends on the reliability of active fire barrier elements such as fire dampers or doors and of the fire extinguishing systems which can be manually actuated.

4. EXEMPLARY IDPSA

4.1 Steps of the Analysis

In the first steps of the analysis, the human actions designated to be applied in case of a fire were investigated in detail [4]. The dynamic model of these actions was constructed on the basis of documents from the reference NPP and walk-talk-throughs at locations relevant for fire fighting. After the model was encoded in the corresponding datasets of the Crew Module, the relevant parameters of the Crew Module which are subjected to epistemic or aleatory uncertainty were identified and the corresponding probabilistic information used to express the uncertainty (distributions and branching information) was specified. The list of relevant parameters included, for instance, time periods needed to execute simple basic actions or discrete parameters referring to the outcomes of basic actions (success/error). If feasible, the probabilities of human errors were derived from the methods ASEP (*A*ccident *S*equence *E*valuation *P*rogram, [6]) and THERP (*T*echnique for *H*uman *E*rror *R*ate *P*rediction, [7]) which are recommended by the technical document on PSA methods [8] supplementing the German PSA Guide. Uncertain parameters and the corresponding probabilistic data were entered as input to the Probabilistic Module of MCDET.

The subsequent simulations were performed without running the FDS code. They were performed just by the combination of the Crew Module, Probabilistic Module and the Scheduler Module of MCDET. That combination ran very fast and provided more than 100 Dynamic Event Trees (DETs) for each of several conditions. Those conditions identified as being decisive to the human actions of the fire fighting process were given by relevant states of the fire detection and alarm system (none, only one or at least two of the detectors operate as required) as well as by the fire progression (e. g. visibility of smoke in front of the door when shift personnel reaches the fire compartment or production of smoke in the fire compartment).

From the DETs resulting from the MCDET simulations, various conditional distributions could be derived by using the corresponding post-processing module of the MCDET tool. The distributions refer to the timing of human actions such as the time period between fire alarm and the arrival of fire fighters at the fire compartment door or the time period between the arrival at the fire compartment door and the start of fire suppression. They express stochastic variability and were used as input to the simulations performed in the following analysis steps.

Those analysis steps dealt with the modelling, simulation and evaluation of the interaction of the fire dynamics with relevant stochastic factors affecting the fire dynamics. The stochastic factors taken into account refer to the timing and outcome (success/error) of tasks of human actions affecting the evolution of the fire, the operability of the fire detection and alarm system as well as to the functioning of active fire barrier elements (i. e. fire dampers and fire doors) and of the fire extinguishing systems to be manually actuated. Modelling information on the dynamics-stochastics interaction was specified as input to FDS as well as to the Probabilistic Module of MCDET. The input of the Probabilistic Module includes the parameters subjected to stochastic variability (aleatory uncertainty) as well as the distributions and branching information expressing the stochastic variability. It also comprises information on the relevant parameters subjected to epistemic uncertainty. Corresponding information is given in Table 1.

The calculations of FDS and those of the Probabilistic Module were supervised by the old version of the MCDET Scheduler Module allowing for calculating each DET in one process. The simulation approach made extensive use of the restart capabilities of FDS. The fire scenario was calculated up to 1800 s (0.5 h) after ignition. The output of the simulations comprises data of approx. 2400 different fire sequences from a sample of 120 individual DETs. These data were evaluated by using corresponding post-processing modules of the MCDET tool.

Table 1: Epistemic Parameters and Specified Probability Distributions

Epistemic Parameter	Reference Value	Distribution	Distribution Parameters
Value of optical density at which emerging smoke is visible under fire compartment door [1/m]	0.3	uniform	Min = 0.2, Max = 0.4
Response time index for activation temperature of fire dampers [$\sqrt{m \cdot s}$]	125	uniform	Min = 50, Max = 200
Threshold value of optical density D below which fire compartment can be entered [1/m]	0.4	uniform	Min = 0.3, Max = 0.5
Fraction of fuel mass (oil) converted into smoke	0.097	uniform	Min = 0.095, Max = 0.099
Time to reach 1 MW heat release rate [s]	425	uniform	Min = 250, Max = 700
Conductivity of cable [W/m*K]	0.275	uniform	Min = 0.15, Max = 0.4
Specific heat of cable [kJ/kg*K]	1.225	uniform	Min = 0.95, Max = 1.5
Depth of cable isolation [m]	0.0016	uniform	Min = 0.0012, Max = 0.002
Cable density [kg/m^3]	1131	uniform	Min = 833, Max = 1430
Specific heat of concrete [kJ/kg*K]	0.65	uniform	Min = 0.5, Max = 0.8
Conductivity of concrete [W/m*K]	1.75	uniform	Min = 1.4, Max = 2.1
Thickness of concrete walls in the fire compartment [m]	0.37	uniform	Min = 0.32, Max = 0.42
Probability that a fire door falsely stays open	0.005	beta	$\alpha = 1.5, \beta = 236.5$
Probability that fire damper fails to close	0.01	beta	$\alpha = 1.5, \beta = 117.5$
Failure temperature of I&C cables [°C]	170	uniform	Min = 145, Max = 195
Critical time periods [s] with temperatures of I&C cables between:			
145 °C - 150 °C	360	uniform	Min = 300, Max = 420
150 °C - 160 °C	240	uniform	Min = 200, Max = 280
160 °C - 170 °C	180	uniform	Min = 150, Max = 210
170 °C - 180 °C	120	uniform	Min = 90, Max = 150
> 180 °C	40	uniform	Min = 20, Max = 60

4.2 Results of the Analysis

The safety related targets which could be damaged and, therefore, were selected to be considered in this analysis, are I&C cables routed below the ceiling of the fire compartment. The fire was regarded as successfully suppressed as soon as the temperature inside the cable jacket fell below 120 °C after the fire fighters started to extinguish the fire.

Figure 2 shows the temporal evolution of the temperature inside the cable jacket for those sequences from all DETs for which the fire detection and alarm system operates as required. Differences between the sequences are due to the combined influence of aleatory and epistemic uncertainties. Distinct colors used in Figure 2 indicate which sequences lead to successful fire suppression (green and red curves) and which not (black curves). Successful fire suppression can be performed either by the fire patrol (green curves) or by the fire brigade (red curves). The fire patrol can extinguish the fire by a portable fire extinguisher or by manually actuating the stationary fire extinguishing system from outside the fire compartment (see section 3.2). If the fire detection and alarm system operates as required, the fire patrol can suppress the fire mostly within 800 s (~ 13 min) after fire ignition. If the patrol fails to suppress the fire, the fire brigade can extinguish the fire with their equipment. If the fire brigade succeeds to suppress the fire, mostly 800 to 1200 s (~ 13 to 20 min) – in some cases 1200 to 1500 s (20 to 25 min) - elapse after fire ignition.

Figure 2: Temperature inside the Cable Jacket

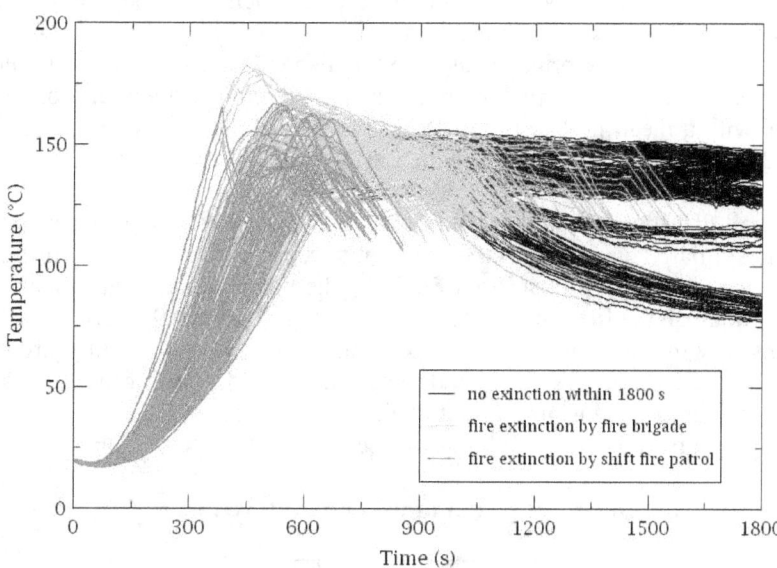

For sequences without any fire extinction within 1800 s (black curves), three distinguished temperature clusters are clearly visible in Figure 2. In the first cluster, the temperature inside the cable jacket decreases below 100 °C within 1800 s after ignition. The associated sequences are characterized by the corresponding fire damper operating as demanded and the fire door being closed (see Chapter 3). The same is true for the sequences of the second cluster with a temperature on a level between 105 °C and 120 °C. Main reasons for a smaller temperature decrease compared to that of the first cluster seem to be higher values of the depth of cable jacket material combined with lower values of the cable density. Both parameters are considered as subjected to epistemic uncertainty (see Table 1). In the third cluster, the temperature remains at a rather high level between 125 °C to 150 °C up to the end of simulation time (1800 s). This is the consequence of an open fire damper or an open fire door.

Figure 3: Distribution of the Time Elapsed from Fire Ignition to Successful Fire Suppression

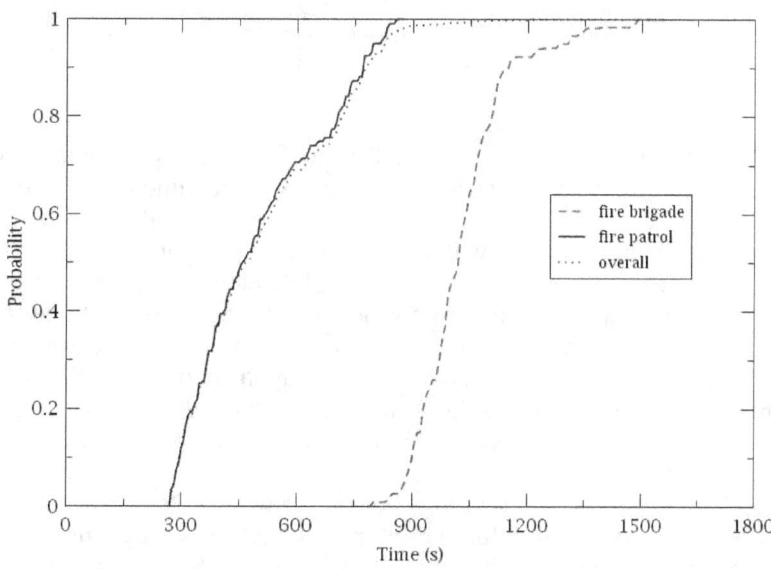

Figure 3 shows the overall distribution and two conditional distributions of the time elapsed from fire ignition to successful suppression, if the fire detection and alarm system operates as required. The conditional distributions refer to the conditions that either the fire patrol or the fire brigade suppresses the fire successfully. The distributions underline the results which could be derived from Figure 2. If the fire patrol is able to extinguish the fire, this happens within 780 s (13 min) after ignition with a mean probability of 0.92. If the fire brigade has to extinguish the fire, suppression is successful 13 min after ignition at the earliest. The fire brigade arrives at the fire compartment approx. 3 to 16 min later than the fire patrol with a mean probability of 0.95. The brigade is able to suppress the fire between 780 and 1320 s (13 and 22 min) after ignition with a mean probability of 0.95.

Obviously, the overall distribution and the conditional distribution referring to fire suppression by the fire patrol are quite similar. This is due to the fact that successful fire suppression (within 1800 s) by the fire patrol occurs with a high probability of 0.994. The mean conditional probability that the fire brigade is able to extinguish the fire within 1800 s after fire ignition is 0.999 given the fire patrol is not successful. Reasons why the fire brigade cannot extinguish the fire within 1800 s are recovery actions performed e.g., if the fire fighters misunderstand the correct fire location, if the equipment is defect or if the water supply for fire extinguishing is not available. All these situations were considered in the model of human actions applied for fire fighting (see section 4.1).

Figure 4: Distribution of the Maximum Temperature inside the Cable Jacket

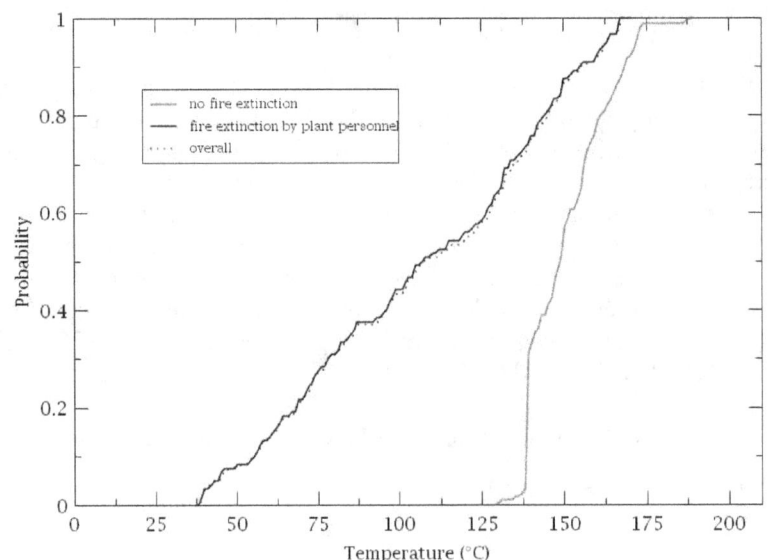

Figure 4 shows the overall distribution and two conditional distributions of the maximum temperature inside the cable jacket. The conditional distributions refer to the conditions, if the fire is successfully extinguished by the plant personnel or not. If the fire can be successfully suppressed, the maximum temperature can be reduced below 150 °C with a mean probability of about 0.87. If the fire cannot be suppressed, the mean probability of a maximum temperature below 150 °C is approx. 0.51. The main effect on the maximum temperature results from the actions of the fire patrol, because he/she can start the fire suppression quite early and therefore avoid a higher temperature maximum inside the cable jacket. The overall distribution and the conditional distribution referring to fire suppression by the plant personnel are nearly identical. This is due to the very high conditional probability for successful fire suppression within 1800 s. The mean value is 1.0 - 5.97 E-06.

Estimates of the probability of I&C cables to be damaged by the fire were derived based on two different failure criteria. According to failure criterion 1, the cables were assumed to be damaged, if the temperature inside the cable jacket exceeds a critical value. With criterion 2, the failure of a cable was supposed to be determined by both a high level of the temperature inside the cable jacket and a critical time period with the temperature on a high level. Reference values and uncertainty quantifications referring to the failure temperature (criterion 1) and the critical time periods (criterion 2) are specified in Table 1. It is emphasized that the specifications are used only for demonstration purposes. They were derived from available experimental data on failure temperatures of I&C cables [9] and should be checked for real applications, in particular, with respect to the critical time periods for given temperature levels.

Figure 5: Epistemic Uncertainty of the Conditional Probability of Damaged I&C Cables

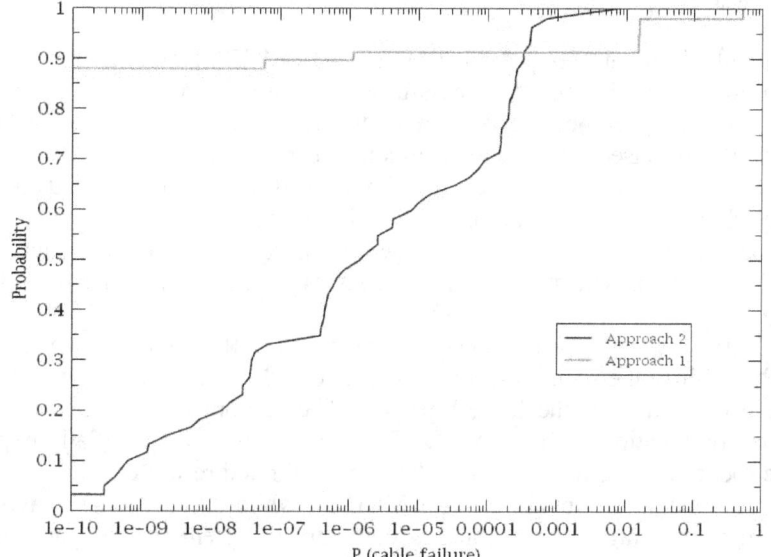

The mean conditional probability that an I&C cable in the fire compartment is damaged during the fire scenario was estimated to be 1.76 E-02 based on failure criterion 1 and 2.08 E-04 based on criterion 2. Figure 5 shows the epistemic uncertainty of the cable damage probability for each criterion. It can be seen that the epistemic probability distributions due to both criteria differ significantly. Since just rather limited sample data were available from the MCDET-FDS simulations (a sample of 60 values is available to express the epistemic uncertainty of the cable damage probability), the one-sided upper (95 %, 95 %)-tolerance limit was calculated as an estimate of the 95 %-quantile of the cable damage probability. With failure criterion 1, the (95 %, 95 %)-tolerance limit is 0.5 which would certainly not be acceptable. The (95 %, 95 %)-tolerance limit based on criterion 2 is 7.5 E-03 which seems to be more reasonable.

The results clearly show the effect of different failure criteria on the calculated cable damage probability. They underline what intuitively seems to be more realistic, namely to consider both high temperature and the exposure time to high temperature as a criterion for assessing targets as deteriorated or damaged. To this purpose, corresponding data should be made available. As demonstrated, the MCDET method can use this data in order to provide corresponding risk assessments.

5. CONCLUSIONS

An exemplary IDPSA of a fire scenario was performed to demonstrate its modelling capacity and evaluation options in the frame of a probabilistic fire risk analysis. The necessary simulations were carried out by a combination of the MCDET tool and the FDS code. This combination allowed for a quite realistic simulation of the interaction between the fire dynamics and relevant stochastic influencing factors over time.

What made the MCDET tool particularly useful for the analysis was its extra Crew Module which permitted to consider the human actions designated to be performed for fire fighting as a dynamic process. MCDET simulations performed just with the Crew Module without running FDS provided a huge amount of data for generating conditional distributions of the timings of human actions. Those distributions were used as input to the IDPSA performed with the combination of FDS and MCDET. Besides the timings of human actions, the individual performances of all systems and components designated to be applied for fire detection, alarm, confinement and suppression were considered as subjected to aleatory uncertainty. Epistemic uncertainties were taken into account as well. Those ones which were considered as potentially important refer to parameters of the FDS code and failure criteria specified for cable targets.

From the huge amount of output data provided by the FDS-MCDET simulations, the distribution of the time elapsed from fire ignition to its successful extinction and - what is even more interesting - distributions referring to the temporal evolution of the temperature inside the jacket of the cable targets could be calculated. Based on these distributions, quantifications useful for fire risk assessment were derived, e.g. the conditional probability of safety related I&C cables to be damaged by the fire. The results clearly showed the effect of different failure criteria on the calculated cable damage probability. They also underlined the necessity of getting more time-dependent data on the reliability of SSC in order to be able to provide more realistic safety assessments by means of an IDPSA method.

The main lesson learned was the importance of having a well validated dynamics model when performing an IDPSA. Since the scenario-specific model did not exist initially, it had to be created as input to FDS just in the course of the IDPSA project. The amount of work necessary to make that model applicable in combination with the MCDET tool was higher than initially expected. A lot of activities had to be spent to find a way of how to handle the limited restart capabilities of FDS in order to make the code running in combination with MCDET while simultaneously avoiding extensive calculation time and data storage. Corresponding commands implemented in the Driver Module of MCDET finally made the communication between FDS and MCDET practicable.

Acknowledgements

The work presented in this paper was sponsored by the Ministry of Economics and Technology (BMWi) within the frame of the Research and Development Project RS1198.

References

[1] M. Kloos, J. Peschke, "*MCDET - A Probabilistic Dynamics Method Combining Monte Carlo Simulation with the Discrete Dynamic Event Tree Approach*", Nuclear Science and Engineering, 153, pp. 137-156 (2006).

[2] K. McGrattan, et al. "*Fire Dynamics Simulator. Technical Reference Guide Volume 1 : Mathematical Model*". NIST Special Pub. 1018, Sixth Edition (2013).

[3] M. Kloos, J. Peschke, "*Consideration of Human Actions in Combination with the Probabilistic Dynamics Method MCDET*", Journal of Risk and Reliability, 222, pp. 303-313 (2008).

[4] M. Kloos, J. Hartung, J. Peschke, M. Röwekamp. "*Advanced probabilistic dynamics analysis of fire fighting actions in a nuclear power plant with the MCDET tool*", Proceedings of European Safety and Reliability Conference ESREL 2013, Amsterdam, Netherlands, 29 Sep -02 Oct 2013, on CD-ROM (2013).

[5] M. Kloos, J. Hartung J. Peschke, J. Scheuer , "*Updates and current application of the MCDET simulation tool for dynamic PSA*", Proceedings of ANS PSA 2013 International Topical Meeting on Probabilistic Safety Assessment and Analysis, Columbia, SC, September 22-26, 2013, on CD-ROM (2013).

[6] A. D. Swain, "*Accident Sequence Evaluation Program Human Reliability Analysis Procedure*", NUREG/CR-4772, Washington, DC (1987).

[7] A. D. Swain, H. E. Guttmann, "*Handbook of Human Reliability Analysis with Emphasis on Nuclear Power Plant Applications. Final Report*", NUREG/CR-1278, Washington, DC (1983).

[8] Facharbeitskreis (FAK) PSA, "*Methoden zur probabilistischen Sicherheitsanalyse für Kernkraftwerke. Stand: August 2005*", BfS-SCHR-37/05. Bundesamt für Strahlenschutz (BfS), Salzgitter, in German (2005).

[9] S. Babst, et al., "*Methoden zur Durchführung von Brand-PSA im Nichtleistungsbetrieb*", GRS-A-3579, Gesellschaft für Anlagen- und Reaktorsicherheit (GRS) mbH, Köln, in German (2011).

Uncertainty Propagation in Dynamic Event Trees - Initial Results for a Modified Tank Problem

Durga R. Karanki[a,*], Vinh N. Dang[a], and Michael T. MacMillan[a]
[a]Paul Scherrer Institute, Villigen PSI, Switzerland

Abstract: The coupling of plant simulation models and stochastic models representing failure events in Dynamic Event Trees (DET) is a framework to model the dynamic interactions among physical processes, equipment failures, and operator responses. The benefits of the framework, as a number of applications show, include, for instance, the capability to account for the aleatory timing of equipment failures or operator actions on sequence outcomes and to consider the impact of the number of available trains (rather than having to identify the bounding cases). The integration of physical and stochastic models may additionally enhance the treatment of uncertainties. Probabilistic Safety Assessments as currently implemented, e.g. for Level 1, propagate the (epistemic) uncertainties in the probability distributions for the failure probabilities or frequencies; this approach does not consider propagate uncertainties in the physical model (parameters). The coupling of deterministic (physical) and probabilistic models in integrated simulations such as the DET allows both types of uncertainties to be considered. The starting point in this work is to consider wrapping an epistemic loop, in which the epistemic distributions are sampled, around the DET simulation. To examine the adequacy of this approach, and to allow different approaches and approximations (for uncertainty propagation) to be compared, a simple problem is proposed as a basis for comparisons. This paper presents initial results on uncertainty propagation in DETs, obtained for a tank problem that is derived from a similar one defined for control system failures and dynamic reliability. An operator response has been added to consider stochastic timing.

Keywords: Epistemic and aleatory uncertainties, Dynamic PSA, Monte Carlo simulation, Dynamic Event Tree Analysis

1. INTRODUCTION

Typical accident scenario in a Nuclear Power Plant (NPP) involves complex interactions between physical process and safety systems (safety equipment and operator response). The response of a safety system is inherently random in nature, which is often referred as aleatory uncertainty [1]. The response of physical process can also have aleatory elements; for example, initial level, break size, break location, etc. Dynamic event tree (DET) analysis provides a framework to simulate the accident scenario considering the dynamic interactions [2], where mathematical models of physical process and safety systems are used. The limitations in assessing the parameters of these models introduce another type of uncertainty, which is often referred as epistemic uncertainty [1]; for example, demand failure probability of safety equipment, human error probabilities, and thermal hydraulic parameters. These epistemic variables can significantly impact the simulated accident dynamics and ultimately the risk estimate; for example, uncertainty in TH parameter or operator response can change the outcome of an accident sequence affecting the final risk estimate. Hence risk quantification must consider both epistemic and aleatory uncertainties in both physical and safety system models along with their dynamic interactions.

In the current PSA practice [3], accident sequence models are first developed and then solved for a cut set equation. A point estimate of risk (e.g. Core Damage Frequency for level-1 PSA) can then be obtained using mean values for the PSA parameters. A Monte Carlo simulation is run to propagate epistemic uncertainty in PSA parameters. The obtained CDF distribution thus accounts for the aleatory and the epistemic uncertainties of safety system responses, e.g. demand failures. However, the current approach does not propagate uncertainties in TH parameters through to the risk model outcomes. The

*durga.karanki@psi.ch

success criteria definitions are the interface between the physical system simulations and the PSA models; they are normally calculated with point estimates of Thermal-Hydraulic (TH) parameters, using bounding parameter values in some cases.

For similar problems in the literature [4, 5], a two-loop Monte Carlo simulation has been used. epistemic variables are sampled in the outer loop while aleatory variables are sampled in the inner loop. In this work, the inner loop dealing with the aleatory response is a DET simulation. There are two DET approaches in the literature to consider dynamic interactions in NPP, discrete DET [6-8] and MCDET [9] approaches; these studies demonstrated the potential of DET approaches in addressing the complex interactions providing insights for risk assessment. Further, DET approaches have been found to be useful to assess the impact of dynamics [10] and the detrimental effects of bounding [11] in the quantification of risk. The DET approach can also provide a framework to consider epistemic and aleatory uncertainties.

In this work, the discrete DET framework along with epistemic uncertainty analysis is applied to quantify risk and identify important contributors in the light of uncertainties and dynamics. The analysis determines the impacts of physical uncertainties and safety system uncertainties on the accident evolution and final risk estimate. Monte Carlo simulation with convergence criteria for epistemic uncertainty analysis and appropriate discretization strategies for DDET are considered. To examine the adequacy of this approach, and to allow different approaches and approximations (for uncertainty propagation) to be compared, a simple problem is proposed as a basis for comparisons. This paper presents initial results on uncertainty propagation in DETs, obtained for a tank problem that is derived from a similar one defined for control system failures and dynamic reliability [12]. An operator response has been added to consider stochastic timing. The results from DDET approach are compared with analytical solution including, important risk contributors and uncertainty importance measures.

The paper is organized as follows. Section 2 explains the methodology and its elements. Section 3 presents application to the tank problem, its analytical solution, and its comparison with DDET approach. The detailed analysis of the results and discussion is presented in Section 4. Finally, conclusions are given in Section 5.

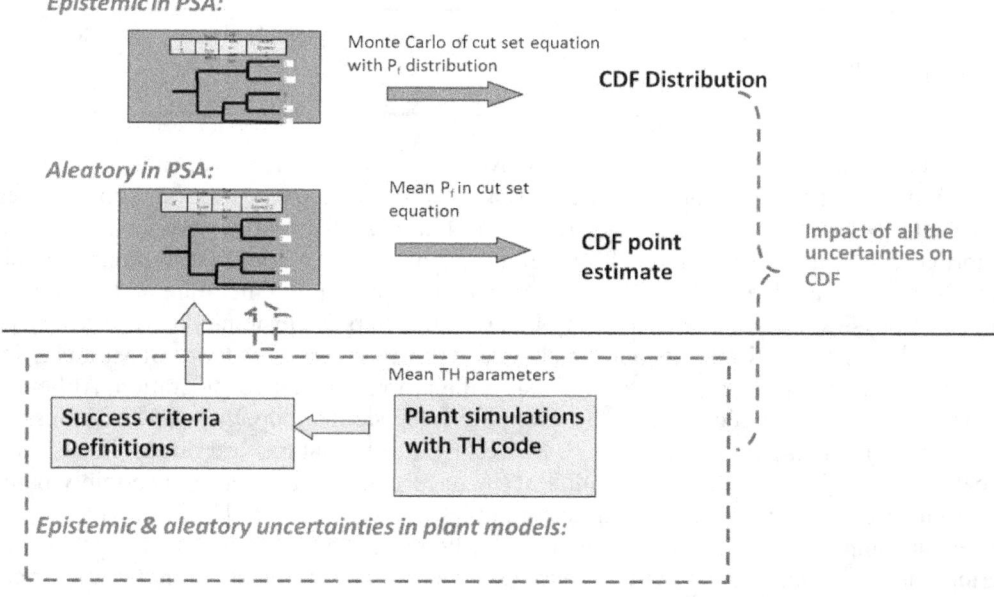

Fig. 1: Propagation of uncertainties in the current PSA practice

2. DYNAMICS AND UNCERTAINTIES IN RISK QUANTIFICATION

The classical combination of fault tree and event tree analyses is used to develop risk models in Probabilistic Safety Assessment of NPPs. Success criteria requirements (for fault tree development) and sequence outcomes (for event tree development) are derived based on plant simulations with thermal-hydraulic codes. PSA model is then solved for a cut set equation, which is subsequently quantified to estimate core damage frequency (CDF) using plant specific or/and generic reliability data. In order to account for epistemic uncertainty in PSA parameters such as demand probabilities of safety systems, HEPs, etc., Monte Carlo simulation is used to quantify the uncertainty in CDF. Fig. 1 shows the current practice of uncertainty propagation in PSA; epistemic and aleatory uncertainties in stochastic elements (safety systems and operator responses) are properly accounted, but these uncertainties are also present in the physical process. For example, epistemic uncertainties in physical process (TH model) parameters or the natural variability of break size or location could change the structure of the event trees or success criteria definitions, subsequently the risk estimate. Any uncertainties related to the success criteria are treated through enveloping / bounding [11, 13]. Thus the uncertainties in PSA models and TH models are separately treated and they are not propagated across the interface. All these uncertainties (both epistemic and aleatory) in the dynamic interactions of physical process and stochastic systems (safety systems and operator responses) must properly be accounted in the risk assessment.

Table 1: Epistemic and aleatory uncertainties in safety and physical models

	Aleatory	Epistemic
Safety equip. & OAs (PSA)	Demand failure probability, failure times, recovery time of safety equipment	Parameter uncertainty of discrete & continuous aleatory variables
	Response time of OA	
Physical process (TH)		TH parameters
	Aleatory variables in TH model (discrete & continuous)	Parameter uncertainty of TH aleatory variables

Table 1 gives a summary of the uncertainties involved in risk calculations. In safety system models, equipment failures on demand, time to failure (during operation), recovery time, and operator response times are aleatory variables, usually characterized with binomial distribution, exponential distribution, and lognormal distribution respectively. The parameters of these aleatory distributions are epistemic variables. Both these variables in PSA studies were well explored in the literature and used in practice [4, 5]. In plant physics models, the uncertainty in TH parameters is epistemic in nature; the examples of aleatory variables are initial levels and break size. Thus there are four types of variables in a full scope risk model and their propagation to final risk is the problem under consideration. Although some studies considered TH epistemic variables by wrapping an epistemic loop, these uncertainties were not propagated up to final risk quantification [9]. Some of these limitations are due to issues with DET quantification of risk [11]. Besides propagation of the uncertainties, ranking of uncertainty parameters and important risk contributors considering uncertainties is also necessary [14, 15], which help to see their individual impact on risk, and further in uncertainty and risk management. In addition to considering epistemic variables of both TH and PSA models in risk quantification, the current study also considers aleatory variables of physical process. A solution is proposed here to this problem of integrated treatment of uncertainties (both epistemic and aleatory in both TH and PSA models) in quantifying risk and its uncertainties.

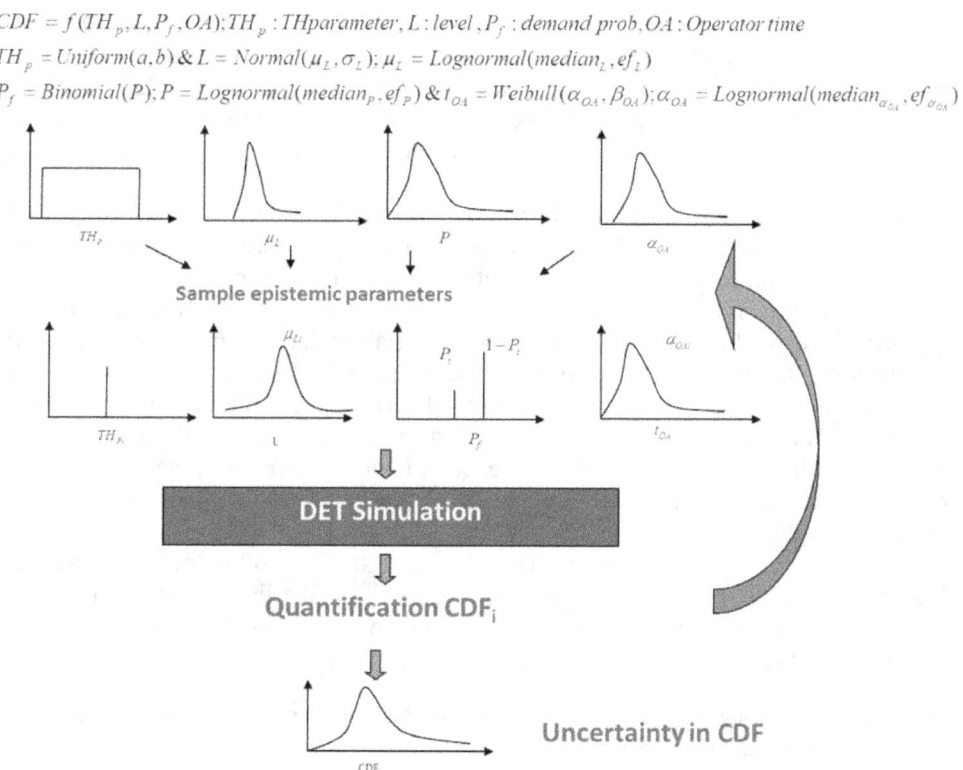

Fig. 2: Computational framework – DDET along with epistemic uncertainty analysis

Computational Framework – DDET along with epistemic uncertainty analysis

The objective is quantifying the risk, identifying accident sequences, and ranking of important parameters in the light of uncertainties and dynamics. The premise of the solution approach, while considering uncertainties and dynamics in accident scenario simulation, is based on the primary distinction of parameters on the nature of uncertainty, and not based on the physical vs. safety systems. The two-loop Monte Carlo simulation strategy, which was used in the literature for similar problems, is adapted for the current problem. The inner loop modeling the aleatory response is replaced by a DDET simulation. In this way, the aleatory response is addressed by the DDET rather than by means of Monte Carlo sampling. Mathematically, risk is a function of epistemic and aleatory parameters of the model, as depicted in equation (1): it has two variables relating physical process and two variables relating to safety systems. 'TH_p' is an epistemic variable (physical) and physical level 'L' is an aleatory variable; P_f and OA are aleatory variables (safety system) and their distribution parameters are epistemic variables. Let us assume that the probability distributions for all these epistemic and aleatory variables are available. The following steps are involved in the computational methodology as shown in Fig. 2.

i. **Epistemic sampling:** The epistemic variables are sampled based on their distribution. In this case, epistemic parameter TH_p and epistemic parameters of aleatory variables, viz. μ_L, P, and α_{OA}, are sampled and they are treated as constants in the next step. A convergence criterion for epistemic Monte Carlo sampling is required to check the accuracy and keep the number of computations to manageable size. A criterion that uses the acceptable percentage of error and confidence levels as inputs is used for the implementation. Details of convergence criterion are discussed in section 3.4.

ii. **Aleatory physical variables:** DDET simulation considers both physical and safety system aleatory variables. The aleatory variables of physical process are different from safety systems. While the initial conditions or boundary conditions of accident initiator depend upon the aleatory physical variables, the aleatory variables of safety systems influence post-accident initiation and drive accident evolution.

Branches are generated for each of the aleatory physical variables. As they are not time dependent, these branches correspond to accident initiating event. If any of these variables is continuous (e.g. level in this case), they are discretized on a logarithmic scale as discussed in section 3.3.

iii. **Aleatory safety system variables:** DDET tool simulates accident scenario with the boundary conditions from steps 1 and 2, branches are generated as and when the safety system are demanded. The response of aleatory safety system variables can be discrete (for example success or failure on demand P_i or $1-P_i$) or continuous such as operator response time. Appropriate discretization is used if the variables are continuous. The logarithmic discretization strategy ensures optimal number of branches with less conservatism at reduced computations. The branches of aleatory physical variables are simulated subsequently.

iv. **Quantification of risk:** Each DDET generated from step iii is evaluated to quantify risk and important contributors. These results correspond to an epistemic sample i. The computation is switched over to next epistemic sample, i.e. step i. The computations continue until the convergence criterion is satisfied.

v. **Quantification of uncertainty in Risk Estimates:** The following measures of risk are obtained from the simulations: Epistemic uncertainty distribution of final risk, important sequences, uncertainty importance measures (ranking of uncertainty parameters), ranking of risk contributors along with their uncertainties.

The next section presents an application of this computational framework to simple tank problem and also its comparison with analytical results.

3. APPLICATION TO A SIMPLE TANK PROBLEM

3.1 Depleting Tank Problem

A tank problem has been derived from a similar one defined for control system failures and dynamic reliability [12]. An operator response has been introduced to consider stochastic timing. There is a cylindrical tank of diameter 'D' with an initial water level of H_i. Tank starts depleting due to a spurious signal that opens a valve, which has a diameter of the leak as 'd'. Alarm is the cue for operator action. Operator has to close the valve before the tank level reaches a critical level H_f. The objective is to estimate the likelihood of the tank reaching a critical level considering all epistemic and aleatory uncertainties in the scenario. Time taken for a depleting tank to reach a level H_f based on Bernoulli's equation is [16, 17]:

$$TW = \frac{A}{aC}\left(\sqrt{H_i} - \sqrt{H_f}\right)\sqrt{\frac{2}{g}} \qquad (1)$$

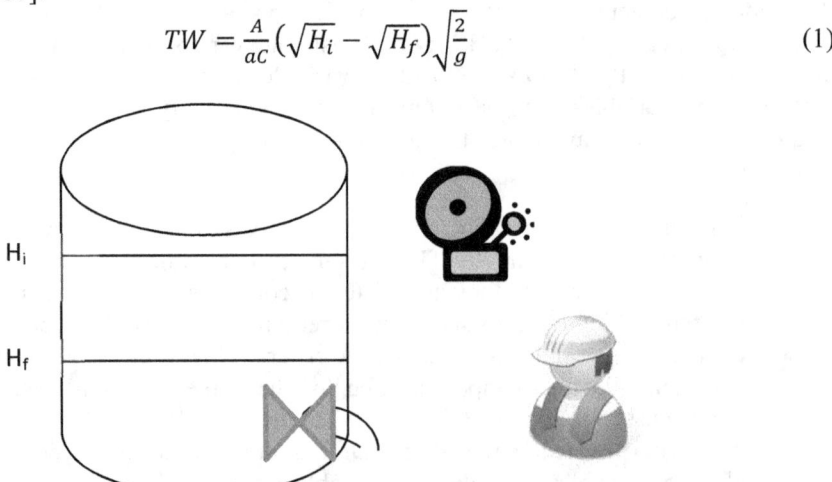

Fig. 3: A depleting tank with an initial level H_i and a critical level H_f

Nomenclature:
TW – Time
a – area of the hole
H_f – Critical tank level
g – gravitational force
P(V) – Valve failure prob
A – area of tank
H_i – Initial tank level
C – discharge coefficient
T_{OA} – Response time of operator

The tank depletes to critical level when the operator does not act before a time, which is the time taken for the tank to reach the critical level. Operator has a cue from an alarm, which is due to fall of level, and in response operator needs to close the valve. The valve needs to function on demand to stop the leak. The tank failure depends on the failures on demand of alarm and valve, and human response. The time dependent (dynamic) element in the problem is human response time competing with the time taken by the tank to reach the critical level, which depends on initial level and other constants.

Table 2: Epistemic and Aleatory uncertainties in the tank model

	Aleatory Variables	Epistemic Variables
Safety System Models	Demand failure probability of Valve (2e-4)	P-Lognormal(1.24e-4, 5)
	Response time of OA – $g(t_{OA})$ Lognormal(360s, 2)	Error factor-Uniform(1.8, 2.2)
Physical Process Models		Discharge coefficient C Uniform(0.72, 0.98)
	Initial tank level Hi Normal(10, 0.3)m	
Other data used in calculations	Diameter of the tank -2m Critical tank level – 2m Diameter of the hole -0.05m g – 9.8m/sec^2	

Table 2 gives the summary of aleatory and epistemic uncertainties assumed in the analysis. In physical process model, tank level is an aleatory variable and discharge coefficient is an epistemic variable. In safety system models, demand failure probabilities of valve and alarm, and operator response time are aleatory variables, where as their distribution parameters are epistemic variables.

3.2 Analytical Solution

This section presents the analytical solution for the system failure probability, as a baseline result with which to compare the DDET solution, in both cases, with uncertainties.

Tank failure probability FP can be expressed as a function of likelihoods of alarm, valve, and human error probability, which is shown in equation (2).

$$FP = f(P(A), HEP, P(V)) \qquad (2)$$

The failure probability of alarm and valve are independent of physical parameters or time dependent elements. But the HEP is the probability of the aleatory variable response time (R) exceeding another aleatory variable time window (W) or time taken for the tank level to reach the critical level (see Fig. 4). The time window is an aleatory variable as it is a function of initial level, which is another aleatory variable. HEP is shown in equation (3), which can be simplified using reliability theory on load-resistance or stress-strength concept [18] as shown below:

$$HEP = P(R > W) \qquad (3)$$

Differential HEP is the probability of response time falling in the interval 'dr' around r and the time window being smaller than the value 'r' simultaneously is

$$d(HEP) = f_R(r)dr \int_0^r f_W(w)dw$$

The HEP is given as the probability of time window 'W' being smaller than the response time 'R' for all possible values of R.

$$HEP = \int_0^\infty f_R(r)dr \int_0^r f_W(w)dw = \int_0^\infty f_R(r)F_W(r)dr \qquad (4)$$

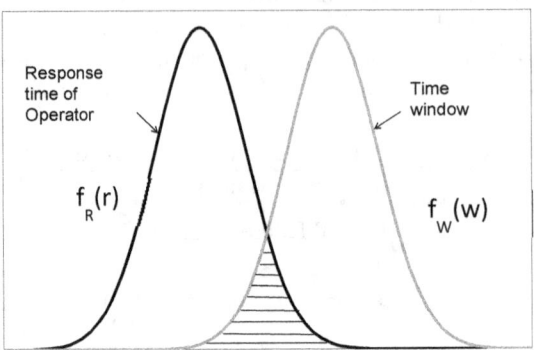

Fig. 4 Operator response time and time window

PDF of response time is known, but PDF of time window is not known; as time window is a function of core level whose pdf is known, we can derive its PDF using transformation of random variables [19] as shown below: Equation (1) can be simplified to

$$W = \frac{A}{aC}\left(\sqrt{H_i} - \sqrt{H_f}\right)\sqrt{\frac{2}{g}} = k_1\sqrt{H} - k_2; \text{ where } k_1 = \frac{A}{aC}\sqrt{\frac{2}{g}} \text{ and } k_2 = \sqrt{H_f} \times k_1 \qquad (5)$$

As mentioned in Table 2, H_i is a normal distribution, we have to find probability density or cumulative distribution function (CDF) of 'W'. The CDF of W can be expressed as

$$F_W(w) = P(W \leq w) \qquad (6)$$

Equation (5) can be rearranged to derive H as a function of w:

$$H = \left(\frac{w + k_2}{k_1}\right)^2$$

Substituting equation (5) in equation (6) and expanding further:

$$F_W(w) = P(k_1\sqrt{H} - k_2 \leq w) = P\left(H \leq \left(\frac{w + k_2}{k_1}\right)^2\right) = \int_0^{\left(\frac{w+k_2}{k_1}\right)^2} f_H(h)dh$$

$$F_W(r) = \int_0^{\left(\frac{r+k_2}{k_1}\right)^2} f_H(h)dh \qquad (7)$$

Substituting equation (7) in equation (4) gives the HEP for final calculations:

$$HEP = \int_0^\infty f_R(r) \int_0^{\left(\frac{r+k_2}{k_1}\right)^2} f_H(h)dh.dr \qquad (8)$$

Numerical integration method has been used to solve equation (8) for HEP and with the data mentioned in Table 2.

3.3 Discrete DET Solution

The discrete DET approach has been applied on the tank problem. DDET is shown in Fig. 5. Continuous aleatory variables, viz., tank level and operator response times are discretized. The alarm and valve have two branches either success or failure.

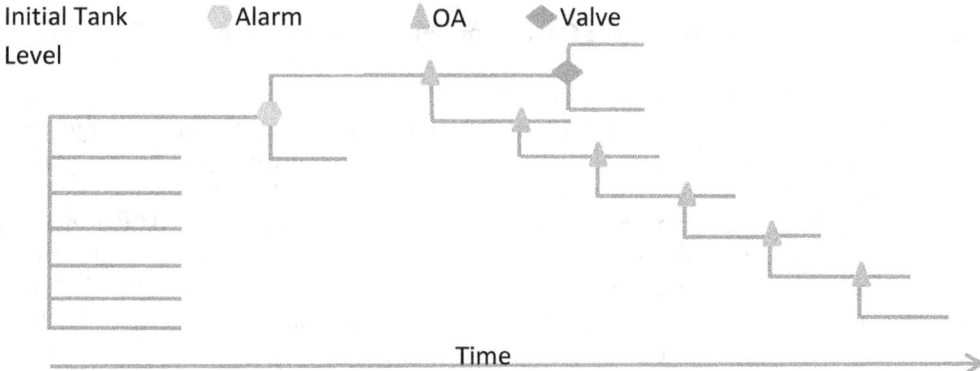

Fig. 5: Discrete DET of the tank problem considering aleatory uncertainties

The initial tank level and operator response time are discretized as they are continuous random variables. The discretization strategies used in the literature [7, 20] are 3 percentiles which normally represent low, median, and high values. This strategy is reasonable for a qualitative understanding of the sequences, but, in quantification, can results in overestimation. It is also important to know if the variable to be discretized is sensitive as a whole or in certain parts (e.g. upper or lower tails) of the distribution. Since the tank level is sensitive for all the values, it was discretized linearly on the whole distribution. The logarithmic discretization strategy ("log strategy") is used in case of operator response distribution on the upper tail (between 0.9 and 1.0 in cum. prob.). The premise for selecting this range is that human error probability (HEP) is assumed to be in the range of 0.0001 and 0.1; the lower values than this range would not contribute significantly compared with other risk contributors and the higher values would make only a marginal error. 5 different discretization strategies (4, 5, 7, 10, and 20 Branches; the last 3 with log strategy, see Table 3) are considered and their results are compared with analytical result. Fig. 6 shows the 7-branch log strategy, where the tail is divided into 3 branches (intervals) in log scale; the remaining 4 branches correspond to 5%, 50%, 90%, and skip, which are necessary to see quick, normal, late, and never actions. Like the 7-branch log strategy, the 10- and 20-branch strategies discretize cumulative probabilities between 0.9 and 1.0 in log scale into 6 and 16 branches respectively. Table 3 shows the discretization strategies for 4, 5, and 7 branches used in the calculations. The percentiles of response time and the branch probabilities are also shown.

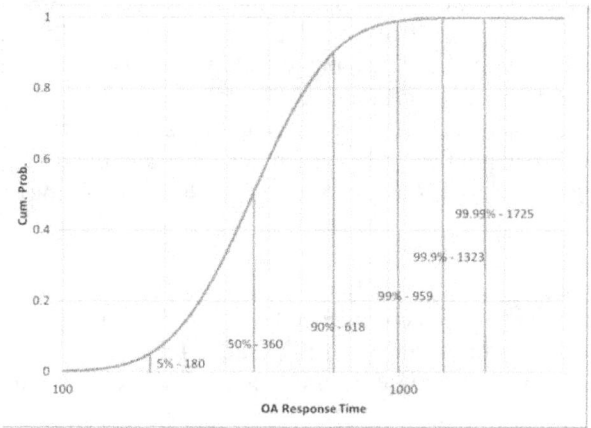

Fig. 6: Log discretization strategy for operator response time

Table 3: Discretization strategies for OA in DET simulations

	4-branch	5-branch	7-branch
Operator response time	(5, 50, 95)tiles, skip	(5, 50, 95, 99.9)tiles, skip	(5, 50, 90, 99, 99.9, 99.99)tiles, skip
Branch probability	0.05, 0.45, 0.45, 0.05	0.05, 0.45, 0.45, 4.9e-2, 1e-3	0.05, 0.45, 0.4, 9e-2, 9e-3, 9e-4, 1e-4

3.4 Uncertainty Propagation by Monte Carlo Simulation with Convergence Criteria

The uncertainty (epistemic) in distribution parameters of aleatory variables are propagated using Monte Carlo simulation which is widely used in current PSA practice [3]. The convergence criterion is based on the specified confidence level and percentage error. The method proposed by Driels and Shin [21] in Monte Carlo simulations of weapon effectiveness is adapted for this problem. Let risk y is a function of epistemic variables whose uncertainties to be propagated. Let 'n' is the initial number of Monte Carlo simulations run (sample size). Sample mean and standard deviation are calculated. The current percentage error and estimate of number of runs required to achieve a specified percentage of error are determined using the equations 9 and 10 [21]. Assuming 'y' as a normally distributed random variable, the percentage error of the mean risk is

$$E = \frac{100 * Z_c * S_y}{\bar{y} * \sqrt{n}} \tag{9}$$

Where Z_c confidence coefficient, S_y standard deviation, and mean of sample is $\bar{y}$.

A relationship between the number of trial runs necessary, confidence interval, and acceptable error is shown in Equation 10.

$$n = \left[\frac{100 * Z_c * S_y}{E * \bar{y}}\right]^2 \tag{10}$$

It was reported that the estimate of number runs convergence quickly after a few initial runs. This convergence method has been applied in the current calculations.

4. RESULTS AND DISCUSSION

The methods discussed in the previous section, i.e. analytical and DDET method, have been applied on the tank problem to determine the failure probability. In the first set of calculations, aleatory uncertainties are only considered and epistemic parameters are kept at their mean values; the second set of calculations considers both epistemic and aleatory uncertainties. The comparison between analytical and DDET aleatory results are shown in Table 4 The analytical method solved with numerical integration technique is the reference result. Several discretization strategies are compared with the reference result. DDET with 3%tile and 4%tile methods that were used in the literature are found to be conservative in estimation. The former overestimates by 83.9 times and the latter depend on the percentile assigned to skip action giving different results. The sensitive to skip percentile indicates it may change from case to case. Although it is obvious that larger the number of discretization levels the better accuracy in DDET calculations, log discretization strategy is found to give satisfactory results with few number of branches; for example, the percentage errors are 98% and 31% for 7 and 10 log branches respectively and the 20 branch (log) case converged with the reference result.

Monte Carlo sampling for epistemic calculations uses the convergence criteria discussed in the previous section. The criterion uses 95% confidence level and 5% error with respect to estimated mean. Comparing the epistemic mean with aleatory results (Table 4) from the analytical method, the former is higher than the latter indicating ignoring epistemic uncertainties could underestimate the risk.

Table 4: Comparison of failure probability without considering epistemic uncertainties

	Analytical	DDET-discretization						
	Numerical Integration	4 Br.	5 Br.			7 Br.	10 Br.	20 Br.
			99%tile*	99.9%tile	99.99%tile*			
Failure Probability	5.98e-4	5.02e-2	1.02e-2	1.19e-3	5.02e-2	1.19e-3	7.83e-4	6.43e-4
Overestimation		83.9	17	1.98	83.9	1.98	1.31	1.07

*Sensitive cases for 5 branch discretization

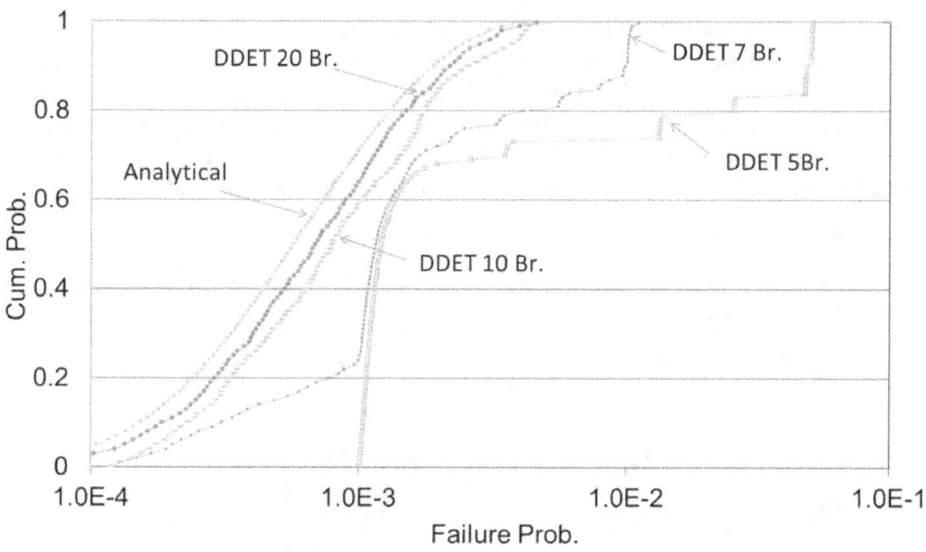

Fig. 7: Cumulative probability functions for epistemic uncertainty in failure probability

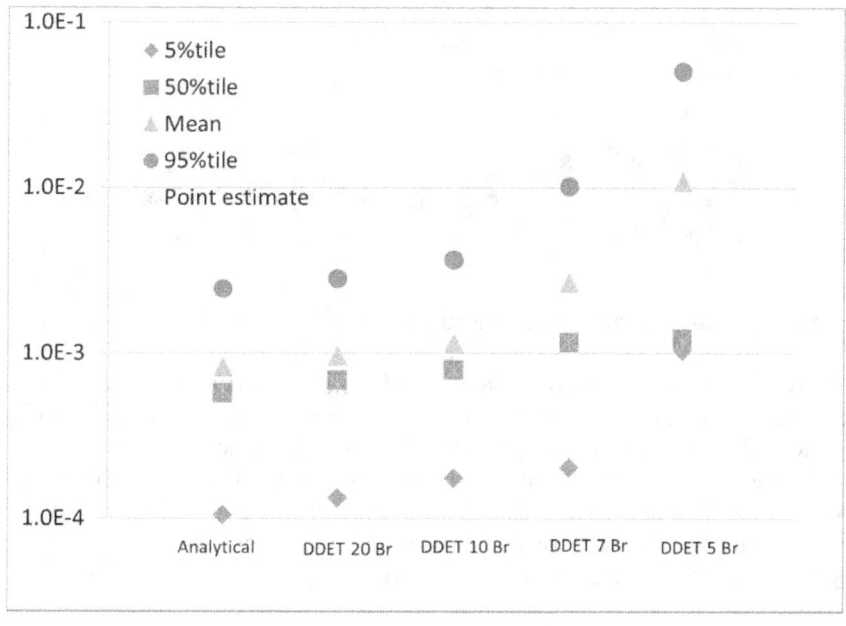

Fig. 8: Comparison of methods - Percentiles and mean of failure probability

In epistemic calculations, the discretization strategies are more thoroughly tested and compared with analytical results. Fig. 7 shows the cumulative probability functions for epistemic uncertainty in failure probability of the methods under consideration; further, the comparison of percentiles and mean among the methods is shown in Fig. 8. The median of all methods are close, but the upper tails of 5-branch and 7-branch are longer than other CDFs. The 20-branch and 10-branch approaches are in fairly good agreement with analytical CDF including the mean and tails. As expected 5-branch CDF is conservative, the log discretization strategies with a few more branches (7br., 10-branch) shifts CDFs close to the analytical result. Provided the percentiles focus on tails of operator response time distribution as in log strategy (e.g. 90, 99, 99.9, 99.99, skip), the results of 5-branch can be close to results of 7-branch.

Importance measures (risk contributors) and their uncertainties, and uncertainty importance measure (Pearson correlation coefficient method) of epistemic parameters have been calculated. The comparison of results among the methods reveals that operator error and its distribution parameter are top contributors to risk and its uncertainty. The ranking order is same among all methods, but the risk and uncertainty contribution of valve and its distribution parameters in 5-branch case are underestimated because of overestimation of operator error in aleatory calculations.

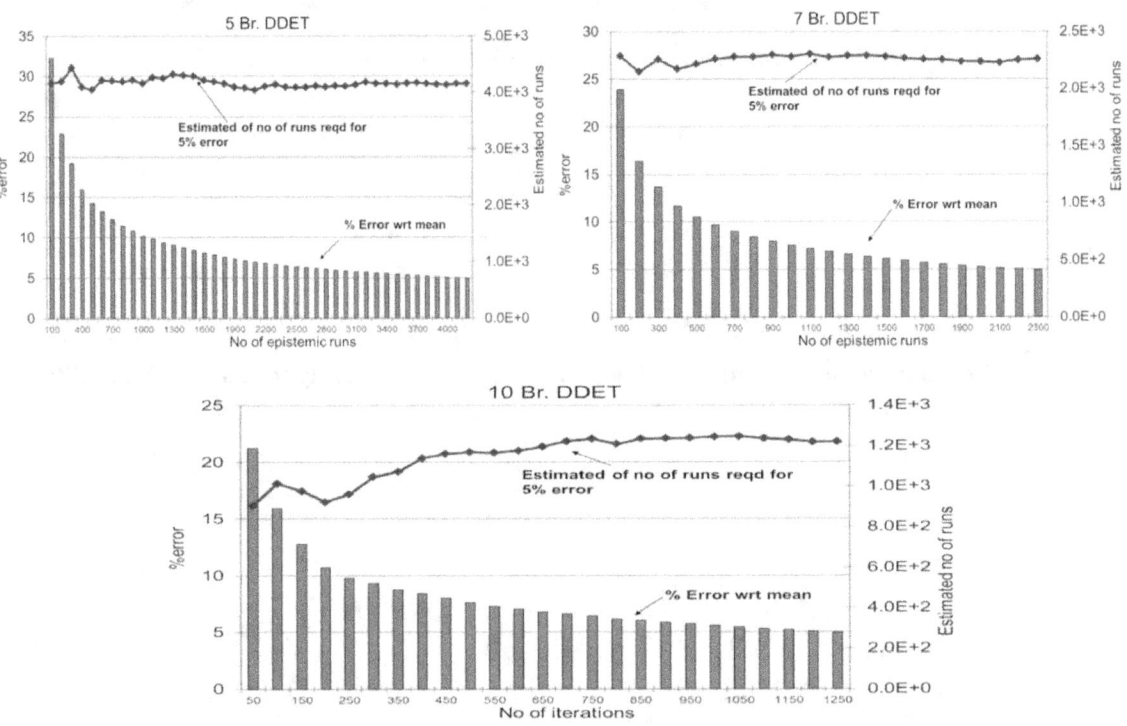

Fig. 9: Convergence criteria and no of runs required – comparison

The DDET approach gives an overestimate of risk in general, but there is no chance of underestimation. Log discretization strategy in DDET reduces conservatism in risk estimate to a great extent and only with few branches involved. Nevertheless, it is important to note that the number of runs gets multiplied as the number of continuous variables to be discretized increases. On the other hand, the number of continuous variables to be discretized in the simulations of nuclear power plant (NPP) accident scenarios are limited, for instance in MLOCA scenario of PWR type NPP there are only two safety functions whose response is continuous.

Convergence criteria and the number of runs

Accuracy costs, but acceptable error in results with limited number of runs is quite important during NPP accident scenario simulations, which challenges even today's computational resources. Convergence criteria used in current calculations monitors the current percentage error and estimates the number of runs required to achieve a given percentage of error with respect to mean. Fig. 9 shows such monitoring plots for DDET with 5, 7, and 10 branch cases. In DDET with 5-branch case, after 500 runs it consistently estimates the required runs to be 4200 runs, and the percentage error after 4200 runs matches 5%. This trend of estimating runs required after a few initial runs is noticed in all other cases as well; only plots for 5, 7, 10-branch DDET cases are shown in Fig. 9. This online convergence criterion could be very useful when the complex nuclear scenarios are explored. Depending on the estimated number of runs and current percentage error after a few initial runs (with initially considered values of percentage error and confidence level), the criterion can be modified to reduce the number of runs or improve the accuracy.

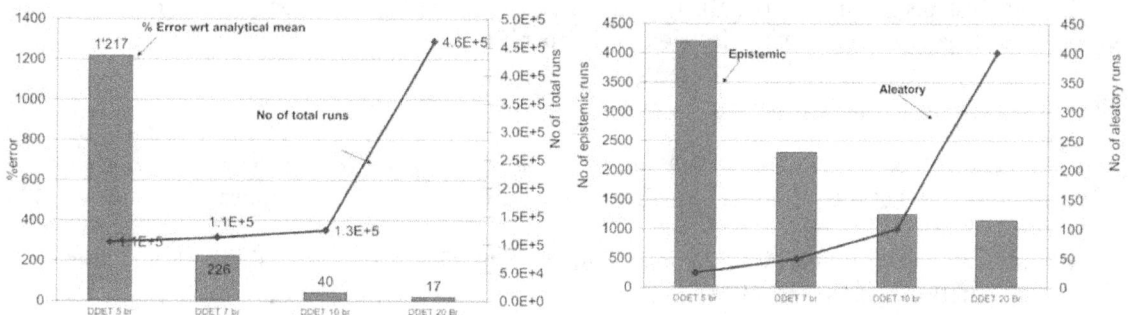

Fig. 10: Comparison of runs and %error among DDET methods

Epistemic Vs Aleatory Runs

An interesting relationship between the number of epistemic and aleatory runs is present. The number of epistemic runs is inversely proportional to the number of aleatory runs before it becomes stable. Their product which gives total number of runs and the percentage error with respect to analytical mean are also dependent. Fig. 10 shows such a relation among the DDET cases. The epistemic runs represent the total number of Monte Carlo simulations sampling epistemic parameters, where each simulation produces a DDET; whereas the aleatory runs represent the number of sequences in each DDET. The number of epistemic runs decreases as the number of aleatory runs increases, and both lines meet in DDET 10-branch case. Looking at the total number of runs versus percentage error with respect to analytical mean, larger number of runs required to reduce the error from 40% to 17%. The optimal point is DDET 10-branch case, which gives less error with a few runs.

Having an optimal number of discretization levels (branches) helps the analyst. However, determining the optimal number of discretization levels is quite challenging in real plant applications and it should not be the aim in such cases. The number of branches may be added in steps and the change in risk estimate among them shall be monitored. When the change is not substantial, it indicates the convergence.

5. CONCLUSIONS

In risk quantification, both epistemic and aleatory uncertainties are present due to inherent physical and safety system variability and their model parameters. Ignoring any one could lead to inappropriate estimation of risk and its uncertainties. DDET along with epistemic uncertainty analysis has been proposed as an approach for solving the problem of integrated treatment of uncertainties and dynamics in quantifying risk. The approach has been demonstrated with the application to modified tank problem. The analytical solution provided as a means of comparison with the obtained results from

DDET cases. The proposed log discretization strategy reduces the conservatism in risk estimate, which is present in discretization strategies used in the literature. The log strategy needs only a few branches to converge with analytical results. Nevertheless, in DDET approach the number of aleatory runs gets multiplied as the number of continuous variables to be discretized increases. The number of epistemic runs depends on the number of aleatory runs. But increasing the aleatory runs beyond a point does not increase the accuracy significantly, but only increases total computational time significantly. Optimal allocation of computational resources between epistemic and aleatory runs ensures accuracy in risk estimate. The convergence criterion in epistemic calculations helps to monitor the current percentage error and estimates number of runs required to achieve a specified accuracy. The computational resources can be used more efficiently to improve accuracy in risk estimate. In more complex problems like NPP accident scenarios, the online convergence criteria will be particularly useful.

Acknowledgements

This work has been performed in the frame of a joint project with PSI's Laboratory for Reactor Physics and Systems Behaviour (LRS). It was partly supported by swissnuclear, the nuclear energy section of the organisation of Swiss electricity grid operators.

References

[1] Scott Ferson, Lev R. Ginzburg, *"Different methods are needed to propagate ignorance and variability"*, Reliability Engineering & System Safety, Volume 54, Issues 2–3, pp. 133-144, (1996).
[2] N. Siu, *"Risk Assessment for Dynamic Systems: An Overview"*, Reliability Engineering and System Safety, 43, pp.43-73, (1994).
[3] IAEA, *"Procedure for conducting probabilistic safety assessment of nuclear power plants (level 1)"*, Safety series no. 50-P-4, International Atomic Energy Agency, 1992, Vienna.
[4] Eduard Hofer et al., *"An Approximate epistemic uncertainty analysis approach in the presence of epistemic and aleatory uncertainties"*, Reliability Engineering and System Safety, 77, pp. 229-238, (2002).
[5] K. Durag Rao et al., *"Quantification of epistemic and aleatory uncertainties in level-1 probabilistic safety assessment studies"*, Reliability Engineering and System Safety 92, pp. 947-956, (2007).
[6] K. S. Hsueh, A. Mosleh, *"The Development and Application of the Accident Dynamic Simulator for Dynamic Probabilistic Risk Assessment of Nuclear Power Plants"*, Reliability Engineering and System Safety, 52, pp.297-314, (1996).
[7] Hakobyan, et al., *"Dynamic generation of accident progression event trees. Nuclear Engg. and Design"*, 238, pp. 3457-3467, (2008).
[8] Izquierdo JM, et al., *"SCAIS (Simulation Code System for Integrated Safety Assessment): Current status and applications"*, ESREL 2008, Safety, Reliability and Risk Analysis – Martorell et al.(eds), Taylor & Francis Group, London, pp. 121-128, (2009).
[9] Martina Kloos et. al., *"MCDET: A probabilistic dynamics method combining Monte Carlo simulation with the discrete dynamic event tree approach"*, Nuclear Science and Eng., 153, pp. 137-156, (2006).
[10] D.R. Karanki, V.N. Dang, T.W. Kim, *"The Impact of Dynamics on the MLOCA Accident Model – An Application of Dynamic Event Trees"*, Proc. 11th Probabilistic Safety Assessment and Management / European Safety and Reliability 2012 (PSAM11/ESREL2012), Helsinki, Finland, 25-29 June 2012, CD-ROM.
[11] D.R. Karanki, V.N. Dang, *"Quantified Dynamic Event Trees Vs PSA – A Comparison for MLOCA Risk"*, ANS PSA 2013 International Topical Meeting on Probabilistic Safety Assessment and Analysis, Columbia, SC, USA, Sep 22-26, 2013, American Nuclear Society, 2013, CD-ROM.
[12] Aldemir, T., *"Computer-assisted Markov failure modeling of process control systems"*, IEEE Transactions on Reliability, R-36, pp. 133-144, (1987).

[13] L. Podofillini, V.N. Dang "*Conventional and dynamic safety analysis: Comparison on a chemical batch reactor*" Reliability Engineering & System Safety, Volume 106, pp. 146-159, (2012).

[14] E. Borgonovo, "*Measuring uncertainty importance measure: Investigation and comparison of alternative approaches*", Risk Analysis, 26, pp. 1349-1361, (2006).

[15] Piero Baraldi, et al., "*Component Ranking by Birnbaum Importance in Presence of Epistemic Uncertainty in Failure Event Probabilities*", IEEE Transactions on Reliability, Vol. 62, No. 1, (2013).

[16] http://www.LMNOeng.com, LMNO Engineering, Research, and Software, Ltd. 7860 Angel Ridge Rd. Athens, Ohio 45701 USA.

[17] Daugherty, R. L., J. B. Franzini, and E. J. Finnemore, "*Fluid Mechanics with Engineering Applications*", McGraw-Hill Inc. 8ed, 1985.

[18] S. S. Rao, "*Reliability-Based Design*", McGraw-Hill Publishers, 1992.

[19] http://www2.econ.iastate.edu/classes/econ671/hallam/documents/Transformations.pdf

[20] Davide Mercurio, "*Discrete Dynamic Event Tree modeling and analysis of NPP crews for safety assessment*", Ph. D. Thesis, Diss. ETH No. 19321, 2011.

[21] M.R. Driels, Y.S. Shin, "*Determining the Number of Iterations for Monte Carlo Simulations of Weapon Effectiveness*", Naval Postgraduate School, Monterey, California, United States, (2004).

An Approach to Physics Based Surrogate Model Development for Application with IDPSA

Ignas Mickus[a*], Kaspar Kööp[a], Marti Jeltsov[a], Yuri Vorobyev[b], Walter Villanueva[a], and Pavel Kudinov[a]

[a]Royal Institute of Technology (KTH), Stockholm, Sweden
[b]Moscow Power Engineering Institute, Moscow, Russia

Abstract: Integrated Deterministic Probabilistic Safety Assessment (IDPSA) methodology is a powerful tool for identification of failure domains when both stochastic events and physical time dependent processes are important. Computational efficiency of deterministic models is one of the limiting factors for detailed exploration of the event space. Pool type designs of Generation IV heavy liquid metal cooled reactors introduce importance of capturing intricate 3D flow phenomena in safety analysis. Specifically mixing and stratification in 3D elements can affect efficiency of passive safety systems based on natural circulation. Conventional 1D System Thermal Hydraulics (STH) codes are incapable of predicting such complex 3D phenomena. Computational Fluid Dynamics (CFD) codes are too computationally expensive to be used for simulation of the whole reactor primary coolant system. One proposed solution is code coupling where all 1D components are simulated with STH and 3D components with CFD codes. However, modeling with coupled codes is still too time consuming to be used directly in IDPSA methodologies, which require thousands of simulations. The goal of this work is to develop a computationally efficient surrogate model (SM) which captures key physics of complex thermal hydraulic phenomena in the 3D elements and can be coupled with 1D STH codes instead of CFD. TALL-3D is a lead-bismuth eutectic thermal hydraulic loop which incorporates both 1D and 3D elements. Coupled STH-CFD simulations of TALL-3D typical transients (such as transition from forced to natural circulation) are used to calibrate the surrogate model parameters. Details of current implementation and limitations of the surrogate modeling are discussed in the paper in detail.

Keywords: IDPSA, Dynamic PSA, Surrogate Model, TALL-3D.

1. INTRODUCTION

Generation IV nuclear reactor designs often incorporate pool type configurations with passive safety features which depend on complex physical interactions and local flow conditions. Lack of operation on large scale means the pre-knowledge of such systems performance, their safety and failure modes is scarce. The IDPSA methodology is proposed in such cases for a comprehensive safety analysis. It is a powerful tool to identify failure domains when both stochastic events and physical time dependent processes are important [1]. This method relies less on expert knowledge of the system and more on the deterministic codes to explore the event space.

Several thousands of calculations need to be run for a successful analysis of a given design. Pool type nature of the designs requires a 3D simulation code to be used as 1D system codes fail to capture complex flow phenomena e.g. mixing and stratification during transient conditions. However, application of a CFD code for the full model of the reactor design is not feasible due to unreasonably long calculation time. To avoid this, approaches for coupling STH and CFD codes have been proposed, which retain accuracy as well as efficiency [2].

However, coupled STH-CFD simulations are still too computationally expensive for IDPSA applications. Here another approach is to use calculation results from a CFD code to create a physics-based Surrogate Model (SM) which would give comparable outcome. Such SM could be used coupled

---
* mickus@kth.se

with an STH code instead of CFD. Using a surrogate model would considerably decrease the calculation time and allow IDPSA methodology to be applied more efficiently.

2. TALL-3D FACILITY DESCRIPTION

Both separate models, calculation codes, and coupled code approaches need to be validated against real-world experimental data. This requires complex and highly specific experimental equipment. Therefore a lead-bismuth eutectic (LBE) thermal hydraulic test facility TALL-3D was built at the Royal Institute of Technology (KTH). The loop is designed to provide experimental data for single STH, CFD and coupled STH/CFD code validation with the aim to study complex feedbacks between the 3D and system scale phenomena [3].

The layout of the facility is shown in Figure 1. The loop is composed of 3 sections: the main heater leg (left); the 3D test section leg (center); and the heat exchanger leg (right). Total height of the facility is 6980 mm whereas the height of the loop piping is 5830 mm. The width of the loop is 1480 mm. The pipes have internal diameter of 27.86 mm and outer diameter of 33.4 mm. The main heater leg contains a 25 kW rod type electrical heater which is 8.2 mm in diameter and the heated part has a length of 870 mm. An expansion tank is installed on the top of the main heater leg. The 3D leg contains a heated pool type test section in its lower part. The test section heater power is 15 kW. The heat exchanger leg contains a heat exchanger in the top part and an electromagnetic pump below it. The secondary loop utilizes DOWTHERM RP as cooling fluid and a fan is used as a secondary heat exchanger. The LBE is stored in the sump tank located at the lower left corner of the loop.

Three ball valves (one per section) are installed for fine tuning of the flow. This enables equalizing the flow rates in the main heater and 3D legs in natural circulation mode, thus allowing the mass flow rates to be brought out of balance by a relatively small effort, leading to flow oscillations in the two hot legs [3].

All parts of the facility in contact with the LBE are manufactured of SS-316L stainless steel to ensure corrosion and erosion resistance. Stainless steel is characterized by a relatively high thermal expansion coefficient, therefore the effects of thermal deformation must be considered. For this purpose TALL-3D was designed with a system of expansion joints, anchored fixation points and pipe guides to preserve the loop geometry, account for different thermal expansion of different legs, damp possible vibrations, and preserve a strict vertical position of the 3D test section.

An extensive amount of instrumentation is installed in both loop type components and the 3D test section to provide sufficient information for single STH, coupled STH/CFD code validation, and investigation of isolated physical phenomena. This includes two Coriolis flow meters on the primary loop; a five-group differential pressure measurement system, enabling measurements of total 15 differential pressures around the loop; and numerous thermocouples positioned in the flow, on the pipe walls as well as on the walls of the insulation. The instrumentation is connected to the Data Acquisition System (DAS) and is operated via a LabView interface.

Most of the facility can be represented as a 1D model since small diameter pipes have negligible 3D flow tendencies. Pool type test section in the middle leg introduces complex axisymmetric geometry and requires full three dimensional flow solution in transient conditions for correct outlet temperature estimation. This means there is a clear separation between the STH code domain and the CFD code domain for the TALL-3D facility.

Figure 1: Schematic view of the TALL-3D facility

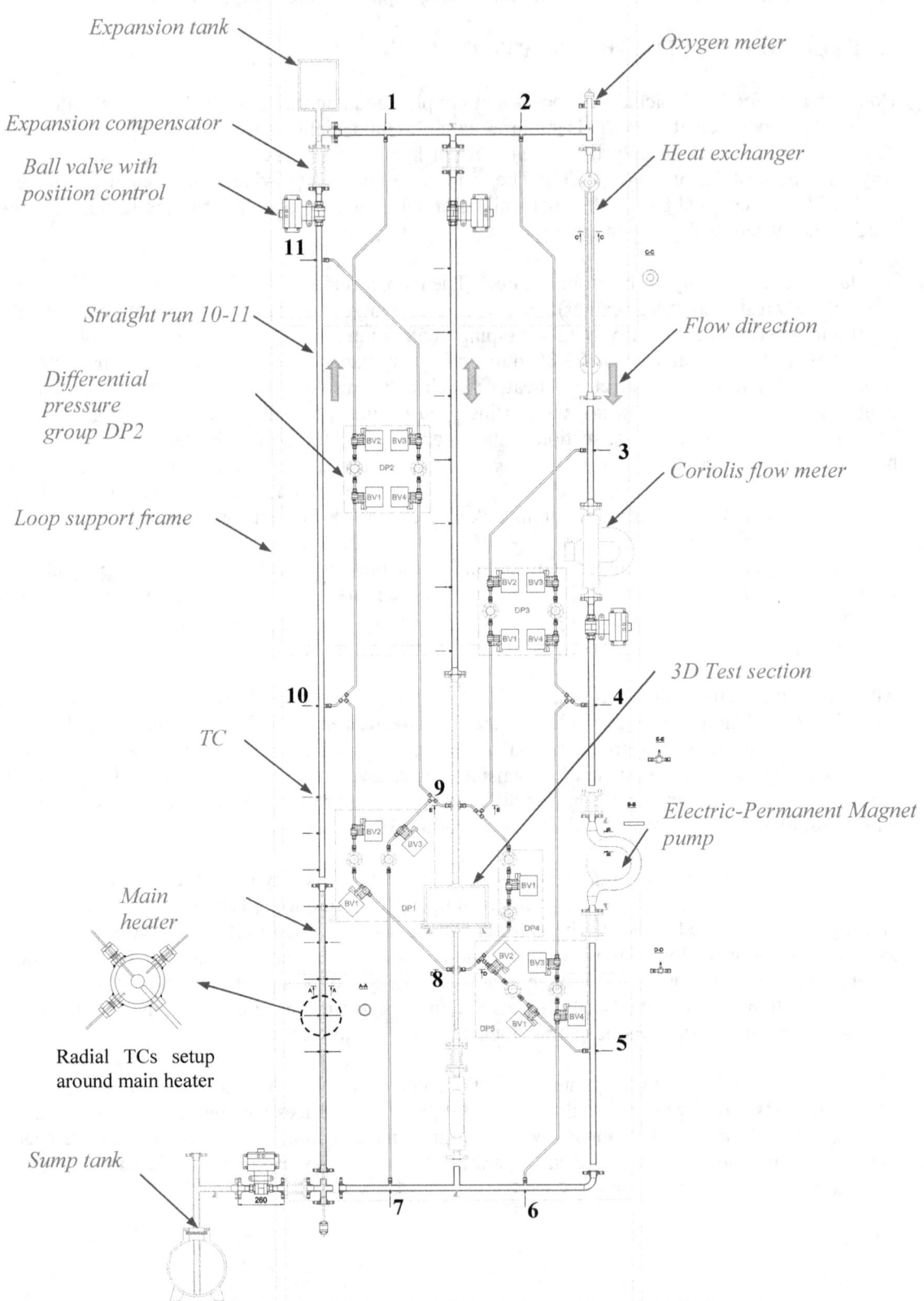

2.1. 3D test section

TALL-3D design incorporates a 3D test section to introduce flow conditions that cannot be properly captured by 1D STH codes. It is an axisymmetric cylindrical vessel with an outlet at the top and an inlet at the bottom. A 15 kW rope heater is installed at the upper part around the lateral wall. The heater is used to facilitate stratification development in the pool, and allows outlet flow temperatures of up to 500°C. The schematics of the test section are shown in Figure 2. Internal diameter and height of the test vessel are 300 and 200 mm respectively. A circular plate of 200 mm diameter is installed inside, orthogonal to the flow direction. The purpose of the plate is to reflect the upward LBE jet towards the walls of the vessel, thus enhancing mixing in the pool at high inlet mass flows. At low flows the jet does not penetrate all the way to the plate and stratified layer forms in the upper section of the pool. The temperature distribution is measured by thermocouples installed in-flow, and on internal and external walls of the test section.

Figure 2: Schematics of the TALL-3D test section vessel

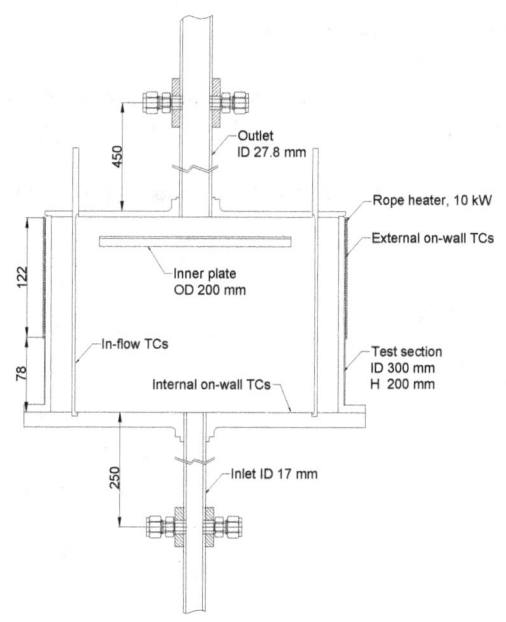

CFD analysis was performed in the 3D test section design process for choosing the proper inlet pipe diameter, test section height, heater power and geometry. The aim of the design is to ensure, that if the whole loop is close to instability in natural circulation conditions, the 3D test section would be the component responsible for 1D/3D feedbacks. Namely the pool should be mixed at high flow rates and stratified at low flow rates. The selection of parameters was done using the scaling analysis developed for buoyant jets in pool-like geometries [4]. The configuration was selected in such way that stratification is allowed to develop in the pool at flow rates up to 0.65 kg/s. Figure 3 shows the results of pre-test CDF modelling. The pool is mixed when the inlet mass flow rate is 1 kg/s and stratified when the inlet mass flow rate is 0.2 kg/s.

Figure 3: CFD simulation results for the 3D test section. Temperature is shown on the left and velocity on the right of the pool for flow rate 1.0 kg/s and 0.2 kg/s

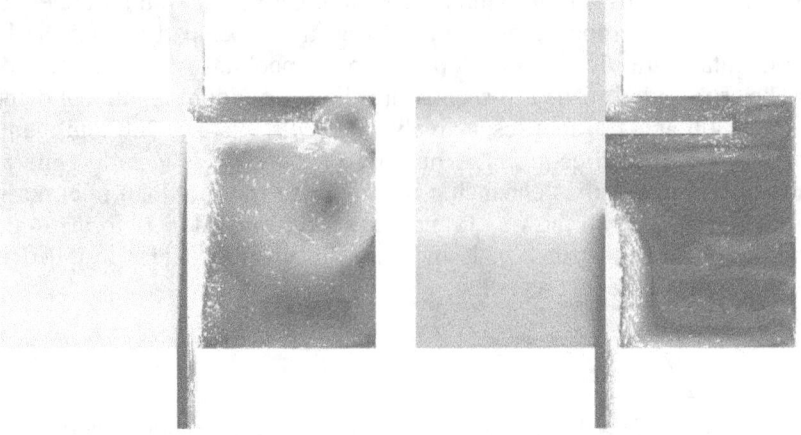

The effects resulting from such 3D component introduction into the loop design were demonstrated in pre-test STH and coupled STH-CFD simulations of the TALL-3D facility. The presence of the test section causes non-linear interactions between 3D and 1D loop components, which result in oscillations, not properly resolved by single STH code, as shown in Figure 4.

Figure 4: LBE temperatures at the bottom and top of 3D test section [2]

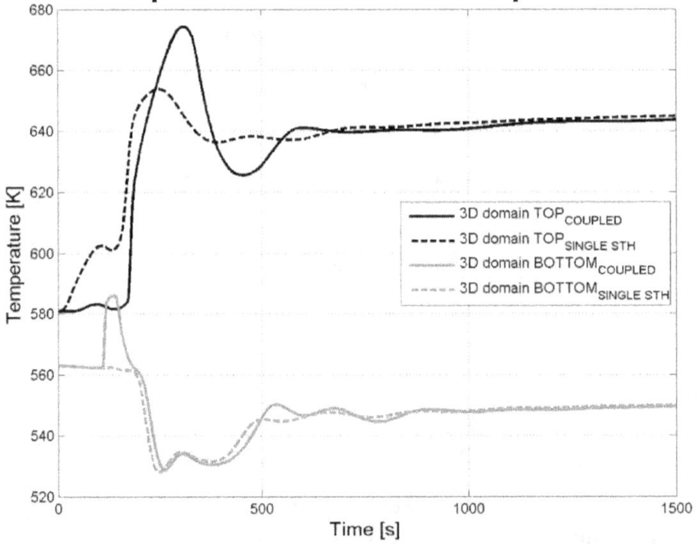

3. SURROGATE MODEL DEVELOPMENT APPROACH

Surrogate Model (SM) is necessary when using original multi-dimensional high resolution codes becomes too computationally expensive. For example, a TALL-3D transient simulation using coupled STH-CFD approach takes several days of computing. Such approach is not suitable in the IDPSA methodology, where large numbers of calculations need to be run to identify failure domains and perform sensitivity and uncertainty analysis. Therefore we seek to develop a surrogate model to replace the CFD part in coupled calculations to decrease the computational time, and make the coupled calculation suitable for IDPSA.

The aim of the SM is to reproduce only a limited set of results with acceptable accuracy, while disregarding the detailed modelling of the actual physical system, thus avoiding the direct consideration of physical processes which would introduce complexity and increase computational effort (ex. solution of 3D conservation equations which leads to a CFD approach). However the SM needs to give better results than currently available models implemented in STH codes to resolve previously identified 3D effects. Thus the criterion of a successful SM development is to obtain a more accurate solution, than achievable by a single STH code, and achieve better computational performance compared to CFD or coupled STH-CFD approaches.

One of the more popular method for surrogate model development depends on an advanced approximation of a complex function which is represented by a database of full model solutions e.g. application Neural Networks (NN). However, this way no actual physics modelling is involved in the SM, and the procedure is essentially a fitting process. This approach requires large number of data calculated by the original model, so the actual gain in computational efficiency can even decline in the process of SM development and application. Furthermore, it results in a SM that cannot be assumed reliable outside of the domain of parameters covered in the database of the original model solutions, since no physical reasoning is applied to the extrapolation [5].

Our approach is to capture the most important physical phenomena in the surrogate model itself. Then, a process of calibration is applied to determine the parameters which represent the physical processes not directly modelled in the SM. Since a limited set of SM output results is desired, only few parameters need to be considered for calibration. These SM closures can be deduced by standard fitting procedures, or by more advanced application of Neural Networks. Therefore our goal is to seek for an intermediate approach in between a pure NN application and a pure analytical solution. The process can be summarized as shown in Figure 5. These steps form an iterative process, which is continued until the desired SM precision is reached.

Figure 5: Surrogate model development procedure

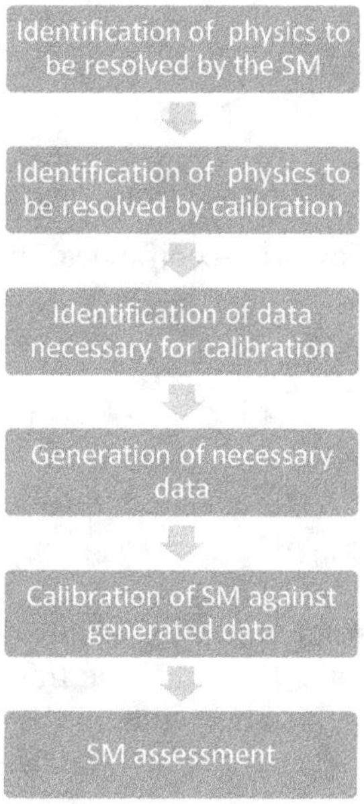

- Identification of physics to be resolved by the SM
- Identification of physics to be resolved by calibration
- Identification of data necessary for calibration
- Generation of necessary data
- Calibration of SM against generated data
- SM assessment

The SM should correctly reproduce the history of the system, so temporal evolution equations have to be solved explicitly. However the exchange rates in these equations are of local nature, and do not depend on history effects. Therefore they can be determined by applying empirical closures. In this work we develop the surrogate model based on pre-calculated CFD solutions. Currently an idealized CFD model is utilized, and no heat losses are considered. We believe that CFD application will give us the general form of closures, which will be later adjusted in accordance with the experimental results.

4. DEVELOPMENT OF SM FOR TALL-3D

The ultimate goal of SM development for the TALL-3D test section is to couple the SM with a STH code to be used for IDPSA. The surrogate model will use the inlet mass flow, inlet temperature and pool heater power values to estimate outlet temperature of the 3D pool at given moment. This value will be used in turn in STH code in the same way as in STH-CFD coupling [2].

Four CDF transients were run to gather insight about the mixing and stratification phenomena (see Table 1). LBE temperature at the test section inlet and the test section heater power were kept constant at 425 K, and 5 kW respectively. The transition from fully mixed to fully stratified states was seen in CFD results in the decreasing mass flow rate transients, while the stratified pool was fully mixed in the increasing mass flow rate cases. The test section outlet temperature and total momentum values were monitored in CFD solution and analyzed.

Table 1: Parameters of CFD transient cases run for mixing/stratification analysis

Transient	Initial $\dot{m}$ (kg/s)	Final $\dot{m}$ (kg/s)	$d\dot{m}/dt$ (kg/s²)
1	1	0.2	-0.005
2	1	0.2	-0.1
3	0.2	1	0.005
4	0.2	1	0.1

The most straight-forward approach to evaluate the outlet temperature comes from a simple energy balance over the 3D test section. Assuming the LBE pool as a homogeneous control volume, the outlet temperature T can be calculated as follows:

$$\frac{d(mc_p T)}{dt} = \dot{m}\bar{c}_p(T_{in}-T) + \dot{Q}_h \tag{1}$$

Here m is the mass of LBE in the test section, $\dot{m}$ is the mass flow, T_{in} is the inlet temperature, $\dot{Q}_h$ is the heater power, and $\bar{c}_p$ is the average isobaric specific heat, which can be evaluated as:

$$\bar{c}_p = \frac{1}{T-T_{in}} \int_{T_{in}}^{T} c_p dT' \tag{2}$$

where the isobaric specific heat correlation is given as [6]:

$$c_p = 159 - 2.72 \cdot 10^{-2} T + 7.12 \cdot 10^{-6} T^2 \tag{3}$$

Solving equation (1) gives good agreement with CFD simulation results when predicting steady state outlet temperature values. However, it fails to predict the transient behavior as shown in Figure 6, as expected. Such discrepancies are caused by 3D effects in the test section, namely the inlet jet – wall jet, and the inlet jet – stratified layer interactions. Therefore the 3D test section cannot be modelled as a homogeneous control volume.

Figure 6: Test section outlet temperature during decreasing mass flow transients (1 kg/s to 0.2 kg/s) at two rates - slow (0.005 kg/s2) and fast (0.01 kg/s2)

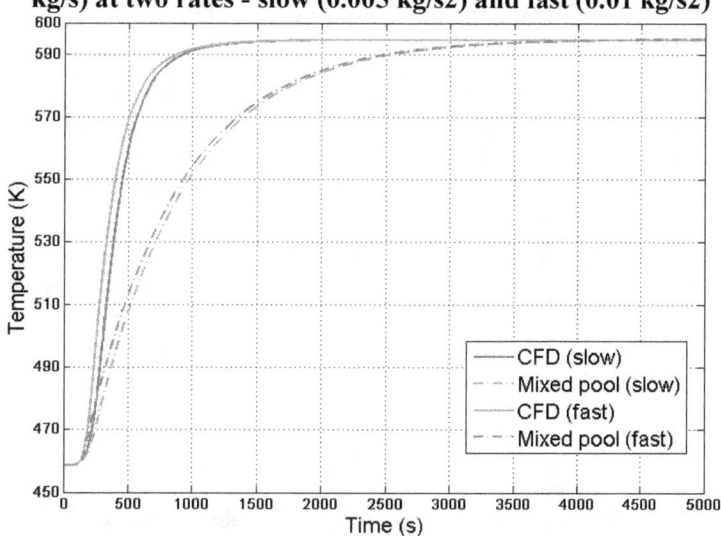

Without going into detailed investigation of the 3D phenomena, the additional effects may be included in equation (1) by adding a "3D term" $\dot{Q}_c$, to account for the observed discrepancies:

$$\frac{d(mc_p T)}{dt} = \dot{m}\bar{c}_p (T_{in} - T) + \dot{Q}_h + \dot{Q}_c \qquad (4)$$

Figure 7: $\dot{Q}_c$ **term correlation with total pool momentum for decreasing mass flow transient**

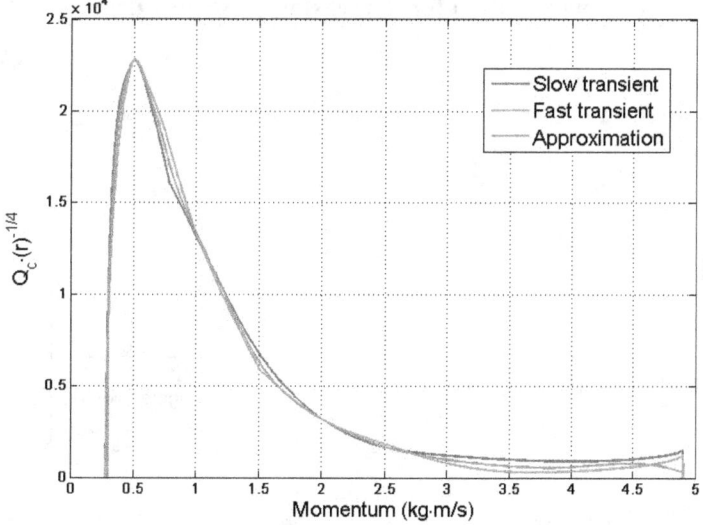

Taking the test section outlet temperature data from the CFD solution, the $\dot{Q}_c$ term can be calculated directly from equation (4). The evolution of this term (adjusted for different rates of change of mass flow $\dot{Q}_c r^{-1/4}$) during the decreasing mass flow transient (stratification development in the pool) was found to correlate with the instantaneous pool momentum values, as shown in Figure 7.

Figure 8: $\dot{Q}_c$ **term as calculated by CFD and from the correlation with the total pool momentum for decreasing mass flow transient**

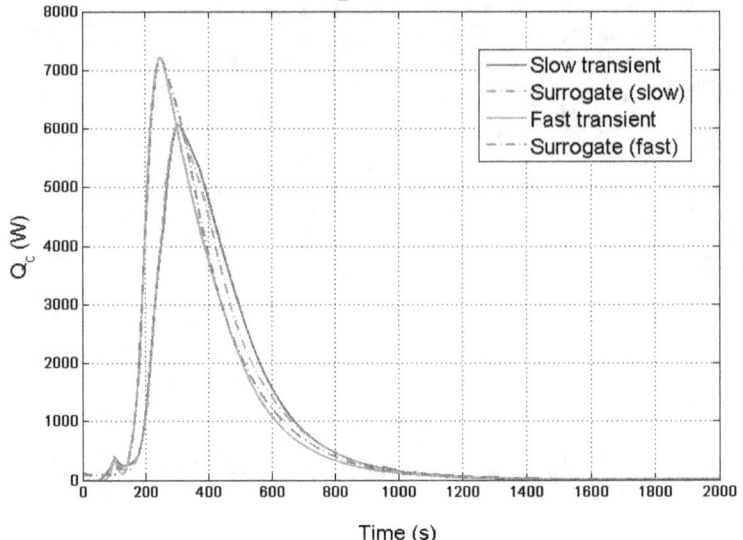

Here r denotes the rate of change of mass flow ($d\dot{m}/dt$). Approximating the observed trend, the $\dot{Q}_c$ term evolution during the transient can be reproduced reasonably well, as shown in Figure 8. Substituting the reproduced $\dot{Q}_c$ values back into eq. (4) a significant improvement over the mixed pool temperature prediction is achieved, as shown in Figure 9.

Figure 9: Test section outlet temperature values calculated by CFD and by solving eq. (4) using the $\dot{Q}_c$ approximation for decreasing mass flow transient

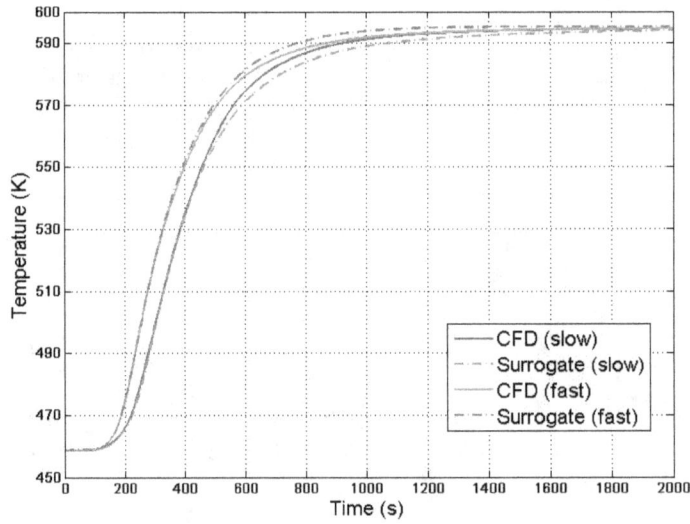

The same approach, however, does not yield a good correlation between $\dot{Q}_c$ and pool momentum values during the increasing mass flow (mixing of the stratified pool) transient, as shown in Figure 10. The correction of the $\dot{Q}_c$ term by $r^{-0.67}$ allows matching the amplitude, but the pool momentum value at which the 3D effects have the highest influence seems to be dependent on the rate of change of mass flow (flow acceleration), contrary to what was found in the stratification development case.

Figure 10: $\dot{Q}_c$ term correlation with total pool momentum for increasing mass flow transient

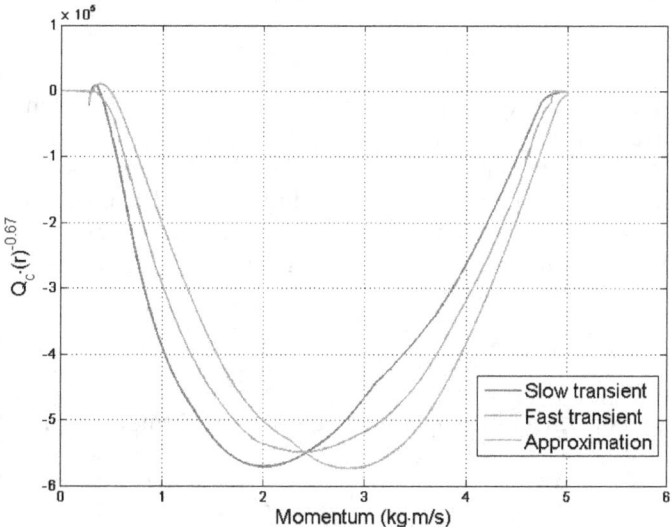

Roughly approximating the observed trend, and substituting the $\dot{Q}_c$ values into eq. (4) results in temperature prediction as shown in Figure 11.

Figure 11. Test section outlet temperature values calculated by CFD and by solving eq. (4) using the $\dot{Q}_c$ approximation for increasing mass flow transient

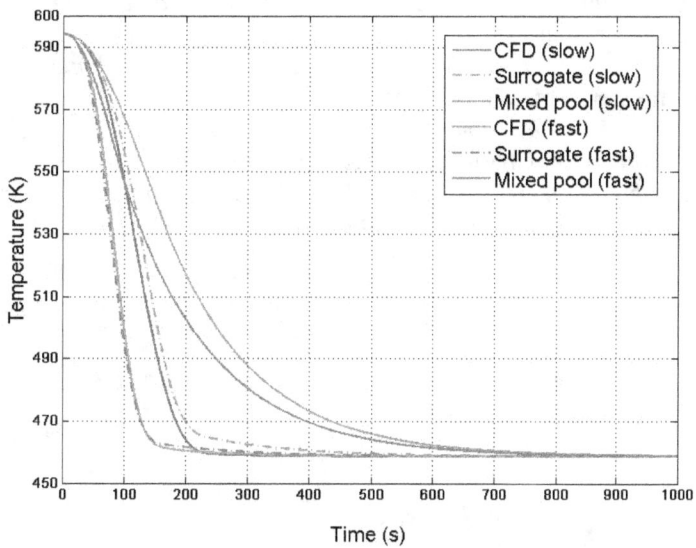

Better agreement could be achieved by directly interpolating the $\dot{Q}_c$ values in the form as shown in Figure 10. This would require running CFD cases of limiting flow acceleration that could be expected in the TALL-3D loop. Furthermore, despite it was shown that 3D effects can be correlated with the pool momentum, the pool momentum itself must be calculated from the known inlet mass flow rate and flow acceleration values, for the surrogate model to be fully self-sufficient. Currently pool momentum values were taken directly from CFD calculations therefore the momentum modelling is still to be resolved. However, after identifying the relation between the 3D effects and the total pool momentum value, Neural Network approximation approach could be applied to link the $\dot{Q}_c$ values with the pool momentum, and the pool momentum with the known parameters of LBE mass flow rate and flow acceleration.

6. CONCLUSION

We can conclude from the results shown in this paper that the 3D effects present in the TALL-3D test section pool can be correlated with the total momentum in the pool and that the test section outlet temperature can be predicted reasonably well in case of transient with mass flow decrease (stratification development), but further effort is needed to link pool momentum with inlet mass flow and flow acceleration. Also, further work including analysis of additional transients, confirmation of the stratification transient SM results as well as development of an approach for momentum prediction is foreseen.

Acknowledgements

This work has been carried out under the support of European 7th framework project THINS.

References

[1] Y. Vorobyev, P. Kudinov. *"Development of methodology for identification of failure domains with GA-DPSA"*, 11th International Probabilistic Safety Assessment and Management Conference and the Annual European Safety and Reliability Conference (PSAM, ESREL) (2012)

[2] M. Jeltsov, K. Kööp, P. Kudinov and W. Villanueva. *"Development of a Domain Overlapping Coupling Methodology for STH/CFD Analysis of Heavy Liquid Metal Thermal-Hydraulics"*, NURETH-15, Pisa, Italy, May 12-17 (2013)

[3] M. Jeltsov, K. Kööp, D. Grishchenko, A. Karbojian, W. Villanueva, P. Kudinov. *"Development of TALL-3D Facility Design for Validation of Coupled STH and CFD Codes"*, NUTHOS-9, Kaohsiung, Taiwan, September 9-13 (2012)

[4] P. F. Peterson. *"Scaling and analysis of mixing in large stratified volumes"* International Journal of Heat and Mass Transfer, **Vol. 37**, no. 1, pp. 97-106 (1994)

[5] P. Kudinov, M. Davydov. *"Development of Surrogate Model for Prediction of Corium Debris Agglomeration"*, ICAPP 2014, Charlotte, USA, April 6-9 (2014)

[6] OECD/NEA. *"Handbook on Lead-bismuth Eutectic Alloy and Lead Properties, Materials, Compatibility, Thermal-hydraulics and Technologies"*, OECD (2007)

A toolkit for integrated deterministic and probabilistic risk assessment for hydrogen infrastructure

Katrina M. Groth[*a], Andrei V. Tchouvelev[b,c]
[a] Sandia National Laboratories, Albuquerque, NM, USA
[b] AVT Research, Inc., Canada
[c] International Association for Hydrogen Safety, HySafe

Abstract: There has been increasing interest in using Quantitative Risk Assessment [QRA] to help improve the safety of hydrogen infrastructure and applications. Hydrogen infrastructure for transportation (e.g. fueling fuel cell vehicles) or stationary (e.g. back-up power) applications is a relatively new area for application of QRA vs. traditional industrial production and use, and as a result there are few tools designed to enable QRA for this emerging sector.

There are few existing QRA tools containing models that have been developed and validated for use in small-scale hydrogen applications. However, in the past several years, there has been significant progress in developing and validating deterministic physical and engineering models for hydrogen dispersion, ignition, and flame behavior. In parallel, there has been progress in developing defensible probabilistic models for the occurrence of events such as hydrogen release and ignition. While models and data are available, using this information is difficult due to a lack of readily available tools for integrating deterministic and probabilistic components into a single analysis framework. This paper discusses the first steps in building an integrated toolkit for performing QRA on hydrogen transportation technologies and suggests directions for extending the toolkit.

Keywords: hydrogen; integrated deterministic and probabilistic; codes and standards; software; QRA;

1. INTRODUCTION

Early market hydrogen fuel cell installations have set a precedent for safe use of hydrogen. As the hydrogen industry shifts toward market deployment and commercialization, safety remains a top priority. In North America, hydrogen infrastructure must meet all applicable codes and standards to demonstrate that they are safe, reliable, and compatible.

Considerations of regulations, codes, and standards (RCS) for use of hydrogen as a transportation fuel are influenced by industrial gas, oil and gas and nuclear power industries. The RCS requirements used in these traditional industrial applications are generally conservative and often barriers to the further promulgation of hydrogen fueling infrastructure. To address these limitations, and to enhance the scientific basis for codes and standards, there has been a push toward using risk information to develop and revise hydrogen-specific standards such as NFPA 2: Hydrogen Technologies Code. During the previous code cycle, various parts of NFPA 2 have been revised using a combination of deterministic and probabilistic analyses. QRA has been performed to address separation distances in the 2011 revision to NFPA 2 [1], and to address size limits of fueling rooms in the 2014 code revisions [2]. Similar work is being planned during the upcoming revision cycle of the Canadian Hydrogen Installation Code (CHIC) within the 2014-15 timeframe.

While general QRA methods are applicable to hydrogen systems, more widespread use of QRA for code revisions is limited due to significant gaps in current data, models, and tools available for applying QRA on hydrogen systems. These gaps have been discussed in the references [2], [3], [4], [5]. Major gaps include the lack of hydrogen-specific data for use in scenario quantification (especially release probabilities) and the lack of physics-based, hydrogen-specific models (for use in

---
[*] Corresponding author. E-mail address: kgroth@sandia.gov

deriving ignition probabilities, and for consequence modeling), and inadequate handling of uncertainty. Several international research teams are actively working to address many of these gaps through data collection, experiments, and analyses.

In June 2013, Sandia and HySafe organized an expert workshop to identify limitations of existing QRA tools [5]. Additional gaps identified during the workshop include oversimplified assumptions regarding leak duration and ignition timing, and a lack of tools for performing QRA with hydrogen-specific models and data.

The results of the workshop showed that use of QRA for hydrogen applications currently suffers from inefficiency, requiring multiple experts and a variety of disparate modeling techniques. These deficiencies prevent the code development committees from self-sufficient operations, as they require multiple experts to consolidate and operate the relevant probabilistic and physical models. The deficiencies also affect the industry as they struggle to design systems compliant with the codes, suggest revisions to the code and apply for variances to codes necessitated by site-specific constraints.

Sandia and HySafe are working to build simplified toolkits to facilitate the use of QRA within the hydrogen fuel cell industry. This paper presents progress on those activities. Section 2 discusses the purpose, basic concept, objectives and targeted audiences for QRA tools applied to hydrogen. Section 3 discusses the underlying hydrogen-specific QRA methodology, which integrates available probabilistic and deterministic models into a single approach. Section 3 also discusses how this methodology is implemented in Sandia's HyRAM (Hydrogen Risk Assessment Models) toolkit, which is intended to support development of hydrogen codes and standards such as NFPA 2. Section 4 discusses the HyRAM interface design, and Section 5 presents the next steps for development of hydrogen-specific QRA tools.

2 TOOLKIT NEED

2.1 Purpose

From the hydrogen safety research community perspective, one of the key tasks of scientific work is to translate fundamental scientific findings into practical formulas, which are easily applied in daily work. While Computational Fluid Dynamics (CFD) codes produce high resolution results for complex physical phenomena, the substantial financial and computational resources required for CFD make it unusable for daily safety decisions. For many safety decisions or for analyses with large numbers of relevant variables (i.e., QRA), validated simplified models, correlations, and statistics data can be used to provide robust results with significantly fewer resources than required by simulations.

The goal of this work is to integrate these simplified pieces into a software toolkit that can be used to assess the hazards and risk in scenarios associated with certain hydrogen system configurations, in a timely manner. Significant and ongoing international efforts, initiated under the auspices of the US DOE, the IEA HIA Tasks 19/31 on hydrogen safety and HySafe, produce various first order models, statistical models, empirical correlations and criteria for the myriad physical and engineering processes relevant to understanding the hazards associated with hydrogen systems. The resulting toolkit will provide practical, efficient access to state-of-the-art models and data required to perform risk assessments of hydrogen systems. In particular subject matter experts, who cannot afford their own safety research or expensive numerical simulations, will benefit from this unique resource. At the same time, the toolkit can serve as a reference basis for the development of risk-informed RCS

2.2 Basic Concept

The primary objective is to develop a library of modern hazard assessment tools that contains best-available models and data relevant to understanding and quantifying risk in hydrogen fuel cell infrastructure. The toolkit must:

- Contain the latest available data and models (ideally, validated for hydrogen infrastructure use) relevant to quantifying the probability of progression various hazard scenarios;
- Contain the latest available data and models (ideally, validated for hydrogen infrastructure use) relevant to prediction of physical properties of hydrogen releases and ignition events, and the consequences of those events;
- Contain risk metrics the represent observable quantities (e.g., physical parameters, losses, number of fatalities) relevant to decision making for safety, codes, and standards
- Facility relative risk comparison, sensitivity analysis, and treatment of uncertainty;
- Be built in a modular configuration;
- Contain user-friendly, graphical interfaces;
- Provide default models, values and assumptions, and provide transparency about those defaults; furthermore, it must allow modification of these defaults to reflect different systems and new knowledge.

The toolkit has two stakeholder audiences: *users* and *developers* [5]. The *users* group includes organizations and analysts interested in developing codes and standards, designing and permitting stations, justifying code variances, etc. The *developers* group includes researchers and agencies who focus on development and validation of the underlying models and data.

As safety is a public concern and a big part of relevant scientific work is funded by public agencies, the toolkit would ideally be an open and free software system, that is well documented and quality assured in a cooperative manner. Ideally, the tools shall be maintained by the hydrogen safety research community itself, after an initial framework is developed by public agencies such as the US DOE. A potential custodian of an international toolkit might be HySafe.

A typical case will consist of a user-defined scenario specified by the inventory or hydrogen flow, geometrical settings (e.g., confinement and/or congestion), system parameters, mitigation measures, up to a maximum release size. For a statistical analysis, including treatment of uncertainty, any of these scenario properties could also be defined by a probability distribution instead. Then the user will select among appropriate models for analysing the scenario. The input required by the selected tool will be input by the associated tool interface. If any of the input parameters lie outside the validity range of the models, an appropriate warning will be displayed for the user.

Each module will have a defined documented set of input parameters and a set of output or result parameters. Each module shall be described in detail, with a defined valid range of input parameters. Literature support and experimental and computational validation exercises relevant to each module should be documented, along with the valid range of the model and key underlying assumptions.

With all requirements defined so far, multiple implementations can be envisaged. Independent custodians of the toolkit could facilitate development and use for the different communities. Many configurations or arrangements of custodianship are possible and require international partnership. In the current approach, Sandia is developing a toolkit targeted toward US DOE domestic stakeholders (such as the NFPA 2 code committee or state fire marshals). In parallel, HySafe is coordinating the international community to "crowd-source" efforts to develop and host an open-source toolkit.

The Sandia version is being developed in C# with an underlying MVC (Model View Controller) framework to enable development of multiple different user views on top of the same underlying models. Sandia is currently pursuing a .NET software framework with a planned Windows and HTML interfaces for codes and standards users. For the international version being developed by HySafe, a WEB2.0 kind of implementation is envisaged; a system which allows for immediate testing and on-the-fly editing of the tools is the Smalltalk dialect Squeak based dynamic web development framework Seaside.

3. QRA APPROACH

The methodology used in Sandia's HyRAM toolkit starts with the hydrogen-specific QRA approach documented in [2]. The approach uses a combination of probabilistic and deterministic models to evaluate the expected fatal accident rate for workers in a warehouse with an indoor hydrogen fuel dispenser. The methodology uses traditional probabilistic approaches to assess the likelihood of various hydrogen release and ignition scenarios, which can lead to thermal and overpressure hazards. Several types of deterministic models are used together to characterize the physical effects of the hazards. Information from the physical effect models is passed into probit functions used to calculate consequences in terms of number of fatalities.

3.1 Risk Metrics

Currently, the methodology focuses on calculating fatality risk. The HyRAM toolkit includes three well-known fatality risk metrics:
- FAR (Fatal Accident Rate) – the expected number of fatalities per 100million exposed hours
- AIR (Average Individual Risk) – the expected number of fatalities per exposed individual
- PLL (Potential Loss of Life) – the expected number of fatalities per system-year.

In addition to the fatality risk metrics, the code also outputs various metrics, which may be relevant to codes and standards users:
- Expected number of hydrogen releases per system-year (unignited and ignited cases)
- Expected number of jet fires per system-year (immediate ignition cases)
- Expected number of deflagrations/explosions per system-year (delayed ignition cases)

3.2 Hazards

The primary hazards related to the use of hydrogen are the release and subsequent ignition of hydrogen. The two main hazards associated with releases of hydrogen are exposure to thermal radiation from jet fires and exposure to overpressures from explosions. Both of these hazards can affect people, property, structures, and the environment directly or indirectly.

Two other hydrogen-related hazards are not included in the approach from [2]: asphyxiation and projectile damage. Hydrogen can cause oxygen displacement, which can lead to asphyxiation. However, since hydrogen is highly buoyant and diffusive, asphyxiation is likely only a concern in extremely small, well-sealed spaces. High-pressure gases escaping from a ruptured containment vessel can propel debris or projectiles at high velocity. However, due to the continuous improvement of gas storage cylinder technology (e.g., leak-before-rupture failure mechanisms, passive protection measures like temperature-actuated pressure relief devices), it is assumed that risk from fires significantly dominates risk from debris. As a result, the risk of projectiles is not modeled in the current framework. Future modules should be designed to address these hazards, or quantitative analyses should be conducted to ensure that these hazards negligibly contribution to risk for all use cases.

3.3 Accident Scenario Models

A jet fire is assumed to result from the immediate ignition of a hydrogen release, while delayed ignition events result in explosions (e.g., deflagrations or detonations). These two hazards can be represented in an Event Sequence Diagram (ESD) model, as shown in Figure 1.

The initiating event for all scenarios is a release of hydrogen gas from a hydrogen system (e.g., damaged piping or a lose fitting or valve on a storage tank, compressor or dispenser,). Hydrogen releases from the dispenser can occur through one of several mechanisms:
- External leaks from individual components, or separation of an individual component
- Shutdown failures
- Accidents

Release characteristics vary widely, and different sized releases are associated with different consequences, and in some cases with different causes. The approach assumes one ESD for each of five releases sizes, based on release areas equal to specific percentages of the pipe cross-sectional flow area: 0.01%, 0.1%, 1%, 10%, 100%. In the HyRAM toolkit, users input the pipe outer diameter and wall thickness, and the toolkit calculates pipe flow area and release size. In future versions, release size may be sampled from a distribution instead of assigned by the algorithm.

3.2.1 Accident Scenario Quantification

Quantification of 100% releases is based on a combined Fault Tree (FT) and parts count approach; release frequencies for the other four release sizes are based solely on the parts count approach. To analyze a system, users input parts counts for each of the nine component types: compressors, tanks, filters, flanges, hoses, joints, pipes, valves, and instruments.

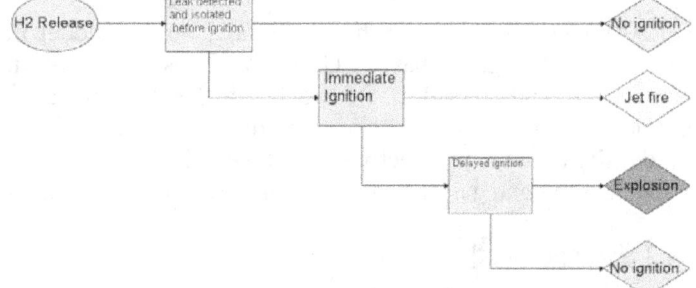

Figure 1: Event Sequence Diagram for scenarios resulting from hydrogen releases.

The parts count approach uses frequency data generated from a Bayesian combination of hydrogen-specific data with data from similar industries, provided in [6]. Each combination of component type and leak size is associated with a lognormal distribution for frequency of leaks (per year). Quantification is achieved by multiplying the number of components by component-specific leak frequencies.

A FT approach is used to incorporate root causes of shutdown failure and accidents. The quantification of 100% releases is achieved by adding the frequencies from the part counts to the release probabilities obtained by multiplying annual system demands by the results of the FT. Users manually enter number of system demands. Currently, cut-sets associated with the system FT must be hard-coded into HyRAM. Component failure probabilities (per demand) for the basic events in the FT were assembled from generic data from non-hydrogen industries, documented in [2]; each of these failures events is associated with either a beta distribution or an expected value.

Hydrogen ignition probabilities used in the ESD were developed by Tchouvelev within the Canadian Hydrogen Safety Program [7]. This approach uses a look-up table to assign mean probability of immediate and delayed ignition as a function of hydrogen release rate. The hydrogen release rate used to assign ignition probabilities is discussed in Section 3.4.1. The leak detection and mitigation event in the ESD is quantified using a point estimate value documented in [2].

In the current HyRAM toolkit, calculations are performed with mean values of the associated probability distributions. The scenario module outputs a vector of expected annual frequencies of jet fires and of explosions resulting for each of the five release sizes.

3.4 Physical Effects Models

Physical effects models are used to characterize the magnitude of hazards associated with ignited hydrogen releases. The primary physical effects of ignited gaseous hydrogen releases are fire effects (impinging flames, temperature, heat flux) and explosion effects such as pressure and impulse waves. The dominant damage mechanism from jet flames is heat flux, and the dominant damage mechanism from explosions is overpressure.

Characterizing physical effects requires modeling a series of aspects of hydrogen behavior: release behavior, flame initiation, flame sustainment, radiation patterns and overpressures. In HyRAM, we use a combination of available first-order deterministic behaviors models and CFD models.

3.4.1 Hydrogen Release Characteristics

The first step in characterizing consequences is to characterize the release of hydrogen and the extent of the flammable envelope. Thermodynamic parameters of releases from high-pressure hydrogen systems can be estimated using notional nozzle models. Papanikolaou et al [9] compare five notional nozzle models for hydrogen jets, and found that the best accuracy was obtained by using either the Schefer et al. [10] or the Birch et al. [11] models. Both models implement conservation of mass and momentum, but the two models implement different equations of state. Birch uses the Ideal gas law, whereas Schefer implements the Abel-Noble equation of state for non-ideal gases.

HyRAM incorporates the Schefer notional nozzle model because high-pressure hydrogen releases exhibit non-ideal gas behavior. User input consists of system temperature and pressure, and ambient temperature and pressure. For each of the five release sizes, the nozzle model outputs key thermodynamic parameters: hydrogen mass flow rate, effective gas temperature at the exit, effective gas density at the exit, effective release velocity, effective leak area and diameter at the exit, effective Mach number, and gas pressure, temperature and density at the nozzle throat.

3.4.2 Hydrogen jet flame effects

Releases from high-pressure hydrogen systems that are ignited immediately produce momentum driven jet flames. Houf and Schefer [12] developed and validated a first-order model for predicting the radiative heat flux and the flammability envelope of a hydrogen jet flame; this model has been used to provide the basis for several parameters specified in NFPA2.

The HyRAM code incorporates the Houf & Schefer model, which is used to calculate the heat flux at a given axial and radial distance from a hydrogen jet flame. This module uses the thermodynamic parameters calculated by the notional nozzle model as input conditions. User input consists of information about the axial and radial positions of interest.

Since the focus of the current approach is on human harm, the positions of interest are places where people are located relative to the flame. In the current version of HyRAM, users input the number of potentially exposed persons (e.g., number of persons in the warehouse or near the facility) and the length and width of the facility (measured from the system). HyRAM includes a module that generates a random position for each person by sampling either a uniform distribution or a normal distribution. The axial and radial position of each person is fed into the Houf and Schefer model. The output is a matrix containing a heat flux value for each person, for each release size.

3.4.3 Overpressure effects

The dominant damage mechanism from explosions is overpressure. Since first-order model for predicting overpressure and impulse effects for QRA purposes are still being developed, the HyRAM toolkit requires user input from the results of simulations to enable the QRA calculations. CFD codes such as FLACS, FUEGO, and FDS have been used to simulate overpressure and impulse effects, and have been validated for aspects of hydrogen behavior [21], [22].

3.5 Harm (Fatality) Models

The results of physical effects evaluations must be translated into a probability of causing damage to an individual, component, or structure for use in a QRA. Probit models are used to establish the probability of injury or fatality for a given exposure. LaChance et al. [13] provides an overview of probit models for thermal exposures and for overpressure exposures, and provides recommendations regarding the applicability of the models to hydrogen hazard scenarios. HyRAM incorporates the four probit models for thermal fatalities: the Eisenberg model [14], the Tsao & Perry model [15], the TNO model [16], and the Lees model [17]. HyRAM also incorporates several overpressure models: lung hemorrhage fatality models by Eisenberg [18] and by HSE [19], as well as the TNO models [20] for head impact fatalities, for structural collapse, and for debris impact fatalities.

For thermal exposures the consequences to the exposed person are a function of radiative heat flux and exposure time. The thermal probit module uses the heat flux vector from the jet flame module. User input consists of a thermal exposure time for the population. The output is a matrix of thermal fatality probabilities; this vector is summed over the population to provide a vector the expected number of fatalities from thermal exposures for each release size.

For overpressure exposures, the consequences to the exposed person are a function of peak overpressure; this value is obtained from the overpressure effects calculations The TNO probit models also require impulse values or projectile fragment mass and velocity. Impulse values are obtained from the overpressure effects calculations. If a projectile model is selected, users must manually enter projectile fragment mass and velocity. As was done for the thermal fatalities, the expected number of fatalities from a given release size was calculated by summing the probability of fatality over the entire exposed population.

4. SANDIA'S HyRAM PLATFORM

4.1 User interfaces

The HyRAM toolkit contains two user-interfaces – one interface designed to facilitate performance-based permitting of hydrogen stations via the alternate design option provided by NFPA2 (the pilot user group), and a second for end-to-end QRA analysis using the approach from [2]. A planned third interface will allow stand-alone implementation of any of the deterministic consequence models.

The interfaces in HyRAM are designed to facilitate user activities such as comparison among various options (e.g., RCS requirements, design tradeoffs) and also comparison with other industries. Because of the focus on relative risk comparison and fast-running calculations, simplified models are be used in lieu of resource intensive simulations or CFD models. While the users have a strong need for defensible underlying physics models in the toolkit, they do not necessarily intend to manipulate the parameters of those underlying models. The development and selection of underlying model parameters fall under the purview of the developer community.

4.2 Code structure

The HyRAM toolkit is being developed in the MVC (Model-View-Controller) framework. The use of MVC enables a single underlying model to be associated with multiple user views. This flexible framework permits development of multiple user interfaces to meet the needs of different groups of users. The target platform is the .NET environment, with planned extension to an HTML interface after completion of the initial .NET interfaces. Both interfaces will contain user views targeted at QRA analysts and other user views targeted for NFPA2 users. The algorithm described in Section 3 has been modularized, as pictured in the flowchart in Figure 2. User input is divided into three categories: system component counts, system design parameters, and facility parameters (see Table 1). Users must also select the probit function to be used in calculating thermal and overpressure fatalities, using the options described in Section 3.5. Currently users must also enter peak overpressure and impulse for each release size. The default ESDs and FT from [2] are programmed as the initial scenarios in the toolkit.

HyRAM contains the frequency data generated from a Bayesian combination of hydrogen-specific data with data from similar industries, as provided in [6]. Each combination of component type and leak size is associated with a lognormal distribution for frequency of leaks (per year). HyRAM also contains generic values for component failure probabilities (per demand) for the basic events documented in [2]. Additionally, the toolkit contains physical constants and parameters relevant to hydrogen gas behavior modeling, including molecular weight, flammability limits, heat of combustion, and others. Data is stored internally in SI units, and the toolkit contains modules to enable conversion to English units.

Figure 2: Flowchart of modules contained in the HyRAM toolkit

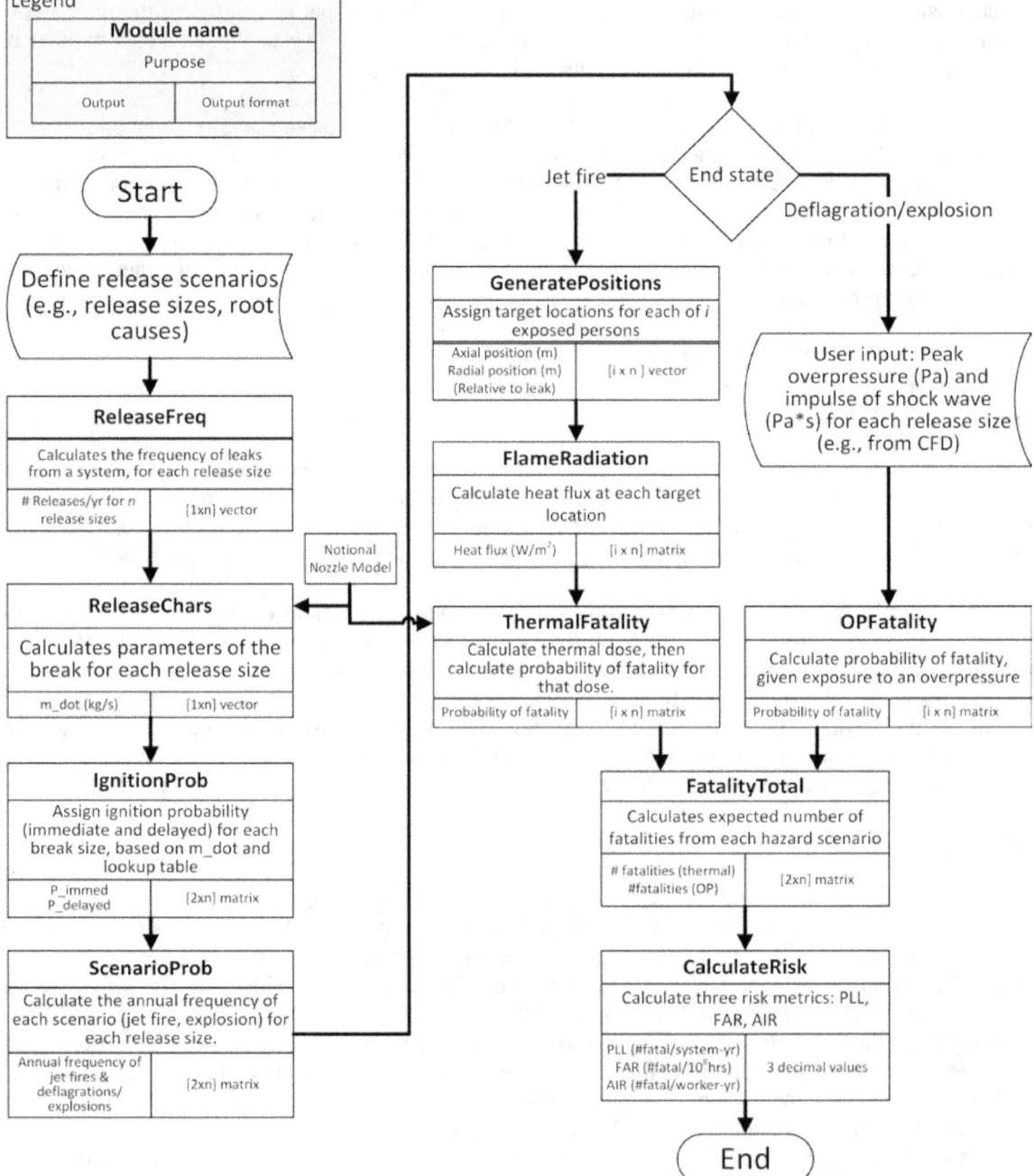

Table 1: Required user input for the HyRAM toolkit

System Component counts	System design parameters
- Compressors (#) - Cylinders (#) - Valves (#) - Instruments (#) - Joints (#) - Hoses (#) - Pipes (m) - Filters (#) - Flanges (#)	- Pipe outer diameter & wall thickness - Internal temperature, pressure - Number of demands (annual) Facility parameters - Facility temperature, pressure - Dimensions: length, width, height - Population: number of persons, locations, exposed (working) hours

5. NEXT STEPS

The HyRAM toolkit is in an early stage of development, and there are many features that must be added or improved to better enable the use of QRA. Filling these gaps requires a broad range of expertise and significant effort from the international hydrogen community. The proposed toolkit provides a framework for unifying the resulting science and engineering models into a tool to support RCS development. There are a number of areas for extension of the toolkit; some require developing or expanding science and engineering models, and others require the implementation of existing models in a compatible computational framework. Some key areas for expansion are:

- **Inclusion of additional hazards.** In previous work, it has been assumed that the risk contribution from fire and overpressure effects will render the risk contribution of debris and asphyxiation to be negligible. However, performance-based design scenarios require consideration of pressure vessel burst scenarios and toxic releases, in addition to fire and explosion scenarios. Future work will identify candidate models suitable for addressing these hazards, which may then be implemented in the HyRAM system.

- **Expanding treatment of uncertainty.** Assumptions about the timing of ignition and duration of exposures both significantly impact the calculated risk results. Future versions of the toolkit should be expanded to include dynamic risk assessment methods or sampling techniques to address uncertainty in timing and duration. In the current toolkit, calculations are limited to use of mean values from the probability distributions for leaks and component failures. Future versions of the toolkit will contain modules that enable sampling over the full distribution range.

- **Development of graphical interface for ESDs and FTs.** Future versions will include a graphical interface that enables users to modify the ESD/FT models and the probabilities associated with basic events, etc.

- **Inclusion of additional risk metrics.** The current toolkit focuses on calculating fatality risk. However, codes and standards users have also expressed interest in estimating risks that can be expressed in terms of cost. Future versions may also include features that enable plotting QRA results alongside various externally defined risk metrics or risk acceptance criteria.

- **Inclusion of additional physics models.** The current toolkit lacks simplified models for predicting overpressures from hydrogen deflagration, detonation events, as well as physical behavior of liquid hydrogen releases. Sandia has developed first-order models relevant to modeling the accumulation of hydrogen in delayed ignition events, and has also identified first order models for overpressure prediction. These models will be incorporated into the next version.

- **Increasing fidelity of included physics models.** Ruggles and Ekoto [8] performed a comparative study on various notional nozzle approaches and conducted experimental work comparing notional nozzle model predictions with measured hydrogen properties in free jets. They suggest a series of improvements to better reflect release thermodynamics, such as near-field jet entrainment. Similarly, the free jet assumption may also need to be adjusted, since many hydrogen leaks occur in close proximity to surfaces and barriers. Benard and Tchouvelev [23] have performed extensive studies of surface effects on hydrogen and methane jets and developed engineering correlations that take into account release pressure, orifice and proximity to surfaces for both horizontal and vertical jets. These and other physics-based model improvements are further discussed in [4].

In addition to these next steps for developing the toolkits, a more general, and omnipresent concern, is development and validation of models and datasets for use in the toolkit. The inclusion of meaningful, representative data in the toolkit is of critical importance to successful, defensible implementation for QRA. Developers must collaborate with data-collection organizations to ensure that data are of sufficient fidelity to be used in the toolkit. Industry groups may be able to facilitate data availability by collecting and sanitizing data for use in the toolkit. Furthermore, science-based models must be continually developed, validated, and integrated into the toolkit, to reduce uncertainty and to ensure that RCS decisions are based on defensible, accurate information.

6. CONCLUSION

QRA is an important tool for maintaining the safety and the commercial viability of the hydrogen industry, and is thus expected to play an increasing role as hydrogen technology shifts toward market deployment. Furthermore, QRA offers a means for using the best science and engineering models to develop codes and standards, as well as to facilitate the design and permitting process for hydrogen fueling stations. While there has been progress in developing defensible probabilistic and deterministic models for hydrogen systems over the past decade, use of QRA is limited due to lack of readily available tools for integrating and implementing these models for use by the hydrogen safety community.

Sandia and HySafe are working to build simplified toolkits to facilitate the use of QRA within the hydrogen fuel cell industry. The toolkit methodology integrates relevant probabilistic and deterministic physics models into a single analysis framework. The toolkits are intended to provide a practical, open access to the state-of-the-art required for hydrogen risk assessments of the hydrogen and fuel cell industry. In particular, industry experts, including codes and standards developers and station designers, who need fast running, high-level insights rather than expensive and resource-intensive numerical simulations, will benefit from this unique resource. At the same time, the toolkit may serve as a reference basis for the development of risk-informed standards and regulations, such as NFPA2. The HyRAM toolkit as well as the tools being developed by HySafe provide a framework for bringing together the expertise of the international hydrogen safety research community and putting it in the hands of the decision makers who ensure the safety of the hydrogen industry.

Acknowledgements

The authors would like to thank the US DOE Fuel Cell Technologies Office for funding and support of the HyRAM tool. Sandia National Laboratories is a multi-program laboratory managed and operated by Sandia Corporation, a wholly owned subsidiary of Lockheed Martin Corporation, for the U.S. Department of Energy's National Nuclear Security Administration under contract DE-AC04-94AL85000.

This work integrates significant international effort initiated under the auspices of HySafe, IEA HIA Tasks 19/31 on hydrogen safety, and the US DOE Fuel Cell Technologies Office. The authors would like to thank their international partners developing and validating hazard assessment tools at the Hydrogen and Fuel Cell Technologies program (Sandia National Laboratories, Livermore CA and Albuquerque NM), at the Insitute for Nuclear and Energy Technologies (Karlsruhe Institute of Technology, Germany) and at the Hydrogen Research Institute (Université du Québec à Trois-Rivières, Canada).

References

[1] J LaChance. "*Risk-informed separation distances for hydrogen refueling stations.*" International Journal of Hydrogen Energy, 34, pp. 5838-5845 (2009).

[2] KM Groth, JL LaChance, & AP Harris. *Early-Stage Quantitative Risk Assessment to Support Development of Codes and Standard Requirements for Indoor Fueling of Hydrogen Vehicles.* SAND2012-10150, Sandia National Laboratories, (2012).

[3] HJ Pasman. "*Challenges to improve confidence level of risk assessment of hydrogen technologies*" International Journal of Hydrogen Energy, 36, 2407-2413 (2011).

[4] A Kotchourko et al. "*State of the art and research priorities in hydrogen safety*" (Draft report) European Commission Joint Research Centre, (2014).

[5] K Groth & A Harris. "*Hydrogen Quantitative Risk Assessment Workshop Proceedings.*" SAND2013-7888, Sandia National Laboratories (2013).

[6] J LaChance et al. "*Analyses to Support Development of Risk-Informed Separation Distances for Hydrogen Codes and Standards.*" SAND2009-0874, 2009.

[7] A Tchouvelev. "*Quantitative Risk Comparison of Hydrogen and CNG Refueling Options.*" Canadian Hydrogen Safety Program Presentation at IEA HIA Task 19 Meeting, (2006).

[8] AJ Ruggles, & IW Ekoto, "*Ignitability and mixing of underexpanded hydrogen jets.*" International Journal of Hydrogen Energy, 37, pp. 17549-17560 (2012).

[9] E Papanikolaou, D Baraldi, M Kuznetsov, A Venetsanos, "*Evaluation of notional nozzle approaches for CFD simulations of free-shear under-expanded hydrogen jets*", International Journal of Hydrogen Energy, 37, pp. 18563-18574 (2012).

[10] RW Schefer, et al. "*Characterization of high-pressure, underexpanded hydrogen-jet flames*", International Journal of Hydrogen Energy, 32, 2081 (Aug, 2007).

[11] AD Birch, DJ Hughes, F Swaffield. "*Velocity decay of high pressure jets.*" Combustion Science and Technology 52, 161-171 (1987).

[12] W Houf & R Schefer. "*Analytical and experimental investigation of small-scale unintended releases of hydrogen.*" International Journal of Hydrogen Energy, Elsevier, 2008, 33, 1435-1444

[13] J LaChance, A Tchouvelev, & A Engebo. "*Development of uniform harm criteria for use in quantitative risk analysis of the hydrogen infrastructure.*" International Journal of Hydrogen Energy, 2011, 36, 2381-2388

[14] NA Eisenberg, CJ Lynch, & RJ Breeding. "*Vulnerability model. A simulation system for assessing damage resulting from marine spills.*" SA/A-015 245, U. S. Coast Guard, (1975).

[15] CK Tsao & WW Perry. "*Modifications to the vulnerability model: a simulation system for assessing damage resulting from marine spills.*" Report ADA 075 231 US Coast Guard; (1979).

[16] G Opschoor, ROM van Loo & HJ Pasman. "*Methods for calculation of damage resulting from physical effects of the accidental release of dangerous materials.*" International Conference on Hazard Identification and Risk analysis, Human Factors, and Human Reliability in Process Safety. Orlando, Florida; January (1992).

[17] FP Lees. "*The assessment of major hazards: a model for fatal injury from burns.*" Transactions of the Institution of Chemical Engineers;72(Part B):127e34 (1994).

[18] Center for chemical process safety. "*Guidelines for chemical process quantitative risk analysis.*" American Institute of Chemical Engineers, (2000).

[19] UK Health & Safety Executive. "*Major hazard aspects of the transport of dangerous substances.*" (1991).

[20] "*Methods for the determination of possible damage.*" In: CPR 16E. The Netherlands Organization of Applied Scientific Research; (1989).

[21] P Middha, OR Hansen, J Grune, J & A Kotchourko. "*CFD calculations of gas leak dispersion and subsequent gas explosions: Validation against ignited impinging hydrogen jet experiments.*" Journal of Hazardous Materials, 179(1-3), pp 84-94; (2010).

[22] OR Hansen, J Renoult, SR Tieszen, & M Sherman. "*Validation of FLACS–HYDROGEN CFD consequence model against large-scale H2H2 explosion experiments in the FLAME Facility.*" ICHS, Pisa 8–10 September, (2005).

[23] P Bénard et al. "*High Pressure Jets in the Presence of a Surface*", 2nd International Conference on Hydrogen Safety, San Sebastian, Spain, Sept 11-13 (2007).